Information Systems Engineering and Management

Volume 21

The book series "Information Systems Engineering and Management" (ISEM) publishes innovative and original works in the various areas of planning, development, implementation, and management of information systems and technologies by enterprises, citizens, and society for the improvement of the socio-economic environment.

The series is multidisciplinary, focusing on technological, organizational, and social domains of information systems engineering and management. Manuscripts published in this book series focus on relevant problems and research in the planning, analysis, design, implementation, exploration, and management of all types of information systems and technologies. The series contains monographs, lecture notes, edited volumes, pedagogical and technical books as well as proceedings volumes.

Some topics/keywords to be considered in the ISEM book series are, but not limited to: Information Systems Planning; Information Systems Development; Exploration of Information Systems; Management of Information Systems; Blockchain Technology; Cloud Computing; Artificial Intelligence (AI) and Machine Learning; Big Data Analytics; Multimedia Systems; Computer Networks, Mobility and Pervasive Systems; IT Security, Ethics and Privacy; Cybersecurity; Digital Platforms and Services; Requirements Engineering; Software Engineering; Process and Knowledge Engineering; Security and Privacy Engineering, Autonomous Robotics; Human-Computer Interaction; Marketing and Information; Tourism and Information; Finance and Value; Decisions and Risk; Innovation and Projects; Strategy and People.

Indexed by Google Scholar. All books published in the series are submitted for consideration in the Web of Science.

For book or proceedings proposals please contact Alvaro Rocha (amrrocha@gmail.com).

Sachi Nandan Mohanty · Suneeta Satpathy ·
Xiaochun Cheng · Subhendu Kumar Pani
Editors

Explainable IoT Applications: A Demystification

Springer

Editors
Sachi Nandan Mohanty
School of Computer Science
and Engineering (SCOPE)
VIT-AP University
Amaravati, Andhra Pradesh, India

Xiaochun Cheng
Department of Computer Science
Swansea University
London, UK

Suneeta Satpathy
Centre for Cyber Security
SoA University
Bhubanesawar, Odisha, India

Subhendu Kumar Pani
Department of Computer Science
and Engineering
Krupajal Engineering College
Bhubanesawar, Odisha, India

ISSN 3004-958X ISSN 3004-9598 (electronic)
Information Systems Engineering and Management
ISBN 978-3-031-74884-4 ISBN 978-3-031-74885-1 (eBook)
https://doi.org/10.1007/978-3-031-74885-1

This Springer imprint is published by the registered company Springer Nature Switzerland AG
The registered company address is: Gewerbestrasse 11, 6330 Cham, Switzerland

Preface

In the rapidly evolving realm of the Internet of Things (IoT), where interconnected devices weave a tapestry of data and insights, the need for transparency and understanding has never been more critical. As the complexity of IoT applications continues to grow, so does the demand for clarity in how these intelligent systems operate and make decisions. *Explainable IoT Applications: A Demystification* embarks on a journey to unravel the intricacies of IoT and shed light on the often opaque processes governing smart devices. This book serves as a guide for both seasoned professionals and curious minds delving into the world of explainable AI, machine learning, deep learning, blockchain, and cloud within the IoT landscape.

The pages unfold a narrative that transcends the technical jargon, providing a comprehensive exploration of the fundamental principles, challenges, and breakthroughs in achieving explainability in IoT applications. From demystifying the inner workings of IoT architectures to dissecting the security considerations that underpin these systems, each chapter strives to empower readers with knowledge and insights that transcend the conventional boundaries of IoT comprehension. The book delves into research studies that illustrate the transformative potential of clarity in the IoT ecosystem exploring how explainable IoT reshapes industries, transforms healthcare diagnostics, and informs decision-making in smart cities, smart homes, smart waste management as well as security aspects. This book is not just a compilation of theories and algorithms; it is a testament to the collaborative efforts of researchers, engineers, and visionaries who strive to make IoT understandable, interpretable, and, most importantly, trustworthy. As we explore the future trends, emerging technologies, and security considerations, we invite readers to contemplate the responsible and transparent deployment of IoT applications.

We hope this demystification of explainable IoT applications sparks curiosity, encourages dialogue, and inspires the next wave of innovations that prioritize clarity and understanding in the interconnected world of IoT.

Amaravati, India — Sachi Nandan Mohanty
Bhubanesawar, India — Suneeta Satpathy
London, UK — Xiaochun Cheng
Bhubanesawar, India — Subhendu Kumar Pani

Contents

Essential Uses of IoT and Machine Learning

IoT Pro-Interventions: Transforming Industries and Enhancing Quality of Life

Anasuya Swain, Suneeta Satpathy, Bijay Kumar Paikaray, and Jitendra Pramanik

Abstract Internet of Technology is a smart approach of communication to access the required data in anytime and at any place. The pro intervention characteristics of IOT help to collect and process data automatically and provide the responses for early decision making for the increment of effectiveness, production growth and conservation of the resources. The buzz word IOT has its popularity all over the globe due to its Big Horizon of spread sheet and smart applications for the smooth and easy going living. With an emphasis on proactive interventions and their wide range of applications, "Internet of Things: Pro-Interventions and its Applications" delves into the ever-changing world of the Internet of Things (IoT). The study explores the benefits of IoT in a future where linked devices rule, highlighting how it can transform a number of businesses and enhance people's quality of life in general. "Pro-Interventions" are proactive steps made possible by IoT that include real-time monitoring, predictive analysis, and preventive actions. This abstract looks at how IoT, with its network of smart devices and sensors, enables interventions to proactively solve problems in industries like industrial processes, healthcare, agriculture, health monitoring, and smart cities that optimize resource utilization predictive maintenance, demonstrating its potential to enhance the enhance the efficiency and sustainability. Furthermore, the chapter also discusses challenges and ethical considerations associated with widespread IoT implementation, highlighting the need for responsible and secure deployment. In total the chapter contributes to a comprehensive understanding of the

A. Swain
IITTM, Bhubaneswar, India

A. Swain (✉) · S. Satpathy
Centre for Cyber Security, Siksha 'O' Anusandhan (Deemed to be) University, Bhubaneswar, Odisha, India
e-mail: swainanasuya@gmail.com

B. K. Paikaray
Centre for Data Science, Siksha 'O' Anusandhan (Deemed to be) University, Bhubaneswar, Odisha, India

J. Pramanik
Department of Mining Engineering, National Institute of Technology, Rourkela, Odisha, India

S. N. Mohanty et al. (eds.), *Explainable IoT Applications: A Demystification*,
Information Systems Engineering and Management 21,
https://doi.org/10.1007/978-3-031-74885-1_1

positive impact of IoT, paving the way for informed decision-making and responsible integration of these technologies into our interconnected world.

Keywords IOT · Smart healthcare · Smart city · Smart home · Smart agriculture · Security issues

1 Introduction

IOT is the collection of number of technology based and network connected devices for connecting and exchanging data. It is an application mechanism which is a collection of services and software for the analysis of the data with the machine learning artificial intelligence program and provides the information for the decisions [1]. The application of this Tool helps to facilitate remote service, automation system remote monitoring, automation system remote health monitoring and energy notification system, smart transportation system, digital control system, smart farming [2, 3]. Nkone Kevin Ashion of Procter and Gamble first introduced the idea of IOT, which he later shared at the MIT Auto-ID Centre in 1999. After this he called it as "IOT". Now—a—days IOT refers to the period of time when almost usable items rather than people have their connectivity with the internet for the assessment of smart technology approach with its multiple application and usage of architecture levels.

Levels of Architecture of IOT

IOT has become a smarter technology with a variety of Levels of Architecture like as Level-1, level-2, Level-3, level-4, level-5, level-6. In level-1 architecture of IOT (Fig. 1), the application host is a single node or device that handles sensing, actuation, data storing, and analysis [3–5]. There are little data requirements and low processing demands for the analysis.

In level-2 architecture of IOT (Fig. 2) a level-2 Internet of Things is a system which has a single node. This single node handles local analysis in addition to sensing and or actuation. This architecture is a cloud–based one with a huge data and all the data are saved in the cloud.

In level 3 architecture of IOT (Fig. 3), an IOT system at level 3 only includes one node. Applications are cloud-based, with data analysis and storage taking place there as well [6–8]. IOT systems of Level-3 are applied for the solutions which are developed after the interpretation of large amount data ad its computation processing. Advanced data gathering, administration, monitoring applications, and cloud computing for data analysis are all at this level. This location houses extensive amounts of information, sometimes referred to as big data. Cloud storage is utilized for quickly gathering large amounts of data that have been found. Cloud-based data analysis can be initiated by using mobile or web applications.

The level-4 IOT systems carry out local analysis. This IOT system makes its application and the saves the data with the help of cloud system. The nodes here have

Fig. 1 Level-1 architecture of IOT

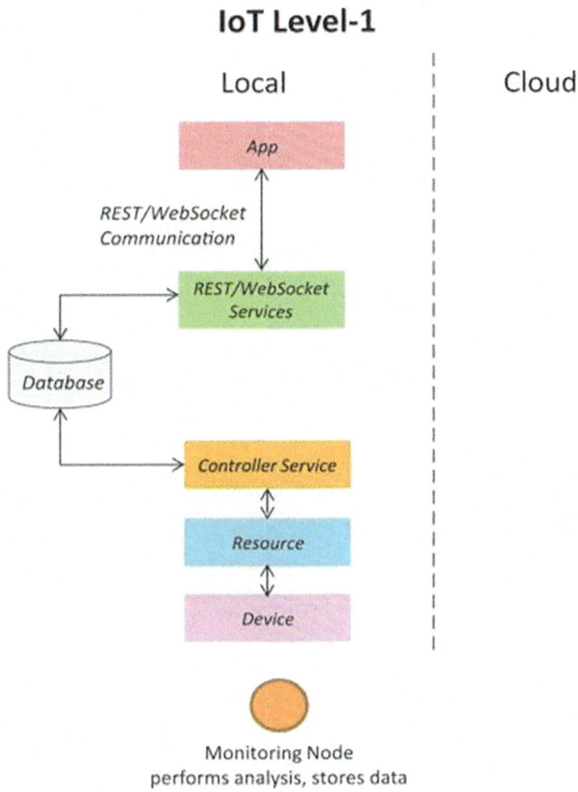

to observe both the local and cloud data and permit the users to receive, subscribe the various data those are collected in the IOT devices. This level of architecture has its several nodes with large amount of data saving with the facility of computational analysis for the solutions towards the problem. Level 4 utilizes multiple sensors that are unable to communicate with each other. Each sensor transmits its data to the cloud individually. The amount of data storage needed requires the utilization of cloud storage. The relevant control action is chosen by accessing data in the cloud and utilizing web or mobile applications (Fig. 4).

The architecture of the IoT can be organized using the five-layer model, which includes the levels: device, communication, data processing, application, and business. IoT layered architecture streamlines the allocation of parts and functions in an IoT system through its five-layer paradigm, ensuring effective management and scalability.

Figure 5 shows IOT Level-5 architecture which has its feature of coordination and multiple end nodes do the actuation and sensing activity. Coordinator node collects the data and gathers in the end nodes and saves in the cloud. This system has its both application of cloud and data analysis [9–11]. Level-5 is important due to its

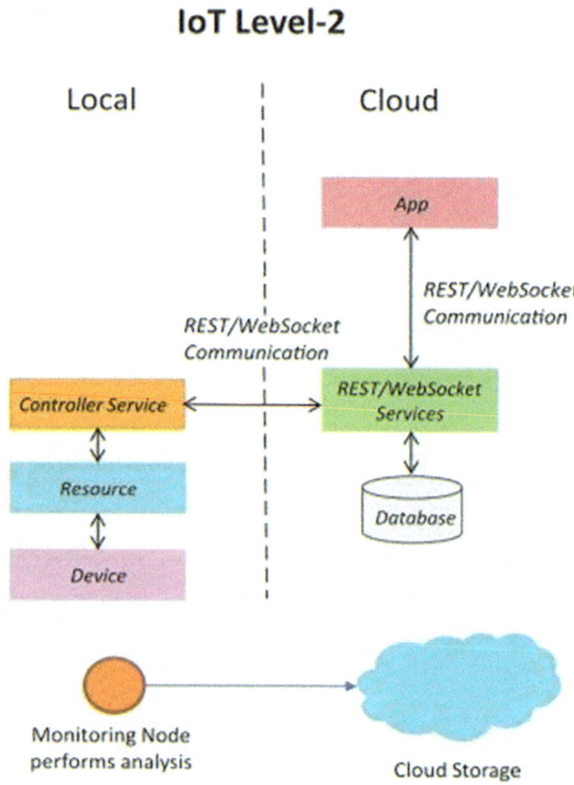

Fig. 2 Level-2 architecture of IOT

effective solutions which is available due to its wireless sensor network and huge amount of data for its computation and analysis

The Level-6 includes both the management and security layers. While it offers administrative and security functions, it is not considered a distinct layer because it interacts with all other layers. However, it is a significant issue that requires attention on a global scale. In level-6 architecture of IOT (Fig. 6), Cloud-based application allows for the visualisation of the results. This system is a centralised controller which is aware about the full status of the entire end node and has its command over all the end nodes [12–14].

2 Applications of IOT

IOT leads to the convenient and connected life style by reducing labour and eliminating the chances of human errors with smart devices and efficiency [15–17]. The most usable IOT applications are as follows.

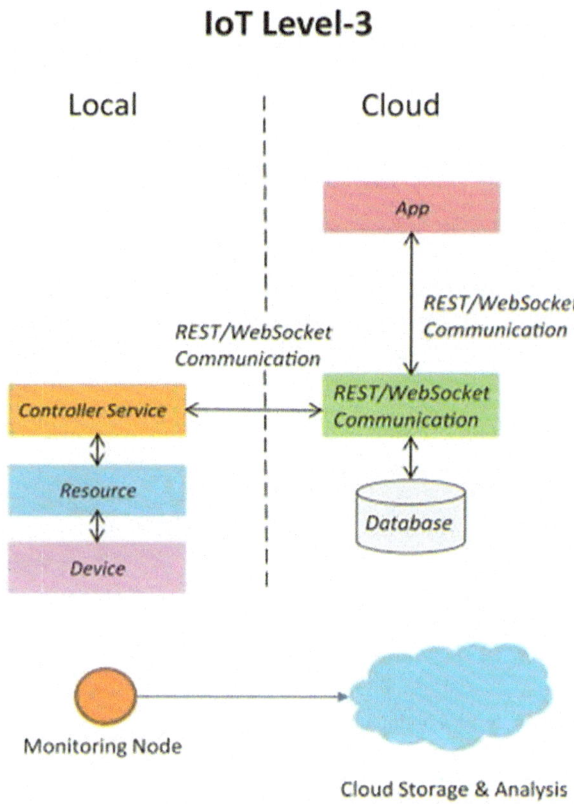

Fig. 3 Level-3 architecture of IOT

2.1 *Smart Homes*

The practical application of IOT makes the smart home facility by providing the convenience and home security [18–20]. The various devices with the IOT for the smart home are the smart door lock, smart bulb, and smart thermostat.

2.2 *Smart Door Lock*

Smart Lock enables the users to entry into the system without the key and to open the door by the remote or by the usage of smart phones or any other device which is connected with the internet. Smart locks develop the usage of IOT-enabled sensors. Smart locks facilitates the owners the virtual keys and unlock their door from anywhere without a key. The connectivity with Internet of things (IOT) and the devices develops the smart assistants and smart home management systems [21–23]. Smart locks provide extra functionality when a door is unlocked, some tasks can

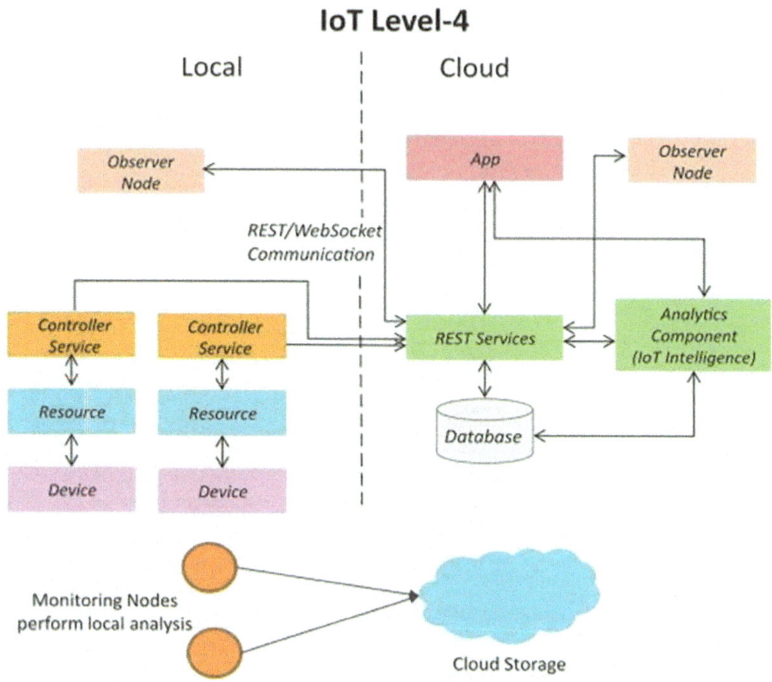

Fig. 4 Level-4 architecture of IOT

be automated, such as turning on your lights and adjusting your temperature, or the security system can be triggered to record and send video if the door is unlocked outside of the normal business hours [24–26].

2.3 *Smart Bulbs*

An internet-connected LED light bulb that can be modified, scheduled, and remotely controlled is known as a smart bulb. Home automation needs Internet of Things (IOT) application with the products in the expanding area at home; smart lights are among the most quickly successful ones. Smart bulbs are Wi-Fi-enabled light bulbs that can be independently controlled by a mobile app or smart assistant. The majority can adjust the colour or brightness. Light switches can manage groups of lights and function as an adapter for conventional light bulbs. Accessibilities of a smart bulb can be the ability to change light temperature, colour, and brightness; Remote Control Option; Choice for Group Control; May be included into Smart Home Routines [27–29]. Voice commands may be used to control it, you can make schedules, and it saves energy.

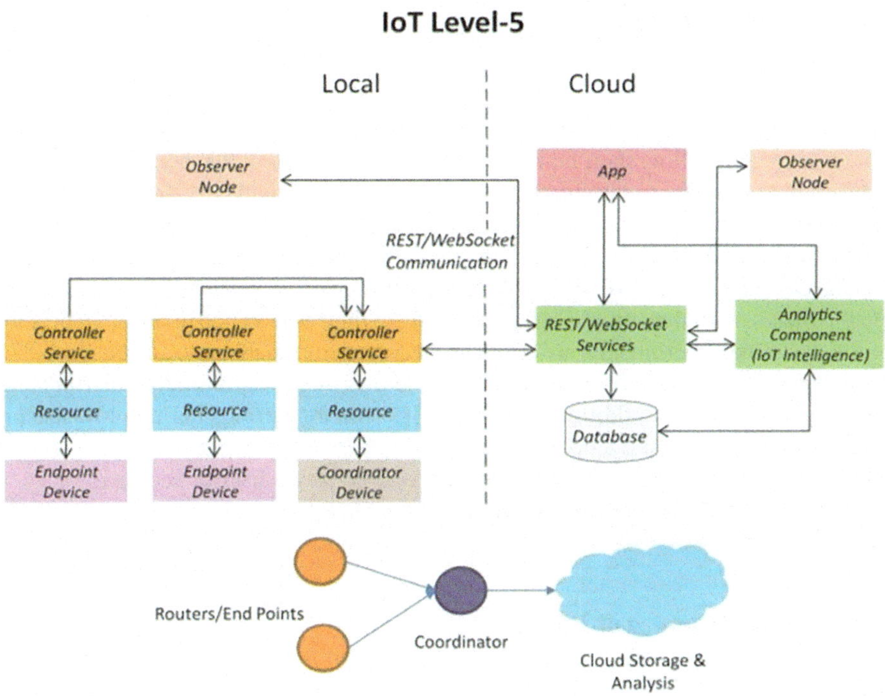

Fig. 5 Level-5 architecture of IOT

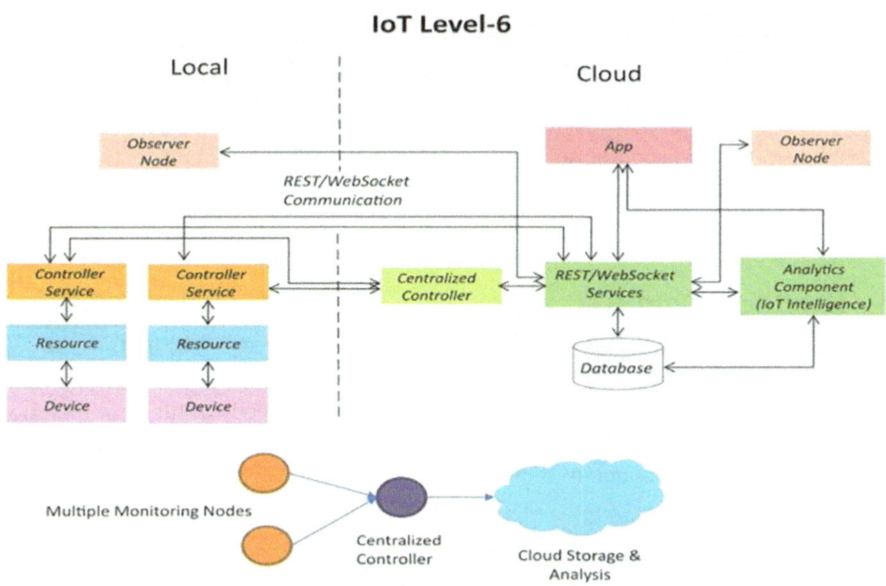

Fig. 6 Level-6 architecture of IOT

2.4 Smart Thermostat

IOT thermostats are a component of home automation systems that are in charge of managing a home's heating or cooling systems. They enable the user to set a schedule for controlling their home's temperature throughout the day, such as setting a different temperature at night. Smart thermostat is automatically adjusted with the temperature settings for heating and cooling and its best performed activity [30–33]. According to the fields study data analysis thermostat saves a lots of energy and has the designation of ENERGY STAR. In that they offer a schedule capability that enables users to set the temperatures according to their preference in different periods of the day. Smart thermostats are comparable to programmable thermostats [34–37]. To further limit the risk of human error while utilising programmable thermostats, smart thermostats use additional technologies. Smart thermostats employ sensors to determine the house occupancy and halt heating and cooling until the occupant comes back. This device is connected Wi-fi and helps the user to access it for 24 × 7 h and also saves the consumers' energy and money.

2.5 Smart Home Appliances

Smart home appliances are famous for their characteristics of saving of time, energy and money and a helping the user to make a successful planning, scheduling, monitoring and controlling the various activities. The technology of the IOT home automation system helps to control the household appliances by using its multiple control system paradigms [38, 39]. Addition of IOT in the appliances of the home makes it be self-controlled Windows, refrigerators, fans, lights, fire alarms, kitchen timers, and other electrical and electronic equipment. IOT devices are administered and used to configure by the usage of software programme which are now assessable in the form of the smartphone app and web integration [40–43].

3 Healthcare and IOT

IOT based healthcare enables real time observer of health and access the data resulting improved patience health, experience and enhanced health care operation [44–46]. Various IOT devices have their array benefits with certain challenges for the healthcare providers and to their patients.

3.1 Health Rate and Blood Pressure Monitor

IOT based health monitoring system helps to record the health status of the patients very quickly with its sensors. Sensor is very less expensive and smaller in size measures the human body's temperature, heart rate and oxygen saturation level and its result is displayed in the webpages. Application of this IOT technology helps to develop the Heart Rate Monitoring system with the objective of sensing the patient's heartbeat in order to track both the regular check-ups and the risk of a heart attack. The pressure sensor (MEMS pressure sensor) utilised primarily in the oscillate metric method to detects the pressure in the cuff utilising air as a pressure transmission medium. The use of wireless body area networks and the internet of things in smart health care are demonstrating improved performance. Although patients must stay in hospitals as part of traditional wellness programmes, WBAN admits that these patients can continue with their regular daily routines. It reduces foundation costs as well as pharmaceutical work costs. The WBANs policies used in cloud computing have more advantages, such as increased effectiveness, greater performances, utilities, and greater dependability, but they are still in the early stages of development and may face numerous obstacles and practical difficulties [2, 3].

4 Transportation

IOT devices and platforms can manage vehicles and drivers speed location; fuel level, mileage, maintenance and driving behaviour. Smart routing scheduling and despatching of the vehicles and drivers are also possible based on traffic, weather and demand patterns.

4.1 Traffic Management

Transportation industry is activated today due to the IOT application in it and sage of the various intelligent gadgets like embedded sensors, actuators, smart objects. This smart technology helps to collect, gather and provide this information at the time of the need of the users in the specified way. Transformation of the transportation industry is held due to the addition of IOT enabled technology and its smart solutions. Today the traffic issues like the more number of vehicle rising, road rising and population rising are making the transportation system a very complex one, which is going to be solved with the help of the incorporate of IOT into transportation with more extensive and secure transportation benefits. An advanced traffic management system (TMS) is a context-aware technology that uses predictive analytics and real-time data from linked road infrastructure to efficiently manage traffic on major thoroughfares. IOT applications not only is a helping ad towards the traffic but also helps to patrolling police to make the effective management of the traffic and reduction of the likelihood of accidents with the facilities of parking, autonomous traffic light systems, smart accident assistance. This technology also helps to reduce the traffic jams by providing timey notifications to drivers and notifications of safety messages towards the traffic management centres.

4.2 IOT Makes Easy Parking

IOT based parking systems is a IOT based devices with sensors and microcontroller which conveys information about vacant and occupied parking spaces. The customer can get the up-to-the-minute information by using the mobile app on the availability of every parking spot and can get the best one after giving their choices. This technology develops the smart parking by its economical and effective approach for the tracking availability of parking spots in real time. Searching the free parking space is possible by the usage of IOT application with the cameras and sensors also helps to point drivers in the direction of the best location with the usage of digital signage like LED displays. The Internet of Things gadget is made up of an ESP8266 microcontroller and an HC-SR04 distance sensor. The microcontroller is linked to the AWS IOT service via the MQTT protocol, and the sensor regularly measures the distance and delivers this data to it. This smart system is also making the parking less complicated and time consuming.

4.3 Vehicle Location Monitoring

The ideal parking spot can be found quickly by using a smart car monitoring system, which also increases the control and safety of parking security guards by handling all the data and lowers management expenses by automating more tasks and reducing human labour. Vehicle detection system alerts drivers when overweight vehicles approach overhead impediments like bridges, tunnels, and other structures. By producing a continuous record of vehicle operation, monitors make it possible to identify drivers and operators who disobey the law. Vehicle monitoring data can also be used to locate cars with disabled or tampered speed limiters.

4.4 Automotive Cars

Automotive IOT refers to the sophisticated network based gadgets such as sensors cameras and GPS trackers which is also linked with the cloud and provide the real time data and helps the manufacturing and transportation processes of cars. This IOT based device helps to prevent accidents and make right time judgement to control the vehicles operation. IOT application is now adding the variety of proximity sensors and improves driving safety and comfort in order to decrease human error and improve driving safety and comfort.

5 Smart Manufacturing

IOT enabled technology helps to connect various machines, devices and sensors to a central network and provides full visibility of assets, processes, resources and products manufacturing.

5.1 Industrial Communication

Managers need the smart collection and analysis of the data for smart manufacturing to make better decisions and maximise the production. IOT connectivity is a solution making technology which is installed at the factory level and transmit data from sensors and machines to the cloud. More and more organisations will aim to embrace these solutions to assist digitise their production processes, expedite data collecting, and enhance overall efficiency as continuing investment in IOT devices has driven innovation and expanded accessibility.

5.2 Production Flow Monitoring

The Internet of Things has made it possible to create a smart factory enables machine communication. Machines that communicate with one another are the bridge that connects 20th-century manufacturing to the intelligent manufacturing of the twenty-first century.

5.3 Improve Field Service Scheduling

IOT can help you save a lot of time and money with the use of a computerised management system. This tool aids in corrective repair of assets in addition to keeping your field service workers up to date on all client information and providing alerts to let them know about the next customer to support [47].

5.4 Engine Management

IOT solutions continuously identify the potential problems and check the condition of the vehicle. The sensors collect real-time information on things like fuel usage, engine temperature, fluid levels, and runtime. After that, the data is analysed to spot potential failure scenarios and warn the driver [48].

6 Smart Cities and IOT

IOT works as a leverage to improve the quality of life and enhance sustainability by connecting sensors lights meters to gather the data and interpret it for the further improvement and action against the deviation. IOT application in smart cities and its various formats are as follows.

6.1 Water Distribution

IOT is a network of physical items and gadgets with electronic sensors and its connection enables them to communicate and carry out tasks effectively and smoothly. An IOT smart water metre keeps tabs on the quantity, quality and pressure of water used in a building or company. One can monitor the water flow through the entire plant through the distribution routes by using an IOT smart water sensor, assisting in leak detection to cut down on water waste. A water distribution system is made up of several components, including tanks, pumps, valves, and pipes. The set-up of the pipelines makes it easier to transport water from the source to each individual dwelling. The most crucial factor for a lifetime of anticipated loading circumstances is how to design and run an effective water distribution system. This system must also be capable of supporting a typical circumstance such as pipe breaks, mechanical failure of pipes, valves, and control systems, power outages, and erroneous estimation of demand. In the modern world, Internet of Things (IOT) technology has had its significant impact on everything being wirelessly connected and makes the process simpler by using the sensors made specifically for things to connect anything.

7 Energy Engagement and IOT

IOT and its application in energy engagement help the business analysis, energy quality control, boost efficiencies and environmental impact [44, 49].

7.1 Electric Grid Automation

IOT smart grid is connected with the various devices and hardware that responds to the human needs and makes an effective two-way communication. IOT based smart grid infrastructure demands less money and has its more durability than the electrical infrastructure. IOT and the sensor gather more information from grid assets to grid operators with better visibility into infrastructure performance [50]. Transmission and distribution system can be controlled with the help of changing grid conditions based on shifting generation mixes and environmental factors [51].

7.2 Wireless Grid Communication Engagement

A new technology called the smart grid is helping to transform industries and social networks all around the world. By implementing smart grid infrastructure, different regional and state grid systems in the energy industry have started the transition from being power consumers to producing, sharing, and storing energy.

However, cyber security and resilience are just two of the numerous difficulties in utilising energy grids in smart grids due to the sometimes remote and harsh settings. A smart grid's communication infrastructure is made up of diverse gadgets like controllers, sensors, and actuators that communicate in real time to track the health of the grid's infrastructure.

8 IOT in Agriculture

Monitoring and direction towards agriculture management is easy due to the IOT involvement in agriculture. IOT is now available in the form of Robots, drones, remote sensors, computer imagery, ever-evolving machine learning and analytical tools. IOT in agriculture is used to monitor the crops, survey the map fields, and give farmers information they may use to make time- and money-saving and firm management decisions. Precision agriculture, crop monitoring, livestock monitoring, irrigation management, smart pest control, fertiliser management, and weather forecasting are a few typical examples of today's IOT-based agriculture.

8.1 Smart Farming and Crop Monitoring

Smart farming is possible due to the help of its IOT approach and association. The smart solutions derived from IOT is helping to make the automation of the irrigation system and effective monitoring of the agricultural process with the addition of sensory aids in the agricultural field for the checking of the different conditions of the light, humidity, temperature, soil moisture and crop health. Farmers can keep an eye on the state of their fields from anywhere. IOT platform "Ubidots" enables farmers to manage all of their connected equipment (weather station, irrigation system, soil moisture sensor, etc.) from a single dashboard and can also check the status of all their equipment and gadgets in real-time with this IOT platform. Information about weather, moisture, temperature, and soil fertility is collected by using IOT technology. Crop online monitoring offers the in the information regarding weed, water level, insect, and animal intrusion detection, crop growth and agriculture detection. IOT-enabled sensors placed everywhere over farms allow for continuous monitoring of crops and environmental parameters for changes in temperature, moisture content, pH level, humidity etc. [46, 52, 53]. The "micro-controller unit (MCU)" board analyses any anomaly found by the sensors, and the farmer is notified right away.

8.2 Climate Monitoring and Forecasting

The ability to manage climate variability and how it manifests in the everyday weather depends on fast and accurate climatic information as well as current knowledge of the frequency and intensity of extreme events, potential effects, and their persistence. For instance, monitoring and reporting efforts pertaining to droughts offer a baseline of data as well as a barometer of change in climatic conditions that may signal the onset of drought. Strategy for the drought avoidance and preventive measures can be done with the help of the drought indicator usage. IOT based Climate monitoring process helps the strategist to make efficient planning and the operations of different events like climate fluctuations in the frequency intensity and location of extreme. The three main categories of climate forecasting either and climate events are as follows: Forecasts made within one to seven days of an event are referred to as short-range forecasts. Typically, medium-range forecasts are released one to four weeks in advance. Long-range predictions are made anywhere from one month to a year in advance.

9 Network for IoT in Agriculture

The current state of IOT research and key technologies engaged with monitoring agricultural machinery operations. Platform for IOT in Agriculture is held through analytics of big data and analytics of cloud models [54, 55].

9.1 Analytics of Big Data

The main task of big data analytics is to collect data from many sources, make it usable for analysts. Crop diseases and crop growth models are built using farm data to ensure that there is little to no loss in agriculture. Additionally, it determines productivity and the best cost analysis Platform for a big data analytics-based IOT agricultural network and there are six main parts that make up this network-Farming knowledge, large data analysis skills, monitoring and sensing expertise, storage services, physical installation and communication protocol. Recently, India's economy has been mostly dependent on agriculture. The producers choose a variety of appropriate fruits, vegetables, and crops. The disease prediction method is accomplished by this system's emphasis on IOT with a machine learning mechanism. Image cloud was employed to find the damaged leaf. Well but they are not in a different kind of job, but, but they are using photos captured in the field and observes the ground at regular intervals. In order to reduce the waste of agricultural products, a challenge has been set up in this chapter to use machine learning to display the afflicted leaf. The chapter can be extended in order to develop a computerised integrated system that will allow for more accurate tracking of the rice both during manufacturing and after operations. Environmental factors have a significant impact on the health risks associated with natural products.

Plant pathogens are often considerably more exposed to the elements than animal pathogens, which may offer their pathogens a stable range of internal heat levels. In general, environmental changes will affect natural product infections, but there are several relationships between the host, the pathogen, and potential vectors.

The impacts of environmental change can occasionally be hidden by land managers' efforts. Some biological agents, such as plant diseases, have an effect on plants either directly or indirectly [15]. Pathogens are the name for the biological agents. Nematodes, bacteria, fungi, and viruses are a few examples of common pathogens. In addition to these pathogens, non-pathogenic diseases can also develop when environmental factors like pH, humidity, and climate change.

9.2 Analytics of Cloud Model

The infrastructure of cloud computing is made in numerous parts, including virtualization, networking, storage, and hardware, all of which are combined into a single architecture to support all kinds of crucial processes [21, 56]. A proper security system is necessary to prevent unauthorised access to any sensitive data.

Contemporary household appliances, tools, toys, and electronics can communicate with servers, clouds, and services using Ethernet or wifi. Many of these "devices" depend entirely on microprocessors for their functionality. Due to their incapacity to manage the intricacies of Internet connectivity, these devices are unsuitable for front-line IoT applications. Efficient and secure connections require a sophisticated

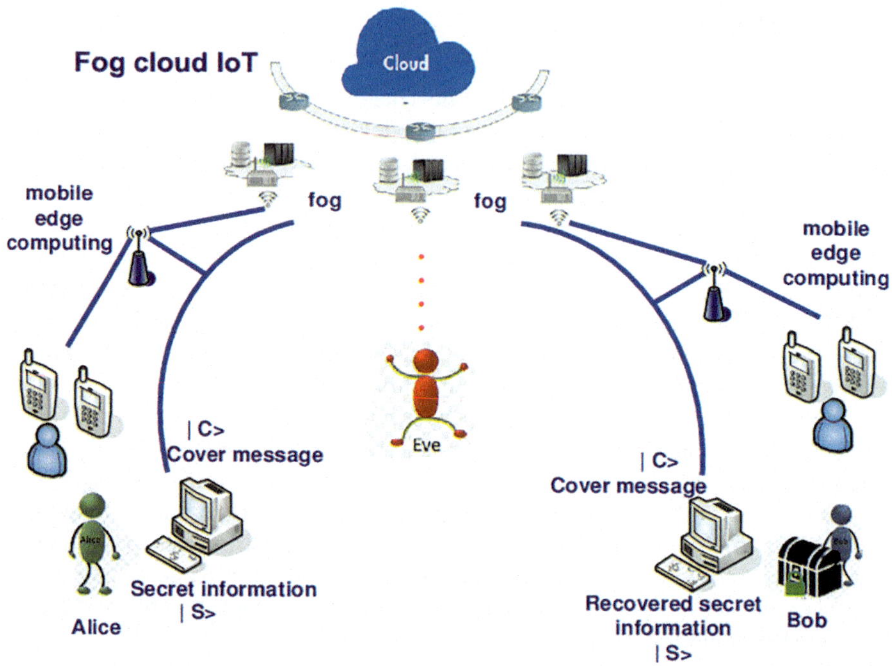

Fig. 7 Architecture of security system

device capable of managing tasks such as authentication, encryption, timestamps, caching, proxies, firewalls, connection loss, and other related functions. Products need to be reliable and able to function autonomously in the field (Fig. 7).

10 Functions of IOT Devices

Today developed global network infrastructure and its self-configuring capabilities built on open and interoperable communication protocols and into the information network. This network system has the identities, physical characteristics, and virtual personalities, and frequently exchange data related to users and their surroundings. IOT devices with distinct identities and the ability to conduct remote sensing, actuation, and monitoring functions are typically referred to as "Things" in the IOT context [57–61]. Few of those functions can be like performing some local chores and other IOT infrastructure functions depending on temporal and spatial constraints; exchanging data with other connected devices and applications (directly or indirectly);collecting data from other devices and process the data locally; sending the data to centralised servers or cloud-based application back-ends for processing; or without delving into the intricate details of the implementation, logical design of an IOT system refers to an abstract representation of the entities and processes. An IOT system is made up of various functional building elements that give it the ability to be identified, sensed, actuated, communicated and managed.

10.1 Wearable Smart Watches

People typically own wearable devices that can be worn on their wrists, such as smart watches, fitness bands, and other kinds of wearable technology. The two most popular wearable, smart watches (60%) and fitness trackers (27%) are by far the most widely owned. The most well-liked manufacturers of wearable are what people pick for their smart watches and fitness trackers.

10.2 Fitness and Activity Monitor

Primarily fitness activity tracker is of three types, i.e. Tracker for basic fitness, Heart rate monitoring tracker for Heart rate, monitoring trackers of Heart rate with GPS. A fitness tracker is a sensor based device that is used to track the orientation, movement, and rotation. This tracker collects data and converts it into steps, calories, sleep quality and general activity you perform through the day.

11 Conclusion

In summary, the Internet of Things (IoT) is a revolutionary force that has changed how we live, work, and interact with the outside world. It offers a plethora of advantages and interventions. Across a range of industries, the pro-interventions and uses of IoT have greatly improved productivity, connectedness, and ease. One of the most significant benefits is in the field of healthcare, where IoT has made it possible to monitor patients remotely, enhance diagnosis, and streamline the delivery of treatment. IoT has improved supply chain management, decreased downtime through predictive maintenance, and optimized manufacturing processes in a variety of industries. The Internet of Things (IoT) has enabled the smart home ecosystem to achieve previously unheard-of levels of automation and control, improving comfort and energy efficiency. Furthermore, precision farming—which allows farmers to make data-driven decisions for higher crop output and resource utilization—has been made possible by the integration of IoT in agriculture. IoT applications have improved traffic control, public services, and general urban planning in smart cities, which has aided in sustainable development. Even though we are amazed by the benefits of IoT, we also need to recognize and deal with its drawbacks, like security and privacy issues. For IoT to remain successful, finding a balance between innovations and protecting sensitive data will be essential. So, Internet of Things has unquestionably become a part of our everyday lives by presenting a rich tapestry of possibilities and developments. The proper development and implementation of IoT technologies will be essential to realizing its full potential and guaranteeing a safe and morally upright digital future as we traverse this interconnected terrain.

References

1. Hasan, M.Z., Al-Rizzo, H., Al-Turjman, F.: A survey on multipath routing protocols for QoS assurances in real-time wireless multimedia sensor networks. IEEE Commun. Surveys Tuts. **19**(3), 1424–1456, 3rd Quart. (2017)
2. Adeel, A., et al.: A Survey on the Role of Wireless Sensor Networks and IoT in Disaster Management, pp. 57–66. Springer, Singapore (2019)
3. Ahmed, R., Malviya, A.K., Kaur, M.J., Mishra, V.P.: Comprehensive survey of key technologies enabling 5G-IoT. SSRN Electron. J. April 2019, 488–492 (2019)
4. Yin, J., Yang, Z., Cao, H., Liu, T., Zhou, Z., Wu, C.: A survey on Bluetooth 5.0 and mesh. ACM Trans. Sens. Netw. **15**(3), 1–29 (2019)
5. Bembe, M., Abu-Mahfouz, A., Masonta, M., Ngqondi, T.: A survey on low-power wide area networks for IoT applications. Telecommun. Syst. **71**(2), 249–274 (2019)
6. Al-Turjman, F., Ever, E., Zahmatkesh, H.: Small cells in the forthcoming 5G/IoT: traffic modelling and deployment overview. IEEE Commun. Surveys Tuts. **21**(1), 28–65 (2019)
7. Andrews, J.G., et al.: What will 5G be? IEEE J. Sel. Areas Commun. **32**(6), 1065–1082 (2014)
8. Agiwal, M., Saxena, N., Roy, A.: Towards connected living: 5G enabled Internet of Things (IoT). IETE Tech. Rev. **36**(2), 190–202 (2019)
9. Ullah, H., Nair, N.G., Moore, A., Nugent, C., Mus-champ, P., Cuevas, M.: 5G communication: an overview of vehicle-to everything, drones, and healthcare use-cases. IEEE Access **7**, 37251–37268 (2019)
10. Conti, M., Dehghantanha, A., Franke, K., Watson, S.: Internet of Things security and forensics: challenges and opportunities. Future Gener. Comput. Syst. **78**, 544–546 (2018)
11. MacDermott, Á., Baker, T., Shi, Q.: IoT forensics: challenges for the IoA era. In: Proc. 9th IFIP International Conference on New Technologies, Mobility and Security (NTMS), Jan 2018, pp. 1–5 (2018)
12. Alenezi, A., Atlam, H.F., Alsagri, R., Alassafi, M.O., Wills, G.B.: IoT forensics: a state-of-the-art review, challenges and future directions. In: Proceedings 4th International Conference on Complexity, Future Information Systems and Risk (COMPLEXIS), May 2019, pp. 106–115 (2019)
13. Lillis, D., Becker, B., O'sullivan, T., Scanlon, M.: Current challenges and future research areas for digital forensic investigation. In: Proceedings 11th ADFSL Conference on Digital Forensics Security Law (CDFSL), Daytona Beach, FL, USA, May 2016
14. Arafat, M.Y., Mondal, B., Rani, S.: Technical challenges of cloud forensics and suggested solutions. Int. J. Sci. Eng. Res. **8**(8), 1142–1149 (2017)
15. Satpathy, S., Pradhan, S.K., Ray, B.B.: A digital investigation tool based on data fusion in management of cyber security systems. Int. J. Inf. Technol. Knowl. Manag. **2**(2), 561–565 (2010)
16. Yaqoob, I., Hashem, I.A.T., Ahmed, A., Kazmi, S.M.A., Hong, C.S.: Internet of Things forensics: recent advances, taxonomy, requirements, and open challenges. Future Gener. Comput. Syst. **92**, 265–275 (2019)
17. Sadineni, L., Pilli, E., Battula, R.B.: A Holistic Forensic Model for the Internet of Things. Springer International, Cham, Switzerland (2019)
18. Lu, Y., Xu, L.D.: Internet of Things (IoT) cyber security research: a review of current research topics. IEEE Internet Things J. **6**(2), 2103–2115 (2019)

19. Balaji, S., Nathani, K., Santhakumar, R.: IoT technology, applications and challenges: a contemporary survey. Wireless Pers. Commun. **108**, 363–388 (2019)
20. Čolaković, A., Hadžialić, M.: Internet of Things (IoT): a review of enabling technologies, challenges, and open research issues. Comput. Netw. **144**, 17–39 (2018)
21. https://www.designworldonline.com/part-1-connectivity-and-iot-in-motion-and-general-automation/
22. Hossain, M.: Towards a holistic framework for secure, privacy-aware, and trustworthy Internet of Things using resource efficient cryptographic schemes. Ph.D. dissertation, April 2018. https://doi.org/10.13140/RG.2.2.33117.72165
23. Al-Sharrah, M., Salman, A., Ahmad, I.: Watch your smart watch. In: Proceedings of International Conference in Computing Science and Engineering (ICCSE), pp. 1–5 (2018)
24. Rondeau, C.M., Temple, M.A., Lopez, J.: Industrial IoT cross-layer forensic investigation. Wiley Interdiscip. Rev. Forensic Sci. **1**(1), Art. no. e1322 (2019)
25. Wang, K., Du, M., Sun, Y., Vinel, A., Zhang, Y.: Attack detection and distributed forensics in machine-to-machine networks. IEEE Netw. **30**(6), 49–55 (2016)
26. Tekeoglu, A., Tosun, A.: Investigating security and privacy of a cloud-based wireless IP camera: NetCam. In: Proceedings 24th International Conference Computing and Communication Networks (ICCCN), Oct. 2015, pp. 1–6 (2015)
27. Ammar, M., Russello, G., Crispo, B.: Internet of Things: a survey on the security of IoT frameworks. J. Inf. Security Appl. **38**, 8–27 (2018)
28. Satpathy, S., Swain, P.K., Mohanty, S.N., Basa, S.S.: Enhancing Security: Federated Learning against Man-In-The-Middle Threats with Gradient Boosting Machines and LSTM. In: 2024 IEEE International Conference on Advanced Video and Signal Based Surveillance (AVSS), pp. 1–8. IEEE (2024, July)
29. Gubbi, J., Buyya, R., Marusic, S.: Internet of Things (IoT): a vision, architectural elements, and future directions. Future Gener. Comput. Syst. **29**(1), 1–19 (2013)
30. Knight, E., Lord, S., Arief, B.: Lock picking in the era of Internet of Things. In Proceedings IEEE CPS Workshop Data Security Privacy Forensics Trust (DSPFT), pp. 835–842 (2019)
31. Al-Turjman, F., Abujubbeh, M.: IoT-enabled smart grid via SM: an overview. Future Gener. Comput. Syst. **96**, 579–590 (2019)
32. Bhoopathy, V., Behura, A., Reddy, V.L., Abidin, S., Babu, D.V., Albert, A.J.: IOT-HARPSECA: a secure design and development system of roadmap for devices and technologies in IoT Space. Microprocessors Microsyst, 104044 (2021)
33. Abassi, R.: VANET security and forensics: challenges and opportunities. Wiley Interdiscip. Rev. Forensics Sci. **1**(2), Art. no. e1324 (2019)
34. Hossain, M., Hasan, R., Zawoad, S.: Trust-IoV: a trustworthy forensic investigation framework for the Internet of Vehicles (IoV). In: Proceedings IEEE 2nd International Congress Internet Things (ICIOT), Oct. 2017, pp. 25–32 (2017)
35. Lee, E.K., Gerla, M., Pau, G., Lee, U., Lim, J.H.: Internet of Vehicles: From intelligent grid to autonomous cars and vehicular fogs. Int. J. Distrib. Sens. Netw. **12**(9), 1–14 (2016)
36. Al-Turjman, F., Lemayian, J.P., Alturjman, S., Mostarda, L.: Enhanced deployment strategy for the 5G drone-BS using artificial intelligence. IEEE Access **7**, 75999–76008 (2019)
37. Renduchintala, A., Jahan, F., Khanna, R., Javaid, A.Y.: A comprehensive micro unmanned aerial vehicle (UAV/Drone) forensic framework. Digit. Invest. **30**, 52–72 (2019)
38. Ramachandra, G., Iftikhar, M., Khan, F.A.: A comprehensive survey on security in cloud computing. Procedia Comput. Sci. **110**(2012), 465–472 (2017)
39. Hossain, M., Hasan, R., Skjellum, A.: Securing the Internet of Things: a meta-study of challenges, approaches, and open problems. In: Proceedings IEEE 37th International Conference on Distributed Computing Systems Workshop (ICDCSW), pp. 220–225 (2017)
40. Jain, U., Rogers, M., Matson, E.T.: Drone forensic framework: sensor and data identification and verification. In: Proceedings IEEE Sensors Applications Symposium (SAS), 2017, pp. 1–6
41. Ferrag, M.A., Maglaras, L.: DeliveryCoin: an IDS and block chain based delivery framework for drone-delivered services. Computers **8**(3), 58 (2019)

42. Satpathy, S., Pradhan S.K., Ray B.B.: A digital investigation tool based on data fusion in management of cyber security systems. Int. J. Inf. Technol. Knowl. Manag. June 2010, **3**(2), pp. 561–565 (2010). http://www.csjournals.com/IJITKM/PDF%203-1/77.pdf
43. Oriwoh, E., Jazani, D., Epiphaniou, G., Sant, P.: Internet of Things forensics: challenges and approaches. In: Proceedings 9th IEEE International Conference on Collaborative Computing: Networking, Applications and Worksharing, October 2013, pp. 608–615 (2013)
44. Behura, A., Kabat, M.R.: Energy-efficient optimization-based routing technique for wireless sensor network using machine learning. In: Progress in Computing, Analytics and Networking, pp. 555–565. Springer, Singapore (2020)
45. Hasan, K., Biswas, K., Ahmed, K., Nafi, N.S., Islam, M.S.: A comprehensive review of wireless body area network. J. Netw. Comput. Appl. **143**, 178–198 (2019)
46. Tahsien, S.M., Karimipour, H., Spachos, P.: Machine learning based solutions for security of Internet of Things (IoT): a survey. J. Netw. Comput. Appl. **161**, 102630 (2020)
47. Cook, A., et al.: Internet of Cloud: Security and Privacy Issues, pp. 271–301. Springer, Cham, Switzerland (2018)
48. Alenezi, A., Zulkipli, N.H.N., Atlam, H.F., Walters, R.J., Wills, G.B.: The impact of cloud forensic readiness on security. In: Proceedings of 7th International Conference on Cloud Computing Services Science, pp. 539–545 (2017)
49. Buvana, M., Loheswaran, K., Madhavi, K., Ponnusamy, S., Behura, A., Jayavadivel, R.: Improved resource management and utilization based on a fog-cloud computing system with IoT incorporated with classifier systems. Microprocessors Microsyst, 103815
50. Sahoo, S., Halder, R.: Blockchain-based forward and reverse supply chains for e-waste management. In: International Conference on Future Data and Security Engineering, pp. 201–220. Springer, Cham (2020)
51. Stoyanova, M., Nikoloudakis, Y., Panagiotakis, S., Pallis, E., Markakis, E.K.: A survey on the internet of things (IoT) forensics: challenges, approaches, and open issues. IEEE Comm. Surv. Tutorials **22**(2), 1191–1221 (2020)
52. Whaiduzzaman, M., Hossain, M.R., Shovon, A.R., Roy, S., Laszka, A., Buyya, R., Barros, A.: A privacy-preserving mobile and fog computing framework to trace and prevent covid-19 community transmission. IEEE J. Biomed. Health Inform. **24**(12), 3564–3575 (2020)
53. Bhoi, S.K., Jena, K.K., Panda, S.K., Long, H.V., Kumar, R., Subbulakshmi, P., Jebreen, H.B.: An Internet of Things assisted unmanned aerial vehicle based artificial intelligence model for rice pest detection. Microprocess Microsyst **80**, 103607
54. Escamilla-Ambrosio, P.J., Rodríguez-Mota, A., Aguirre-Anaya, E., Acosta-Bermejo, R., Salinas-Rosales, M.: Distributing computing in the Internet of Things: cloud, fog and edge computing overview. In: Maldonado, Y., Trujillo, L., Schütze, O., Riccardi, A., Vasile, M. (eds.) NEO 2016. Studies in Computational Intelligence, vol. 731. Springer, Cham (2018). https://doi.org/10.1007/978-3-319-64063-1_4
55. Abd El-Latif, A.A., Abd-El-Atty, B., Hossain, M.S., Elmougy, S., Ghoneim, A.: Secure quantum steganography protocol for fog cloud internet of things. IEEE Access **6**, 10332–10340 (2018)
56. Behura, A., Priyadarshini, S.B.B.: Assessment of load in cloud computing environment using C-means clustering algorithm. In: Intelligent and Cloud Computing, pp. 207–215. Springer, Singapore (2019)
57. Khan, M.A., Salah, K.: IoT security: review, blockchain solutions, and open issues. Future Gener. Comput. Syst. **82**, 395–411 (2018)
58. Chung, H., Park, J., Lee, S.: Digital forensic approaches for Amazon Alexa ecosystem. In: Proceedings of 17th Annual DFRWS USA, pp. S15–S25 (2017)
59. Nieto, A., Rios, R., Lopez, J.: A methodology for privacy-aware IoT-forensics. In: 16th IEEE International Conference on Trust, Security and Privacy in Computing and Communications, 11th IEEE International Conference Big Data Science Engineering, 14th IEEE International Conference Embedded Software Systems, 2017, pp. 626–633
60. Basahel, S.B., Bajaba, S., Yamin, M., Mohanty, S.N., Lydia, E.L.: Teamwork optimization with deep learning based fall detection for IoT-enabled smart healthcare system. Comput.

Mater. Continua. **75**(1), 1353–1369 (2023). ISSN: 1546–218. https://www.techscience.com/cmc/v75n1/51539
61. Potluri, S., Mohanty, S.N.: An efficient scheduling mechanism for IoT based home automation system. Int. J. Electron. Bus. **16**(2), 147–156 (2021). ISSN: 1470-6067. https://doi.org/10.1504/IJEB.2021.115719

A Comprehensive Review of Machine Learning Approaches in IoT and Cyber Security for Information Systems Analysis

G. Prabhakar Reddy, P. Deepan, M. Arsha Reddy, R. Santhoshkumar, and B. Rajalingam

Abstract Sophisticated software analysis techniques are necessary for the security of current IoT systems. Static analysis is one such technique that has consistently proven useful. The labor of human specialists needs to be automated and intellectualized since the connections between IoT systems are becoming more complicated, larger, and more heterogeneous. Therefore, we postulate that machine-learning techniques can be useful for static analysis of IoT systems. The study's ontology is reflected in the research plan, which seeks to validate the hypothesis. The most important things that this work has accomplished are: Streamlining the process of static analysis for IoT systems and formalizing model decisions for ML problems; reviewing and analyzing a large body of literature in the field; validating that machine learning tools are appropriate for each step of static analysis; and proposing a concept for an intelligent framework to aid in static analysis of IoT systems. The findings are groundbreaking because they formalize the processes and solutions as "Form and Content," examine each stage from the viewpoint of the complete suite of machine-learning solutions, and account for the entire static analysis process (beginning with the research of IoT systems and ending with the delivery of the results).

Keywords Cyber security · Static analysis · Machine learning · Analytical model · IoT techniques and formalization

G. Prabhakar Reddy (✉) · P. Deepan · M. Arsha Reddy
St. Peter's Engineering College, Hyderabad 500043, India
e-mail: prabhakar.sp17@gmail.com

R. Santhoshkumar · B. Rajalingam
St. Peter's Engineering College, Hyderabad 500100, India

S. N. Mohanty et al. (eds.), *Explainable IoT Applications: A Demystification*,
Information Systems Engineering and Management 21,
https://doi.org/10.1007/978-3-031-74885-1_2

1 Introduction

The lack of protection for Internet of Things (IoT) systems from malicious actors is a pressing global concern. Data availability, confidentiality, and integrity can all be jeopardized by software errors, which significantly affect IoTS security [1, 2]. Whereas in the past only certain data or software components were tested to ward against such invasions, it is now essential to conduct both static and dynamic analyses of the entire information system [3]. A great deal of information, articles, and software with multiple applications make up IoTS. Every one of these components could be harmful in its own way: PHP files, which contain code for web servers, can have backdoors; exe files, which contain code for programs, can have program bookmarks; and JPG files, which contain Stego attachments, can set up a covert channel for the leak of private and malicious information [4]. The information security (IS) condition of a system can be better understood through an in-depth analysis of its objects, which is an integral part of IoTS analysis [5, 6]. Improper time management prevents the production of expert security analyses due to factors such as an excess of IoTS files, large amounts of data included within them, high levels of data heterogeneity, document container architectures, etc.

The bulk of human effort goes into creating and fixing rigid rules in classical automation, so they don't provide the intended results. Machine learning (ML) is an interesting and potentially useful strategy since it entails shifting resources from human cognition, which is plainly resource-intensive, to AI, which is software-implementable and so less resource-intensive. Distinct applications that can be executed repeatedly under the same conditions are more suited for DA. Covering the code thoroughly during the investigative process will require making use of all possible program execution scenarios. Hence, reports of errors, execution logs, or program results can be produced for further examination, either automatically or by hand. It may take a long time to cover even a tiny percentage of their code using this method, and the system components work in a dynamic environment that is difficult to simulate virtually. Therefore, this approach will be a significant difficulty for large IoTS. Which is why the authors think static analysis is the best bet for making analysis applicable to modern IoTS.

We use the data presented above to provide a working hypothesis for the current investigation: One area where machine learning might improve information security for IoTS systems is in SA. To verify it, this research is being conducted. In their article, the authors express their belief that future research should concentrate on the DA task, and they propose limiting consideration to just the SA task. Although machine learning applications are still in their infancy in the realm of information security, they have already proven their worth in a number of areas, including but not limited to: finding vulnerabilities in source and machine code [7–9], detecting attacks in networks [10], forecasting balances on large distributed system components [11, 12], detecting anomalies in real-time data [13], and many more. Since a direct search for vulnerabilities is just one of the subtasks of IoTS analysis, it is crucial to evaluate applying machine learning (ML) to the entire cycle of system examination, not just for executable files. This is because this subtask is typically represented in scientific publications, but there are many others.

2 Research Literature Works

IoT security is especially important as it directly impacts the actual world, including how systems work and how people live their lives. For instance, mistakes made when operating smart home security systems may cost owners money; mistakes made while operating the transportation system [14, 15] can result in collisions and gridlock; and mistakes made when using medical IoT devices can cause patients to pass away quantitative analysis. Due to the distinct functional aims of IoTS, the unique challenge of executing SA is evident. Devices therefore run different OSs, distributions, and CPU architectures. In specifically, every one of these choices is chosen according to the devices' tasks. Certain Internet of Things (IoTS) have specific requirements, such as high operating speed, battery life, ultra-precise complicated computations, etc. Because of this, a more comprehensive and standardised set of SA tools and a technique is required for the whole range of IoTS variety.

The existence of a whole interconnected Internet of Things system implies several of limitations on security measures. Consequently, dynamic analysis in particular will be quite limited since it is rather challenging to simulate the environment of a device that is always interacting with the outside world. However, security issues can occasionally arise specifically when several IoTS devices are interacting rather than just one. Machine Intelligence. The abundance of IoT vendors and the range of available solutions used forbid the implementation of hand-crafted SA regulations or those that call for the professional manual labour of specialists. If not, SA will require a lot of resources. A vast variety of cyber-physical interfaces (sensors, impact sources, etc.) make the manual process of creating a test case quite difficult. This means that some of a person's creative talents must be partially replaced by technology. This is what makes using ML justified. Reviewing the presumptions that already exist for such an application should be the first step in confirming if a new set of procedures may be applied to the tasks at hand. To validate the hypothesis, we create a continuous research challenge consisting of two subtasks:

1. To demonstrate that the changes made for information security at every level of IoTS SA can be supported by machine learning (ML)-based solutions.
2. To provide a summary of machine learning (ML)-based solutions that are suitable for every phase of IoTS security assurance.

3 Research Discussion on Existing Works

3.1 Machine Learning for SA of IoTs

Numerous well-organized and comparative reviews of scholarly articles cover the topic of machine learning's applications to information systems and the Internet of Things (IoT) in particular. But, compared to, instances, network security, just a small fraction of these concerns are unique to IoTS security. Here is a brief overview of some reviews that caught my eye. According to Xue et al. [8], a number of methods for binary code analysis have been thoroughly examined. Taxonomy is used to list the four components of the corresponding framework: feature embedding, applications, analytical approaches, and feature extraction. Machine learning approaches to code analysis are getting a lot of attention lately. As a result, over a hundred publications have used machine learning (ML) to classify code clones, text tokens, cryptographic methods, viruses, and related topics.

Clustering is elucidated from multiple angles, including data preparation for other classifiers, authorship recognition, clone detection, base-level program resemblance to a virus, and identification of auxiliary features in the code, such as virus traits and entrance points in functions. Malware classification, function entrance site identification, and function recognition are some of the works that address the use of machine learning (ML) to binary code analysis. The following tasks are the primary emphasis of the work [7]: Recommender Systems: tools that aid programmers (such as code auto completion) in their work; Inferring Formatting, variable naming, and other code style standards must be followed. Problems with Code: Finding Inconsistencies That Might Point to Security Flaws; The procedures involved in writing "smart" program code, as well as documentation, traceability, and information retrieval, are known as programme synthesis.

In these publications, the topic of analyzing code for errors and anomalies using machine learning (ML) is covered. This means that the evaluation only covers a subset of possible security threats. In the first group, we have vulnerability prediction tools that go beyond SA and use a variety of source metrics, binary code, and code commits to make their predictions. The second group's goal is to detect code anomalies that might be the consequence of security flaws. This is achieved by looking for deviations in API requests from relevant patterns and missing standard checks. In order to find code vulnerabilities directly, the third group use pretrained patterns. Clustering and classification are employed in the works to achieve this objective. The fourth category describes approaches that were not addressed in the first three.

Visualization as a tool for securing software interactions in complex information systems is the main area of concentration. To enhance the intellectualization of interaction study, it is recommended to employ clustering, dimensionality reduction, anomaly detection [16], classification, and regression. It is clear that most attempts, despite the great amount of work, are inefficient because they only apply to the source code. The study concludes that there has been insufficient progress in the area of using ML to find code vulnerabilities. Although these evaluations are comprehensive (within the range of 100–200 articles reviewed), all of the works that were considered are solely related to software security testing in its direct form.

Since most human effort goes into designing and debugging these rules, classical automation's strict rules don't produce the expected outputs. Therefore, a fascinating and potentially effective strategy is machine learning (ML), which entails shifting the resource-intensive human intellectual work to the software-implementable artificial intelligence. It is more acceptable to use DA when dealing with separate programs that may be run again given the same conditions. During the inquiry, it will be essential to use all possible scenarios in which the program runs in order to thoroughly cover the code. As a result, it is possible to acquire program results, execution logs, or error reports for further examination, either automatically or by hand. The problem with this approach is that it may take a long time to cover even a tiny fraction of the code for large IoTs systems, and the system components work in a dynamic environment that is difficult to simulate realistically. The authors conclude that this is why static analysis is the most promising method for developing modern IoTS-relevant analyses.

Based on the facts provided, we propose a hypothesis for this study: Machine learning has the potential to enhance information security in state-of-the-art IoTS systems. This study aims to provide confirmation of that. The paper recommends limiting consideration to the SA task alone, and the authors feel that future studies should concentrate on the DA task. Machine learning applications are still in their infancy in the realm of information security, but they have already proven their worth in a number of areas, including but not limited to: finding vulnerabilities in source and machine code [7, 9], detecting attacks in networks [10], forecasting balances on large distributed system components [11, 12], detecting anomalies in real-time data [13], and many more. Though it is most often depicted in scientific publications, a direct search for vulnerabilities is just one of many subtasks of IoTS analysis. As a result, it is crucial to think about applying machine learning (ML) to the whole cycle of system examination, not just for executable files.

3.2 Machine Learning for IoT Security

We also examine the most recent publications on the subject of applying machine learning to improve IoT security [15, 17–21]. The challenge of recognising IoT devices is the focus of the effort. An overview of machine learning-based identification techniques is given in the article. There are four main types of identification processes: pattern recognition, deep learning, unsupervised learning, and anomalous device detection. As a result, this work has some relevance to the present topic. Nevertheless, SA Stage 1 is the only one that can be linked to the suggested fixes. The IoT device security trends are evaluated. Out of the 20 surveys that are analysed, only 25% take machine learning into account. Nonetheless, the primary taxonometries deal with classifying assaults rather than countering them. The identification of dynamic assaults rather than static flaws in device implementation is the primary setting in which machine learning (ML) is used to enhance security (i.e., in programme code, algorithms, architecture, etc.).

The study highlights how important it is to look for harmful IoT software and how underdeveloped current solutions are. A special focus is placed on the Unux like programmes that are run in the ELF format. According to data, Windows accounts for 22% of IoT file formats (PE) and Linux for 70% (ELF); less than 20% of IoT users frequently use other operating systems.

The taxonomy divides characteristics into two categories: logs and resource utilisation for DA, and metrics, graphs, and trees, sequences, and dependencies for SA. Although this survey's results are somewhat related to the ongoing study, they are mostly attributable to Stage 3 (with some Stages 1 and 2 taken into account as well), not the entire SA process. Only machine learning issues like classification—and its particular instance, detection—are emphasised at the same time. Both ML-based IoT security models and IoT threats are described in. An IoT model with layers is suggested. Based on this approach, a Q work flow for threat detection is explained. All countermeasures, however, fall under the category of dynamic analysis as they all entail examining device network behaviour at different network levels. However, IoTS has not been directly analysed as a software system. Just three ML techniques are mentioned: regression, clustering, and classification. Some techniques are featured individually, such as reinforcement learning and rule-based systems.

The use of Federated Learning (FT) for the Internet of Things is taken into account. Utilising this method will eliminate the requirement to transmit training data. The FL-IoT system is introduced, outlining the exchange procedures between FT-server, local training, and both. It is suggested that FT may be used to identify malicious software in a static manner. The primary uses of machine learning are in anomaly detection, regression, and classification. The primary means of thwarting Internet of Things attacks is through dynamic device analysis. Simultaneously, the survey is focused on the use of FT in IoT rather than systematising SA. Keep in mind that using FT can improve safety and efficiency when doing the SA (at all stages and tasks of ML).The focus of the work is on IoT security in 5G networks, which include enormous numbers of devices and fast data transmission speeds. An analysis is conducted on the current physical layer authentication techniques. ML technologies, including autoregressive random process, Kalman filtering prediction, AdaBoost classifier, kernel machine, reinforcement learning (especially Q-learning and Dyna-Q), and reinforcement learning, are studied. Schemes are seen from the standpoint of implementing ML.

Initially, as the solutions under consideration process the external consequences of malicious devices while they are operating, they primarily fall under the category of dynamic analysis. Our assessment, however, concentrates on SA. Only at a specific moment in time and without the actual deployment of IoTS is such a study possible.

Second, although providing countermeasures partially based on ML, the studied methods primarily aim to introduce taxonometry for systematising devices. Our evaluation presents the use of fundamental ML tasks to systematise all SA steps. We study the challenge of finding strategies to fight risks during SA, rather than the difficulty of recognising threats themselves.

Thirdly, the options under consideration mostly involved systematisation based on current dangers and remedies. This may result in the potential loss of some of the components (for instance, if solutions of this kind have not yet been suggested). To begin, we synthesise all of the answers as a 4×5 matrix, or most. Next, we demonstrate that each of these decisions is legitimate. Our method is therefore more methodologically sound.

Fourth, none of the assessments took into account all of the primary tasks—classification, anomaly detection, regression, clustering, and generalization—that are accomplished using machine learning. IoT security typically just addresses the answers to the first two issues. Classifiers are essentially the sole tool used for SA of binary codes. As a result, new uses of more uncommon ML solutions for the Internet of Things may be "missed." Fifth, the solutions under consideration are mostly focused on analysing specific IoT devices or how they interact. We provide the review as an IoTS-wide study. To that end, the notion of an intelligent framework is put forth.

4 SA Model

In order to develop subject-specific models, we take a closer look at the SA domain from the perspective of machine learning application there. We do this by drawing on both established theoretical prerequisites and real-world knowledge from the published investigations. First, the SA as a whole can be separated into several time periods or stages (with a specific goal) due to its heterogeneity and frequent nonlinearity. Second, ML has to be used in the solutions. As a result, it is feasible to identify the jobs that machine learning is suitable for. Consequently, the stages of SA and the ML solutions relevant to them may be systematised in a single model. Classical and generalised time segments were utilised for the stages, with the exception of balancing. The Model aims to do the following. Firstly, the fact that it was formed explicitly validates that different solutions may be used to ML problems at every step of SA. Simultaneously, this evidence will be constructed based only on one foundation (form and content, as explained below). The module therefore functions as a theoretical demonstration of the hypothesis. Second, techniques for addressing issues for stages may be practically implemented using the theoretical model.

4.1 Different Stages of Static Analysis

SA is typically broken down into many phases since it is a complicated process that involves a wide range of activities that must be completed and the use of qualitatively distinct data. To make things easier, we assign each activity to a stage of SA using the formalisation below. Any data is represented by its look, or Form (F), and the information it contains, or Content (C). A line-by-line record of execution errors, for instance, might have the following entries: Content—errors and Form—text strings. Subsequently, a tuple <F | C> may be used to represent specific data that is both altered in the SA process and saved in the IoTS. From this vantage point, we view every step as a change of the incoming data's shape and content. After summarising the techniques and resources for SA software that have been discussed, we may integrate them according to the data transformation technique. There are four stages of IoTS analysis that can be distinguished with enough abstraction without going against accuracy.

4.1.1 Collecting the Data

The goal of this first step is to separate the subsystem (also known as the "vertical component") or level (also known as the "horizontal component") from the Internet of Things so that it may be processed further. This stage is required since there are several factors to consider while analysing the Internet of Things, just like any other complicated and diverse item. At this point, a specific aspect needs to be selected. Information about the Internet of Things itself serves as the stage's input. Only a portion of the IoTS data chosen for processing is sent to the stage's output. While important, this phase might not directly connect to IoTS duties. The following are typical stage actions: choosing the scope and direction of the study; organising executable code; typing and tagging files; and unpacking data containers (archives).

4.1.2 Preparation of Data

In order to make the data about a portion of the IoTS acceptable for the processing procedures, this stage is meant to transform the data. Without this stage, all processing techniques would have to modify to fit the IoTS data's format, which is inherently unproductive. The IoTS data point chosen in Stage 1 serves as the stage's input. A portion of the IoTS data is transformed into an appropriate format by the stage's output. This stage, of course, has to transform the IoTS data into a format that works for all of the processing methods if there are several ones. Instead than solving particular IS issues, this phase transforms the data into a format that most accurately captures the characteristics connected to those problems.

4.1.3 Processing the Data

It is the most important and difficult step from both a theoretical and practical perspective. The current state of data processing is based on IS (intrusion security, vulnerabilities, malware, stego, bugs, backdoors, etc.). In this stage, the results are defined by the sum of all of its methods. The data is prepared for processing when it enters the stage, which is made possible by Stage 2. Each method typically makes use of a number of complex algorithms, the output of which are often entirely novel and high-quality data sets. Consequently, the data we receive at the end of the stage may have a completely different Form and Content.

Here, information security tasks are handled directly: finding vulnerabilities, clones of viruses, evaluating code security metrics, fixing vulnerabilities, anticipating threats, etc. Now is a good time to construct and maintain an infiltration model, find and predict vulnerabilities, search for clones, convert code to a human-readable format, determine authorship, etc. So, for example, the sum of all function calls can reveal that this function is harmful at the stage. The exact match between the database virus and the function instructions is not critical.

4.1.4 Formation of Results

The last step, known as "result formation," provides all of the IoTS analysis's findings. At this point, the data collected during Stage 3 processing have been transformed into a single form (or, less frequently, a set), making them suitable for additional analysis by software and humans. The outputs of every data processing technique are at the stage's input. Depending on the goal of the analytic application, the stage's output may be seen as problem-oriented.

4.1.5 Transformation of Form and Content

The stage modifies IS task outcomes to fit the desired perspective specified in the particular SA aim. Systematisation of IoTS object properties and features; categorization of viruses and vulnerabilities; presentation of the discovered virus and vulnerability list; presentation of security metrics and data; choice of IS recommendations, etc. are typical stage activities. For instance, the result of this step might be the list itself, as infection statistics can, when it comes to the obtained list of harmful routines. Table 1 presents their Form and Content transformations for an easier to understand presentation of the steps of the IoTS.

In the process of allocating each of the following activities to the SA stages, the formalization that is presented in Table 1 will be applied in an implicit manner.

5 Systematization of SA Stages and ML Solutions

We offer the model that systematises the current study based on the task, which takes the form of the following table: tasks completed using machine learning are represented by rows, and phases of IoTS analysis are represented by columns. The research articles relevant to IoTS analysis are then taken into consideration. Based on its qualities, we allocate each job to one or more phases and one or more machine learning tasks. Therefore, theoretically, we should be able to fill every cell in the table using the whole collection of such research. The notation work groups in the

Table 1 Form and content transformations in information system static analysis

Transformations	Data collection	Data preparation	Data processing	Result formation
Form and content	$< F_2 \| C_2 >=$ $Stage_1(< F_1 \| C_1 >)$	$< F_3 \| C_3 > j =$ j $Stage_2(< F_2 \| C_2 >)$	$< F_4 \| C_4 >=$ $Stage_3(< F_3 \| C_3 >)$	$< F_4 \left(\bigcup_{j=1...M} < F_4 \| C_4 > F_5 \| C_5 >= Stage \right.$
Form	$F_2 = F_1$	$F_36 = F_2$	$F_46 = F_3$	$F_56 = F_4$
Content	$\|C_2\| < \|C_1\|$	$SC3j \equiv C2$ $j = 1 \ldots N$	$C_46 = C_3$	$C5 = SC4j$ $j = 1 \ldots M$

cells are introduced as Sx_Ty, where x denotes the analysis stage number (x = 1…4) and y denotes the ML task number (y = 1…5).A scientific publication, for instance, would fall under group S3_T1 if it describes how to look for code vulnerabilities (Stage 3) using categorization (Task 1).

The following factors led to the consideration of many papers. First, more credible evidence for the hypothesis will be evaluated if several studies have been conducted on the application of each ML problem's solution at every step of the SA. Second, in order to complete each MO assignment, it is important to ascertain the existence and extent of research for each stage of the SA. This will make it possible to determine which regions are relatively undisturbed by the others and give them more attention.

Thirdly, the success of each impacted region may be predicted based on the statistical distribution of work characteristics throughout time (e.g., the stage of SA, the job of MO, bringing to the experiment).

5.1 Collecting the Data

The authors of this work presented and evaluated the use of ML classifiers for the purpose of typing basic information systems files, including IoTS files. You can classify these pieces as S1_T1 works. In a similar vein, the authors used an SVM to tag files according to their location in the file system; this could be useful for forensic investigations. You can classify this piece as S1_T1. System crashes, data interception by secret services, storing information in "impersonal" files to prevent leaks, and other similar events can leave data files without a description. A previous effort addressed this issue. Part of the process involves comparing files by classifying their attributes into sets. Text document clustering using the time-honored TF-IDF method is demonstrated.

This method may be used for the first round of document collecting. Since they will all share characteristics, they will all be studied using comparable techniques. Similarly, clustering techniques are covered in the study [88] in order to get information ready for forensic processing by professionals in the relevant field. The S1_T4 group is the owner of the work. A classification-based approach for finding and locating machine-printed text and signatures in photos is presented. For this, multiple instance learning, or MIL, is employed. Although the technique is unrelated to IS, it may be applied to commercial and forensic document indexing. The piece is a part of the S1_T1 group is employed for clustering in MIL. As a result, S1_T4 can be applied to the job. The SVM classifier is used in a traditional manner to detect packaged executable files (PE format). This allows you to specify which items should be treated once they are unpacked. The S1_T1 group is the owner of the work. A technique for determining which documents include packaged executables is explained. On the other hand, this is accomplished via an anomaly detection technique. The work is therefore a part of the S1_T2 group.

The research shows how important it is to find abnormalities in file integration systems. The self-learning concepts form the foundation of the suggested remedy.

The S1_T2 group is the owner of the work. The lifespan of files is predicted using absolute path symbols in the applied problem of file optimisation. The lifespan of a file refers to the period between its creation and its most recent reading. Prediction is done using regression techniques based on the Random Forest and CNN Models. Files may be sorted by lifespan using this approach, and only those files that fall under a specified range can be processed. This can be applied to forensics in order to locate files pertaining to the timeline of cybercrime. The work is therefore a part of the S1_T3 group. Stego attachment detection is a traditional problem that has been addressed. A universal solution based on multiple linear regression is suggested for this reason. Furthermore, the nested message's length is acquired. The information that is disclosed can be used to further process the photos that are found in this manner. As a result of the partial localization of the stego field, the work is associated with both the S1_T3 group and, to a lesser degree, the S2_T3 group.

The use of Principal Component Analysis (PCA) to reduce dimensionality in the traditional SVM-based document classification approach is explained. The S1_T1 group is the owner of the work. It is explored how to identify handwritten writings in graphical graphics. This is accomplished using a convolutional and recurrent neural network. Sensitive information, such as passwords or personal information, may be contained in the inscriptions; so, IS may benefit from their discovery and verification. The work falls into the S1_T1 and S2_T1 categories because, in addition to the presence of a handwritten inscription, it has been translated into a particular text. The goal of the effort is to improve forensics operations. It is suggested that documents and mobile apps be processed using machine learning for categorization and clustering purposes. For this, the authors suggest using kNN and SVM. The work is therefore a part of groups S1_T1 and S1_T4.

5.2 *Preparation of Data*

The evaluated study [8] states that the works and are categorised as S2_T1.The determination of variable types, a secondary decompilation challenge, has been solved. It is suggested to employ the Random Free and SVM classifiers for this purpose, as they performed better than the others. The piece is a part of the S2_T1 group. A machine learning (ML)-based categorization technique is presented that enables very accurate identification of Function Entry Points, or the first byte of each function. In order to disassemble function instructions for additional examination, this is required. The S2_T1 group is the owner of the work. The progress of architecture reconstruction techniques is covered in the review. It includes a description of and citations for studies that recreate programme architecture using clustering. The acquired architecture may be analysed to find high-level vulnerabilities later on. The S2_T4 group is the owner of the work.

The software refactoring problem is resolved by the effort. The HASP clustering technique is suggested as a way to organise software classes into packages. The focus of work is on a novel method for regaining binary malware code that is active on embedded devices—such as Smart cards—that Sykipot is considering. This is accomplished by gathering information about the device's power usage via the side channels. The dimensionality of the feature space is reduced using PCA and LDA (Linear Discriminant Analysis) techniques, and then kNN (K-Nearest Neighbours) classification is utilised. It is demonstrated that this method is not just theoretical but also applicable in real life. This work and the one described are comparable. Equal credit for the effort is due to the S2_T5 and S3_T5 groups.

An executable binary instruction sequence is transformed into a grayscale picture. After that, dimensionality reduction via LDA is used to both compress the picture and produce a training sample that is more ideal. As a result, the work is associated with the groups S2_T5 and S4_T5, which are next to Stage 3.A suggested approach for predicting row and column separators in tabular data is based on a logistic-regression classifier. When preparing data in files for processing using the Stage 3 techniques, this chore may be frequent. The work is therefore a part of the S2_T3 group. A deep neural network-based approach to log text analysis is provided. The most important information on unsuccessful application launch attempts will be contained in the anomalies found in the text. This may be used to isolate sizable data files containing the most dubious information in the interest of SA, such as the breakdown of vital IS services. In Stage 4, the method's application will contribute to the formation of more significant outcomes. The work is therefore a part of groups S2_T2 and S4_T2.Comparable to the article, which details the tests conducted to find abnormalities in OpenStack system logs. An SVM with several cores is utilised for this. The S4_T2 group is the owner of the work. The work that has been previously detailed is a part of group S2_T3, and group S2_T1 work.

5.3 Information Processing

The following works can be categorised as S3_T1 in accordance with the reviewed review [8] as they use categorization to identify vulnerabilities. We also discuss a method that uses regression analysis to forecast security holes in upcoming Test Cases based on current ones for a single programme set. The work falls within the S3_T3 category. Review [7] states that the efforts are part of group S3_T2 and identify code abnormalities. Review [9] indicates that the works are categorised as S3_T1, S3_T2, S3_T3, and S3_T4.

The goal of the endeavour is to use machine learning to detect harmful code in software. The introduction of taxonometry highlights several process phases, including the display of the code-containing file, the detection of indicators, and the straightforward classification of the harmful code. In the last stage, machine learning classifiers including kNN, DT, Boosted Algorithms, SVM, ANN, Bayesian Networks, Naive Bayes, and OneR are employed. All rights to the work are reserved for the S3_T1 group.

Using n-grams of binary code, a method for text classification is applied. The next step is to determine whether the sample is malicious or not by applying one of several classification algorithms. The task should be labeled as S3_T1.How to identify harmful Android applications is detailed in the developed program HOSTBAD. The challenge of finding irregularities using ML is resolved by utilizing the features of received/sent SMS, received/sent calls, device activity status, and active applications/processes. An algorithm based on DCA that was similar to was published in (Dendritic Cell Algorithm). The S3_T2 group includes both components.

The usage of many binary data feature detection techniques based on n-grams for malware detection and classification is covered. These techniques include CFsSubset, Principal Components, InfoGain Attribute, Correlation AttributeEval, GainRatio Attribute, and Symmetrical UncertAttribute. SVM and PCA are applied to get the best results. Equal credit for the work is due to groups S3_T1 and S3_T5.explains the following technique for detecting malware in programmes that need user authorization. First, PCA is used to reduce the Android application's permission dimensionality. Second, malware is detected using the SVM classifier. Credit for the work is shared evenly between groups S3_T1 and S3_T5.It is explored if machine learning may be used to thwart assaults without requiring the direct use of files. Perceptron is used to find abnormalities in the command lines of common Windows operating system programmes in order to achieve this goal. The S3_T2 group is the owner of the work. A method for identifying malicious content in PDF files is explained. For this, artificial neural networks and PCA are employed. The piece is a part of the S3_T1 group. It is suggested that malevolent Android applications be viewed, and the resulting photos should be categorised. For this, SVM, KNN, and Random Forest are employed. The S3_T1 group is the owner of the work. Security research is focused on Internet of Things devices that use command line interpreters that are common to Linux shells. Malicious software is thought to be capable of both system hacking and further infection through the use of shell commands. The software programme ShellCore, which uses static code analysis to identify malware by analysing shell instructions, is the suggested remedy. Because the answer is dependent on categorization, the work may be divided into groups S3_T1 and S3_T5.The S3_T5 category includes the works that have been previously described.

5.4 Result Formation

It is suggested to use the o-glasses approach, which visualises document files—which aren't always executable—in order to look for shellcode. For $\times$ 86 binary code, a high F-measure of around 99.95% is stated. For this, a unique type of one-dimensional convolutional neural network (1d-CNN) is employed. The technique handles malware visualisation even though it applies the entire code analysis cycle. As a result, S4_T1 should be the classification for this job. This study addresses two malware detection related problems: (1) accurately identifying groups of mutant malware and (2) detecting malware signatures from logs (e.g., the xml made while executing an exe file in a sandbox) for further classifier training. Ensembles of classifiers are presented as a solution to the first difficulty. The suggestion is to use clustering to address the second issue. An approach is used to reduce the dimensionality of t-SNE (t-distributed Stochastic Neighbour Embedding) space in order to visualise the acquired findings in 2D space. Thus, the work is referred to the groups S4_1, S4_4, and S4_5, respectively, by the two methods to issue resolution and the method for visualising the malware.

The paper tackles the problem of how visualisation techniques and anomaly detection techniques (like outliers) might operate together effectively. The author's algorithm, called "hdoutliers," which differs from specific techniques (explained in several publications, etc.), is suggested for this purpose. The work is therefore a part of the S4_T2 group. It is suggested to use a method that combines numerical data with natural language text in the system log to find abnormalities. This may be used for the data gathered using the Stage 3 procedures in the last stage of SA. Consequently, the work is a part of the S4_T2 group. A machine learning approach for automatically categorising vulnerabilities based on their linguistic descriptions is presented. This work falls within the S4_T1 category as it may be used to apply an approach like this to the vulnerabilities found during the SA result formatting step, making them easier to convey to the expert.

Based on the vulnerabilities found in NVDs, an attempt is made to forecast 0-day vulnerabilities in goods. For this, regression models that are linear and quadratic are employed. Evidently, the technique may be used to forecast the emergence of new IP vulnerabilities in light of those discovered (during Stage 3). A deep neural network and logical regression may be used to estimate an image's attractiveness based on its quality. Many experiments were carried out. As a result, techniques for visualising SA findings can be modified for additional processing. The S4_T3 group is the owner of the work. A system that is comparable to this one, but categorises dangers to a smart city's transport infrastructure, is explained in. To do this, threats are divided into clusters, each of which is mapped to a certain class, using machine learning without the need for a teacher. The work is therefore a part of the S4_T4 group.

6 Conclusion

A theoretical and practical demonstration of the hypothesis regarding the use of ML in SA was presented in the study. When discussing the first formal proof, each SA Stage's execution as well as the completion of the stage's ML tasks were taken into account. From a theoretical perspective, the existence of this model and the accuracy of its description support the idea. This scientific conclusion is significant because it validates the formal proof equipment used by particular judgements to solve certain issues. Intelligent SA algorithms may be practically implemented using the model as a foundation. In relation to the second proof, an assessment of previous research relevant to the IoTS SA phases from the standpoint of the tasks resolved by ML was conducted in order to achieve this. The establishment of a single, consistent basis of intelligent solutions—that is, employing ML—in the best interests of SA is where the scientific relevance of this discovery resides. Developers of IoTS analysis systems may thus, if needed, make an educated decision on which solution to use for this SA Stage. They are able to evaluate the level of technical implementation and elaboration of the selected course of action. In order to provide IS for complicated IoTS, the resultant models that connect SA and ML enable the theoretical and practical formulation of methodological solutions.

Naturally, this necessitates the development of a suitable framework to guarantee that all phases are carried out with the entire range of machine learning techniques for Big Data and heterogeneous data. The intellectualization that the ML techniques enable will be a key component of the framework. Such systems are highly sought after in the IoTS IS domain, as several studies highlight.

Although evaluations of ML applications for IoT security exist (e.g., from the perspective of combating threats), there are some notable distinctions between the review and taxonomy suggested in this article. First off, the focus of the study is mostly on IoTS analysis; attack detection and neutralisation are not covered. Second, the analysis is precisely static (not dynamic), allowing for the early detection of system faults. Thirdly, a thorough examination of the SA phases and ML tasks enables us to make assumptions about both current and potential analytic techniques. There are currently no such reviews in any published scientific works.

References

1. Kucherova, K., Mescheryakov, S., Shchemelinin, D.: Using predictive monitoring models in cloud computing systems. In: Distributed Computer and Communication Networks. Springer International Publishing: Cham, Switzerland, pp. 341–352 (2018)
2. Buinevich, M., Izrailov, K., Vladyko, A.: Metric of vulnerability at the base of the life cycle of software representations. In: Proceedings of the 2018 20th International Conference on Advanced Communication Technology (ICACT), Chuncheon, Korea, 11–14 February 2018 (2018)
3. Komashinskiy, D., Kotenko, I.: Malware detection by data mining techniques based on positionally dependent features. In: Proceedings of the 2010 18th Euromicro conference on parallel, distributed and network-based processing, Pisa, Italy, 17–19 February 2010, pp. 617–623 (2010)
4. Ageev, S., Kopchak, Y., Kotenko, I., Saenko, I.: Abnormal traffic detection in networks of the Internet of things based on fuzzy logical inference. In: Proceedings of the 2015 XVIII International Conference on Soft Computing and Measurements (SCM), St. Petersburg, Russia, 19–21 May 2015, pp. 5–8 (2015)
5. Desnitsky, V.A., Kotenko, I.V., Nogin, S.B.: Detection of anomalies in data for monitoring of security components in the Internet of Things. In: Proceedings of the 2015 XVIII International Conference on Soft Computing and Measurements (SCM), St. Petersburg, Russia, 19–21 May 2015, pp. 189–192 (2015)
6. Kotenko, I., Saenko, I., Skorik, F., Bushuev, S.: Neural network approach to forecast the state of the Internet of Things elements. In: 2015 XVIII International Conference on Soft Computing and Measurements (SCM), St. Petersburg, Russia, 19–21 May 2015, pp. 133–135 (2015)
7. Allamanis, M., Barr, E., Devanbu, P., Sutton, C.: A survey of machine learning for big code and naturalness. ACM Comput. Surv. **51**, 36 (2017)

8. Xue, H., Sun, S., Venkataramani, G., Lan, T.: Machine learning-based analysis of program binaries: a comprehensive Study. IEEE Access **7**, 65889–65912 (2019)
9. Ghaffarian, S., Shahriari, H.R.: Software vulnerability analysis and discovery using machine-learning and data-mining techniques: a survey. ACM Comput. Surv. **50**, 1–36 (2017)
10. Kotenko, I., Saenko, I., Kushnerevich, A., Branitskiy, A.: Attack detection in IoT critical infrastructures: a machine learning and big data processing approach. In: Proceedings of the 27th Euromicro international conference on parallel, distributed and network-based processing (PDP), Pavia, Italy, 13–15 February 2019, pp. 340–347 (2019)
11. Mescheryakov, S., Shchemelinin, D., Izrailov, K., Pokussov, V.: Digital cloud environment: present challenges and future forecast. Future Internet **12**, 82 (2020)
12. Pattnaik, L.M., Swain, P.K., Satpathy, S., Panda, A.N.: Cloud DDoS attack detection model with data fusion & machine learning classifiers. EAI Endorsed Trans. Scalable Inform. Syst. **10**(6) (2023)
13. Deepan, P., Sudha, L.R.: Comparative analysis of remote sensing images using various convolutional neural network. In: EAI endorsed transaction on cognitive communications (2021). ISSN: 2313-4534. https://doi.org/10.4108/eai.11-2-2021.168714
14. Deepan, P., Sudha, L.R.: Remote sensing image scene classification using dilated convolutional neural networks. Int. J. Emerg. Trends Eng. Res. **8**(7), 3622–3630 (2020). ISSN: 2347-3983
15. Satpathy, S., Swain, P.K., Mohanty, S.N., Basa, S.S.: Enhancing Security: Federated Learning against Man-In-The-Middle Threats with Gradient Boosting Machines and LSTM. In: 2024 IEEE International Conference on Advanced Video and Signal Based Surveillance (AVSS), pp. 1–8. IEEE (2024, July)
16. Satpathy, S., Pradhan S.K., Ray B.B.: A digital investigation tool based on data fusion in management of cyber security systems. Int. J. Inf. Technol. Knowl. Manag. June 2010, **3**(2), pp. 561–565 (2010). http://www.csjournals.com/IJITKM/PDF%203-1/77.pdf
17. Buinevich, M., Izrailov, K., Stolyarova, E., Vladyko, A.: Combine method of forecasting VANET cybersecurity for application of high priority way. In: Proceedings of the 2018 20th International Conference on Advanced Communication Technology (ICACT), Chuncheon, Korea, 11–14 February 2018, pp. 266–271 (2018)
18. Deepan, P., Sudha, L.R.: Deep learning and its applications related to IoT and computer vision, artificial intelligence and IoT: smart convergence for ecofriendly topography. Springer Nature, p. 223–244 (2021)
19. Raju, A.D., Abualhaol, I.Y., Giagone, R.S., Zhou, Y., Huang, S.: A survey on cross-architectural IoT malware threat hunting. IEEE Access (2021)
20. Mohanty, S.N., Singh, T., Goel, R., et al.: A study on building awareness in cyber security for educational system in India using interpretive structural modellings. Int. J. Syst. Assur. Eng. Manag. (2024). https://doi.org/10.1007/s13198-024-02273-3
21. Mohapatra, S., Yesubabu, M., Sahoo, A., Mohanty, S., Mohanty, S.N.: IoT-based ubiquitous healthcare system with intelligent approach to an epidemic. In: Recent Patents on Engineering. https://doi.org/10.2174/0118722121240884230926092316

Empowering Industries with IoT and Machine Learning Innovations

Application of Machine Learning in the Internet of Things

Shantanu Bhattacharya, Shubham Kumar, Aditi Bhattacharya, Anjanna Matta, and G. Sucharitha

Abstract In the current world data is more important to analyze the surroundings. All the new inventions or future predictions are made using IoT. These data to make our life easy. These days the maximum of data collection is done through IoT, because it is easy and the data can be stored in an organized way. Many countries are working on how to protect this data. The major issue in the world is data security or protection and the topic of research that is going on IoT devices is that these devices should be made that intelligent, so that they can detect useful information and store it on a cloud and storage of inadequate information can be avoided. So, how can we make our IoT devices intelligent. Here machine learning comes into the picture. Machine learning (ML) helps that device to analyze things from the given instructions and store the information on the cloud. ML helps developers to classify the data and clustering/finding pattern in it. With this help, IoT devices can scan the product and analysis of the product using past data or in-build functions. Using Machine learning IoT devices will not only increase the performance of the devices but it will also help in the security of the devices and the data. Whenever a product is related to social welfare or any security-based system, the security of that particular device is given priority because the data that the device is transmitting or receiving may be important or highly confidential. So, if the device is secured, then the leakage of data can be avoided. Apart from this machine learning can be used to detect malicious attacks, analyzing mobile endpoints, a combination of Ml & AI to learn human behavior and to automate wearisome security tasks, and many more. There are many other applications of machine learning in IoT like advancing smart city projects, smart transportations, managing Big Data, thread detection, any kind of accident tracker, and many more smart devices that will help to optimize the remote areas of a country.

S. Bhattacharya · S. Kumar · A. Bhattacharya · A. Matta (✉)
Department of Mathematics, Faculty of Science and Technology (IcfaiTech), ICFAI Foundation for Higher Education, Hyderabad, Telangana 501203, India
e-mail: anjireddyiith@ifheindia.org

S. Bhattacharya
e-mail: shantanu.bhattacharya18@ifheindia.org

G. Sucharitha
Institute of Aeronautical Engineering, Hyderabad, India

S. N. Mohanty et al. (eds.), *Explainable IoT Applications: A Demystification*, Information Systems Engineering and Management 21,
https://doi.org/10.1007/978-3-031-74885-1_3

Keywords Internet of Things · Big Data · Machine learning · Smart devices

1 Introduction

In this chapter, it will be discussing the application of machine learning in the Internet of Things based projects. We all know that our world population is very huge and it is increasing day by day by which the data that is produced per day is very large and it is becoming quite challenging in managing and securing these data. On the other side, the task is not only to protect data, but the data should also be segregated in a particular format, i.e. splitting the information in a different section which will make data analysis easy [1]. All these sounds very good, but the implementation was not possible by it devices only. Here, used Machine Learning algorithms here to detect and analyze the information that IoT device is transmitting so that the useful information is stored successfully. But still, there are lots of challenges that IoT devices are still facing which are solved by Device management, Data integrations, and many more [2]. Here one more thing that needs to be specified is that Machine learning is not only helping it to analyze data (or handling back-end), but it has also sent the filtered information to devices back for decision making [3, 4]. For example, a lot of research is going on "Self-Driving Cars" [5–7]. This mechanism is completely based upon sensors and rapid decision making. These decisions are made possible with the help of ML algorithms which helps the sensors to verify the objects that are coming in front of the car and process it to define the path that cars should follow. In further sections we will discuss in detail, how we can apply a machine learning algorithm to solve a problem in IoT.

In short, we can tell that involvement of the Machine learning algorithm is now reducing human interaction with devices with is making our life easier and more comfortable, not only these sensors are integrating their performance by adopting computational methodologies they are also becoming more efficient and consistent in data or object description.

2 Introduction to Internet of Things (IoT)

IoT refers to the Internet of things that can be referred as "Connecting things to the internet that can enable them to collect and exchange data between each other. An IoT ecosystem consists of many embedded systems such as processors, sensors, and other hardware components used to collect and store information in them. This information or data are used for the analysis of the product or can also be used to check the performance. All IoT based devices are independent of human interaction with them. The connectivity and communication between any IoT devices depend upon the instruction deployed in them [8–12]. These days all the companies are taking the help of its devices to make them work smarter and easier [13, 14]. Importance of Internet of Things in organizations/Industries is,

- Monitoring their total business processes
- Develop their customer experience
- Improve employee productivity
- Analyzing and integrating various models.

Importance of Internet of Things (in general) is,

- Access data from everywhere at any time on any device.
- Developers the communication between connected devices, whether it can be 2 or many devices connected to each other.
- Easy transformation of "Data packets" over the connected devices.
- Automation of daily tasks to improve human intervention.

2.1 Internet of Everything (IoE)

The data can do many things. Before doing the action, it is needed proper network connections. Current days Internet of Everything (IoE). Simply, it can be defined as an "Intelligent connection between people, process, data, and things. We have observed in definition that IoE is based on Data, Process, Things, and People, let us discuss them in some detail (these four are generally called as Pillars of IoE).

- **Data**: Devices gather information based upon the task or area they are made for, by which this data is used by data Analysists to analysis the data and use them accordingly.
- **Process**: Every data that we store in these devices are not required for every person present in their connection. So, delivering the appropriate information to appropriate the person is the most important thing carried out in this part.
- **Things**: All the physical devices that are connected to the internet and with each other for creating decisions are called Internet of things.
- **People**: With all these interconnected devices people connect or use these technologies to complete their work.

What is the difference between Internet of Things (IoT) and Internet of Everything (IoE)?

The major difference between these two technologies is that IoE is based on pillars where as IoT focusses on physical objects only, it is the interconnectivity of physical objects that transfers and receives data whereas IoE is combining many other technologies for wider usage.

Although these two are different, two similarities should know, they are

- These two systems are decentralized i.e. there is no single control point, they connect different devices with the help of nodes, so that they need not to depend

on each other to perform the task. With the help of this technology, we can avoid delays in transformation of instructions.
- As we know that these two systems are distributed by which they possess a highly vulnerable to penetration and cyberattacks. This can also be said as "The higher the devices are connected to a network, the higher the susceptibility to breaches".

Advantages of Internet of Everything at workplace are,

- Connecting roads with hospitals.
- Connecting people and food in the supply chain.
- Connecting all home appliances for comfortable living.

2.2 Architecture of IoT

The architecture of the Internet of things is divided into 4 layers i.e.

Layer-1: Sensors and Controllers

As we all know that IoT based devices are completely dependent on hardware components because these things are first responsible for transferring the data from their memory all the connected databases or data centers. Now, you may have a question that from where and how we are collecting this information? To answer this question, we have to first know about sensors. A sensor is a module or embedded device that detects the changes happening in the environment and send that changes to data centers. These sensors are arranged or grouped according to their purpose and data types and the main reason for using sensors in a maximum of the appliances is that they consume very less power and can connect to other modules in low connectivity also. Another important part of this layer is actuators. An actuator is a module that is responsible for the movement of a model or system. Many types of actuators are most commonly used are Electrical actuators (i.e. Converts electric energy to mechanical torque), Mechanical linear actuator (i.e. Converts rotatory motion to linear motion), and Hydraulic actuator (i.e. Converts a physical compression to a mechanical motion). It is also that connected systems or modules should not be capable of interacting with each other in their gateway, but they should also be capable of recognizing and talk to each other to improve the whole development.

Layer-2: Gateways and Data Acquisition

The functionality of this layer is also similar to layer-1 because we get data from sensors and actuators. But it is essential to describe this IoT layer because we process the data, filter it, and transfer to edge infrastructure and cloud-based platforms. Here the data that we get from sensors converted to other formats so that it makes the work easier to filter, control, and reject the unnecessary data, by which the volume of information can be minimized and which will positively affect network transmission cost and response time. Another aspect is security because data transfer is bidirectional, so to prevent data leakage IoT systems are deployed with proper encryptions and security tools. This will also reduce malicious attacks on IoT devices.

Layer-3: Edge Analytics

When we have a large IoT project or a large architecture then there is a chance of a loophole. And in the face of limited accessibility of data and data transfer these edge modules can provide a faster response to any changes in the system or their environment. These edge computing is becoming more popular in the industry because of its high performance in data transmission. If we consider a real-time project these edge infrastructures are located close to a data source so that it will be easy to react to any change in the environment and produce the output in the form of instant actionable intelligence. By this, the data which require the power of the cloud can be processed and forwarded. In this process, we also minimize network exposure to enhance security and reduce bandwidth consumptions for better performance.

Layer-4: Data Center or Cloud Platform

Do you agree that everything that can sense the situation and react to it have a central body or a component that acts as a brain for that? If we consider sensors as the neurons and gateways as a backbone of an IoT system, the cloud acts like a brain for the complete system because on contrary to edge technologies, cloud computing or data centers are designed to store, perform filtering process and to have the ability for deep analysis of massive data with the help of data analytics engines and Machine learning algorithms. When all these advantages were accepted by industries the cloud computing started to contribute to higher production rates in the Internet of Things projects. This cloud computing has made the analysis work easier and the problem of monitoring data is resolved.

3 Application of Machine Learning (ML)

Machine learning (ML) was presented in the last bit of the 1950s as a procedure for artificial intelligence (AI). After some period, it is stimulated more to algorithms that are computationally appropriate and amazing. In the most recent era, machine learning systems has been utilized thoroughly for a wide degree of tries, including depictions, lose the faith, and thickness evaluation in a collection of utilization areas, for example Bioinformatics, talk confirmation, spam affirmation, computer vision, intimidation zone and propelling associations [15–22]. The counts and techniques start from Recent Machine Learning Applications in the Internet of Things (IoT) different fields including bits of information, number shuffling, neuroscience, and programming designing and utilized even visitor, or most areas identified with machines these days [13].

The accompanying two outdated definitions catch the substance of machine learning:

(1) The improvement of computer models for learning estimates that offer responses to the issue of data acquirement and overhaul upgrade the performance of advanced systems.
(2) The gathering of computational procedures for enlightening machine execution by distinguishing and depicting textures then models in preparing data.

3.1 Algorithms

Algorithms or science accepts the most important capacity in machine learning, for this is the gadget to deal with the data [11]. Here, some fundamental concepts of machine learning are talked about just as the oftentimes useful machine learning algorithms for smart data analysis [12].

4 Classifications

4.1 K-Nearest Neighbors

It is broadly dispensable, in actuality, situations since it is a non-parametric algorithm, which implies it doesn't make any presumption on principal data. K-Nearest Neighbors is quite possibly the most fundamental yet basic classification algorithms in Machine Learning. It has a place with the coordinated learning domain and discovers extraordinary application in model affirmation, data mining, and interference area. One obstruction of KNN is that it requires a limit to the entire arranging set, which makes KNN unscalable to immense illuminating collections.

4.2 Support Vector Machines (SVM)

SVMs are popular for their ability to handle high-dimensional data and their effectiveness in finding optimal decision boundaries, especially in cases where the classes are not linearly separable. Additionally, SVMs can be extended to handle non-linear decision boundaries through techniques such as kernel methods, which map the input features into a higher-dimensional space where the classes become separable by a hyperplane. Overall, SVMs are widely used in various machine learning tasks, including classification, regression, and outlier detection. SVMs are among the best immediately available overseeing learning models that are set up to do sufficiently overseeing high-dimensional enlightening assortments and are capable to the extent memory use, inferable from crafted by help vectors for desire. One immense detriment of this model is that it doesn't clearly give probability measures as mentioned in Fig. 1.

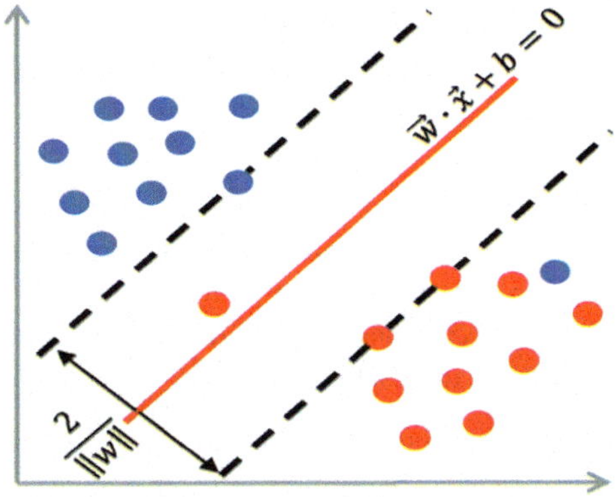

Fig. 1 Linear support vector machine

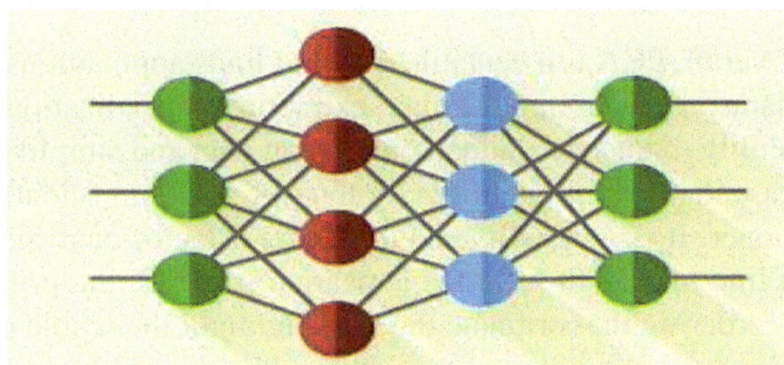

Fig. 2 Node localization in WSNs in 3D space using supervised neural networks

4.3 Neural Network

A neural network is an interrelated gathering of clear dealing with segments, units or centers, whose value is roughly established on the animal neuron. Neural networks are regularly used for statistical analysis and data illustrating, in which their occupation is viewed as a choice as opposed to standard nonlinear regression or cluster analysis techniques. Accordingly, they are ordinarily used in issues that may be outlined with respect to classification, or forecasting. Figure 2 describes about how neural networks are used in 3D visualization of data or node localization.

4.4 Bayesian Statistics

Bayesian Statistics are a strategy that assigns "levels of conviction," or Bayesian probabilities, to ordinary statistical modeling. In this interpretation of estimations, the probability is resolved as the reasonable craving for an event happening subject to as of now known triggers. Or on the other hand, thusly, that probability is a novel cycle that can change as new information is aggregated, instead of a fixed worth subject to repeat or affinity.

4.5 Decision Tree (DT)

Decision tree algorithms are significant, entrenched machine learning procedures that have been utilized for a wide scope of utilizations, particularly for classification problems. Decision tree gives a nonparametric strategy to parceling datasets. Alternative data mining procedures incorporate regression or ANOVA models that speak to associations between factors as cross-product between them. Decision tree algorithms incorporate negligible necessities for data preparation and robust performance enormous data sets.

4.6 Principle Component Analysis (PCA)

Overall, PCA is a versatile tool that finds applications in various domains, including data preprocessing, feature extraction, visualization, and pattern recognition. Its ability to uncover hidden patterns in data and simplify complex datasets makes it an essential technique in the toolbox of any data scientist or machine learning practitioner. It is generally called a general factor analysis where regression determines a line of best fit [20]. As appeared in Fig. 3, the primary parts utilize the included reordering that ordinate the first highlight the whole data.

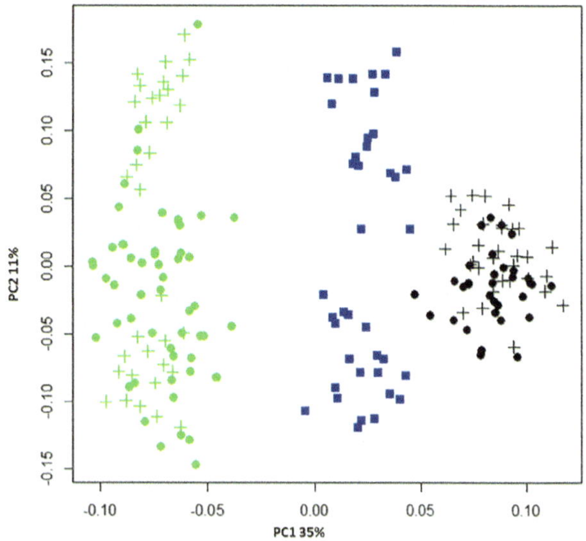

Fig. 3 A simple 2D visualization of the principal component analysis algorithm

4.7 K-Means Algorithms

K-means clustering is widely used in various fields such as machine learning, data mining, image analysis, and pattern recognition for tasks such as segmentation, compression, and data reduction. It's a simple and efficient algorithm, but its performance can depend on the choice of initial centroids and the inherent structure of the data. This results in a separating of the data space into Voronoi cells. It is renowned for cluster analysis in data mining [21]. These alternative clustering algorithms can be particularly useful when dealing with non-Gaussian data distributions or when the presence of outliers is a concern. However, it's important to note that each algorithm has its own advantages and limitations, and the choice of algorithm should be made based on the specific characteristics of the dataset and the objectives of the analysis.

4.8 Reinforcement Learning

Reinforcement learning is a region of Machine Learning. It is connected to making a fitting move to intensify grant in a particular condition. Figure 4 describes a sample working tree of Reinforcement Learning. It is used from various programming and machines to find the best lead or way it should take on a specific condition. Reinforcement learning taking in fluctuates from the controlled learning in a way that in managed learning the arrangement data has the fitting reaction key with it so the model is set up with the privilege react to itself while in help learning, there is no answer aside from the stronghold administrator picks what to never really out the given endeavor. Without a training data set, it will without a doubt pick up from its experience.

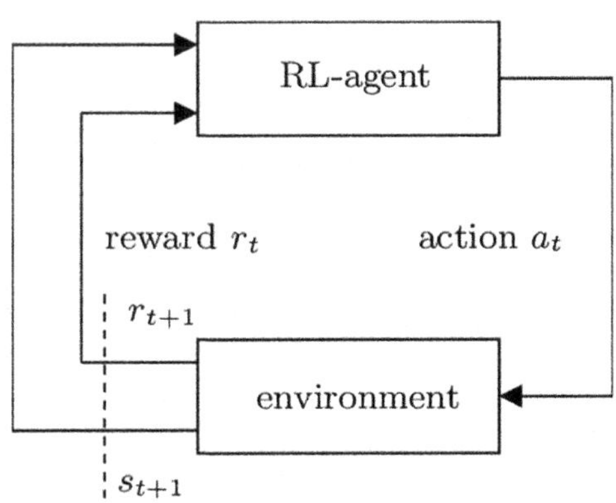

Fig. 4 Reinforcement learning

5 ML Application to IoT

ML is the rule procedure among these computational applications to it. Likewise, there are stacks of utilization both in investigation and industry.

5.1 Cost Savings in Industrial Applications

Prescient capacities are extraordinarily useful in a mechanical setting. By attracting data from various sensors or on machines, machine learning estimations can "apprehend" what's ordinary for the machine and thereafter distinguish when something bizarre begins to happen. Predicting when a machine needs upkeep is unbelievably significant, changing over into countless dollars in saving expenses. Companies are presently utilizing machine learning to foresee with over 90% exactness when machines will require support, which means gigantic cost cuttings.

5.2 Embellishment Experiences to Individuals

We're when in doubt all acquainted with machine learning applications in our standard ordinary existences. Both Amazon and Netflix use machine figuring out some approach to take in our propensities and give a superior trouble than the client. That could mean proposing things that you may like or giving important suggestion to films and TV shows. So also, in IoT machine learning can be commonly enormous in outlining our condition to our own propensities [19]. Overall, the Nest Thermostat demonstrates how machine learning can be integrated into everyday devices to create smart, energy-efficient homes while enhancing user comfort and convenience.

5.3 Smart Transportation

These days, everything from toothbrush to beds are getting smarter and smarter using IoT and Machine learning models. With this the introduction to IoT in field of transportation makes them to "feel" and "think" which leads in the development of Intelligent Transportation Systems (ITS) [3]. The most significant topic of research is to work on transport navigation and route optimization. Smart transportation will counter many day-to-day problems like route optimization, parking, lights, accident detection, road anomalies, infrastructure [5]. Let's discuss these in brief subsections.

5.4 *Route Optimization*

The purpose of Route Optimization is to find the best route to his destination in order to minimize the traffic jams by which time and fuel consumption can he saved which will make it economically friendly for the client [1]. The main ideology that we want to present is to explore the capabilities of mobile Crowed-sensing for ITS by using a swarm intelligent algorithm development of Route Optimization.

5.5 *Parking*

There are many companies working in Smart parking systems and are still under trials. This is developed to track availability of parking lots and sort parking slots as per customer requirements. Not only this some parking mechanism are developed to send notification bills to the user based on "How much time they have parked their vehicle?". These IoT devices are connected to a centralized server which manages the functionality of these device data [17]. These parking systems use ultrasonic sensors to check parking space and WIFI module to share data with users and the servers as well.

5.6 *Accident Detection*

Accident cases are increasing day by day in the world as well as the death cases which made this section as the major part of research in Smart transportation which will help many countries to control the traffic speed. Basically, this system will have rebuilt accident-prone areas and will notify the driver about this which will prevent accidents. Now the question comes is there are many accidents prone areas still to be updated or in the future if we want to add a new area then? How will it work? There are certain steps that are followed in accident detection as mentioned in Fig. 5, if not, the answer is this system will not only contain the inbuilt data, it will also gather information from the sensors present on roads which will continuously integrate the build-in data [9]. Here we can use DRAM method which is used in object detection from image data.

6 Future Scope

Machine learning is research locale that has pulled in a huge load of astonishing characters and it can unveil further [23, 24]. Regardless, the three most critical future sub-issues are picked to be discussed.

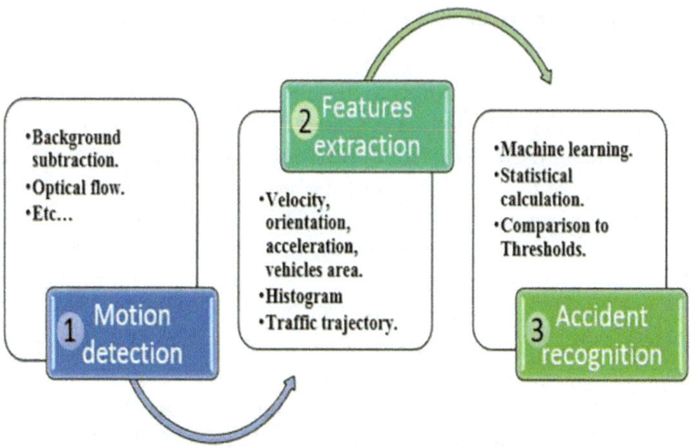

Fig. 5 Steps involved in accident detection

6.1 Explaining Human Learning

A referred to previously, AI hypotheses have been perceiving fitting to comprehend features of learning with people and creatures. Support learning calculations gauge the dopaminergic neurons incited exercises in creatures during remuneration-based learning with astonishing precision. ML algorithms for enlightening sporadic delineations of normally showing up pictures anticipate visual highlights distinguished in creatures' underlying visual cortex. All things considered, the significant drivers inhuman or creature learning like incitement, repulsiveness, critics, hunger, intuitive activities, and learning by experimentation throughout various time scales, are not yet considered in ML algorithms. This a potential instance to find a more summed up idea of discovering that entails both creatures and machines.

6.2 Programming Languages Containing Machine Learning Primitives

Indeed, the integration of machine learning (ML) algorithms with traditional programming languages is commonplace in modern applications. Often, ML algorithms are used as components within larger software systems, where they perform tasks such as data analysis, pattern recognition, or predictive modeling [18]. By defining a set of data sources and desired outputs, developers can explore the available machine learning methods in these languages and select the most appropriate ones for their specific problem domain and objectives. This approach enables them to quickly prototype and iterate on machine learning solutions without having to implement algorithms from scratch, ultimately accelerating the development process and improving the effectiveness of their programs already using this thought in a more unassuming degree [10]. Regardless, a charming new request is raised to develop a model to define relevant learning experience for each subroutine named as "to be dominated", timing, and security in the example of any unforeseen modifications to the program's function.

6.3 Perception

A summed-up idea of computer perception discernment that can connect ML algorithms that are used in various types of computer perception, insight today, including yet not restricted to exceptionally progressed vision, discourse acknowledgment, etc. is another potential research zone [15, 16, 25–30]. One idea provoking problem the coordination of different senses (e.g., sight, hear, contact) to set up a framework that utilizes self-regulated figuring out how to gauge one tactile knowledge using the others. Researches informative brain research have noted more powerful learning in humans when various input modalities are provided and concentrates on co-preparing techniques indicate similar results.

7 Summary

This study highlights the facts on application of Machine Learning algorithms on the Internet of Things projects. Machine learning gives a simple IoT project or system to think and analysis the situation and react to it based on the output required. We have also discussed about few ML algorithm majorly used to integrate. This integration has helped developers in developing systems like smart transportation, smart home, smart city, even smart classes for students, google analytics and many daily usage tools which are helping the users in their day to day life.

References

1. Yang, J., et al.: Optimization of real-time traffic network assignment based on IoT data using DBN and clustering model in smart city. Fut. Gener. Comp. Syst. **108**, 976–986 (2020)
2. Hashem, I.A.T., et al.: The role of Big Data in smart city. Int. J. Inform. Manage. **36**(5), 748–758 (2016)
3. Talari, S., et al.: A review of smart cities based on the Internet of Things concept. Energies **10**(4), 421 (2017)
4. Jain, B., et al.: A cross layer protocol for traffic management in Social Internet of Vehicles. Fut. Gener. Comp. Syst. **82**, 707–714 (2018)
5. Zantalis, F., et al.: A review of machine learning and IoT in smart transportation. Fut. Internet **11**(4), 94 (2019)

6. Keni, P., et al.: Automated street lighting system using IoT. Int. J. Adv. Res. Ideas Innov. Technol. **4**(3), 1970–1973 (2018)
7. Chowdhury, D.N., et al.: A vehicle-to-vehicle communication system using IoT approach. In: 2018 Second International Conference on Electronics, Communication and Aerospace Technology (ICECA). IEEE (2018)
8. Sucharitha, G., Tannmayee, B., Dwarakamai, K.: Revolution in IoT: smart wearable technology. In: Internet of Things and its Applications, pp. 407–425 (2022)
9. Cui, L., et al.: A survey on application of machine learning for Internet of Things. Int. J. Mach. Learn. Cybern. **9**, 1399–1417 (2018)
10. Andročec, D., Vrček, N.: Machine learning for the internet of things security: a systematic. In: 13th International Conference on Software Technologies, vol. 4120 (2018)
11. Sucharitha, G., Mandeep Sai, M.: Developments in agriculture technology using Internet of Things. In: Internet of Things and Its Applications, pp. 341–360 (2022)
12. Sucharitha, G., et al.: Custom manufacturing using Industry 4.0: cost-effective industry revolution model. In: Cloud Analytics for Industry 4.0, pp. 17 (2022)
13. Sucharitha, G., et al.: The impact of green manufacturing in Industry 4.0 for future ecosystems. In: Cloud Analytics for Industry 4.0, pp. 1 (2022)
14. Desai, A.M., Jhaveri, R.H.: The role of machine learning in Internet-of-Things (IoT) research: a review. Int. J. Comp. Appl. **179**(27), 0975–8887 (2018)
15. Miettinen, M., et al.: IoT sentinel: Automated device-type identification for security enforcement in IoT. In: 2017 IEEE 37th International Conference on Distributed Computing Systems (ICDCS). IEEE (2017)
16. Meidan, Y., et al.: ProfilIoT: a machine learning approach for IoT device identification based on network traffic analysis. In: Proceedings of the Symposium on Applied Computing (2017)
17. Kotenko, I., et al.: Neural network approach to forecast the state of the Internet of Things elements. In: 2015 XVIII International Conference on Soft Computing and Measurements (SCM). IEEE (2015)
18. Baldini, G., et al.: Physical layer authentication of Internet of Things wireless devices through permutation and dispersion entropy. In: 2017 Global Internet of Things Summit (GIoTS). IEEE (2017)
19. Parmaksız, H., Karakuzu, C.: A review of recent developments on secure authentication using RF fingerprints techniques. Sakarya Univ. J. Comp. Inf. Sci. (2022)
20. Jeong, H.-J., Lee, H.J., Moon, S.M.: Work-in-progress: cloud-based machine learning for IoT devices with better privacy. In: 2017 International Conference on Embedded Software (EMSOFT). IEEE (2017)
21. Jincy, V.J., Sundararajan, S.: Classification mechanism for IoT devices towards creating a security framework. In: Intelligent Distributed Computing. Springer International Publishing (2015)
22. Nobakht, M., Sivaraman, V., Boreli, R.: A host-based intrusion detection and mitigation framework for smart home IoT using OpenFlow. In: 2016 11th International Conference on Availability, Reliability and Security (ARES). IEEE (2016)
23. Mahmood, M.T., Ahmed, S.R.A., Ahmed, M.R.A.: Using machine learning to secure IOT systems. In: 2020 4th International Symposium on Multidisciplinary Studies and Innovative Technologies (ISMSIT). IEEE (2020)
24. Do, V.T., et al.: Strengthening mobile network security using machine learning. In: Mobile Web and Intelligent Information Systems: 13th International Conference, MobiWIS 2016, Vienna, Austria, August 22–24, 2016, Proceedings 13. Springer International Publishing (2016)
25. Liu, G., Mao, S., Kim, J.H.: A mature-tomato detection algorithm using machine learning and color analysis. Sensors **19**(9), 2023 (2019)
26. Srinivas, M., Sucharitha, G., Matta, A. (eds.): Machine Learning Algorithms and Applications. John Wiley & Sons (2021)
27. Samudrala, S.: Machine Intelligence: Demystifying Machine Learning, Neural Networks and Deep Learning. Notion Press (2019)

28. Stockheim, T., Schwind, M., Koenig, W.: A reinforcement learning approach for supply chain management. In: 1st European Workshop on Multi-agent Systems. Oxford, UK (2003)
29. Jena, P.K., Khuntia, B., Palai, C., Nayak, M., Mishra, T.K., Mohanty, S.N.: A novel approach for diabetic retinopathy screening using asymmetric deep learning feature. Big Data Cogn. Comput. **7**(1), 25 (2023). ISSN: 2504-2289. https://doi.org/10.3390/bdcc7010025
30. Sharma, N., Mangla, M., Mohanty, S. N., Gupta, D., Tiwari, P., Shorfuzzaman, M, Rawashdeh M.: A smart ontology-based IoT framework for remote patient monitoring. Biomed. Sig. Process. Control (ELSEVIER) **68**(2), 102717–102729 (2021). ISSN: 1746-8094. https://doi.org/10.1016/j.bspc.2021.102717

A Framework for Sustainable Smart Healthcare Systems in Smart Cities

Chandrakant Mallick, Parimal Kumar Giri, and Bijay Kumar Paikaray

Abstract In the last few years, there has been a rise in worldwide interest in smart cities as a solution to urban difficulties, notably in healthcare, which uses technology to improve accessibility and efficiency of healthcare systems. The chapter discusses the role of smart cities on healthcare, highlighting their potential for improved global services through real-time monitoring, personalized treatment through predictive analytics, IoT devices, sensors and smart mobile apps. As an outcome of this study, it suggests a technological framework for smart city platform to transform healthcare practices to build specialized, adaptive healthcare systems in light of Sustainable Development Goals. The key elements of this framework emphasize on privacy and security measures, environmental health monitoring, collaboration initiatives, scalability, and sustainability strategies. Smart healthcare solutions can enhance accessibility, reduce costs, optimize resource allocation, and streamline patient care, leading to faster medical response times and reduced hospital workload.

Keywords Smart city · IoT · Smart health · Artificial Intelligence · Information and communication technology · Mobile health

1 Introduction

A smart city is an urban infrastructure that makes use of ICT and IoT devices to collect and analyze data, improving quality of life, sustainability, and urban services [1]. Smart cities nowadays use technology like the IoT [2], Big data analytics [3], cloud computing [4], AI and Machine Learning [5], smart grids [6], urban mobility

C. Mallick (✉) · P. K. Giri
Department of Computer Science and Information Technology, Gandhi Institute of Technological Advancement (GITA) Autonomous College, Bhubaneswar, India
e-mail: ckmallick@gmail.com

B. K. Paikaray
Department of Computer Science and Engineering, Siksha 'O' Anusandhan (Deemed to be) University, Bhubaneswar, Odisha, India

S. N. Mohanty et al. (eds.), *Explainable IoT Applications: A Demystification*, Information Systems Engineering and Management 21,
https://doi.org/10.1007/978-3-031-74885-1_4

solutions, and civic engagement platforms [7] to improve urban life. These systems offer real-time monitoring, data analysis, energy efficiency, traffic control, and public interaction, thereby improving smart city services and quality of life of people. The main components of smart city infrastructure are shown in Fig. 1 and include the following [8].

Internet of Things (IoT)—based wearable and ubiquitous devices enable remote patient monitoring, providing real-time health status information to healthcare providers, enhancing healthcare services delivery, and improving response to patients through seamless connectivity and data exchange [9].

Smart wearable devices monitor the health conditions including blood pressure level, heart bit rate, and oxygen saturation levels etc. This facilitates medical professionals to make prompt, well-informed decisions that guarantee patients receive the best care possible [10].

***Artificial intelligence** (AI)*—based advanced healthcare systems using data analysis provides crucial insights for timely disease diagnosis and treatment, enabling healthcare professionals to make immediate decisions and improve treatment outcomes [11].

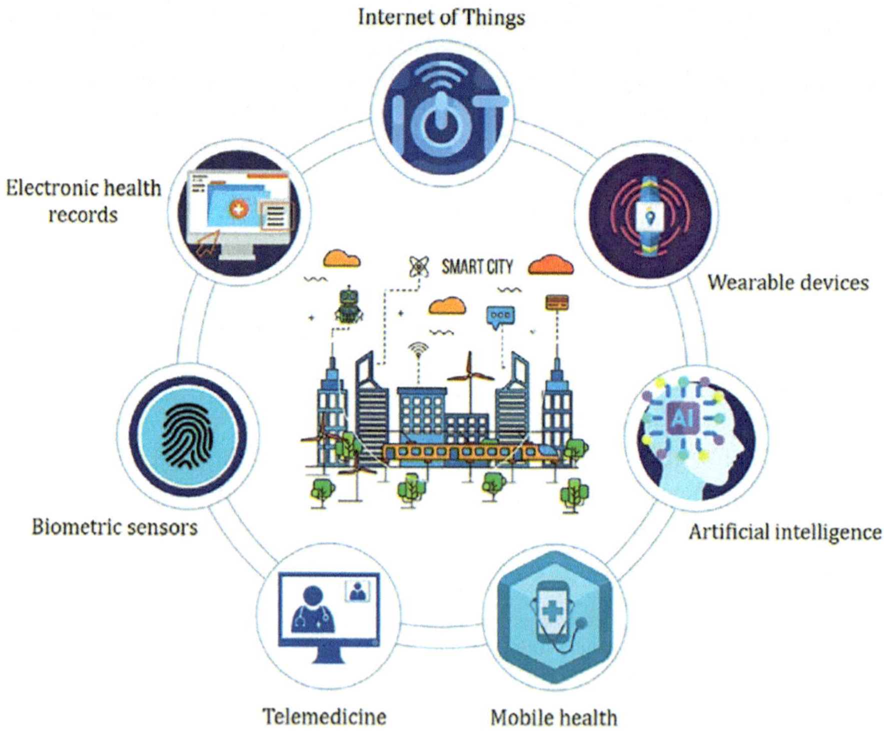

Fig. 1 Smart city infrastructure [9]

Telemedicine offers patients remote consultations with healthcare providers, reducing travel costs and infection risks, and enhancing convenience and accessibility in healthcare [12].

Electronic Health Records (EHRs) enhance healthcare by enabling healthcare providing organizations to retrieve patient data anytime and anywhere, promoting high-end healthcare services, informed decision-making and optimal patient outcomes [13].

Mobile health applications enable people to take control of their chronic conditions, track their health, and communicate with healthcare professionals, promoting personalized care and improved health outcomes through closer healthcare connections [14].

Biometric sensors monitor patients' movements, alerting healthcare providers of unusual activity, improving patient safety and facilitating quick actions to health emergencies for improved patient care [15].

Smart cities utilize ICTs in various applications such as city management, real estate, utilities, healthcare, education, public safety, economy, governance, mobility, environment, social systems, services, resources, infrastructure, and the natural environment [16, 17]. This study explores the impact of integrating the Sustainable Development Goals (SDGs) on the development of smart healthcare systems in smart and sustainable cities. It is evidenced that in recent years, smart cities are transforming the citizens' quality of life with technology, focusing on "smart health" to improve healthcare delivery, promote well-being, and meet the SDGs [18, 19]. The rest part of this text throws light at the nexus of smart cities, smart health, and sustainability, with an emphasis on how smart health projects help to achieve the SDGs and finally suggest a technological framework for the smart healthcare systems in smart cities in the developing countries.

2 Smart Healthcare Systems

Smart health refers to a comprehensive approach to healthcare that uses digital infrastructure, linked devices, and data-driven insights [20]. The key components of smart health are:

2.1 Telemedicine and Remote Monitoring

Telehealth services allow for remote consultations, diagnostics, and monitoring. Wearable gadgets capture health data in real time that facilitates healthcare practitioners to monitor the patients' health conditions and respond early. IoT is transforming healthcare with better patient care enabling remote monitoring, telemedicine, and collecting patient data for better support and check-up recommendations [21].

2.2 Health Information Systems

Healthcare providers can more easily communicate data by integrating electronic health records (EHRs) with health information exchanges. Interoperability improves care coordination and avoids duplication of services [22].

2.3 Predictive Analytics

Machine learning algorithms use health data to anticipate disease outbreaks, identify high-risk groups, and optimize resource allocation. Predictive models assist in averting epidemics and enhance public health planning [23].

2.4 IoT-Enabled Healthcare Infrastructure

Smart cities use IoT devices to monitor air, water, and sanitation. These environmental elements have a huge influence on health outcomes, and real-time data helps policymakers make informed decisions [24].

2.5 Cloud Computing and Healthcare

Healthcare organizations utilize cloud computing for real-time data collection, storage, and sharing, enabling effective decision-making and patient treatment. Cloud infrastructure offers high storage volume and throughput, which is crucial for a large population of patients [25].

2.6 Bigdata and Healthcare

Big data is generated and analyzed by public and private sectors, including healthcare, to improve services, including scientific research, internet of things gadgets, patient medical records, and hospital records [26].

3 Achieving Sustainable Development Goals

Smart health and smart cities can contribute to the SDGs by utilizing technology and data-driven techniques to address social, economic, and environmental issues, fostering innovation, efficiency, and sustainability in urban development and healthcare delivery [27]. The smart city initiatives can address some of the SDGs in the following directions.

3.1 Access to Healthcare

Smart health directly contributes to SDG 3 (Well-being and good health) by increasing access to excellent healthcare, lowering death rates, and enhancing overall well-being. Telemedicine crosses geographical distances, particularly in underprivileged regions, while data-driven preventative interventions combat disease [28]. Telemedicine, remote health monitoring, and mobile health apps can all help to enhance access to healthcare service platforms, especially in underprivileged communities. This contributes to attaining SDG 3 by establishing universal health coverage and lowering maternal and child mortality rates [29].

3.2 Preventive Healthcare

Smart health systems, which use data analytics and IoT devices, can enable early illness identification and promote preventative healthcare actions. This is consistent with SDG 3, since it reduces illness burdens and promotes healthy lifestyle choices [30].

3.3 Technology and Innovation

Smart cities invest in robust healthcare infrastructure, including telecommunications networks, data centers, and medical facilities. Mobile health apps and wearable technology enable people to engage in and manage their own health. Smart cities can achieve SDG 9: Industry, Innovation, and Infrastructure in Healthcare by investing in advanced technologies, building resilient infrastructure, and fostering a culture of innovation [31]. This can improve access to quality care, enhance efficiency, and contribute to progress towards SDG3 [32].

3.4 Digital Inclusion

Ensuring fair access to internet connectivity and digital technology is critical for smart health and smart city initiatives to benefit all segments of society, promoting improved information and educational resource accessibility in order to support SDGs 4 (quality education) and 10 (reduced disparities) [33].

3.5 Enhanced Mobility and Accessibility

Smart transportation solutions, such as intelligent traffic management and public transit optimization, have the potential to cut congestion, pollution, and travel times while also supporting sustainable urban mobility (SDG 11: Sustainable Cities and Communities) and decreasing disparities (SDG 10: Reduce inequality within and among countries) [34, 35]

3.6 Efficient Resource Management

Smart cities use technology like IoT sensors and data analytics to optimize resource distribution, including electricity, water, and transportation. This helps SDG 11 by encouraging sustainable urban development and making better use of resources. Smart health supports SDG 11 by promoting healthier urban settings. Monitoring air quality, creating green areas, and providing active transportation alternatives all enhance physical and mental health. Sustainable healthcare facilities decrease both energy usage and waste [36].

3.7 Resilience and Disaster Management

In line with SDG 11, smart cities can improve their ability to withstand calamities and natural disasters by implementing early warning systems, real-time monitoring, and adaptable infrastructure and SDG 13 by promoting climate resilience and disaster risk reduction [37].

3.8 Environmental Sustainability

In order to help accomplish SDGs 14 (Life below Water), 15 (Life on Land), and 13 (Climate Action), smart cities may monitor and regulate environmental issues like waste management, energy consumption, and air and water quality. They can also reduce pollution and conserve natural resources [38].

3.9 Community Engagement and Empowerment

Smart city efforts frequently include public involvement platforms and digital governance technologies, which support transparent, accountable, and inclusive decision-making procedures (SDG 16: Encourage inclusive, peaceful societies for sustainable development, guarantee everyone's access to justice) [39].

3.10 Public–Private Partnerships (PPPs)

Smart health projects are driven by collaboration between non-government organizations, the private sector, and governments. Public–private partnerships promote innovation, information sharing, and resource mobilization. In the context of smart city healthcare initiatives, SDG 17 (Partnerships for the Goals) calls for collaboration between public and private sectors, as well as academic institutions. This cooperation includes telemedicine, multi-sector planning, and evidence-based policymaking. Successful healthcare innovation requires cooperation between the public and commercial sectors, non-governmental organizations, and other stakeholders. To combine resources, expertise, and technology for development and execution, public–private partnerships are employed [40].

4 Smart Healthcare Initiatives

Smart health initiatives, utilizing technologies like artificial intelligence, the IoT, and telemedicine, are crucial for the advancement of smart cities. They enable proactive, personalized healthcare, streamline processes, minimize costs, and improve quality of life. These initiatives drive innovation and sustainability, promoting a healthier, more equitable future where healthcare is accessible to all [41]. Here are some instances of smart health program that use technology to improve healthcare delivery and well-being.

4.1 Telemedicine and Remote Consultations

During the pandemic, healthcare experts have welcomed remote consultations, allowing them to diagnose problems, analyze radiological imaging, including CT and X-rays in great detail, and even conduct collaborative consultations with doctors in other regions. This transformational strategy also reduces the need for needless hospital visits, democratizing healthcare, especially for marginalized and impoverished groups [42].

4.2 Emergency Response Optimization

Real-time connection is vital in emergency response during the "golden hour"—the key period when effective medical intervention can save lives. In times of emergency, smart technology provides for speedier communication, coordination, and resource allocation [43]. Innovative healthcare systems are required in light of climate change in order to provide safe, high-quality care in unfavorable environments, especially in isolated or neglected locations. Digital health technologies that increase patient information portability, decrease inefficiencies, and improve access can help adapt to climate change. Digital health tools were quickly adopted during the COVID-19 epidemic, but it is still unclear how resilient and effective they are in natural catastrophes [44].

4.3 Wearable Devices and Health Monitoring

Wearable devices provide real-time health data, enabling patients and healthcare practitioners to assess conditions, monitor vital signs, and identify abnormalities early. These technologies allow people to take control of their health and implement preventive interventions [45].

4.4 Predictive Analytics for Disease Prevention

Machine learning algorithms use health data to anticipate disease outbreaks, identify high-risk groups, and optimize resource allocation. Smart analytics improves health outcomes by averting epidemics and enhancing public health planning [23].

4.5 Health Information Systems and Interoperability

Healthcare providers can communicate data more easily when they integrate electronic health records (EHRs) with health information exchanges. Interoperability increases care coordination, eliminates service duplication, and leads to better patient outcomes [46].

4.6 Environmental Monitoring for Health

Smart cities use IoT devices to monitor air, water, and sanitation. Environmental variables have a considerable influence on health outcomes, and real-time data guides policy choices and interventions [47]. These initiatives promote excellent health, fairness, and sustainable communities, which are all in line with the SDGs. As technology progresses, smart health will play an increasingly vital role in shaping how healthcare is delivered in the future.

5 Opportunities and Challenges of Smart Healthcare

Smart cities offer a promising future for healthcare with advanced technologies. These technologies can tailor treatments to individual patients' needs, promoting seamless coordination among providers. We need to discuss the potential opportunities and associated challenges.

5.1 Opportunities

Advanced healthcare systems in smart cities, backed by digital infrastructure and advanced technology, present significant opportunities for innovation and transformation in healthcare delivery. Some of the potential opportunities of smart health platforms can be mentioned as follows:

Technology and Innovation: Smart cities act as centers of research innovation and the adoption of new technologies, offering chances to leverage cutting-edge tools like wearable devices, telemedicine, remote monitoring, and predictive analytics to enhance patient outcomes and healthcare delivery [48, 49].

Preventive Healthcare: Smart health initiatives enable proactive and preventive healthcare by utilizing real-time data analytics and predictive modeling to recognize health hazards, spot early illness symptoms, and implement targeted interventions, lowering healthcare costs while improving population health. [50]

Personalized Medicine: Individual individuals can receive personalized medical treatments and interventions based on their unique features, preferences, and health requirements by integrating health data from wearable devices, electronic health records, and genetic testing [51].

Community Engagement and Empowerment: Smart health initiatives help communities engage and empower themselves in controlling their health and well-being. Individuals can use digital health platforms, Smartphone applications, and community health programs to obtain health information, track their health indicators, and engage in collaborative decision-making processes [52].

Integrated Care and Coordination: Smart health technology enables seamless coordination and collaboration among healthcare practitioners, resulting in integrated care delivery across multiple care settings and specializations. This improves care coordination, eliminates medical mistakes, and ensures continuity of care for patients in smart cities [53].

Opportunities exist for utilizing new technologies (AI, Blockchain, and 5G) and encouraging multidisciplinary cooperation. Deep learning can be effectively integrated into smart healthcare systems in smart cities to improve various aspects of healthcare delivery [54]. By incorporating smart health into the fabric of smart cities, we can build healthier, more resilient communities that can live indefinitely.

5.2 Research Challenges

While smart cities provide several prospects for innovation and improvement in healthcare delivery, addressing issues such as data security, interoperability, regulatory compliance, and ethical concerns is critical to realizing their full potential. A few of such challenges are:

Privacy and Security: Protecting health data from cyber threats and maintaining patient privacy are significant considerations. It is a matter of great concern in providing strong security protections and privacy-preserving strategies to protect sensitive health data in networked smart healthcare systems [55].

Equity Inclusion**:** Equitably accessing smart health technologies is crucial, especially for disadvantaged populations. Addressing disparities in technology, digital literacy, and healthcare resources is essential for advancing smart cities. Prioritizing inclusivity in smart health initiatives can bridge healthcare access gaps, empowering marginalized communities to utilize transformative technologies [56]. Understanding and correcting discrepancies in access to smart health technology and digital health literacy among diverse demographic groups is critical for ensuring equitable healthcare delivery. Fostering equal opportunity and accessibility can create a healthcare system that serves all individuals.

Ethics**:** Balancing technology and ethics is crucial in healthcare systems, especially in AI-based diagnoses. Transparency and objectivity are essential for accountability and preventing biases. Upholding ethical principles can improve healthcare outcomes while safeguarding patient trust and autonomy [57]. Addressing moral quandaries, avoiding fake information and legal problems around the implementation of emerging technologies such as Machine Learning and Deep Learning, IoT, and big data analytics in smart healthcare systems is also a major concern [58].

Health Outcomes and Effectiveness: There is a strong need for assessing how smart health efforts affect health outcomes, patient happiness, and healthcare quality, as well as finding best practices and successful treatments [59].

Cost effectiveness and Sustainability: The process that evaluates the cost-effectiveness and sustainability of smart health technologies and interventions for healthcare system decision-making and resource allocation is a matter of great concern in the current scenarios [59, 60].
User acceptance and Adoptions: There is a need of the study explores the influencing factors of user acceptance and soft adoption of smart healthcare technologies, aiming to design user-friendly and effective healthcare solutions [61].
Community Empowerment and Engagement: Exploring strategies for promoting community engagement and empowerment in smart healthcare initiatives, including participatory design approaches and community-based interventions is another challenge during the crisis [62].
Regulatory and Policy Considerations: Analyzing regulatory frameworks, policy implications, and governance models for smart healthcare systems to ensure compliance with rules and ethical principles [63].
Long-term Health Impacts: Researching the long-term health consequences of smart health interventions, such as their influence on health behaviours, illness prevention, and health inequities is also utmost important [64].

6 Design Goals of Smart Healthcare Platforms

Several essential goals must be addressed while creating a sustainable smart healthcare system to ensure its efficacy and durability. To begin, prioritizing security and privacy measures is critical for protecting sensitive health information and maintaining patient confidentiality. Furthermore, encouraging interoperability among various healthcare systems and devices facilitates smooth data interchange and complete patient care. Equity and inclusion should be essential aspects, with healthcare services and technology available to all people, regardless of socioeconomic position or geographic location. Efficiency and effectiveness are crucial, with the goal of optimizing resource utilization and improving health outcomes. User-centered design concepts should lead the creation of intuitive and user-friendly interfaces, hence increasing adoption and usability. Community participation promotes teamwork and enables individuals to manage their health. Furthermore, sustainability considerations guarantee that the system is both environmentally benign and economically feasible in the long run. To maintain patient trust and autonomy, ethical factors such as openness and fairness must be included into AI-based diagnosis. Finally, regulatory compliance guarantees that all applicable rules and regulations are followed, therefore ensuring patient rights and safety. By combining these design objectives, a sustainable smart healthcare system may efficiently solve the complex difficulties of modern healthcare delivery while also promoting health equity and increasing patient outcomes.

7 Proposed Framework

In the framework of smart health in smart cities with sustainable development objectives, a strong technology foundation is required to smoothly incorporate health programs integrated into the fabric of smart cities. The proposed framework for a sustainable smart healthcare system focuses on security, privacy, data integrity and interoperability, telemedicine, community engagement and collaboration, predictive analytics and AI, scalability and sustainability, and environmental monitoring etc. This suggested technological framework is presented in Fig. 2 and its key components are illustrated as follows.

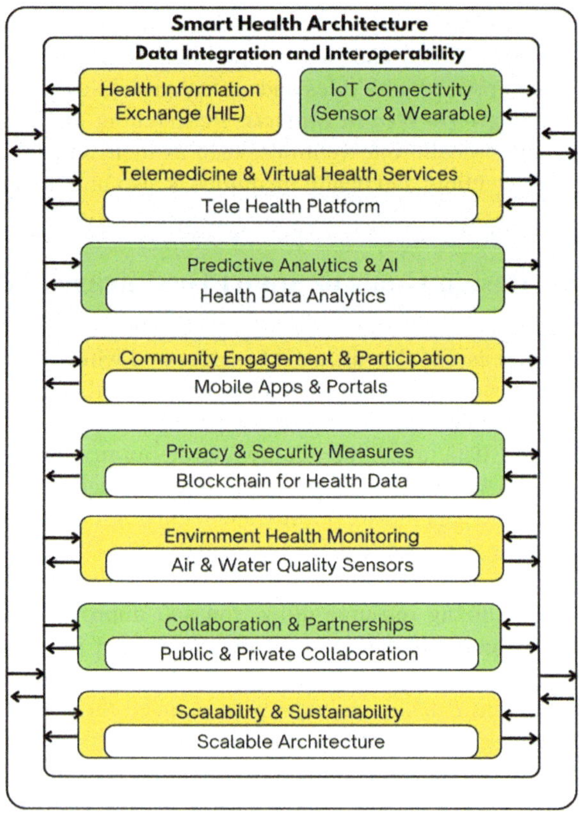

Fig. 2 Proposed framework for smart healthcare system

7.1 Data Integration and Interoperability

The Health Information Exchange (HIE): HIE is a standardized framework for sharing electronic health information (EHRs) across healthcare providers. Ensure compatibility across systems to allow for easy data sharing.

IoT Connectivity: Use dynamic sensors and smart wearable devices to gather real-time patient health information. Integrate this data with existing municipal infrastructure (e.g., air quality sensors, traffic control systems) to gain full insights.

7.2 Telemedicine and Virtual Health Services

Telehealth Platforms: Create secure telehealth solutions for remote consultations, diagnostics, and follow-ups. Utilize video conferencing, chatbots, and AI-powered triage tools.

Augmented Reality (AR) and Virtual Reality (VR): Discover AR/VR applications for medical training, patient education, and mental health therapies.

7.3 Predictive Analytics and AI

Health Data Analytics: Use machine learning algorithms to predict disease outbreaks, identify high-risk groups, and optimize resource allocation.

Early Warning Systems: Create models that detect health crises (e.g., epidemics, heat waves) using environmental and health data.

7.4 Community Engagement and Citizen Participation

Mobile applications and portals: Develop user-friendly apps for health monitoring, appointment scheduling, and wellness programs. Encourage citizens to participate in health surveys and feedback channels.

Community Health Dashboards: Make real-time health indicators (such as air quality and illness prevalence) available to inhabitants.

7.5 Privacy and Security Measures

Blockchain for health data: Investigate block chain technology to protect health information, improve privacy, and enable consent-based data exchange.

Cybersecurity Protocols: Implement strong security measures to safeguard sensitive health information.

7.6 Environmental Health Monitoring

Air and Water Quality Sensors: Use IoT devices to monitor environmental elements that impact health. Integrate this data with health-care systems to make more informed decisions.

Green Space and Active Transportation: Encourage walking, cycling, and access to parks for physical health.

7.7 Collaboration and Partnership

Public–Private Collaboration: Foster collaboration among the government, health-care providers, technology businesses, and research institutes.

Research and innovation hubs: Establish centers for health-technology research, testing, and innovation.

7.8 Scalability and Sustainability

Scalable Architecture: Create a flexible framework to meet population growth and changing health demands.

Resource optimization: Optimize energy use, infrastructure upkeep, and resource allocation to maintain long-term viability.

8 Conclusion and Future Directions

Smart city healthcare technology has the potential to completely transform healthcare systems by improving quality of healthcare, efficiency, accessibility, advancing towards Sustainable Development Goals, and enabling real-time monitoring, personalized treatment, and optimized resource allocation, promoting well-being and global health outcomes. The suggested framework for intelligent healthcare systems in smart cities, aligned with the SDGs, uses advanced technologies like AI, IoT sensors, and mobile applications to enhance accessibility, efficiency, and innovation. This approach aims to build healthier communities, prioritize universal health coverage, and achieve global health equity. It aims to address modern healthcare challenges, promote equity, and enhance patient outcomes by ensuring security, privacy, interoperability, equity, inclusivity, efficiency, effectiveness, usability, community engagement, sustainability, and ethical compliance.

Future research in smart healthcare systems in smart cities will focus on ethical considerations, more efficient data security and privacy, equitable access, interoperability, community engagement, user collaboration, environmental data integration, and sustainable models. This will advance smart health, promote technological innovation, and improve healthcare outcomes.

References

1. Sánchez-Corcuera, R., Nuñez-Marcos, A., Sesma-Solance, J., Bilbao-Jayo, A., Mulero, R., Zulaika, U., Almeida, A.: Smart cities survey: technologies, application domains and challenges for the cities of the future. Int. J. Distrib. Sens. Netw. **15**(6), 1550147719853984 (2019)
2. Alavi, A.H., Jiao, P., Buttlar, W.G., Lajnef, N.: Internet of Things-enabled smart cities: state-of-the-art and future trends. Measurement, **129**, 589–606 (2018)
3. Satpathy, S., Mallick, C., Pradhan, S.K: Big data computing application in digital forensics investigation and cyber security. Int. J. Comp. Sci. Mobile Appl., 129–136 (2018)
4. Massobrio, R., Nesmachnow, S., Tchernykh, A., Avetisyan, A., Radchenko, G.: Towards a cloud computing paradigm for big data analysis in smart cities. Program. Comput. Softw. **44**, 181–189 (2018)
5. Mallick, C., Mishra, S., Giri, P.K., Paikaray, B.K.: Machine learning approaches to sentiment analysis in online social networks. Int. J. Work Innov. **3**(4), 317–337 (2023)
6. Rangarajan, S.S., Raman, R., Singh, A., Shiva, C.K., Kumar, R., Sadhu, P.K., Senjyu, T.: DC microgrids: a propitious smart grid paradigm for smart cities. Smart Cities **6**(4), 1690–1718 (2023)
7. Richter, M.A., Hagenmaier, M., Bandte, O., Parida, V., Wincent, J.: Smart cities, urban mobility and autonomous vehicles: how different cities needs different sustainable investment strategies. Technol. Forecast. Soc. Chang. **184**, 121857 (2022)
8. Mohammadzadeh, Z., Saeidnia, H.R., Lotfata, A., Hassanzadeh, M., Ghiasi, N.: Smart city healthcare delivery innovations: a systematic review of essential technologies and indicators for developing nations. BMC Health Serv. Res. **23**(1), 1180 (2023)
9. Kamel Boulos, M.N., Al-Shorbaji, N.M.: On the Internet of Things, smart cities and the WHO Healthy Cities. Int. J. Health Geogr. **13**(1), 1–6 (2014)
10. George, A.H., Shahul, A., George, A.S.: Wearable sensors: a new way to track health and wellness. Partners Univ. Int. Innov. J. **1**(4), 15–34 (2023)

11. Sanjeev, S., Ponnekanti, G.S., Reddy, G.P.: Advanced healthcare system using artificial intelligence. In: 2021 11th International Conference on Cloud Computing, Data Science & Engineering (Confluence), pp. 76–81. IEEE (2021, January)
12. Tan, A.J., Rusli, K.D., McKenna, L., Tan, L.L., Liaw, S.Y.: Telemedicine experiences and perspectives of healthcare providers in long-term care: a scoping review. J. Telemed. Telecare **30**(2), 230–249 (2024)
13. Anshari, M.: Redefining electronic health records (EHR) and electronic medical records (EMR) to promote patient empowerment. IJID (Int. J. Inform. Dev.) **8**(1), 35–39 (2019)
14. Baig, M., GholamHosseini, H., Connolly, M.J.: Mobile healthcare applications: system design review, critical issues and challenges. Australas. Phys. Eng. Sci. Med. **38**, 23–38 (2015)
15. Angelov, G.V., Nikolakov, D.P., Ruskova, I.N., Gieva, E.E., Spasova, M.L. :Healthcare sensing and monitoring. In: Enhanced Living Environments: Algorithms, Architectures, Platforms, and Systems, pp. 226–262. Springer International Publishing, Cham (2019)
16. Ahad, M.A., Paiva, S., Tripathi, G., Feroz, N.: Enabling technologies and sustainable smart cities. Sustain. Cities Soc. **61**, 102301 (2020)
17. Silva, B.N., Khan, M., Han, K.: Towards sustainable smart cities: a review of trends, architectures, components, and open challenges in smart cities. Sustain. Cities Soc. **38**, 697–713 (2018)
18. Visvizi, A., & del Hoyo, R. P. (Eds.). : Smart cities and the UN SDGs. Elsevier.
19. Grossi, G., Trunova, O.: Are UN SDGs useful for capturing multiple values of smart city? Cities **114**, 103193 (2021)
20. Al-Azzam, M.K., Alazzam, M.B.: Smart city and smart-health framework, challenges and opportunities. Int. J. Adv. Comp. Sci. Appl. **10**(2) (2019)
21. Gera, S., Mridul, M., Sharma, S.: IoT based automated health care monitoring system for smart city. In: 2021 5th International Conference on Computing Methodologies and Communication (ICCMC), pp. 364–368. IEEE (2021, April)
22. Holmgren, A.J., Esdar, M., Hüsers, J., Coutinho-Almeida, J.: Health information exchange: understanding the policy landscape and future of data interoperability. Yearb. Med. Inform. **32**(01), 184–194 (2023)
23. Ibrahim, M.S., Saber, S.: Machine learning and predictive analytics: advancing disease prevention in healthcare. J. Contemp. Healthc. Anal. **7**(1), 53–71 (2023)
24. Javaid, M., Khan, I.H.: Internet of Things (IoT) enabled healthcare helps to take the challenges of COVID-19 Pandemic. J. Oral Biol. Craniofacial Res. **11**(2), 209–214 (2021)
25. Rajabion, L., Shaltooki, A.A., Taghikhah, M., Ghasemi, A., Badfar, A.: Healthcare big data processing mechanisms: the role of cloud computing. Int. J. Inf. Manage. **49**, 271–289 (2019)
26. Dash, S., Shakyawar, S.K., Sharma, M., Kaushik, S.: Big data in healthcare: management, analysis and future prospects. J. Big Data **6**(1), 1–25 (2019)
27. Kolesnichenko, O., Mazelis, L., Sotnik, A., Yakovleva, D., Amelkin, S., Grigorevsky, I., Kolesnichenko, Y.: Sociological modeling of smart city with the implementation of UN sustainable development goals. Sustain. Sci. **16**(2), 581–599 (2021)
28. Papa, A., Mital, M., Pisano, P., Del Giudice, M.: E-health and wellbeing monitoring using smart healthcare devices: an empirical investigation. Technol. Forecast. Soc. Chang. **153**, 119226 (2020)
29. El-Rashidy, N., El-Sappagh, S., Islam, S.R., El-Bakry, H.M., Abdelrazek, S.: Mobile health in remote patient monitoring for chronic diseases: principles, trends, and challenges. Diagnostics **11**(4), 607 (2021)
30. Takian, A., Raoofi, A., Haghighi, H.: COVID-19 pandemic: the fears and hopes for SDG 3, with focus on prevention and control of noncommunicable diseases (SDG 3.4) and universal health coverage (SDG 3.8). In: COVID-19 and the Sustainable Development Goals, pp. 211–234. Elsevier (2022)
31. Tavanti, M.: Technology innovations for the SDGs. In: Developing Sustainability in Organizations: A Values-Based Approach, pp. 389–404. Springer International Publishing, Cham (2023)

32. Howden-Chapman, P., Siri, J., Chisholm, E., Chapman, R., Doll, C.N., Capon, A.: SDG 3: ensure healthy lives and promote wellbeing for all at all ages. In: A Guide to SDG Interactions: From Science to Implementation, pp. 81–126. International Council for Science, Paris, France (2017)
33. Islam, Q.T., Ahmed, J.U., Sayed, A.: Digitization and integration of sustainable development goals (SDGs) in emerging economies. In: Fostering Sustainable Businesses in Emerging Economies: The Impact of Technology, pp. 23–38 (2023)
34. Domínguez, J.C., Chao, C.O., Lucatello, S.: Back to the future: smart technologies and the sustainable development goals. In: The Routledge Handbook of Smart Technologies, pp. 537–554. Routledge (2022)
35. Pandey, U.C., Kumar, C., Ayanore, M., Shalaby, H.R.: SDG10–Reduce inequality within and among countries. Emerald Publishing Limited (2020)
36. Fattahi, M., Keyvanshokooh, E., Kannan, D., Govindan, K.: Resource planning strategies for healthcare systems during a pandemic. Eur. J. Oper. Res. **304**(1), 192–206 (2023)
37. Mosadeghrad, A.M., Isfahani, P., Eslambolchi, L., Zahmatkesh, M., Afshari, M.: Strategies to strengthen a climate-resilient health system: a scoping review. Glob. Health **19**(1), 62 (2023)
38. Egbende, L., Helldén, D., Mbunga, B., Schedwin, M., Kazenza, B., Viberg, N., Alfvén, T., et al.: Interactions between health and the sustainable development goals: the case of the Democratic Republic of Congo. Sustainability **15**(2), 1259 (2023)
39. Erku, D., Khatri, R., Endalamaw, A., Wolka, E., Nigatu, F., Zewdie, A., Assefa, Y.: Community engagement initiatives in primary health care to achieve universal health coverage: A realist synthesis of scoping review. PLoS ONE **18**(5), e0285222 (2023)
40. Dionisio, M., de Souza Junior, S.J., Paula, F., Pellanda, P. C.: The role of digital social innovations to address SDGs: a systematic review. Environ. Dev. Sustain., 1–26 (2023)
41. Jiwani, N., Gupta, K., Whig, P.: Machine learning approaches for analysis in smart healthcare informatics. In: Machine Learning and Artificial Intelligence in Healthcare Systems, pp. 129–154. CRC Press (2023)
42. Mroz, G., Papoutsi, C., Greenhalgh, T.: From 'A rapid and necessary revolution'to 'Telemedicine killed the PE teacher': changing representations of remote GP consultations in UK media during the COVID-19 pandemic. In: Communicating COVID-19: Media, Trust, and Public Engagement, pp. 125–144. Springer International Publishing, Cham (2024)
43. Usoro, A., Mehmood, A., Rapaport, S., Ezeigwe, A.K., Adeyeye, A., Akinlade, O., Razzak, J.: A scoping review of the essential components of emergency medical response systems for mass casualty incidents. Disast. Med. Public Health Preparedness, 1–12 (2023)
44. Lokmic-Tomkins, Z., Bhandari, D., Bain, C., Borda, A., Kariotis, T.C., Reser, D.: Lessons learned from natural disasters around digital health technologies and delivering quality healthcare. Int. J. Environ. Res. Public Health **20**(5), 4542 (2023)
45. Iovanel, G., Ayers, D., Zheng, H.: The role of wearable technology in measuring and supporting patient outcomes following total joint replacement: review of the literature. JMIR Perioperative Med. **6**(1), e39396 (2023)
46. Deepa, V.V., Thamotharan, B., Mahto, D., Rajendiran, P., Sriram, A.L., Chandramohan, K.: Smart embedded health monitoring system and secure electronic health record (EHR) transactions using blockchain technology. Soft. Comput. **27**(17), 12741–12756 (2023)
47. Ramírez-Moreno, M.A., Keshtkar, S., Padilla-Reyes, D.A., Ramos-López, E., García-Martínez, M., Hernández-Luna, M.C., Lozoya-Santos, J.D.J.: Sensors for sustainable smart cities: a review. Appl. Sci. **11**(17), 8198 (2021)
48. Umair, M., Cheema, M.A., Cheema, O., Li, H., Lu, H.: Impact of COVID-19 on IoT adoption in healthcare, smart homes, smart buildings, smart cities, transportation, and industrial IoT. Sensors **21**(11), 3838 (2021)
49. Dwivedi, R., Mehrotra, D., Chandra, S.: Potential of Internet of Medical Things (IoMT) applications in building a smart healthcare system: a systematic review. J. Oral Biol. Craniofacial Res. **12**(2), 302–318 (2022)
50. Razzak, M.I., Imran, M., Xu, G.: Big data analytics for preventive medicine. Neural Comput. Appl. **32**, 4417–4451 (2020)

51. Goetz, L.H., Schork, N.J.: Personalized medicine: motivation, challenges, and progress. Fertil. Steril. **109**(6), 952–963 (2018)
52. Belliger, A., Krieger, D.J.: The digital transformation of healthcare. In: Knowledge Management in Digital Change: New Findings and Practical Cases, pp. 311–326 (2018)
53. Javed, A.R., Sarwar, M.U., Beg, M.O., Asim, M., Baker, T., Tawfik, H.: A collaborative healthcare framework for shared healthcare plan with ambient intelligence. HCIS **10**, 1–21 (2020)
54. Mallick, C., Mishra, S., Senapati, M.R.: A cooperative deep learning model for fake news detection in online social networks. J. Ambient Intell. Human. Comput. **14**(4), 4451–4460 (2023)
55. Sahi, M.A., Abbas, H., Saleem, K., Yang, X., Derhab, A., Orgun, M.A., & Yaseen, A. (2017). Privacy preservation in e-healthcare environments: state of the art and future directions. IEEE Access, 6, 464–478 (2023)
56. Badr, N.G.: Learning healthcare ecosystems for equity in health service provisioning and delivery: smart cities and the quintuple aim. In: The Proceedings of the International Conference on Smart City Applications, pp. 237–251. Springer International Publishing, Cham (2022, October)
57. G Kontra, P., Quaglio, G., Garmendia, A.T., Lekadir, K.: Challenges of machine learning and AI (what is next?), responsible and ethical AI. In: Clinical Applications of Artificial Intelligence in Real-World Data, pp. 263–285. Springer International Publishing, Cham (2023)
58. Mallick, C., Mishra, S., Giri, P.K., Paikaray, B.K.: A meta heuristic optimization based deep learning model for fake news detection in online social networks. https://doi.org/10.1504/IJESDF.2024.10057139 (In press)
59. Blonigen, D., Hyde, J., McInnes, D.K., Yoon, J., Byrne, T., Ngo, T., Smelson, D.: Integrating data analytics, peer support, and whole health coaching to improve the health outcomes of homeless veterans: study protocol for an effectiveness-implementation trial. Contemp. Clin. Trials **125**, 107065 (2023)
60. Bhadula, S., Sharma, S.: Personalized self adaptive Internet-of-Things enabled sustainable healthcare architecture for digital transformation. In: 2023 Advanced Computing and Communication Technologies for High Performance Applications (ACCTHPA), pp. 1–6. IEEE (2023, January)
61. Negash, Y.T., Sarmiento, L.S.C.: Smart product-service systems in the healthcare industry: Intelligent connected products and stakeholder communication drive digital health service adoption. Heliyon **9**(2) (2023)
62. Sahoo, K.C., Sahay, M.R., Dubey, S., Nayak, S., Negi, S., Mahapatra, P., Pati, S.: Community engagement and involvement in managing the COVID-19 pandemic among urban poor in low- and middle-income countries: a systematic scoping review and stakeholders mapping. Glob. Health Action **16**(1), 2133723 (2023)
63. Lauri, C., Shimpo, F., Sokołowski, M.M.: Artificial intelligence and robotics on the frontlines of the pandemic response: the regulatory models for technology adoption and the development of resilient organisations in smart cities. J. Ambient Intell. Human. Comput., 1–12 (2023)
64. Norman, A.A., Marzuki, A.H., Faith, F., Hamid, S., Ghani, N.A., Ravana, S.D., Arshad, N.I.: Technology dependency and impact during COVID-19: a systematic literature review and open challenges. IEEE Access (2023)

Cloud Computing Applications in Digital Health: Challenges Related to Privacy and Security

Tripti Rathee, Minakshi Tomer, and Ishneet Kaur Chadha

Abstract Examining the complex interplay between cloud computing applications and digital health, the main goal is to mitigate privacy and safety-related issues. Delving into the three service models in cloud computing, the analysis includes their correlation with and impact on digital health services, while also scrutinizing potential threats faced by cloud-enabled digital health systems and proposing methods to ensure robust security and confidentiality. In addition, a thorough examination of how cloud computing applications are incorporated into the digital health environment is done. Real-world implementations are examined to shed light on the transformative potential of these applications, emphasizing both their benefits and potential pitfalls. The discussed applications, such as telemedicine and teleconsultation, medical imaging, hospital management, clinical information systems, use of smart devices for medical services, and the IoMT (Internet of Medical Things), exemplify the tangible impact of this integration.

Keywords Cloud computing · Data security in healthcare · Privacy in healthcare · Digital health · Cloud computing applications

1 Introduction

The infusion of digital advancements in healthcare has led to major breakthroughs, with healthcare organizations increasingly migrating their data to cloud computing for more versatility and optimized data access. Despite these advantages, there are difficulties both during and after the switch to cloud-based systems. Robust safety measures are required because cloud computing data is susceptible to vulnerabilities, which is a serious concern.

T. Rathee (✉) · M. Tomer · I. K. Chadha
Department of Information Technology, Maharaja Surajmal Institute of Technology, Delhi, India
e-mail: rathee.tripti@gmail.com

M. Tomer
e-mail: minakshi.tomer@msit.in

S. N. Mohanty et al. (eds.), *Explainable IoT Applications: A Demystification*, Information Systems Engineering and Management 21,
https://doi.org/10.1007/978-3-031-74885-1_5

Protecting against data breaches and providing data privacy has become essential in the world of digital health. Cloud computing and cloudlet adoption have become essential for healthcare organizations; they're no longer optional. These advances in technology have become crucial for the healthcare industry in order to accomplish flexibility and seamless data access [1].

User error during technology implementation as well as a lack of awareness regarding data capture and storage are two of the most significant barriers to healthcare data security. Physicians prioritize taking care of patients over other concerns, commonly disregarding digital data collection. Potential security risks are further aggravated by the difference in knowledge between patients and healthcare providers [2]. Patients might not handle data with the same caution as healthcare providers, which could leave them vulnerable to cybercrime threats.

The adoption of cloud computing and cloudlet technologies presents healthcare organizations with plenty of challenges, which are looked into in this chapter. It analyzes the effects of user error and discrepancies in data security awareness, highlighting the immediate need for holistic approaches to deal with these problems in the rapidly changing field of digital health.

2 Navigating Threats and Ensuring Security and Confidentiality of Digital Health

Digital health data encompasses personal health data which can be generated in digital format by medical gadgets. This includes personal data like a person's weight, height, and blood group identification alongside findings from investigations like the fasting plasma glucose test (FGT). Digital representations of these records are maintained on laptops, PCs, and databases employed by health information systems [3].

As patient medical records are converted to electronic health records (EHRs), medical professionals produce staggering quantities of digital health data. Because they are electronic, EHRs possess perks over conventional written records [4]. This is because software may save and manipulate data effortlessly. Many hospitals, educational institutions, and diagnostic labs store these inestimable datasets, many of which are classified as protected health information (PHI), across a variety of health information systems (HIS).

Individually identifiable health information that is obtained from people and is governed by national or international regulations regarding data breach disclosure is known as protected health information (PHI). PHI includes data about an individual's history, present, and possibly future mental or physical well-being, the administration of healthcare, and any corresponding monetary transactions. When combined with medical records, common identifiers like name, date of birth, place of residence, and mobile phone number turn into PHI. Laboratory findings, health care records, and hospital invoices that include a patient's identifying information are a few instances of PHI.

One of the most significant components of Healthcare Information System (HIS) security is guaranteeing the integrity and confidentiality of patient records. Keeping medical data protected from fraud, malware, and hackers is crucial to keeping HIS secure. Several reasons why Hackers attack HIS are listed in Fig. 1. When sensitive or personally identifiable data is gathered and kept on the system, privacy issues become more prominent. Protecting privacy when sharing such data amongst medical professionals is a major challenge. A variety of information confidentiality techniques, such as data masking, authentication, and encryption, are employed to restrict access to authorized personnel in order to mitigate these concerns.

A major concern is the growing possibility that hackers will use PHI as their target. Hackers put patients' privacy at risk by trying to profit from their personal information. Healthcare providers' shift from maintaining paper-based records to digital records has made their concerns worse.

Although building a national health data warehouse (NHDW) that incorporates a variety of HIS data is essential for improving research and health care, there are significant risks to privacy and data security. Sensitive patient data is contained within a single organization before being integrated into the NHDW, as required by law. To guarantee the protection of patient privacy within the NHDW, particular safeguards must be put in place along with the establishment of a national warehouse.

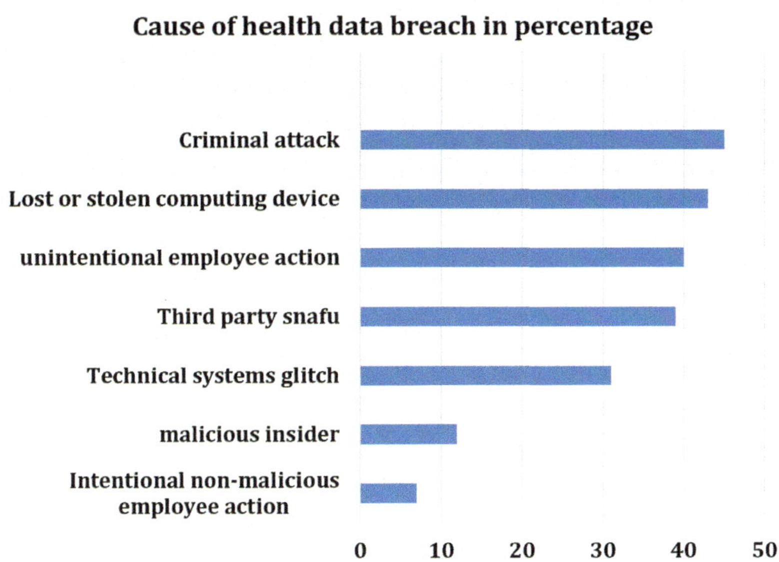

Fig. 1 Reasons of data breach in healthcare industry [3]

3 Defining Cloud Computing

Cloud computing is defined as an on-demand access model that encompasses a variety of deployment and service models for providing hardware and software services over the Internet. Cloud computing encompasses the real-time management and provisioning of programs, software, resources, and information over the Internet [5].

Numerous advantages of cloud computing include easier deployment, reduced expenses, scalability, and better use of hardware resources. Cloud computing is therefore used by all of the big businesses, including Microsoft, Google, and Amazon. Additionally, more and more users are transferring information to cloud-based platforms like Dropbox, Google Cloud Storage, iCloud, Slack, and IBM Cloud on a daily basis. Different security risks arise because the cloud often poses challenges in implementing a variety of enterprise-level security measures [6]. Even though cloud security has gained significant attention over the past decade, there are still unanswered questions. It is critical for researchers, programmers, service providers, and consumers to comprehend cloud security risks in order to implement current security measures or create new ones, as well as to take the necessary safety precautions.

4 Service Models and Design of Cloud

A thorough analysis of the architecture and service models of the cloud is necessary to comprehend confidentiality in it. In order to understand the security challenges, We shall dive deep into the three service models and how they are related to digital health.

4.1 Service Models

As illustrated in Fig. 2, there are three service models in cloud which are:

(a) Platform as a Service (PaaS)
(b) Software as a Service (SaaS)
(c) Infrastructure as a Service (IaaS).

Examples of cloud providers with cloud services can be seen in Table 1.

Software as a Service (SaaS). SaaS refers to a cloud computing service model wherein clients can utilize applications offered by the service provider without the need for local installation on their devices. This approach leads to cost savings by eliminating the requirement for individual software licenses. Typically accessed through web browsers or other client interfaces, SaaS is advantageous for end users as it grants access to cloud-developed applications, freeing users from the complexities of managing the foundational infrastructure. The provider bears the responsibility

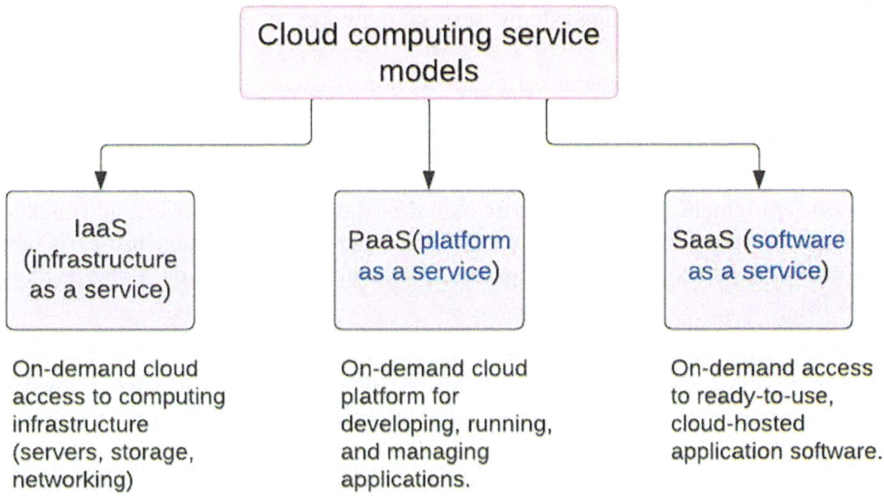

Fig. 2 Service models of cloud computing

Table 1 Cloud service providers offering various cloud services [7]

SaaS	PaaS	IaaS
Zoho	Microsoft Azure	Dropbox
Salesforce.com	Google App Engine	Cisco Cloud
Adobe creative cloud	Heroku	Mozy
		Amazon Web Services (AWS)

of infrastructure management, with only a limited number of customers having the ability to customize configurations. While SaaS works well for lightweight apps like media players or Microsoft Word, slower network speeds can cause processing delays for more resource-intensive apps. Pricing models for SaaS applications vary, with some providers adopting flat-rate charges regardless of usage, while others base charges on actual usage metrics [8].

Platform as a Service (PaaS). Platform as a Service (PaaS) is a category within cloud computing that provides developers with a platform to program, deploy, and maintain applications, eliminating the need for them to handle the intricacies of managing the underlying infrastructure. Under the PaaS model, developers do not need to install or maintain development tools locally in order to create and deploy applications. Instead, they can access a development environment via the internet. The cloud service provider is in charge of protecting the underlying infrastructure and application services [7]. Customers can run existing applications or create and test new ones by renting virtualized servers and related services from PaaS. Customers have control over the deployed apps and their configurations, however, they lack authority over the operating systems, servers, networks, or storage within the cloud environment.

Metrics such as data transfer per gigabyte, hourly usage, I/O requests per million, storage utilization per gigabyte, and data storage requests per thousand are frequently employed to determine the expenses associated with PaaS offerings. Some benefits of using PaaS models include more flexibility in the development process, streamlined version deployment, provided security including data security, recovery, and backup, reduced costs by eliminating the need for expert personnel to manage infrastructure, adaptability to changing circumstances, and the ability to work on a one-to-many basis [8].

Infrastructure as a Service (IaaS). The remarkably flexible model for cloud computing, known as Infrastructure as a Service (IaaS), offers virtualized physical computational assets via the Internet. When it comes to infrastructure management and control, IaaS offers clients the greatest degree of autonomy over PaaS and SaaS, which provide more managed service layers. Under the IaaS model, a cloud service provider accommodates the infrastructure elements commonly found in an on-site data center, including storage, servers, communication devices, and the hypervisor or virtualization layer. Customers are usually billed on a per-use basis, much like they would be for electricity or water at home.

Some of the key advantages of IaaS include the capacity to swiftly scale up and down in response to an enterprise's requirements, greater flexibility and control over computing resources, and avoiding the the cost and intricacy associated with the purchase and administration of physical servers and additional infrastructure within a data center. Each customer's data remains isolated from other customers even when they're stored on the same physical server. However, with this model, clients are responsible for managing aspects such as applications, data, runtime, and middleware. IaaS clients are also responsible for handling security aspects related to their applications and data content, unlike PaaS where the cloud provider also manages some security features, or SaaS where almost everything is managed by the provider.

4.2 How the Three Cloud Service Models Can Correlate with and Benefit Digital Health

The three cloud service models are useful for many aspects of digital health, including processing and storing massive datasets for clinical and research purposes as well as offering scalable platforms for patient management systems. Ensuring compliance, enhancing scalability, lowering overhead, and improving patient-provider collaboration are all achieved while upholding strict security and privacy protocols. Detailed explanations of these applications are shown in Fig. 3.

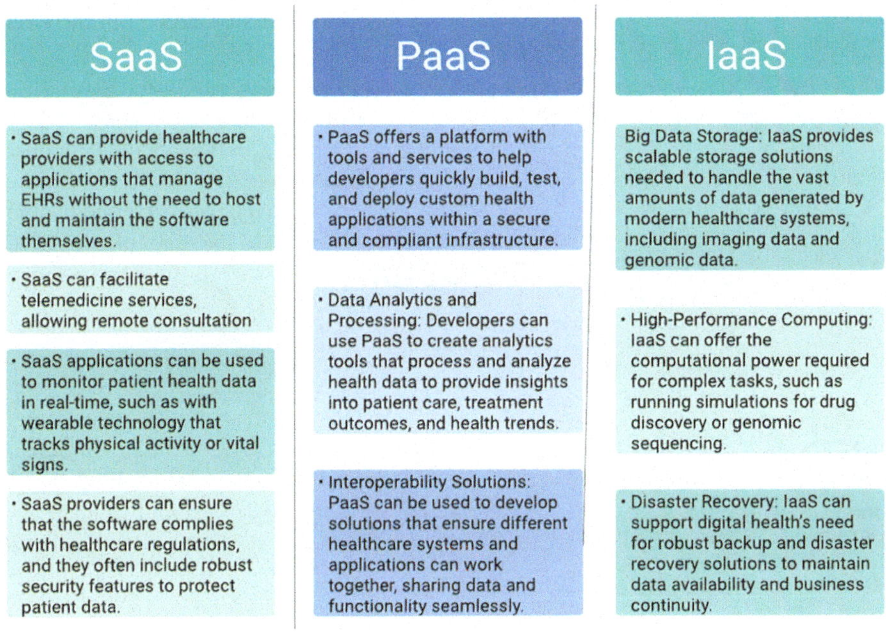

Fig. 3 Use of the various cloud service models in digital health

5 Methods to Ensure Security in the Cloud

The incorporation of cloud computing has transformed healthcare information management, preservation, and accessibility. As discussed in the previous section, the cloud presents benefits in terms of scalability, flexibility, and collaboration. However, it also brings with it special security challenges, especially when handling sensitive medical data. Ensuring patient data confidentiality, integrity, and availability is critical to preserving credibility in healthcare systems. Because of this, the cloud infrastructure must have strong security put into action.

5.1 Role Based Access Control (RBAC)

Personnel working for the healthcare provider, including server management technicians and patients themselves, have access to patient health records. This covers doctors, medical personnel, and employees of cloud service providers. It is essential to implement a role-based access system in order to protect patient data privacy. This is so because various roles—such as medical professionals and technical staff—need varying degrees of access to patient data. The solution to this problem is to give each authorized person a special ID code or number. Users are grouped according to this

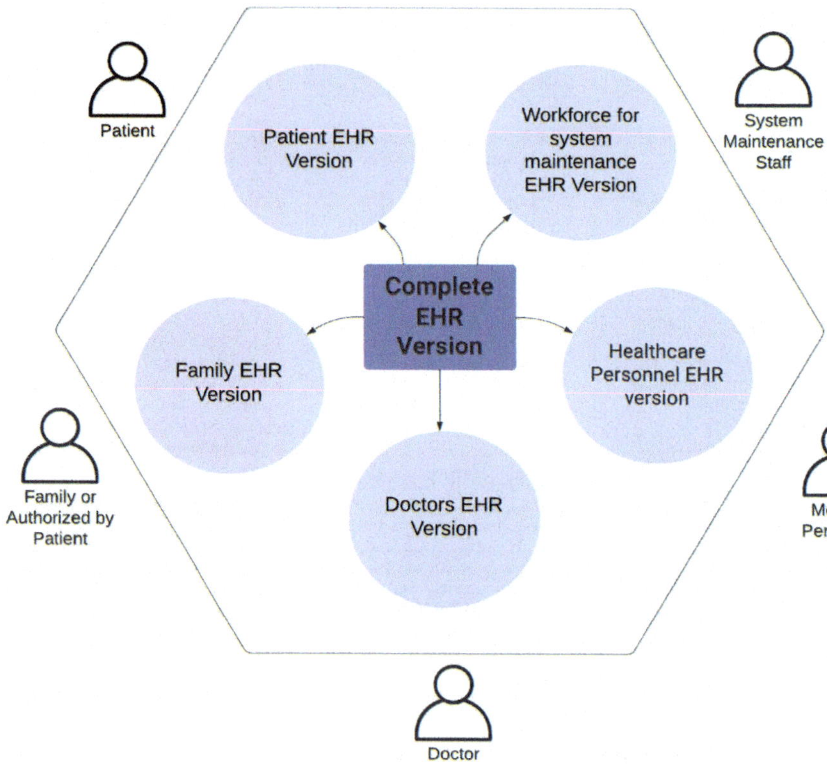

Fig. 4 Different roles that can take part in cloud based digital health [9]

ID, and each group is granted particular access rights to specific segments of the patient data [9]. Such a system is explained in Fig. 4. For example, platform maintenance staff would only have access to facts that are required for the system to function properly, but healthcare providers and patients would have complete access to the entire health record. This role-based structure operates seamlessly.

5.2 Authentication System

The verification of an individual's identity is commonly referred to as user authentication. The user authentication requirements established by the cloud provider must be fulfilled by cloud users. Depending on the integrity and dependability of the mechanism, the provider may offer a variety of authentication methods, each with a different level of security. Additionally, maintaining data integrity and confidentiality depends on authentication mechanisms working properly. The increasing prevalence of hybrid cloud environments has made it even more important to make sure that user authentication is interoperable [10].

Password Authentication. Although password authentication is simple and easy to use, maintaining security requires regular updates and a certain level of complexity. This authentication method is well-known for its vulnerabilities because it can be challenging to verify the legitimacy of a user even when they have the correct username and password. Many users tend to use the same passwords for several cloud services, which puts their account information at serious risk of security breaches. Password authentication is still the most common method, accounting for over 90% of transactions, despite these disadvantages. Password authentication appears in different challenge-response protocols used in cloud deployments today.

Trusted Platform Module-based Authentication. A secure cryptoprocessor is used by the Trusted Platform Module (TPM), a hardware-based security component, to store cryptographic keys and safeguard data. A related variation established by the Trusted Computing Group (TCG), a consortium of AMD, Intel, Hewlett Packard, Microsoft and IBM, is the Mobile Trusted Module (MTM). Due to the increasing use of smartphones, the MTM—which was first created for telecoms terminal authentication—is now being evaluated as a cloud computing authentication technique, especially when combined with Subscriber Identity Module (SIM) integration [10].

Cloud subscribers' devices can use special hardcoded keys for encryption, decryption, and software authentication. Moreover, TPM chips are used in a number of security-related technologies, including biometric authentication, firewalls, and antivirus programs. But TPM technology has flaws of its own, like vulnerabilities to cold boot attacks that let attackers get around disk encryption and use social engineering to find the master password. One major obstacle in a cloud environment is the "Bring Your Own Device" (BYOD) philosophy, which makes it more difficult to integrate TPM devices into business networks.

Public Key Infrastructure-based Authentication. When it comes to building the necessary degree of trust and providing the best possible approach to protect the security, integrity, and credibility of data and communication, using Trusted Third Party (TTP) services in the cloud is essential. A method for implementing strong authentication and authorization that is both legally acceptable and technically sound is the Public Key Infrastructure (PKI) in conjunction with TTP. Public-key cryptography, which PKI uses for authentication, enables users to confirm another party's identity based only on certificates and does not require shared secret data. Single Sign-On is an illustration of TTP authentication in the cloud (SSO). With SSO, a user can rely on assertions and easily access other sites without going through the authentication process once they have received authentication from one site. But depending on a reliable outside party to serve as an authentication server or certification authority presents problems because it can cause a system's fault tolerance and security to deteriorate [10].

Multifactor Authentication. By combining several authentication techniques, multi-factor authentication confirms the identity that a user has claimed. Increasing the number of authentication factors strengthens the validity of the user's identity. ID, password, biometrics, and certificates are the foundation of traditional single-factor authentication. Second-factor authentication now uses SMS, email, phone OTPs, PUSH Notifications, and mobile OATH Tokens thanks to the development of mobile networks. These second-factor approaches are expensive, unfavorable, and logistically challenging, especially when it comes to administration, distribution, support, and management in a cloud environment, even though they work well in closed communities like enterprise clouds [10].

Implicit Authentication. Mobile devices are especially well-suited to the use of user behavior observations for authentication since they can collect a wide variety of client information, such as location, movement, communication, and application usage. To cater a suitable solution for client and personal profile data in the mobile cloud environment, several profiling techniques have been explored. Nevertheless, there is currently no formal model for this method, and it is still difficult to get around technical limitations caused by constrained device resources. Comprehensive research on intelligent mobile authentication services is still lacking [10].

Context-Aware Cloud Authentication. While it is currently impractical to achieve a comprehensive profile of every user and device, a context-aware cloud can be an essential component of a dynamic and adaptive authentication system. Improved authentication services, adaptable access control, and a responsive security subsystem that adapts to the environment's conditions are important elements of this strategy. Consistently collecting structured data from users, the context-aware cloud builds a model of the legitimate user with the ability of accessing systems and resources by actively classifying and inferring.

Adaptive authentication is made possible by the context-aware cloud's dynamic nature, which enables it to change in real-time in response to a user's risk factor. This kind of risk-based authentication presents a dynamic system that considers the user profile and matches it with the related risk profile for a specific transaction. It is closely related to implicit authentication. While lower-risk profiles might be content with a static username and password, elevated risk profiles necessitate more extensive authentication procedures. By adjusting authentication levels according to risk, security can be improved over traditional secret-knowledge techniques, enabling smooth authentication without causing users any inconvenience. To preserve trust in their identity, users actively engage in ongoing re-authentication with the cloud service [10].

5.3 Encryption Methods in EHR (Electronic Health Record) Systems

Strong security measures must be put in place before moving healthcare data to a cloud system, with data encryption being the main focus. Standard encryption techniques are not, however, flawlessly supported by every Electronic Health Record (EHR) system. The fact that different users play different roles makes this problem worse and could result in access overlaps based on job specializations. As a result, various access points to the EHR are required.

Symmetric-Key Encryption (SKE) is one well-known encryption method that is renowned for both its complexity and efficiency. SKE mandates that a single key be used for both encryption and decryption by all healthcare providers. This creates a risk because all EHR data could be compromised at once if there is an encryption breach. An additional encryption technique called Public-Key Encryption (PKE) is costly and requires a public-key infrastructure (PKI) to use for storing electronic health records (EHRs). Ciphertext Policy Attribute-Based Encryption (CP-ABE) has been suggested as a superior choice as a substitute for healthcare institutions.

CP-ABE uses user roles to provide secure data access and management, addressing the drawbacks of conventional encryption techniques. Because of its reasonable performance and low storage requirements, CP-ABE can be used in cloud-based EHR systems as a viable alternative to traditional encryption techniques.

5.4 Digital Signatures

Digital signatures affirm authenticity, integrity, and non-repudiation, which makes them a valuable instrument in the field of digital health. This serves as a layer of protection and is especially useful in the Health Cloud environment to prevent possible instances of fraudulent data transactions. Using digital signatures makes sure that digital records are authentic, which is important while dealing with cloud computing in the digital health space. The digital signature confirms that a message or file really came from the intended sender, giving the recipient peace of mind when it travels over insecure channels. This security tool can be implemented efficiently using a variety of cryptographic algorithms.

6 Applications of Cloud Computing in Digital Health

This section explores the multifaceted applications of cloud computing in the realm of digital health, revolutionizing the landscape of healthcare services. Delving into key domains such as telemedicine and teleconsultation, medical imaging, hospital management, clinical information systems, and the integration of smart devices through the Internet of Medical Things (IoMT), we will explore the unprecedented impact of cloud technology on healthcare delivery. By examining the synergies between cloud computing and various facets of digital health, we shall gain insights into how these innovations are reshaping patient care, improving operational efficiency, and fostering a new era of accessible and personalized healthcare services.

6.1 Telemedicine and Teleconsultation

Technology is rapidly and successfully integrating itself into medical treatments as it reaches its pinnacle, giving rise to "telemedicine"—a technological innovation in the medical field. The days of tedious tasks that are prone to mistakes are over, as exact and faultless procedures take their place, made possible by fast internet connections. These cutting-edge methods provide secure authentication and real-time access to data. The idea is based on live video streaming that is seamless and cloud computing. Authorized medical personnel can access information in a convenient format on the internet. A lot of medical professionals are already using modern telehealth applications, which are having a revolutionary effect on telemedicine thanks to cloud computing [11].

Telemedicine, the remote delivery of healthcare services, is supported by this cloud-based model. Cloud computing makes it simpler to store, process, and share medical data easily, which improves the effectiveness and accessibility of telemedicine services. A conceptual solution for a fully cloud-integrated teleconsultation service is shown in Fig. 5. In this case, the doctor at the urban health center can use their smart card to authenticate themselves inside the cloud-based system. They will then be granted safe access to the database that holds details about the patient who is receiving treatment. This model—as well as one similar to it that was made specifically for rural health centers—was developed as part of research that was published in the International Journal of Clinical & Medical Informatics [12].

6.2 Medical Imaging

The process of viewing the human body using varying technologies in an effort to detect, track, or treat medical conditions is known as medical imaging. Imaging modalities vary depending on the technology employed and the image's information about the body part being studied. X-ray, CT, MRI, Mammography, ultrasound imaging, and Positron Emission Tomography are examples of imaging modalities.

The Picture Achieve and Communication System (PACS) was founded with the intention of storing, processing, retrieving, and visualizing medical images in response to a shift in emphasis toward patient-oriented care [13]. PACS is made up of four components: workstations for visualizing images, imaging devices, storage archives, and a network that connects all of the PACS components. The Digital Imaging and Communication in Medicine (DICOM) standard is used for the distribution and transfer of medical imaging data. DICOM allows medical imaging data to be shared between devices within the same institution, but not between institutions. Healthcare organizations have greatly profited from the transition of medical images to digital systems in terms of reduced expenses, increased productivity, and encouraged practitioner collaboration.

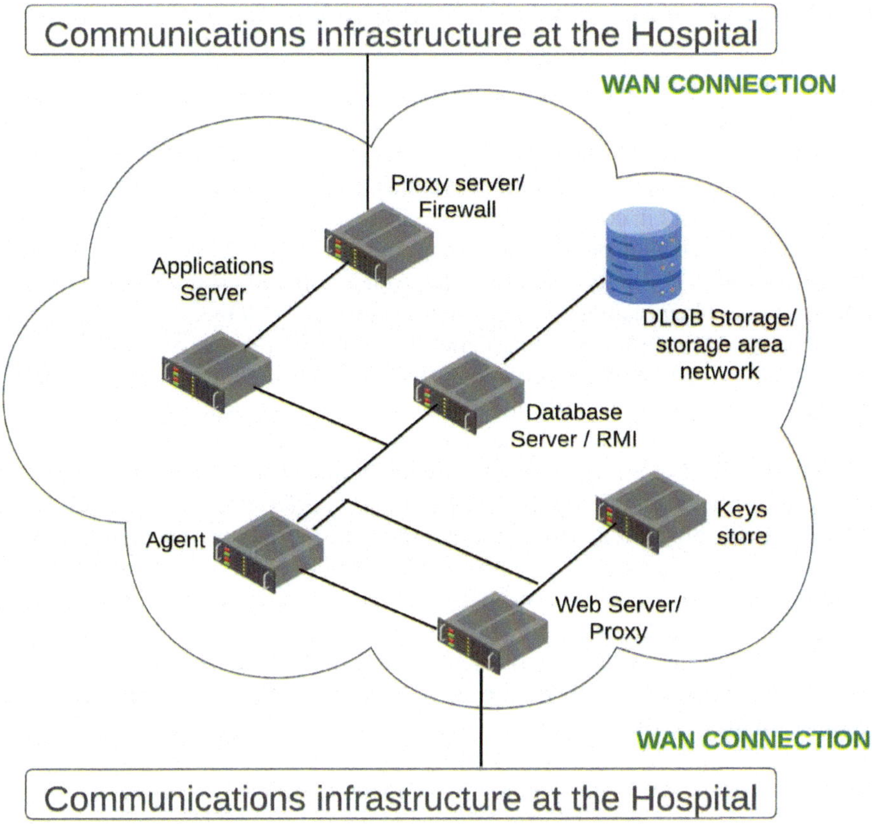

Fig. 5 Cloud-based infrastructure enabling comprehensive deployment of teleconsultation services at a municipal health facility [12]

Radiologists can use cloud computing to log into their local workstation and access reconstruction software that runs on the cloud. Only commands and ordering are made at the workstation level by radiologists processing images; cloud servers handle the computational part of the process, and users' devices preview the processed images. Additionally, the cloud makes it possible to access information from any location as long as there is a strong enough internet connection, which minimizes the need for setting up and maintaining software and hardware and helps to reduce IT costs. According to estimates, cloud computing could reduce IT costs by 20% a year [14].

6.3 Hospital Management and Clinical Information Systems

Hospital Management and Clinical Information Systems have witnessed a paradigm shift with the implementation of cloud computing technologies. The integration of cloud computing brings forth numerous advantages, transforming the way healthcare institutions manage data, streamline operations, and deliver patient care.

One notable trend in contemporary research involves implementing Health Information Systems (HIS) on public cloud computing platforms. This choice is motivated by the scalability, elasticity, and cost-effectiveness offered by the cloud. However, concerns arise when medical services share resources with unknown neighbors on the public cloud, potentially jeopardizing the security of sensitive healthcare data. To address this, private clouds with specialized mechanisms, such as virtual private networks, are recommended for safeguarding sensitive medical information.

Cloud computing facilitates the upload of medical data from cyber medical devices to private clouds (Fig. 6), creating a secure environment where doctors and patients can access services with desirable medical Quality of Service (QoS). The challenge lies in achieving efficient dynamic resource allocation within private clouds, as these resources are limited and managed by independent corporations. Neural network-based resource provisioning controllers have been proposed to optimize resource utilization while meeting specific QoS requirements [15].

The dynamic and unpredictable nature of workload in Hospital Information Systems (HIS) poses a challenge, especially with the sudden appearance and disappearance of flash crowd workloads. Traditional auto-scaling strategies based on thresholds struggle to adapt, necessitating dynamic workload-aware auto-resource scaling strategies. Leveraging control theory and feedback mechanisms, cloud-based HIS can proactively adjust resources based on predictive control algorithms, mitigating the impact of unpredictable workloads.

Furthermore, the diversity of data in HIS, comprising text, images, and other formats, requires a fine-grained approach to resource allocation. Cloud computing allows for both horizontal scaling, adjusting the number of virtual machines, and vertical scaling, modifying the configuration of original VMs. This flexibility enables healthcare institutions to tailor resource allocations to the specific demands of medical services, preventing resource inefficiencies.

6.4 Use of Smart Devices for Healthcare

The swift evolution of cloud computing and the Internet of Things (IoT) indicates that cloud-based s-health systems will soon be necessary for delivering the finest medical care [16]. Due to the constrained storage and processing capabilities of IoT devices, it is essential to routinely transmit diverse information to cloud data centers for analysis and processing. For IoT-based applications, the ideal union of cloud computing and

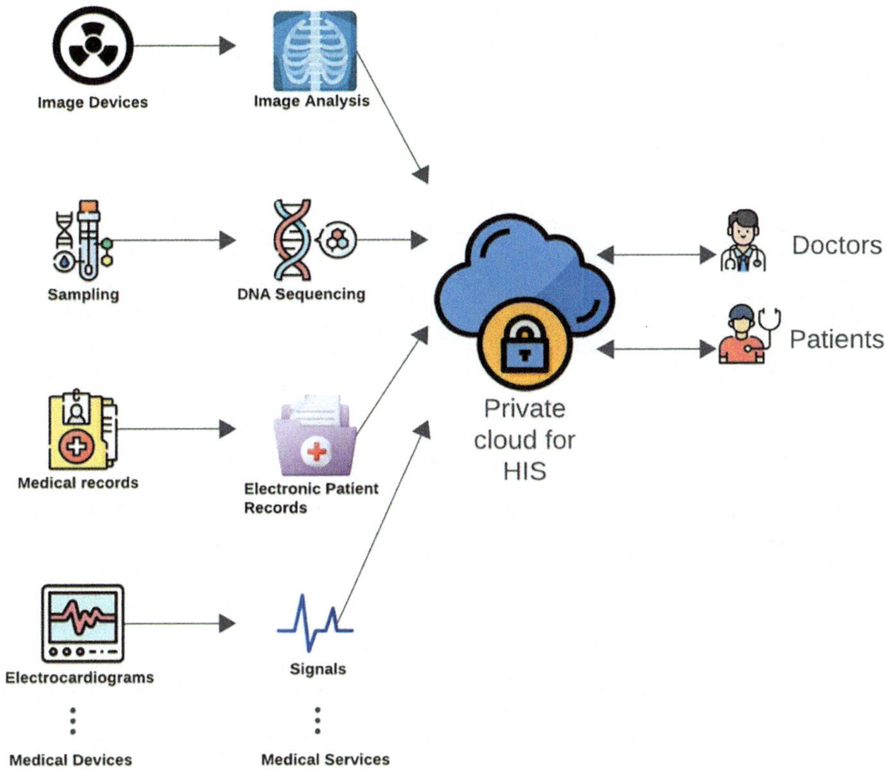

Fig. 6 Private cloud computing structure for HIS [15]

IoT offers a multitude of advantages [17]. In a standard configuration of a cloud-based s-health system, smart devices gather health records, known as s-health records (SHRs), and subsequently transmit them to a cloud server for prolonged storage and smooth collaboration with healthcare professionals. Such a system serves several advantages as shown in Fig. 7. But because patients can no longer access SHRs directly, this arrangement raises concerns about user privacy and data security [18].

Three main security considerations come into play: protecting patient privacy, authenticating data owners, and safeguarding the security of outsourced SHRs. Patients have a right to expect that licensed healthcare providers will have access to accurate SHRs. Healthcare professionals should also confirm that a SHR is authentically issued by an authorized organization and not a fraudulent representation before using it for research or treatment. Patients also anticipate that information about their identities and medical conditions will not be disclosed to uninvited parties and that their personal medical records will be stored in a confidential manner. Nevertheless, complete trust in the cloud service provider is not guaranteed, and the cloud is not obligated to disclose occurrences of data loss. As an example, in 2011, specific data outsourced by users was permanently deleted from Amazon's extensive Elastic

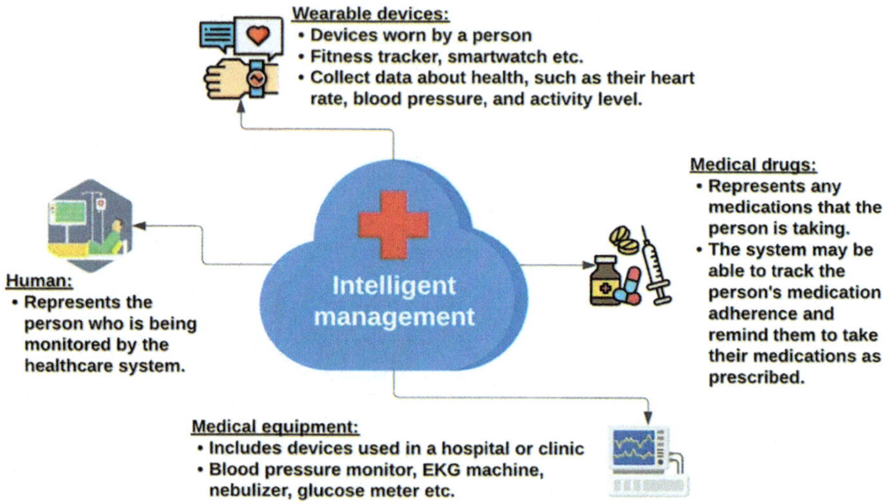

Fig. 7 Cloud architecture for smart healthcare systems [19]

Compute Cloud (EC2), highlighting the crucial need to regularly validate the integrity of outsourced s-health records (SHRs) [16].

On a different note, patients or potentially malicious entities may attempt to falsify SHRs for various motives, including medical insurance fraud. Therefore, the establishment of a security-conscious solution that simultaneously addresses these privacy concerns becomes a vital requirement for the advancement of s-health. The most prevalent tools for ensuring integrity and authenticity verification are digital signatures and Remote Data Integrity Checking (RDIC) protocols.

6.5 Internet of Medical Things

The Internet of Things (IoT) has quickly advanced, linking a vast number of users and devices. Remote health monitoring systems at home are required to reduce the total healthcare expenses and optimize healthcare processes and work-flows [20]. A significant application of the Internet of Things (IoT) is its widespread integration in at-home healthcare; this is sometimes called the Internet of Medical Things (IoMT). Chronic health problems (e.g., heart or lung diseases) are on the rise due to the challenges of modern life combined with the difficulty of getting timely medical examinations and advice. Furthermore, limitations like the dearth of spectrum resources and the volume of medical data impede the continued development of IoMT, especially in light of the requirement for real-time performance. In the context of IoMT, there's growing interest in in-home health monitoring to keep diseases from progressing and to relieve the burden on the healthcare infrastructure [21].

Through integration with conventional medical equipment, IoMT extends the sensing and processing capabilities of IoT. More applications of IoMT are shown in Fig. 8. IoMT makes it possible to use several body sensors on different patients, which makes remote in-home monitoring possible. Patients can move around freely in ubiquitous health monitoring networks, and IoMT can meet a range of healthcare needs thanks to the adaptability of heterogeneous sensors. Despite IoMT's potential to offer a wide range of health monitoring services, medical centers are facing pressure from an increasing number of patients, which is impeding its progress. Local devices, such as laptops and cell phones, cannot keep up with the demands of time-sensitive medical information analysis. The emergence of Mobile Edge Computing is one paradigm that shows promise in addressing this problem (MEC). Through the transfer of the medical analysis task to a nearby edge server, MEC reduces the load on local devices and boosts the capabilities of IoMT by providing abundant computational resources.

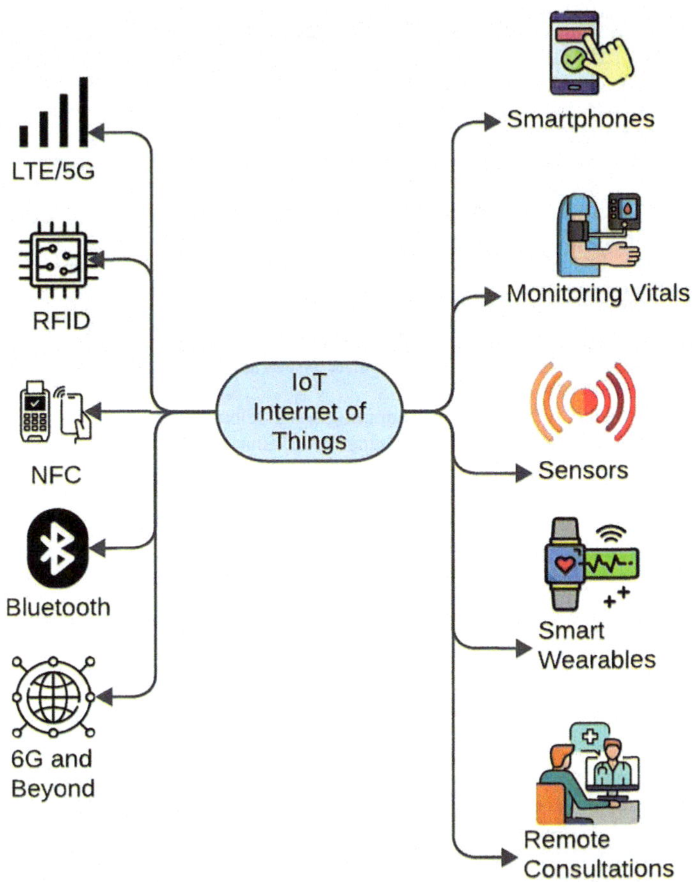

Fig. 8 IOT, enabling technologies and devices in healthcare [22, 23]

Cloud computing can provide a strong platform for the processing, analysis, and storage of the enormous amounts of health-related data produced by various sensors in the context of IoMT. IoMT systems can overcome local device limitations by utilizing the computing power of the cloud, guaranteeing accurate and timely medical information analysis. The cloud offers the scalability required to meet the demanding requirements of time-sensitive tasks while managing the growing number of patients.

References

1. Mittal, A.: Digital health: data privacy and security with cloud computing. Issues Inf. Syst. **21**(1), 227–238 (2020)
2. Bartoletti, I.: AI in healthcare: ethical and privacy challenges. In: Conference on Artificial Intelligence in Medicine in Europe, pp. 7–10. Springer, Cham (2019, June)
3. Khan, S.I., Latiful Hoque, A.S.M.: Digital health data: a comprehensive review of privacy and security risks and some recommendations. Comp. Sci. J. Moldova **24**(2), 71 (2016)
4. Burke, H.B., Sessums, L.L., Hoang, A., Becher, D.A., Fontelo, P., Liu, F., Stephens, M., Pangaro, L.N., O'Malley, P.G., Baxi, N.S., Bunt, C.W.: Electronic health records improve clinical note quality. J. Am. Med. Inform. Assoc. **22**(1), 199–205 (2015)
5. Nicho, M., Hendy, M.: Dimensions of security threats in cloud computing: a case study. Rev. Bus. Inform. Syst. **17**(4) (2013)
6. Kazim, M., Zhu, S.Y.: A survey on top security threats in cloud computing. Int. J. Adv. Comput. Sci. Appl. (IJACSA) **6**(3) (2015)
7. El-etriby, S., Mohamed, E.M., Abdul-Kader.: Modern encryption techniques for cloud computing. In: ICCIT, pp. 800–805 (2012)
8. Bokhari, M.U., Shallal, Q.M., Tamandani, Y.K.: 6 International Conference on Computing for Sustainable Global Development (INDIACom) (2016)
9. Rodrigues, J.P.C., et al.: Analysis of the security and privacy requirements of cloud-based electronic health records systems. J. Med. Internet Res. (2013)
10. Lim, S.Y., Mat Kiah, M.L., Ang, T.F.: Security Issues and Future Challenges of Cloud Service Authentication (2017)
11. Matlani, P., Londhe, N.D.: A cloud computing based telemedicine service. In: IEEE Point-of-Care Healthcare Technologies (PHT), Bangalore, India, pp. 326–330 (2013). https://doi.org/10.1109/PHT.2013.6461351
12. de la Torre-Díez, I., Hamrioui, S., Sainz-de-Abajo, B., Cruz, E.M., López-Coronado, M.: Designing secure cloud-based solutions for several teleconsultation scenarios. Int. J. Clin. Med. Info. **1**(2), 66–73 (2018)
13. Kagadis, G.C., Kloukinas, C., Moore, K., Philbin, J., Papadimitroulas, P., Alexakos, C., Nagy, P.G., Visvikis, D., Hendee, W.R.: Cloud computing in medical imaging. Med. Phys. **40**, 070901 (2013). https://doi.org/10.1118/1.4811272
14. Akkaya, M., Sari, A. Dr., Al-Radaideh, A.T.: Factors affecting the adoption of cloud computing based-medical imaging by healthcare professionals. Am. Acad. Scholar. Res. J. (2016)
15. Gong, S., Zhu, X., Zhang, R., Zhao, H., Guo, C.: An intelligent resource management solution for hospital information system based on cloud computing platform. IEEE Trans. Reliab. **72** (2023)
16. Ali, M., Sadeghi, M.-R., Liu, X., Vasilakos, A.V.: Anonymous aggregate fine-grained cloud data verification system for smart health. IEEE Trans. Cloud Comput. **11**(3) (2023)
17. Awaisi, K.S., Hussain, S., Ahmed, M., Khan, A.A., Ahmed, G.: Leveraging IoT and fog computing in healthcare systems. IEEE Internet of Things Mag. (2020)
18. Liu, Y., et al.: Novel cloud-based framework for the elderly healthcare services using digital twin. In: IEEE Access Special Section on Healthcare Information Technology For the Extreme and Remote Environments (2019)

19. Yang, Z., Liang, B., Ji, W.: An intelligent end–edge–cloud architecture for visual IoT-assisted healthcare systems. IEEE Internet of Things J. **8**(23) (2021)
20. Yang, G., Jiang, M., Ouyang, W., Ji, G., Xie, H., Rahmani, A.M., Liljeberg, P., Tenhunen, H.: IoT-based remote pain monitoring system: from device to cloud platform. IEEE J. Biomed. Health Inform. **22**(6) (2018)
21. Ning, Z., Dong, P., Wang, X., Hu, X., Guo, L., Hu, B., Guo, Y., Qiu, T., Kwok, R.Y.K.: Mobile edge computing enabled 5G health monitoring for Internet of Medical Things: a decentralized game theoretic approach. IEEE J. Sel. Areas Commun. **39**(2) (2021)
22. Razdan, S., Sharma, S.: Internet of Medical Things (IoMT): overview, emerging technologies, and case studies. IETE Tech. Rev., **39**(4), 775–788 (2022). https://doi.org/10.1080/02564602.2021.1927863
23. Aziz, H.A., Guled, A.: Cloud computing and healthcare services. J Biosens Bioelectron **7**, 220 (2016). https://doi.org/10.4172/2155-6210.1000220

An IoT-Based Blockchain-Enabled Secure Storage for Healthcare Systems

Mohd. Harish, Ishita Sharma, Meet Singh, Anushka Gupta, Mustafa Asad, and Rohit Saxena

Abstract The Internet of Things (IoT) is transforming technology by connecting wireless devices, RFID tags, and sensors into a seamless network. Its potential is most evident in healthcare, where it allows for real-time monitoring of patients, tracking of equipment, and extensive collection of health data. This technology has the potential to revolutionize patient care and operational efficiency. However, it is crucial to ensure that sensitive healthcare data is safeguarded. The purpose of this paper is to investigate the integration of blockchain technology into the IoT landscape. In particular, we focus on the use of Ethereum-based blockchain to enhance data security in IoT applications, which is critical in the healthcare industry. This paper proposes a comprehensive approach to utilizing the potential of IoT in healthcare while addressing the unique needs of healthcare data. It prioritizes data security through blockchain, smart contracts, decentralized storage, and encryption. The paper advocates for storing healthcare data in IPFS, which is a decentralized storage solution. Encryption adds an extra layer of security by ensuring that data remains unreadable without the correct keys. This research aims to contribute to the broader discourse on leveraging IoT for transformative healthcare innovation.

Keywords IoT · Blockchain · IPFS · Encryption

1 Introduction

The "Internet of Things" (IoT) has rapidly transformed our world, creating a network of interconnected objects capable of sharing data and information via the Internet. These IoT devices, often high-tech gadgets equipped with sensors, applications, and electronics, possess the ability to sense their surroundings and respond intelligently, albeit with limited power. The impact of IoT, driven by its automation, optimization,

Mohd. Harish (✉) · I. Sharma · M. Singh · A. Gupta · M. Asad · R. Saxena
Department of Computer Science and Engineering, Pranveer Singh Institute of Technology, Kanpur, Uttar Pradesh, India
e-mail: harishaa827@gmail.com

S. N. Mohanty et al. (eds.), *Explainable IoT Applications: A Demystification*, Information Systems Engineering and Management 21,
https://doi.org/10.1007/978-3-031-74885-1_6

and analytical capabilities, has left a profound mark on numerous sectors. Smart devices, equipped with sensors and connectivity, have simplified data collection, management, and processing across industries, including wearables, smartphones, smart automobiles, and more. These devices form interconnected networks that facilitate seamless information exchange, benefiting sectors like inventory management, smart farming, retail, smart cities, smart healthcare, and security systems [1]. Notably, the healthcare sector has undergone revolutionary changes due to IoT's rapid growth [2]. As IoT-connected gadgets are used, the challenge of finding adequate storage space for the volume of generated data emerges. Many businesses and academic organizations opt to store this data in cloud environments. However, the security of cloud-stored data heavily relies on the strategies employed by cloud service providers [3]. Users, lacking control over cloud data storage, face potential threats from hackers and malicious users. Whether data is stored in the cloud or another environment, maintaining robust security is imperative to safeguard user-sensitive information from compromise. Additionally, the substantial data volumes generated by IoT applications can lead to unacceptable transmission delays, potentially impacting the quality of service for real-time applications. To bridge the digital gap and connect IoT devices to the digital world, communication networks are indispensable. Whether data is transmitted via a cloud or alternative means, ensuring a high level of security is paramount to protect user data from potential adversaries. This dual challenge of data storage and secure transmission is central to the IoT landscape (Fig. 1).

IoT sensors and devices play a vital role in the healthcare sector, helping monitor and record patients' vital signs and health data (Fig. 2). Ensuring the privacy of this data is crucial before designing any healthcare infrastructure [4]. For instance, patients can use smart, connected devices to track their health data. However, these devices often need to transmit data to remote servers like the cloud because local storage isn't sufficient [5]. IoT sensors and gadgets are used to collect this health data [6]. Again, it's essential to prioritize data privacy in healthcare infrastructure planning [7]. Imagine a patient using smart IoT devices to monitor their vital signs. To store this data, they must upload it to the cloud or another remote storage solution since on-site storage isn't practical. However, during data transit, there's a risk of exposure to a malicious environment where it might be misused for financial gain, compromising patient data security. This situation could lead to patients losing control over their medical records [8]. To address these concerns, experts strongly recommend encrypting healthcare data before transmission [9]. This approach offers a novel solution to enhance data security in the IoT healthcare domain, incorporating blockchain technology [10]. Blockchain technology continues to evolve and finds various applications in today's world. It has proven effective in the field of cybersecurity [11]. Blockchain's infrastructure is particularly well-suited to address security concerns in IoT devices, networks, and data transmission and storage. Additionally, the complete flow of data in our system is showcased in Fig. 3.

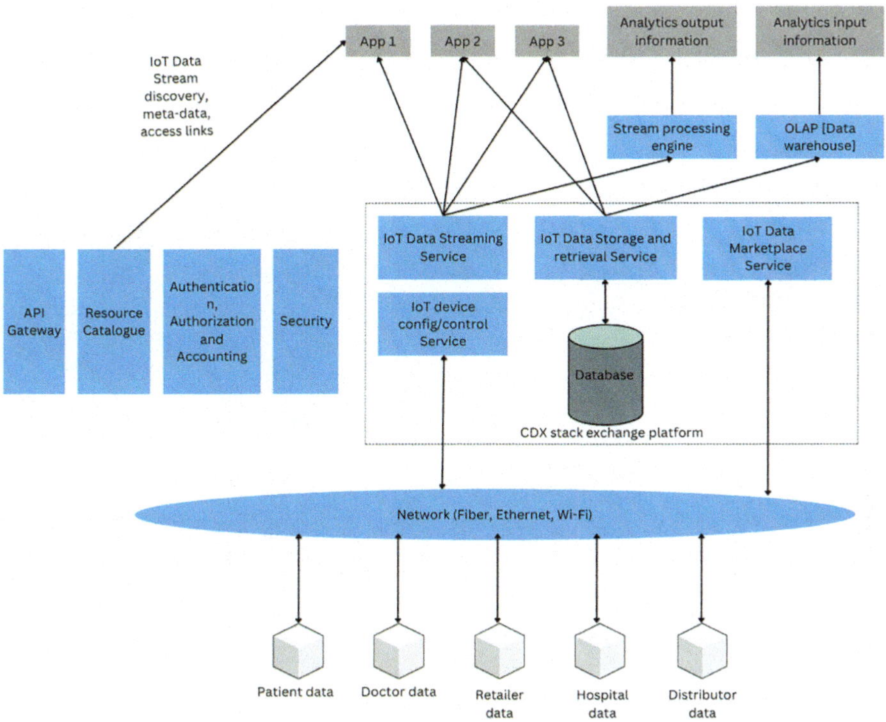

Fig. 1 Illustrating data exchange in the IoT environment

1.1 Contributions

The following major contributions are highlighted in this paper:

- Ethereum-based blockchain integration has strengthened healthcare IoT data security through smart contracts and decentralized IPFS storage, ensuring robust protection for sensitive information.
- The paper stresses IoT security, employing encryption and access control, notably on Raspberry Pi, to ensure data confidentiality and prevent unauthorized access.
- The study contrasts Firebase and IPFS, finding Firebase faster for larger files but highlighting IPFS as the superior choice for data integrity in healthcare IoT decision-making.
- Ethereum's blockchain ensures secure, transparent collaboration among healthcare stakeholders, maintaining data integrity and privacy in the process.

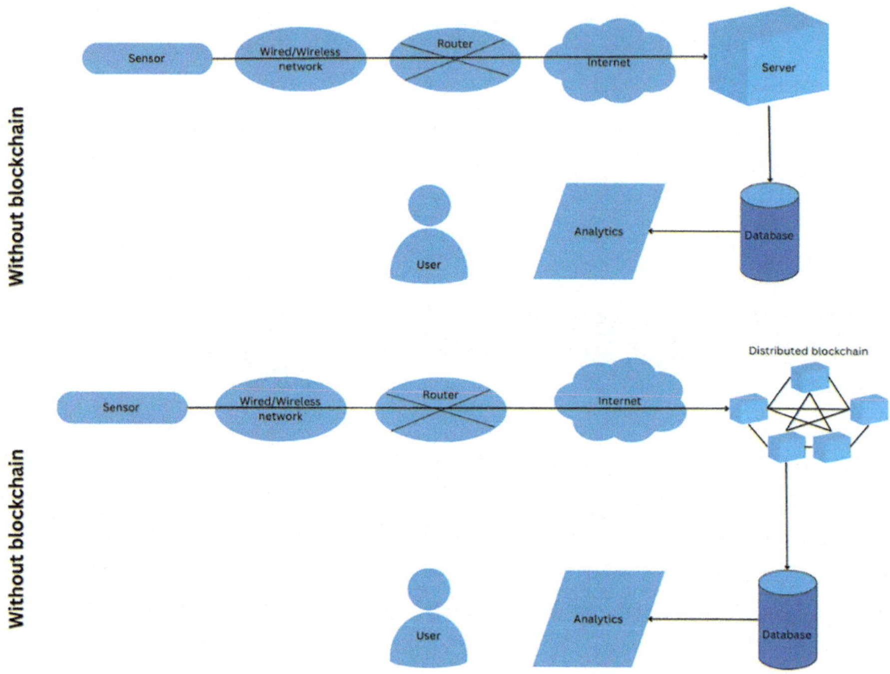

Fig. 2 Data storage of patient's health data with and without blockchain

1.2 Structure of the Work

This study is divided into five sections for clarity. In the introductory section (Sect. 1), we present an overview of IoT and blockchain technology, and their applications in healthcare, and address data security challenges while proposing blockchain solutions. The subsequent section (Sect. 2 delves into the authors' previous work related to medical healthcare systems, IoT, and blockchain, along with background information and problem formulation. Section 3 outlines our proposed methodology and research objectives. Moving on to Sect. 4, we discuss the various techniques employed in the study, the proposed methodology, results, and the workflow of the IoT-blockchain-based medical healthcare system. Finally, in Sect. 5, we provide conclusions drawn from the study's findings and outline potential avenues for future research.

2 Related Work

The following section provides a brief overview of the reviewed literature.

Wu et al. [12] discovered that medical information often travels through insecure channels, putting patients at risk of unauthorized access and tampering. To

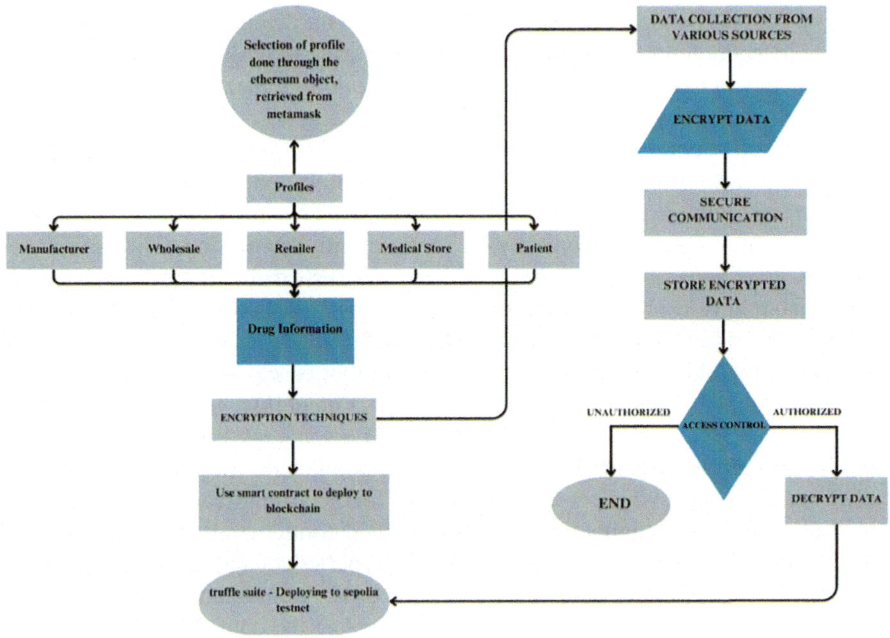

Fig. 3 Complete flow on information in the blockchain

address this problem, they proposed a strategy that combines blockchain technology and general-purpose edge computing. This approach ensures secure and reliable data transfer between medical devices (known as D2D or device-to-device communication). They also used advanced encryption techniques to protect medical data during transfer and storage. In the end, they added blockchain for an extra layer of security. Their simulations proved that this method is both effective and safe, promising greater data security in healthcare. Recognizing the importance of secure device connectivity, Ali et al. [13] proposed an approach by introducing Binary Spring Search (BSS) and combining it with blockchain to enhance IoT network security. BSS is rooted in deep neural networks and group theory. The researchers also developed mechanisms for key revocation and policy changes. This resulted in the creation of a secure patient healthcare data operating system. This system uses blockchain and trust chains to address the efficiency and security challenges faced by digital health information exchange. The goal of Almaiah et al. [14] was to identify effective preventive measures against security risks in healthcare resources using a combination of heuristics, signatures, and voice recognition techniques. They built a system where the central node combines the functions of three identification systems with blockchain to detect malicious sensor nodes. They also used the central node to identify malicious nodes in Wireless Sensor Networks (WSNs) based on criteria like sensor node hash value, node signature, and voting. Through simulations, they achieved a remarkable 94.9% identification rate for malicious messages.

In their research, Panda et al. [15] developed a decentralized IoT architecture using blockchain technology. Their approach involved the use of hash chains for encryption key management to ensure user privacy and secure communication. This method not only enhanced privacy but also improved the security framework for IoT communication. Their study demonstrated that this approach outperformed traditional techniques significantly. Velmurugadass et al. [16] designed a blockchain architecture to collect evidence and maintain data provenance in the cloud. This involved user registration, authentication, data encryption, storage, and activity monitoring. They employed Elliptic Curve Integrated Encryption Scheme (ECIES) technology to encrypt data packets from IoT devices before transmitting them to the cloud server. The system's public safety significantly improved through distributed scheduling. Their innovative approach yielded excellent results with minimal processing costs, and it could help trace data changes, which is crucial for accountability. In their study, Chenthara et al. [17] explored how to protect patients' personal information in healthcare. They used blockchain technology and within the blockchain, they created something called Healthy Chain, which acted as a guardian for electronic health records (EHRs). This framework ensured that patient data remained confidential, available, scalable, and trustworthy. They used a system called Hyperledger Fabric, which is like a very secure foundation, to store these health records. Additionally, they added extra security layers by using SHA-256 hashing to protect the data. The results of their work showed that this model greatly improved security, privacy, scalability, and how different systems could work together when it came to sharing health information. Mohanta et al. [18] talked about the Internet of Things (IoT). They emphasized the importance of using IoT devices while making sure data was protected and secure. The authors found that blockchain can prove to be secure. They even suggested that you could use blockchain to build secure IoT applications. Blockchain's decentralized nature, which means it's spread out and not controlled by one person, adds an extra layer of security. Their research showed that blockchain could be a solution for privacy and security problems related to IoT. Werder et al. [19] explored how blockchain and artificial intelligence (AI) could help with medical research and patient health management. They wanted to make it easier for patients to manage their health information and even get rewards for doing so. They introduced a new concept, for organizing and understanding medical data. They imagined a system where the health information is stored on a blockchain. This blockchain uses consensus algorithms, which are like rules everyone agrees on, to support new ways of discovering medicine, developing markers for diseases, and preventing health issues. When health information is scattered, blockchain and deep learning technology can help create a secure and open system. The main focus of Li et al. [20] was on keeping medical information secure, exchanging it reliably, controlling who has access to it, and protecting people's privacy. They developed something called EHR Chain, a system for electronic health records (EHRs) based on blockchain. This system uses special techniques to make sure medical data is reliable and can be traced back to its source, like a digital detective. They also connected blockchain to something called IPFS, which is like super-secure cloud storage. The primary goal of Zaabar et al. [21] study was to use blockchain technology to make electronic health records

(EHRs) secure and private. They wanted to move away from centralized storage and use distributed databases. These distributed EHRs were stored using IPFS, which is incredibly durable and scalable. They also used Hyperledger Composer to save summaries of data and control who can access it. The purpose of all this was to improve the stability of healthcare management systems and make the data safe.

3 Problem Formulation

In the past decade, the IoT has brought about significant changes. IoT involves various smart devices connected to the internet. These smart devices can collect and share data across different applications. However, creating successful IoT applications comes with challenges such as network traffic, limited capacity, mobile device compatibility, security, and privacy concerns. To make IoT work effectively in the real world, we need to address these issues, including worries about data security, privacy, and governance. One solution to tackle the privacy and security problems of IoT devices is blockchain technology [22]. Blockchain can offer features like smart contracts, digital signatures, and data mining to enhance security [23]. Platforms like Hyperledger Fabric or Ethereum can be used to build IoT applications that are secure and resistant to forgery. Before developing and deploying an IoT application, it's crucial to address security and privacy concerns. Threats to user security and privacy are among the challenges that can arise during IoT implementation. To address these issues, this study combines IoT and blockchain to enable smart devices to operate independently without relying on a central authority. It also ensures that data files are encrypted using advanced encryption technology before storing them in IPFS. With this blockchain-based approach, only authorized parties can access patient files, offering a solution to the current security and privacy problems in IoT applications.

4 Preliminaries

This work focuses on enhancing security for health information systems using blockchain technology. It also aims to address cybersecurity issues and privacy concerns, especially when connecting IoT devices [24]. One aspect involves combining blockchain and IoT to make these systems more secure and private [25]. To create secure data modules, we use IPFS, which generates cryptographic output based on hash values.

4.1 Techniques

Blockchain: Blockchain technology enables individuals and organizations to engage in data transmission and financial transactions directly, bypassing the need for a trusted intermediary. Specific types of nodes are tasked with verifying and validating these transactions [26]. Think of a blockchain as an all-encompassing, verified financial ledger containing authenticated copies of every transaction. This distributed ledger operates on a vast network of computers, or nodes, numbering in the millions. Nodes autonomously join this network, each assuming the role of a network administrator. The inherent design of a blockchain, decentralized and lacking a central repository of information, renders it virtually impervious to hacking attempts. The architecture of a blockchain accommodates a continually expanding list of ordered documents referred to as blocks, with each block carrying a timestamp and a link to its predecessor. Figure 4, provides an overview of the fundamental framework of blockchain technology.

Blockchain information is securely stored in a distributed ledger maintained by computers known as nodes. This technology serves as a robust method for safeguarding sensitive data within the system, allowing for the secure exchange of such information between parties. This versatile tool enables the consolidation of relevant files in a single, secure location. Leveraging blockchain expedites the search for trial participants meeting specific criteria. In essence, blockchain operates as a peer-to-peer (P2P) network of nodes, ensuring secure storage and recording of transaction histories without reliance on any central authority. Figure 5, offers a step-by-step illustration of how blockchain technology operates. The network's data

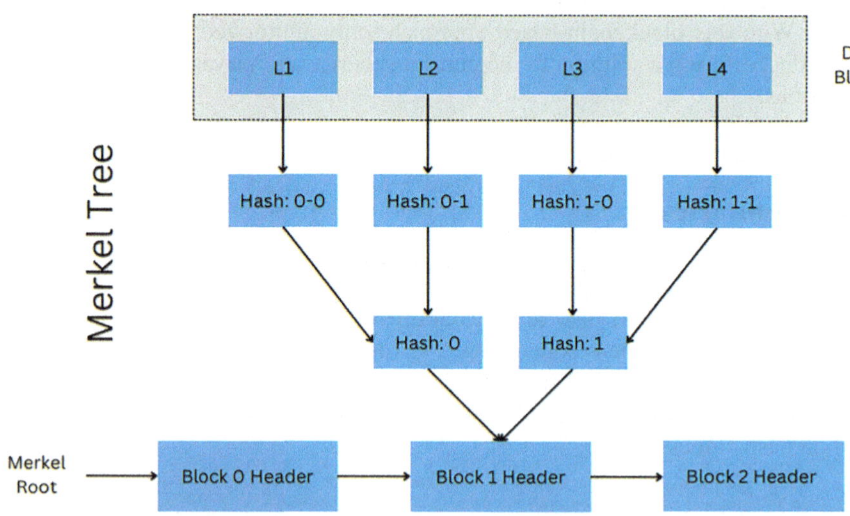

Fig. 4 General framework of blockchain

storage and transmission capabilities foster dependable collaboration by continuously documenting past and current events. This tool has the ability to connect multiple systems, offering valuable insights into a patient's treatment journey. The immutability and security of blockchain technology are widely recognized. Notably, information isn't stored in a singular repository but is instead distributed and duplicated across a network of computers. As new blocks are added, every computer in the world automatically updates the blockchain [27].

IPFS: IPFS, a decentralized file-sharing network, uniquely identifies files based on their content by utilizing cryptographic hashes, which are conveniently stored on a blockchain. However, direct file sharing among users through IPFS is not possible [28]. Instead, it employs a distributed hash table to gather information about file locations and node connections. IPFS stands as a peer-to-peer, decentralized file-sharing system, and its notable innovation lies in replacing traditional geolocation with content-based addressing. In simpler terms, it relies on hash functions rather than

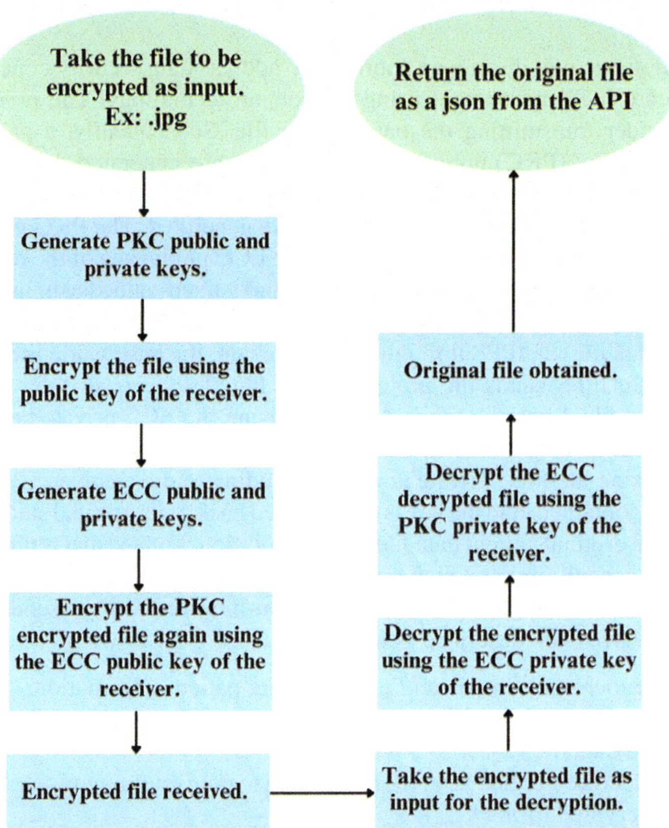

Fig. 5 Flowchart of encryption and decryption

file addresses [29]. Each file uploaded to IPFS is assigned a unique hash, making it easily retrievable if known. IoT data is generated in a secure environment and then preserved within IPFS to ensure that only authorized parties can access it. To safeguard sensitive medical information, the study employed the Elliptic Curve cryptographic method, prioritizing privacy, data integrity, and effective management [30, 31].

Ethereum: Ethereum, a blockchain platform, stands as an open-source, decentralized system distributed across numerous computers. Conceived in 2014 by Vitalik Buterin, Ethereum drew inspiration from the Bitcoin cryptocurrency [32–34]. Ethereum employs the elliptic curve digital signature algorithm, much like Bitcoin, with this algorithm rooted in the discrete logarithm problem, yielding a pair of cryptographic keys. To secure its network, Ethereum relies on an elliptic curve known as Secp256k1.

5 Proposed Methodology

In this section, we will discuss an approach to address data security concerns within the healthcare IoT environment using blockchain technology. The process begins with the sender transmitting the patient data file. Subsequently, a pair of Public Key Cryptography (PKC) public and private keys are generated. Using the public key of the intended receiver, the file is encrypted. Following this, Elliptic Curve Cryptography (ECC) public and private keys are generated. The PKC-encrypted file is then encrypted once more, this time using the ECC public key of the recipient. The resulting protected data file is stored on IPFS, and a fixed-value hash, in conjunction with the transformed key, is recorded on the blockchain. When the receiver requests the patient data file via IPFS through the blockchain, the hash value is checked, and upon a match, IPFS sends the protected data file. The receiver then decrypts the protected data file. First, the file is decrypted using the ECC private key specific to the receiver. Once this is accomplished, the ECC-decrypted file undergoes another decryption using the PKC private key corresponding to the receiver. Ultimately, the original file is obtained, and the key is transformed back to its original state, ultimately extracting the original patient data file for transmission. For a visual representation of this process, consult Fig. 4, which outlines the workflow of our proposed approach.

The steps for showing the process of data transmission from the sender end to the receiver end have been described below:

Step 1: Sender sends the data ($\mathcal{D}_p$) containing patient information.
Step 2: Patient's data file ($\mathcal{D}_p$) is encrypted using a symmetric key (K_{sym}) creating a protected data file ($\mathcal{D}_p$').
Step 3: Receiver's public key ($_{Pub}\ K_R$) is used to convert the symmetric key (K_{sym}) into an encrypted form (K'_{sym}).
Step 4: The protected data file ($\mathcal{D}_p$') is sent to IPFS file storage.
Step 5: A fixed-value hash (H_D) is generated by IPFS as an acknowledgement.

Step 6: The hash value returned from IPFS is stored on the blockchain along with K′.
Step 7: Receiver submits the request for the patient's data file ($\mathcal{D}_p$) to IPFS through blockchain.
Step 8: The data file $\mathcal{D}'_p$ is sent to the receiver by matching the hash (H_D) as a response by IPFS.
Step 9: Receiver's private key ($_{Pri}\ K_R$) is used to decrypt the protected ($\mathcal{D}_p$) data file. The encrypted key (K'_{sym}) is again decrypted to its original form (K_{sym}).
Step 10: The symmetric key (K_{sym}) is used to extract the original data ($\mathcal{D}_p$) from the protected data file ($\mathcal{D}'_p$).

6 Results and Discussion

The proposed methodology underwent initial validation on the Ganache platform, followed by simulating the behaviour of the primary network. A standalone instance of Ganache was employed to simulate a blockchain network. The front end, responsible for record addition and viewing, was developed using ReactJS, while the back end was implemented in JavaScript. Additionally, patients' cryptographic keys were securely stored at MongoDB. To recreate a blockchain network, we utilized the Ethereum platform in conjunction with Solidity for testing purposes, and Web3-JS was employed to interact with the blockchain. INFURA was harnessed to offer reliable, secure, and scalable access to the IPFS gateway and to comprehensively test the IPFS network during the testing phase. The proposed solution was constructed using Firebase and IPFS platforms, enabling us to assess the speed of uploading and downloading data files. Five images (jpg format) were used of sizes—50 KB, 100 KB, 150 KB, 300 KB and 800 KB, both on a decentralized database hosted on IPFS and on Firebase, with subsequent recording of upload and download times. This entails a comparative analysis of the proposed approach's confidentiality, data integrity, and access control aspects in contrast to established blockchain techniques.

The results depicted in Fig. 6 showcase the upload and download speeds of data files in the recommended approach utilizing the Firebase platform. In this scenario, 5 different JPG files employed from the Firebase database, and the timestamps for upload and download were recorded.

The outcomes presented in Fig. 7 illustrate the speed at which data files can be uploaded and downloaded using the IPFS platform.

Figure 8 underscores the contrast in download times for the .jpg data file between the Firebase and IPFS systems. Our analysis reveals that the IPFS platform outperforms the centralized Firebase platform in terms of download speed.

Our system ensures data security for IoT-based healthcare data by processing all transactions (linked to IPFS) on an Ethereum-based blockchain network. Figure 9 presents the difference in download times for the .jpg data file between Firebase and IPFS. As anticipated, the files with larger sizes require more time for uploading and downloading on both IPFS and Firebase.

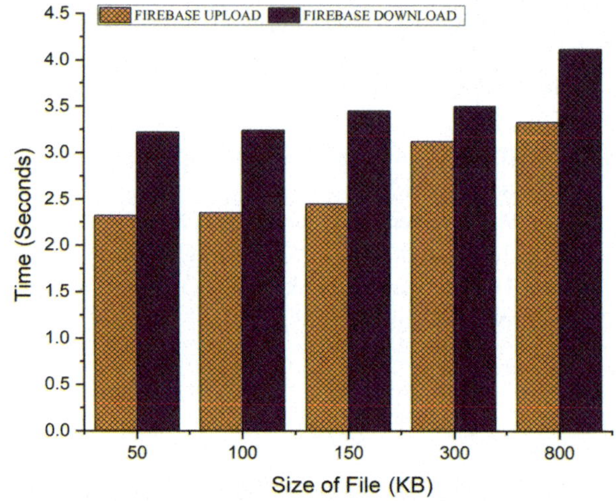

Fig. 6 Upload–download times on the firebase platform

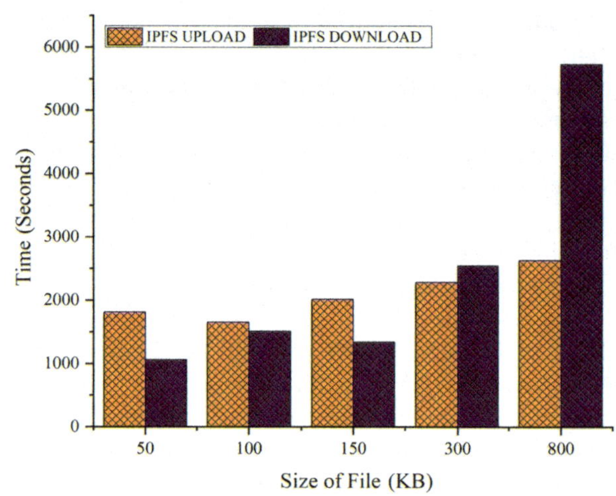

Fig. 7 Upload–download times on the IPFS platform

7 Conclusion and Future Scope

This study explored how blockchain could help address challenges in the medical field related to IoT. It did so by focusing on encryption and access control. We utilized Ethereum's structure to enhance data security and enable seamless collaboration among different stakeholders. To collect and secure data, we employed a Raspberry Pi IoT device, which generated and encrypted the information before storing it in IPFS for decentralized storage. We thoroughly analyzed upload and download speeds for both Firebase and IPFS, highlighting the differences between them. Our findings suggested that while Firebase could perform better in terms of speed for larger files, IPFS significantly outshone Firebase when it came to data integrity. Future research

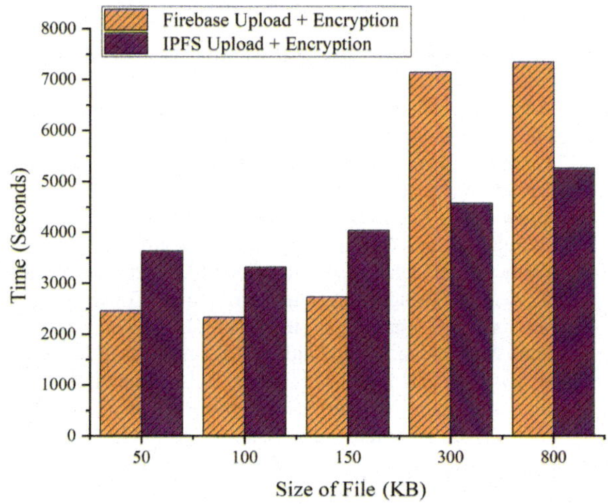

Fig. 8 Comparison of uploading for .jpg file on Firebase and IPFS platform

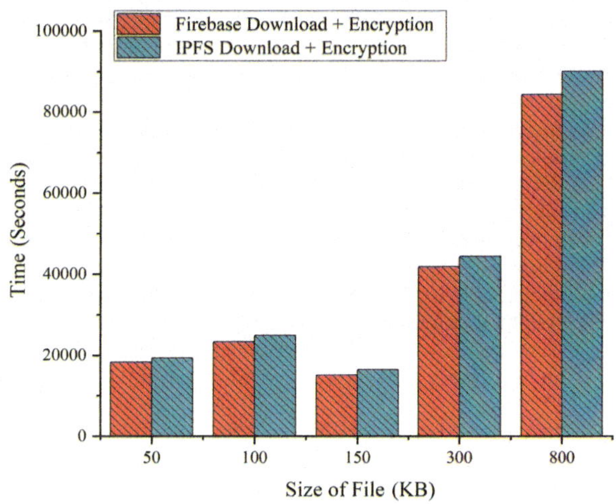

Fig. 9 Difference of download times of .jpg files on Firebase and IPFS platforms

can explore large-scale simulations and different blockchain platforms for managing agreements among system participants.

References

1. Tripathi, S., Chaurasia, B.K.: Lightweight communication in IoT using MQTT. In: IEEE ISKCON, GLA University, Mathura, India, March 3–4, pp. 1–6 (2023)
2. Tripathi, G., Singh, V.K., Chaurasia, B.K.: An energy-efficient heterogeneous data gathering for sensor-based internet of things. Multimedia Tools Appl., 1–24 (2023)

3. Popović, K., Hocenski, Ž.: Cloud computing security issues and challenges. In: The 33rd International Convention MIPRO, pp. 344–349 (2010)
4. Pradhan, N.R., Singh, A.P., Verma, S., Kavita, Kaur, N., Roy, D.S., Shafi, J., Wozniak, M., Ijaz, M.F.: A novel blockchain-based healthcare system design and performance benchmarking on a multi-hosted testbed. Sensors **22**(9), 3449 (2022)
5. Awotunde, J.B., Jimoh, R.G., Folorunso, S.O., Adeniyi, E.A., Abiodun, K.M., Banjo, O.O.: Privacy and security concerns in IoT-Based Healthcare Systems. In: Siarry, P., Jab-bar, M., Aluvalu, R., Abraham, A., Madureira, A. (eds) The Fusion of Internet of Things, Artificial Intelligence, and Cloud Computing in Health Care. Internet of Things. Springer, Cham (2021)
6. Misra, G., Hazela, B., Chaurasia, B.K.: Zero knowledge based authentication for Internet of Medical Things. In: 14th International Conference on Computing, Communication And Networking Technologies (ICCCNT), IIT—Delhi, Delhi India, July 6–8, pp. 1–6 (2023)
7. Talal, M., Zaidan, A.A., Zaidan, B.B., Albahri, A.S., Alamoodi, A.H., Albahri, O.S., Alsalem, M.A., Kim, C.K., Tan, K.L., Shir, W.L., Mohammed, K.I.: Smart home-based IoT for real-time and secure remote health monitoring of triage and priority system using body sensors: multi-driven systematic review. J Med Syst **43**, 42 (2019)
8. Islam, M.M., Bhuiyan, Z.A.: An integrated scalable framework for cloud and IoT based green healthcare system. **11** (2023)
9. Elhoseny, M., Ramírez-González, G., Abu-Elnasr, O.M., Shawkat, S.A., Arunkumar, N., Farouk, A.: Secure medical data transmission model for IoT-based healthcare systems. IEEE Access **6**, 20596–20608 (2018)
10. Singh, A.P., et al.: A novel patient-centric architectural framework for blockchain-enabled healthcare applications. IEEE Trans. Industr. Inf. **17**(8), 5779–5789 (2020). https://doi.org/10.1109/TII.2020.3037889
11. Saxena, R., Arora, D., Nagar, V., Chaurasia, B.K.: Privacy provisioning on blockchain transactions of decentralized social media. In: Blockchain Technology for Social Media Computing, IET (2023)
12. Wu, H., Liu, X., Ou, W.: A novel blockchain-MEC-based near-domain medical resource sharing model. In: Machine Learning for Cyber Security, pp. 40–56 (2023)
13. Ali, A., Almaiah, M.A., Hajjej, F., Pasha, M.F., Fang, O.H., Khan, R, Teo, R., Zakarya, M.: An industrial IoT-based blockchain-enabled secure searchable encryption approach for healthcare systems using neural network, pp. 572–572 (2022). https://doi.org/10.3390/s22020572
14. Almaiah, M.A.: A new scheme for detecting malicious attacks in wireless sensor networks based on blockchain technology, pp. 217–234 (2021). https://doi.org/10.1007/978-3-030-74575-2_12
15. Panda, S.S., Jena, D., Mohanta, B.K., Ramasubbareddy, S., Daneshmand, M., Gandomi, A.H.: Authentication and key management in distributed IoT using blockchain technology. IEEE Internet Things J. **8**(16), 12947–12954 (2021). https://doi.org/10.1109/jiot.2021.3063806
16. Velmurugadass, P., Dhanasekaran, S., Anand, S.S., Vasudevan, V.: Enhancing blockchain security in cloud computing with IoT environment using ECIES and cryptography hash algorithm (2020). https://doi.org/10.1016/j.matpr.2020.08.519
17. Chenthara, S., Ahmed, K., Wang, H., Whittaker, F., Chen, Z.: Healthchain: a novel framework on privacy preservation of electronic health records using blockchain technology. **15**(12), e0243043 (2020). https://doi.org/10.1371/journal.pone.0243043
18. Mohanta, B.K., Satapathy, U., Panda, S.S., Jena, D.: A novel approach to solve security and privacy issues for IoT applications using blockchain. IEEE Xplore (2019). https://ieeexplore.ieee.org/abstract/document/9031931?casa_token=6z7JgJ1yUvgAAAAA:1MedpG22tuvURa_SCmCkXcC8mDud1fkwTn_wIV7FGSEhPebB4U4LRtMHXdklnhIiIYBCBOpXBw
19. Werder, K., Ramesh, B., Zhang, R. (Sophia): Establishing data provenance for responsible artificial intelligence systems. ACM Trans. Manage. Inf. Syst. **13**(2), 1–23 (2022). https://doi.org/10.1145/3503488
20. Li, F., Liu, K., Zhang, L., Huang, S., Wu, Q.: EHRChain: a blockchain-based EHR system using attribute-based and homomorphic cryptosystem. IEEE Trans. Serv. Comput., 1–1 (2021). https://doi.org/10.1109/tsc.2021.3078119

21. Zaabar, B., Cheikhrouhou, O., Jamil, F., Ammi, M., Abid, M.: HealthBlock: a secure blockchain-based healthcare data management system. Comput. Netw. **200**, 108500 (2021). https://doi.org/10.1016/j.comnet.2021.108500
22. Sharma, A.K., Peelam, M.S., Chaurasia, B.K., Chamola, V.: QIoTChain: quantum IoT-blockchain fusion for advanced data protection in industry 4.0. Wiley IET Blockchain (2023). https://doi.org/10.1049/blc2.12059
23. Pradhan, N.R., Singh, A.P.: Smart contracts for automated control system in blockchain based smart cities. J. Amb. Intell. Smart Environ. **13**(3), 253–267 (2021)
24. Saxena, R., Arora, D., Nagar, V.: Classifying blockchain cybercriminal transactions using hyperparameter tuned supervised machine learning models. Int. J. Comput. Sci. Eng. (IJCSE) **26**(6) (2023)
25. Satpathy, S., Mahapatra, S., Singh, A.: Fusion of blockchain technology with 5G: a symmetric beginning. In: Tanwar, S. (eds.) Blockchain for 5G-Enabled IoT. Springer, Cham (2021). https://doi.org/10.1007/978-3-030-67490-8_3
26. Filippi, D.P.: The interplay between decentralization and privacy: the case of blockchain technologies. J. Peer Prod. (7). Alternative Internets (2016). Available at SSRN: https://ssrn.com/abstract=2852689
27. Saxena, R., Arora, D., Nagar, V.: Efficient blockchain addresses classification through cascading ensemble learning approach. Int. J. Electron. Secur. Digit. Forensics **15**(2), 195–209 (2023)
28. Yuan, Y., Wang, F.Y.: Towards blockchain-based intelligent transportation systems. In: IEEE 19th International Conference on Intelligent Transportation Systems (ITSC), Rio de Janeiro, Brazil, pp. 2663–2668 (2016). https://doi.org/10.1109/ITSC.2016.7795984
29. DuPont, Q.: Cryptocurrencies and Blockchains. John Wiley & Sons: Hoboken, NJ, USA, pp. 1–224 (2019)
30. Rai, S., Chaurasia, B.K., Gupta, R., Verma, S.: Blockchain-based NFT for healthcare system. In: 12thIEEE International Conference on Communication Systems and Network Technologies (CSNT), pp. 700–704 (2023)
31. Trautwein, D., Raman, A., Tyson, G., Castro, I., Scott, W., Schubotz, M., Gipp, B., Psaras, Y.: Design and evaluation of IPFS: a storage layer for the decentralized web. In: The SIGCOMM '22: ACM SIGCOMM 2022 Conference, Amsterdam, pp. 739–752 (2022). https://doi.org/10.1145/3544216.3544232
32. Amara, M., Siad, A.: Elliptic curve cryptography and its applications. In: International Workshop on Systems, Signal Processing and their Applications, WOSSPA, Tipaza, Algeria, 247–250 (2011). https://doi.org/10.1109/WOSSPA.2011.5931464
33. Mohanty, S.N., Ramya, K.C., Sheeba Rani, S., Gupta, D., Shankar, K., Laksmanaprabu, S.K., Khanna, A.: An efficient Lightweight integrated Block chain (ELIB) model for IoT security and privacy. Fut. Gener. Comp. Syst. (SCI), **102**(2),1027–1037 (2020). ISSN:0167-739X. https://doi.org/10.1016/j.future.2019.09.050
34. Ganachari, S., Nandigam, P., Daga, A., Mohanty, S.N., Sudha, S.V.: Machine learning based malware analysis in digital forensic with IoT devices. In: Nandan Mohanty, S., Garcia Diaz, V., Satish Kumar, G.A.E. (eds.) Intelligent Systems and Machine Learning. ICISML 2022. Lecture Notes of the Institute for Computer Sciences, Social Informatics and Telecommunications Engineering, vol. 470. Springer, Cham (2023). https://doi.org/10.1007/978-3-031-35078-8_15

Block-Chain Technology in Smart Telemedicine Using IoT

B. Ravi Chandra, B. Tanusree, Y. Sri Ramani, Y. Sowjanya, and Amjan Shaik

Abstract Modern IoHT combines health-related items like sensors and remotely monitored healthcare equipment for patient data evaluation and management. The integration of 5G and blockchain enhances this system's capabilities. Blockchain ensures secure data storage and exchange, benefiting various healthcare aspects such as monitoring equipment, clinical trial data, mobile health apps, electronic health records, and insurance data. The proposed framework utilizes medical devices to collect patient data, which is kept on an online database along with pertinent health information. 5G and blockchain technology facilitate secure, high-speed data transmission. An Artificial Neural Network (NN) is employed to predict diseases and their severity. Various classifiers are compared, with the NN classifier achieving a 98.98% accuracy rate. This trained NN provides more accurate and intelligent predictions compared to traditional classifiers.

Keywords Artificial neural network · Blockchain technology · 5th generation

1 Introduction

Telemedicine employs real-time video and store-and-forward modes for remote medical consultations. The former offers immediate video assessments but relies on good internet and equipment. Store-and-forward involves data storage and forwarding, useful in regions connectivity challenges, typically with a 24–48 h response time. Challenges arise from time and distance, with CDSS and 5G integration mitigating these issues. CDSS reduces response times by providing treatment recommendations based on symptoms. 5G enables faster data transmission and remote vital sign monitoring through smart devices, improving access to care.

B. Ravi Chandra (✉) · B. Tanusree · Y. Sri Ramani · Y. Sowjanya
G. Pullaiah College of Engineering and Technology, Kurnool, India
e-mail: chandrabrc11@gmail.com

A. Shaik
CSE, R&D, St. Peters Engineering College, Maisammaguda, India

S. N. Mohanty et al. (eds.), *Explainable IoT Applications: A Demystification*, Information Systems Engineering and Management 21,
https://doi.org/10.1007/978-3-031-74885-1_7

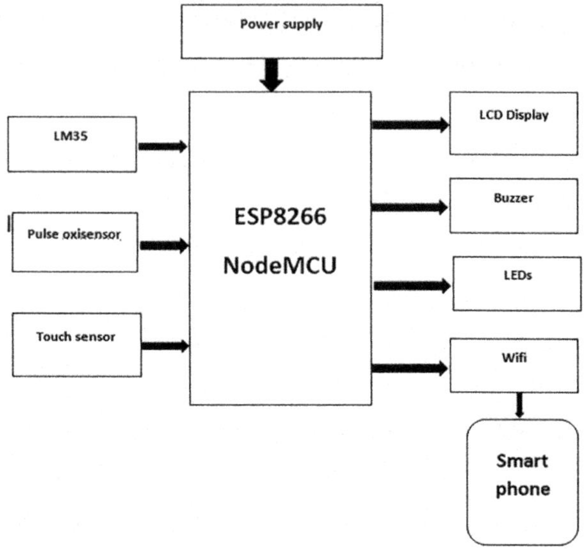

Fig. 1 Block diagram

Blockchain enhances security and data integrity while facilitating data sharing. By addressing these factors and integrating IoT, blockchain, and 5G technologies, telemedicine enhances healthcare delivery, reducing treatment time from days to minutes, and improving the healthcare system's quality and efficiency [1–3].

Block diagram

See Fig. 1.

System requirements

Hardware:

- MAX30100
- ESP8266 NodeMCU
- Touch sensor
- LM35 sensor
- LCD 16 × 2 Display
- Buzzer
- Wifi module
- LEDs
- Jumper wires
- PCB

Software:

- Arduino IDE
- Embedded C
- Blynk Cloud

Working process

The brains behind the system are NodeMCU. Via the I2C interface, the MAX30100 and the touch screen are linked to the nodeMCU. The serial bus interface connection protocol is called I2C (Inter-Integrated Circuit). Given that it only requires two wires for communication, it is also known as a TWI (two-wire interface). SDA (serial data) and SCL (serial clock) are the two connections. I2C is an acknowledgment-based communication protocol, meaning that after sending data, the transmitter waits for the receiver to acknowledge the transmission to be able to determine if the data was successfully received by the receiver. I2C functions in two modes: 1. Master mode 2. In a parallel connection, SDA and SCL wires are used for data exchange and devices on the master and slave aligned clocks. LCD panels and MAX30100 have connections this way, while the temperature sensor is single-channel, connected to a digital pin separately. Blynk is a free IoT platform available on Google Play, requiring an active Wi-Fi connection and a NodeMCU. NodeMCU code should include "BlynkSimpleEsp8266.h" and an authentication code for app access. The Blynk App displays data uploaded to the Blynk App Server under a registered login ID.

Block-Chain Technology

Blockchain is widely applied in healthcare for secure record transfers, medicine supply chain management, and genetic research. It offers transparency, security, and privacy through decentralized storage and authentication. Blockchain enables quick and safe information sharing among patients, doctors, and providers, overcoming privacy and regulatory challenges [4, 5]. This integration enhances patient care and data integrity. In online healthcare monitoring, sensors continuously observe patients, ensuring security. Data collection, disease investigation, and diagnosis are facilitated through sensing and monitoring devices, with communication and security ensured by 5G and blockchain [6, 7].

NodeMCU Firmware:

Based on the ESP8266 Api 0.9.5, NodeMCU is is a freely available Network of Things framework that uses the Python programming language. It contains free applications such as spiffs and lua-cjson. Originating from the ESP8266 release in December 2013, NodeMCU development began in October 2014, with the first firmware file committed by Hong. It expanded to open-hardware with Huang R's contribution of the gerber file for devkit 1.0. Tuan PM then ported the MQTT client library to ESP8266, enabling MQTT IoT support with Lua. A significant update in January 2015 incorporated u8glib, allowing NodeMCU to drive various displays like LCD, OLED, and VGA screens.

NodeMCU ESP8266:

NodeMCU originated after the introduction of ESP8266 in December 2013, a popular Wi-Fi SoC for IoT applications. Development of NodeMCU began in October 2014, with Hong's first commit to the GitHub repository. Developer Huang R contributed the gerber file for the ESP8266 devkit v0.9 in December. Tuan PM ported the MQTT client library, enabling MQTT support with Lua. In January 2015, Devsaurus added u8glib support for various displays. In the summer of 2015, the original creators abandoned NodeMCU, but independent contributors took over, offering over 40 modules for customized firmware (See Fig. 2).

IO index	ESP8266 pin	IO index	ESP8266 pin
0 [*]	General purpose input output 1	7	General purpose input output 13
1	General purpose input output 5	8	General purpose input output 15
2	General purpose input output 4	9	General purpose input output 3
3	General purpose input output 0	10	General purpose input output 1
4	General purpose input output 2	11	General purpose input output 9
5	General purpose input output 14	12	General purpose input output 10
6	General purpose input output 12		

The "ESP8266 Core for the Arduino IDE" emerged to support ESP8266-based modules, including NodeMCU. Notable projects include "The Button" by Peter R Jennings, "Node USB," and "ijWatch," an open Wi-Fi smart watch. NodeMCU pins provide access to GPIO (General Purpose Input/output) for development purposes.

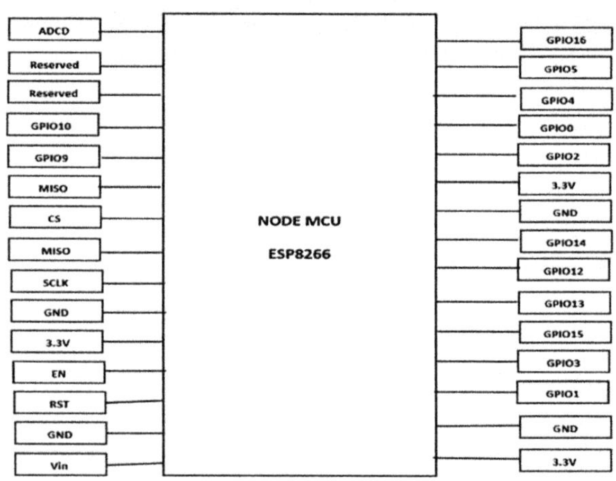

Fig. 2 Pins of NodeMCU

LM35 Temperature Sensor:

See Fig. 3.

Features of LM35 logic module Regulators:

LM35 is a cost-effective, small temperature sensor with a voltage output proportional to temperature from − 55 to 150 °C.

Pin number	Pin name	Description
1	Emitter follower	Typical voltage of the input for 1 Vcc uses is + 5 V
2	Input output	For every 1 °C rise, there ought to be a 10 mV increase. Can vary between – 1 V (– 55 °C) and 6 V (150 °C)
3	Ground	Three Base linked to the circuit's ground

It's accurate to ± 0.5 °C, has low power consumption (less than 60 μA), and comes in various package types. You can easily interface it with microcontrollers or platforms like Arduino by connecting the ground while offering a precisely controlled + 5 V signal for the input terminal bit. Within the range from 35 to – 2 V, the output voltage goes up by 0.01 V for each one degree Celsius rise in warmth.

Pulse oximeter:

Oximeter is crucial for monitoring SpO_2 during the pandemic. They use LEDs and a photodiode to measure SpO_2 levels, which should ideally be above 95%. The MAX30100 sensor, commonly used in oximeter, provides pulse oximeter and heart

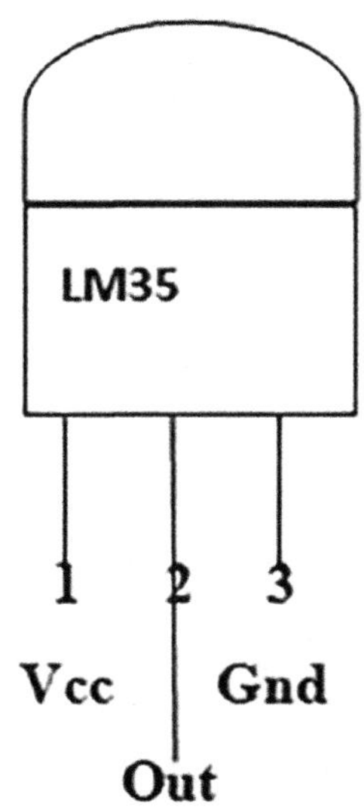

Fig. 3 LM-35 temperature sensor

rate data through I2C but can be affected by factors like motion, temperature, and pressure (Fig. 4).

Pin Configuration of MAX30100 Pulse Oximeter Heart Rate Sensor Module:-

S. No.	Pins	Definition of pins
1	RD	MAX30100 R_DRV
2	VIN	Input voltage (1.8–5.5 V)
3	INT	MAX30100INT
4	SDAIIC-SDA	
5	IRD	MAX30100 IR_DRV
6	SCL	IIC-SCL
7	GND	Base

The MAX30100 is an analog front-end, photo sensor, LED, and pulse oximeter and heart rate sensor module combined into one. It has ambient light cancellation and resistance to motion artifacts and it communicates through I2C. Operating at 3.3 V, it's suitable for battery-efficient wearable devices.

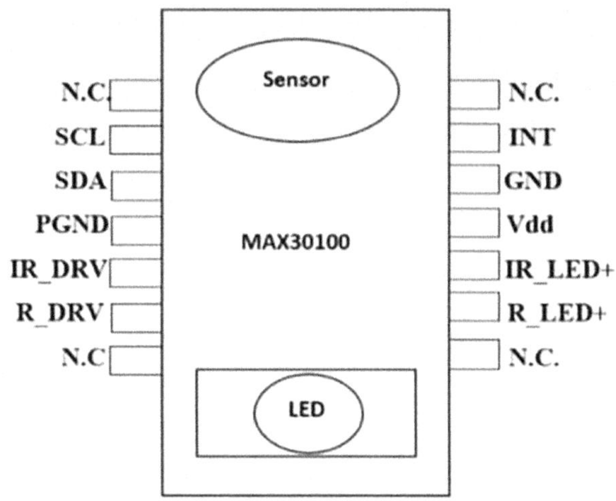

Fig. 4 Pulse oximeter

LCD display:

The 16 × 2 LCD screen is commonly used in various electronic devices due to its cost-effectiveness and programmability. It uses a 5 × 7 pixel grid for each character; consequently it can show sixteen separate characters on every single one of its two lines. Command and Data are the two registers on the LCD (Fig. 5).

Instructions are stored in the Command record. For tasks like initialization, screen clearing, and cursor control, while the Data register holds the ASCII values of characters to display.

Pin No.	Function	Name
1	Surface (0 V)	Surface
2	5 V (4.7–5.3 V) is the power source energy	Vcc
3	Enhancing the brightness with an adjustment circuit	VEE
4	When low, chooses the operation register; while high, selects the information value	Register Select
5	Down to retrieve data from the record; Higher for adding to it	Read/write
6	Transmits information to data pins upon exposure to a high-to-low pulse	Enable
7	Connectors for 8-bit information	DB0
8		DB1
9		DB2
10		DB3

(continued)

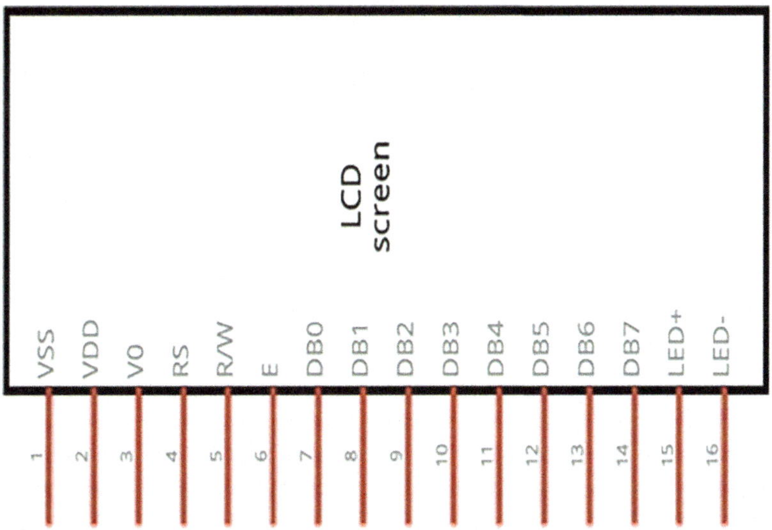

Fig. 5 LCD Display 16 × 2

(continued)

Pin No.	Function	Name
11		DB4
12		DB5
13		DB6
14		DB7
15	Dimmer VCC (5 V)	Led+
16	Dimmer Ground (0 V)	Led−

Protocol for Touch Sensor:

Three pins make up the control interface: GND is used to connect to the surface, VCC is used to give power, and SIG is used for generating digital signals. Electricity Indicator: Shiny green LED with power on the right Touch area: You may touch the trigger hand, much like a hand icon inside the region. Whole placement: four M2 screws the module may be easily installed and set with its 2.2 mm placement hole diameter to accomplish inter-module coupling.

2 Literature Survey

The global population strain on urban healthcare, exemplified by the COVID-19 pandemic, emphasizes the need for better patient monitoring. IoT health monitoring systems, including pulse oximeter, play a vital role in ensuring patient safety during surgeries and ventilation. These systems continuously monitor oxygen levels and are applied to high vascular density areas [1, 6, 8]. IoT in healthcare enhances patient care, reduces errors, and enables remote monitoring through microcontrollers and gateways like Arduino and Raspberry Pi. Wireless protocols, like the HC-06 Bluetooth module, connect sensors for improved data-driven decision-making in densely populated areas [3, 9].

Proposed system:

The proposed e-Health system employs blockchain technology for enhanced patient health monitoring and data security. Blockchain addresses challenges in healthcare data security, privacy, sharing, and storage, ensuring transparency. It integrates healthcare information among various service providers, promoting interoperability and data integrity [4, 5]. The model continuously observes patients through a compact sensor network, ensuring security. It collects health data and facilitates diagnosis and medication communication. The system utilizes 5G for fast data transmission and blockchain for security and data integrity, enhancing healthcare delivery and patient outcomes [6, 7].

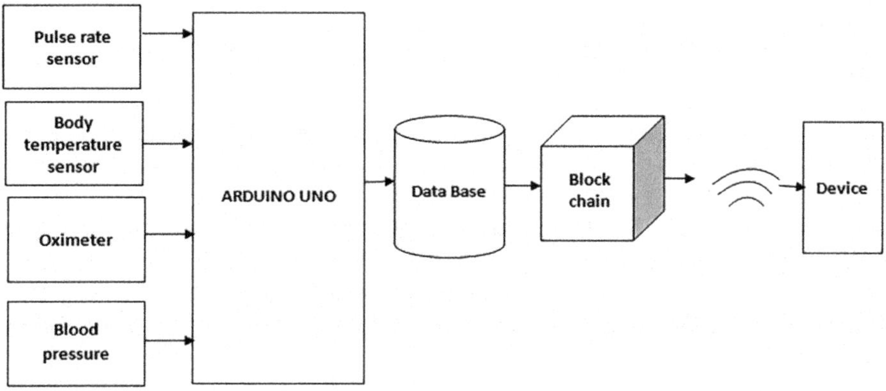

Fig. 6 Proposed block diagram

Proposed methodology:

The system's objective is to record sensor data and display it in a user-friendly way. Users can access information through an app and web dashboard to check if readings are within normal limits. It also notifies users and contacts about medication requirements. The model is smart, predicting and identifying diseases and aiding in treatment. It collects ambient and body sensor data, sends it to a cloud server for analysis, uses a Neural Network for prediction, and integrates a CDSS system for treatment information. Blockchain technology ensures security, while 5G technology makes the system faster and more efficient. Voltage surges from sources like lightning and electrostatic discharge can damage electronic equipment. We use the PC817 opto-isolator to prevent surges and interface it with an ARM Processor and relay PCB. When the relay receives 5 V, the opto-isolator connects; otherwise, it disconnects. This 2-channel relay board allows controlling 240 V appliances from 5 V ARM Processors (Fig. 6).

3 Results and Discussion

Hardware deployment:

This board is known to contain the incorrect value for the 1.5 kΩ pull up resistor on D + (R10), despite the USB specification requiring this. Either a 1.5 kΩ resistor or a suitable resistor value (such as 1.8 kΩ) to be installed between PA12 and 3.3 V in place of the resistance that comes with it, since it is either a 10 kΩ or a 4.7 kΩ resistor. It's also true that certain PCs can tolerate erroneous values, so you can test to see whether it works in your situation before changing the resistance.

Running software:

It is necessary to flash a bootloader using Bluetooth to Sequential or ST-Link (SWD). Reference to Bootloader Flashing. It should be noted that before downloading a sketch, you might need to put the board in "perpetual firmware" mode after flashing the bootloader for the first time. To do this, put a capacitor across pin PC14 and 3.3 V and reset the board. After illuminated a blank sketch, extracting the resistor, and restarting the board, posting new sketches ought to function as intended. In the event that the IDE resets your board successfully but dfu-util reports that no DFU devices are at the moment, you might need to make changes to the tools-folder's maple-upload script. Locate the line that calls upload-reset and adjust its value to a higher value.

Application Blynk:

Blynk was created with the World Wide Web of Things in heart. Furthermore to many additional remarkable attributes, it can store and analyze data, present sensor data, and wirelessly operate devices. The structure of the platform consists of 3 primary components: Blynk Library services, Blynk application, and Blynk Host.

Features:

- Compatible with all supported hardware and devices; same API and UI
- Establishing a cloud connection through Bluetooth, wireless LAN, BLE, Ethernet, USB (Serial), and GSM
- A group of user-friendly widgets Using virtual pins, it's simple to integrate and add new functionality. You can monitor history data via the Super Chart widget. You can communicate between devices using the Bridge widget.

4 Result Analysis

See Fig. 7.

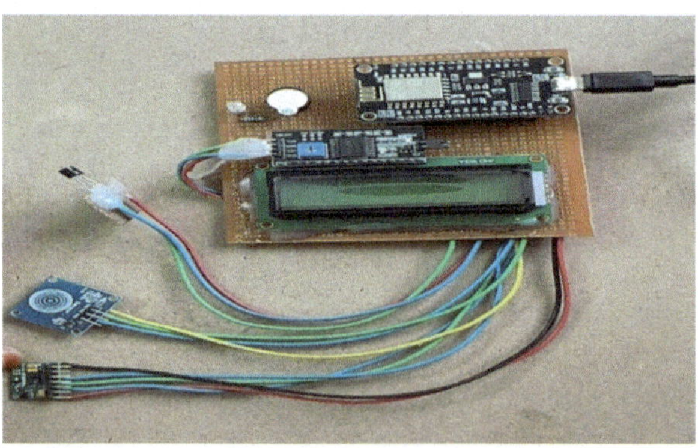

Fig. 7 Experimental kit

Scenario-1

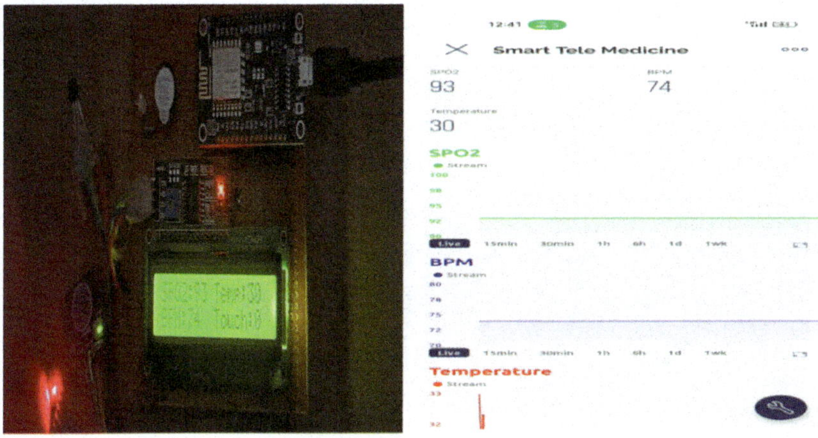

Scenario-2

Scenario-3

Scenario-4

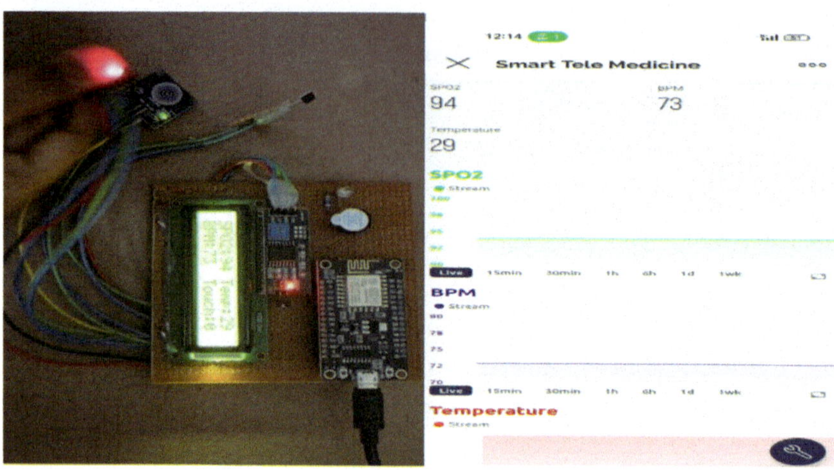

Parameters	Scenario-1	Scenario-2	Scenario-3	Scenario-4
SPO_2	93	95	94	94
BPM	74	70	77	73
Temperature	30	29	29	29
Touch sensor	0	0	0	0

Utilization:

1. It lessens the burden for medical experts as well as the workforce in clinics [8, 10].
2. Players can utilize it to keep an eye on their physical well-being.

3. Remote human health analysis.
4. All health details of patient will be stored in cloud computing system [6, 7].
5. Patients are monitored from distinct places by knowing the patient health records.
6. It is also utilized to assess the well-being of an individual struggling from any ailment that impacts oxygen levels in their blood, like: Cardiovascular attack, Cardiovascular attack, A plastic anemia; Persistent restrictive, breathing disorder (COPD), Retinal cancer, Inflammatory reactions [8, 9].

Benefits:

1. Provides peace of mind through easy health monitoring [1, 4].
2. Lessens the amount of labor and activity needed in institutions.
3. Enables monitoring of multiple patients in healthcare facilities.
4. Creates a comprehensive database of patient health parameter changes overtime, aiding doctors in treatment decisions [2, 6].
5. Minimizes hospital stays with remote patient monitoring.
6. Reduces the need for frequent hospital visits for routine checkups.
7. Ensures data reliability through cloud storage, minimizing the risk of data loss compared to paper records or single-device digital storage [5, 6].

Negative aspects:

1. Readings that are inaccurate could occur from inadequate fingertip position [9].

5 Conclusion

Blockchain technology ensures transparency and security for healthcare data, forwarding it to cloud storage for treatment and future records. 5G technology enhances system efficiency. This IoT-based health monitoring prototype measures and stores data on parameters such as respiration, blood oxygen level, and circulation with accuracy rate above 95%. Medical staff can access real-time data remotely. The system aids in epidemics and crises, including COVID-19 treatment, potentially saving lives. Future enhancements could incorporate more parameters, advanced sensors, processors, and machine learning. The system collects data from various sensors and employs a multi-layer architecture for connectivity, management, and user features.

References

1. Almotiri, S.H., Khan, M.A., Alghamdi, M.A.: Mobile health (m-health) system in the context of IoT. In: 2016 IEEE 4th International Conference on Future Internet of Things and Cloud Workshops (FiCloudW), pp. 39–42, Aug 2016
2. Riazulislam, S.M., Daehankwak, M.H.K.M.H., Kwak, K.S.: The Internet of Things for health care: a comprehensive survey. IEEE Access (2015)

3. Darshan, K.R., Anandakumar, K.R.: A comprehensive review on usage of Internet of Things (IoT) in healthcare system. In: Proceedings International Conference on Emerging Research in Electronics, Computer Science and Technology (2015)
4. Choudhury, T., Gupta, A., Pradhan, S., Kumar, P., Rathore, Y.S.: Privacy and security of cloud-based Internet of Things (loT). In: 2017 3rd International Conference on Computational Intelligence and Networks (CINE)
5. Chhabra, A.S., Choudhury, T., Srivastava, A.V., Aggarwal, A.: Prediction for big data and IoT, In: 2017 International Conference on Infocom Technologies and Unmanned Systems (Trends and Future Directions) (ICTUS), pp. 181–187 (2017)
6. Chavan, P., More, P., Thorat, N., Yewale, S., Dhade, P.: ECG–Remote patient monitoring using cloud computing. Imperial J. Interdiscip. Res. **2**(2) (2016)
7. Mihat, A., Saad, N.M., Shair, E.F., Rahim, R.A., Aslam, A.B.N.: Smart health monitoring system utilizing Internet of Things (IoT) and Arduino. **2**(1) (2022)
8. Kumar, S., Agrawal, P.: Real time IOT based detection of oxygen saturation level and bpm. **6**(6) (2019)
9. Patil, S., Pardeshi, S. Dr.: Health monitoring system using IoT. **5**(4) (2018)
10. Wasson, T., Choudhury, T., Sharma, S., Kumar, P.: Integration of RFID and sensor in agriculture using lOT. In: 2017 International Conference on Smart Technologies For Smart Nation (SmartTechCon)

Securing the Future of IoT-Based Smart Healthcare: Challenges, Innovations, and Best Practice

Iswar Kumar Patra and Saanvi Panigrahi

Abstract In this exploration of the IoT-based smart healthcare system, this chapter explores how smart healthcare and the Internet of Things (IoT) are coming together to reshape patient care. It talks about the good things, like how IoT helps monitor patients and improve healthcare. But there are challenges too, like data breaches that can expose private health info. The chapter also highlights the need for strong security to protect patient data from cyber threats. It covers different topics, from current issues to future trends, making it a journey into the world where technology meets healthcare. The chapter's main message is clear: while IoT can make healthcare better, we must also keep patient info safe in this tech-driven era.

Keywords Healthcare · IOT · Data breaches · Technology · Cyber threats

1 Introduction

With the Immense advancement of the technology in the field of Internet of Things (IOT) and the rapidly evolving landscape of healthcare, the integration of both these fields results a revolutionary force that reshape the traditional paradigms of patient care. The Internet of Things (IoT) involves connecting smart physical devices. These devices come equipped with embedded software, sensors, and network connectivity, facilitating the secure collection and exchange of data among them. Various Fields like Smart cities, traffic congestion, emergency services, health care sector etc. problems are being provided with appropriate solution by the IOT. The application of technology in IoT offers numerous advantages to the healthcare system, covering the integration of IoT in healthcare, ranging from remote monitoring systems and smart sensors to the integration of medical devices. It improves the quality of life, reduces the cost of medical treatment, improves the quality of treatment of cure, gives accurate diagnostic details of the patient and many more advantages both to

I. K. Patra (✉) · S. Panigrahi
Siksha 'O' Anusandhan, Bhubaneswar, Odisha, India
e-mail: iswarpatra100@gmail.com

S. N. Mohanty et al. (eds.), *Explainable IoT Applications: A Demystification*, Information Systems Engineering and Management 21,
https://doi.org/10.1007/978-3-031-74885-1_8

the health care providers and to the patient. The IOT has a large potential in order to serve in the health care. Not only it provides efficient, real time correct analysis of the patient but also provides effective ease of cost interactions between individual patients, clinics, and healthcare through seamless and secure connectivity which is much more beneficial in the serious cases of the patient. The device keeps the patient safe and healthy as well as improves the health care providers towards the patient. Smart Health care devices collect numerous data from a set of real-world cases with great accuracy, less time consuming and increases the size of the diagnosis data. Therefore, Different types of medical devices, sensors, diagnosis devices can be viewed as an integral and core part of IoT.

Further in this chapter we are going to discuss about few topics related to IOT based healthcare innovation, benefits, applications, security threats, vulnerabilities, privacy and security challenges as well about its future trends.

1.1 Importance of Healthcare Innovation

In the ever evolving landscape of the healthcare system, the rising of the chronic diseases leads to get attention in the innovation in the health care system [1]. As the demands on healthcare system grows, the old and traditional method facing challenges of inefficient, resources allocation, including operational efficiency and the quality of patient care, so the need of the innovation in the technology is much more important and allows the healthcare providers as well as the patient to obtain information at a speed they have never seen before.

Innovation in IOT plays in crucial role in Overcoming operational inefficiencies in which manual process and outdated communication methods and introducing automation in order to enhance efficiency of the workflow, in optimizing allocation of resources in order to provide real time insights into utilization of the resource and enabling informed decision making and efficient resource management. Enhancing the care of the patient and elevating the quality of patient care through personalized treatment plans and addressing the burden of chronic diseases by offering solutions for remote patient monitoring and proactive diseases management. In the Innovation of Healthcare, the integration of IoT based smart healthcare system providing pivotal force and promising transformative changes across the various aspects of healthcare delivery.

1.2 IoT Health Care Benefits

The use of IOT within healthcare transform the entire healthcare industry from improving patients' health and increasing efficiency of healthcare workers and facilities. IOT within healthcare able to extract more data which leads to better health care to patients. Even the health institutes also estimated that the integration of IOT

(EHR systems) in hospitals improves profitability. Obviously while utilizing IOT, healthcare providers must take care of privacy and security rules. Typically, the IT departments of medical institutions and hospitals employ secure methods for data storage; however, certain challenges still need addressing. The Internet of Things is consistently and steadily expanding within the healthcare industry [2]. Implementing IoT in healthcare presents numerous benefits outlined below:

1. Cost Reduction:

Healthcare providers can monitor patients in real time using IoT-connected devices. This leads to a decrease in unnecessary hospital visits, saving time, reducing hospital stays, and minimizing unnecessary re-admissions.

2. Enhanced Patient Experience:

Patients remain consistently connected to IoT devices, contributing to more accurate diagnoses as doctors have access to crucial and necessary data.

3. Improved Medication Management:

IoT solutions streamline drug management, reducing the time medical staff spends searching for medications.

4. Positive Outcomes:

By integrating healthcare solutions with technologies such as cloud computing, big data, and IoT, caretakers can track real-time data. This capability aids in making informed decisions, ultimately leading to more effective and efficient healthcare for patients.

5. Efficient Treatment Practices:

Accessing and collecting real-time data provided by IoT networked devices allows healthcare providers to offer personalized and efficient treatments.

6. Enhanced Patient Care Experience:

Proactive treatments, accurate diagnoses, timely interventions by doctors, and superior treatments contribute to an improved overall patient experience.

In the fast-paced contemporary world, individuals often lack the time for consecutive visits to clinics and medical facilities. Healthcare mobile apps offer a multitude of conveniences and play a crucial role in society. Presently, medical practices can seamlessly operate on a unified platform, effectively managing various aspects of business without the need for multiple software implementations. It is now an opportune moment to embrace change and incorporate technology. The healthcare sector presents numerous opportunities and benefits. Gratitude is owed to evolving technology, which empowers us with impactful IoT applications for managing and enhancing health experiences.

So, in the sections of this chapter dive into the specific aspects of IOT in healthcare including its applications in the healthcare facilities, patient monitoring and diagnosis, remote patient monitoring system, cyber security challenges and the future directions of this rapidly evolving field.

The following literature survey will provide depth and breadth of existing research in the field of IoT and its applications in healthcare field.

2 Application in Remote Patient Monitoring

Due to the emergence of highly infectious diseases such as COVID-19, the shift in demographics towards an aging population, and the increase in health complications, several remote patient monitoring systems (RPMS) are gaining popularity. Technological advancements have significantly expanded the usage of RPMS. In these systems, specific focus is placed on patients with conditions like chronic diseases, infectious diseases, mobility challenges, and other disabilities. The primary objective is to enhance the comfort of patients in their daily lives. Patient monitoring systems typically involve wired sensors connected to computers within medical facilities. However, a drawback of this system is that it limits patient mobility. The devices utilized are often large and costly, allowing monitoring for only a limited number of patients. RPMS was introduced as healthcare facilities sought to provide home-based healthcare. The healthcare sector's notable improvement has prompted researchers and companies to employ remote patient monitoring systems to enhance quality, leading to a rapid growth in the industry.

Examples of remote patient monitoring technology:

1. Heart rate monitoring devices.
2. Blood pressure monitoring devices.
3. Remote fertility treatment and monitoring systems.
4. Programs for logging caloric intake.
5. Glucose meters designed for patients.

Benefits of Remote Patient Monitoring:

1. Enhances patient engagement—RPM devices empower patients to play a crucial role in understanding their health conditions.
2. Improves the quality of care—RPM provides patients and healthcare providers with access to relevant patient data, thereby enhancing the overall quality of care.
3. Enhances access to healthcare—RPM enables patients to manage basic healthcare, allowing professionals to serve more patients.
4. Provides a higher level of education and support—RPM offers daily information to patients about their personal conditions, fostering education and support.
5. Ensures patient assurance—Constant monitoring through RPM devices gives patients peace of mind.

2.1 Applications of Remote Patient Monitoring

Vital Monitoring:

Vital monitoring constitutes a routine check-up that assesses fundamental physiological parameters such as body temperature, heart rate, blood pressure, and breathing rate. In remote patient monitoring systems (RPMS), this monitoring occurs outside the hospital setting, significantly enhancing efficiency within hospital premises. Consequently, there is a reduction in hospital visits, with more patients being monitored from the comfort of their homes. Depending on the patient's needs, monitoring can take place in real-time or on a daily basis. For instance, vital monitoring through RPMS can promptly identify a patient's deteriorating condition. Consequently, researchers have implemented vital monitoring RPMS to detect worsening conditions or irregularities in a patient's body. Vital monitoring RPMS is instrumental in identifying disease development, leading to the creation of RMPS tailored for disease diagnostics.

Disease Diagnostic Monitoring:

Disease diagnostic RPMS involves the examination of physiological indicators and other parameters to analyze specific diseases. Researchers have designed diagnostic systems to identify one or multiple diseases, including the capability to track conditions such as bipolar disease. Applications in disease diagnosis are crucial as they provide patients with insights into their current health status, helping them understand whether they are in good health. Additionally, these applications often incorporate a chat functionality that facilitates communication between doctors and patients.

Types of Sensors in RPMS:

According to the authors, sensors in remote patient monitoring systems (RPMS) can be categorized as either contact-based or contactless. Contact-based sensors are wearables that can be applied to the surface of the human body. For example, a wearable sensor can be attached to clothing, an elastic band, or directly onto the skin, connecting to systems or devices that offer smart functionalities. These sensors can be placed at various locations on the human body. On the other hand, contactless sensors encircle the human body but are not directly attached to it. They commonly gather patient RFID or optical data, as they are generally less precise than wearable devices. In contrast, implantable sensors can be positioned inside the heart, brain, stomach, etc., for continuous monitoring of patients' conditions.

3 IOT and Chromic Disease Management

The Internet of Things (IoT) gives us many solutions to make healthcare better and cheaper [3]. It helps improve the quality of healthcare and is important for keeping healthcare services going. It is massive experience of bringing and connecting process core and data-driven decision making. It plays an incredible role in managing chronic diseases management highlight the need or restricting daily activities or both which include heart disease, cancer and diabetes are the primary cause of death and disability in the us.

Industry plays a very vital role in chronic diseases management concurrently reporting and monitoring-It enables to gather information on patient health in real time in such a way that is warble including a fitness tracker. It helps in tracking chronic diseases infected patients health status in order to make preventative measures. A easier way to maintain reports—Tracking on critical conditions—It helps in sending alerts on critical conditions though increasing lifespan of patient by detecting it a right time from normal to trigger alert in healthcare provider if there is critical condition Remote Medical assistance—It helps in management to automate the delivery of medication and provide remainders to patient to take their medications on time by improving adhere and reducing the rise of complications.

In healthcare, IoT allows devices with sensors to measure a specific aspect of a patient's body, keeping an eye on chronic diseases [4]. It helps improve healthcare quality at a lower cost by offering effective treatment and introducing new capabilities.

Process of getting started with chronic diseases management are

1. **Identifying the patient**—Making a list of patients can help in easy access of the entire process. Healthcare providers can identify it based on their medical history, current status and risk factors
2. **IOT devices selection**—After identification, selection of appropriate devices or network for monitoring and collecting data like glucose monitors, blood pressure monitors, smart scale such as fitness traders
3. **Data collection and Analysis**—Medical devices are designed to gather data. It is the process of gathering, measuring and analysing accurate data.
4. **Care plan development**—After collection, the health care provides a care plan to the patients.it includes lifestyle changes improvements and follow up appointment
5. **Remote Monitoring**—It allows healthcare providers to intervene early in case of any critical conditions and male appropriate prevention moves before any serious complication occurs. It also generates alerts if there is a need for urgent medical attention
6. **Patient education and Engagement**—For better outcomes and effective improvement they need to be complete aware of the entire process. It facilitates by providing real time reminders to patients about medication, lifestyle changes and follow up appointments.

7. **Continuous Improvement**—It is an effective healthcare plan and modification for a patient's health improvement

Concluding, the chronic diseases addressed in this book don't encompass all potential illnesses; instead, they spotlight the most widely discussed ones. Additionally, the identification of innovative technologies is grounded solely in the selected literature, acknowledging the potential presence of other technologies in different studies.

4 Security Threat Landscape in IoT Healthcare

The increasing volume of IOT based health care systems has led to an immense rise in complexity and challenges of security of medical devices. These devices require specialized security frameworks and technologies to address them effectivity as well as efficiently. Healthcare organizations must implement these framework and technologies to protect patient data and maintain integrity.

4.1 Data Breaches and Privacy Concerns

In the world of smart healthcare gadgets, like those using the Internet of Things (IoT), technology helps make things better for patients and doctors. But sometimes, there are problems, especially with keeping our private health information safe. Imagine if someone who shouldn't know about your health got hold of your information—that's what we call a data breach. These breaches can happen when the smart devices we use to keep track of our health aren't protected well enough. Although there has been concerns regarding around security on health care devices, there has yet to be many cases where the data has been manipulated. Medical devices in hospital are vulnerable and can be exploited by the hackers in order to gain access of the hospital data due to the weak configuration of the devices.

Two security tools made fake devices with pretend information, kind of like traps [4]. When the researchers looked into it, they found that the information collected by these traps and the attackers were able to get access to these fake medical devices. The attackers successfully took advantage of the same weakness that had been used before by the Conficker infections. In these situations, the attackers didn't realize what they had hacked, and the infected machine became a part of their controlled network without them knowing. Therefore, there has been a rise in drawbacks in healthcare sectors because of IOT. The IOT brings many privacy and security challenges, but as the devices act automatically it has greater risk. Doctors in this scenario can program the medical devices in order to monitor a heart condition of the patient so that the doctor can send right level of electric shock which correct the heartbeat of the patient but if that medical devices will be exploited by the hackers, then the malicious hackers can use the device to deliver a high rate of shock to the patient.

4.2 Unauthorized Access and Authentication Issues

Healthcare organizations are the prime targets for cybercriminals due to the presence of sensitive information of patient and the potential for financial gain. Let envision your health data as a well-protected space. You and your trusted healthcare providers possess a special key, known as authentication, granting access to this exclusive information. This key ensures that only authorized individuals, like you and your doctors, can enter and comprehend the details of your health, maintaining a secure and private environment.

Now, consider the concept of "Unauthorized Access" as someone attempting to infiltrate this health space without the proper key—essentially, an unauthorized entry. In the pages ahead, we'll explore the significance of safeguarding your health information from such attempts. Our journey will also delve into the strategies to fortify your special key, ensuring that only those with rightful access can manage and attend to your health information.

Device Vulnerabilities and Exploits:

Devices and networks needed to be secure, and important health information had to be protected from unauthorized access [5]. Challenges were also faced with authentication, ensuring only the right people could use the system, and making different devices work well together, called interoperability. This subtopic was like looking closely at possible weak points that bad people might use to get into our digital healthcare world. The significant cyber security attacks in healthcare systems faced commonly are:

DDoS Attacks (Distributed Denial of Service)

DDoS attacks, known as Distributed Denial of Service, have the potential to overpower the system, leading to disruptions in critical healthcare application services. This may affect timely access to patient data and interrupt communication channels between devices and healthcare providers.

Phishing Attacks

Phishing attempts may deceive users into sharing sensitive information. This deceptive practice could result in unauthorized entry to patient records, jeopardizing confidentiality and potentially influencing treatment plans.

Device Tampering

Exploiting weaknesses in device security could grant unauthorized individuals the ability to manipulate medical devices. This poses significant risks, as manipulation might lead to inaccurate health readings or the unauthorized access of real-time health data for patients.

Data Interception

Weaknesses in data transmission could expose health data to interception. Unauthorized entry to patient information during transmission may compromise privacy and confidentiality, potentially leading to the inappropriate use of sensitive health details.

Malware Infections

If malicious software enters the system, it could mess up the accuracy of health data and allow unauthorized access to patient records. This could seriously impact how patients are cared for.

Man-in-the-Middle Attacks

During a man-in-the-middle attack, hackers sneak into communication between devices and healthcare providers. This breach could let them access sensitive health info, threatening patient confidentiality.

Insufficient Authentication Measures

If authentication processes are weak, unauthorized individuals might get into medical devices or patient records. It's super important to make authentication stronger to prevent unauthorized access and keep patient data safe.

Zero-Day Exploits

Zero-day exploits target unknown weaknesses in the system. Hackers can use these undiscovered issues to potentially get unauthorized access to the IoT healthcare network.

Eavesdropping

Eavesdropping is when unauthorized individuals listen in on communication between devices. This invasion of privacy might expose sensitive patient info, making us worry about keeping data confidential.

Physical Security Risks

If IoT devices aren't properly secured, there's a risk of unauthorized physical access. This could mean messing with medical equipment or stealing devices that hold sensitive health data.

Weak Encryption Practices

If encryption isn't strong enough, health data sent over networks might be intercepted. Strengthening encryption methods is crucial to making sure patient information stays private during data transmission.

Fixing these weak points was super important to keep our health information safe. Alongside these challenges, healthcare systems often dealt with big cyber-attacks. These attacks tried to find and use the problems in our devices and networks, especially where we checked if users were allowed in. As healthcare kept advancing, strong security rules were needed to protect our health services from these cyber threats. Some significant risks that health care systems can face when there will be system downtime and service disruption, hardware software failure, vulnerabilities in systems, third party interferences and many more making cybercriminals easy to attack.

4.3 Privacy Challenges in IoT

Due to the immense rise in health care systems and technologies data privacy is very important to protect the collections of sensitive data and finance details of patient [6]. Another privacy challenge in IoT-based health care systems is lack of control over data sharing and low management over cybersecurity frameworks.

To solve these risks, health care organizations must implement robust data privacy controls, including encryptions, access controls, and consent management. They must also make sure that systems which are made with privacy in mind and must be aware about how the patients' data are used and shared. Therefore, Regular employee training on data privacy and security is also important. They must be able to identify risks and prevent them. They should be trained on how to respond to risks.

4.4 Challenges and Future Directions

The rapid growth of technological advancements means that security and privacy technologies continuously evolve to keep up with new threats and vulnerabilities. Health care organizations must stay up to date with latest technologies to make sure that patient data must be protected. In order to scope up with these challenges researchers must focus on developing standardized security and privacy on systems [7]. Furthermore, efforts are needed to train and educate the employees, workers and trainers for future. Finally, advancement in AI ML and block chain technology can be used in future to improve the privacy and security of IOT-based health care systems.

4.5 Providing Improved Patient Care

IOT in health care is also known as IoMT (internet of Medical Things). The Internet of Medical Hardware infrastructures that connects health and technology together. IoT uses cloud storage to get connected with it and to keep the data from the patient to analyses later or for future analysis [8]. An example of telehealth is "The Body-Gaudian Monitoring System" is a system that helps the physicians enhance the care to patients instead of denying the overly burden consumers. Requirements of the security was provided by the system in different ways, at first it identifies the information of the patient and data, during transmission it encrypts the data on the devices [8]. Health care organizations must implement strong and firm cybersecurity measures in order to protect patient data, maintain data integrity and prevent financial losses. Regular cyber security audits, staff training and taking up best practices to reduce risks in health care systems.

Cybersecurity Measures in IoT Healthcare:

Securing health data is achieved through encryption, a method of encoding information to ensure only authorized individuals can decipher it. This cryptographic practice acts as a protective shield for our data as it traverses through interconnected devices [9]. Consider IoT healthcare as a network fortress, where network security plays a pivotal role in fortifying its boundaries. Employing segmentation, the network is compartmentalized, restricting unauthorized access to different sections [10]. In this way, even if one segment is compromised, the integrity of the entire system remains intact. Authentication and access controls act as digital gatekeepers, regulating entry into the healthcare network. Just as a castle gate requires the right credentials for access, only authorized users, such as authenticated personnel, can navigate specific areas within the IoT healthcare network. Maintenance of the digital infrastructure is upheld through security patching and updates, akin to reinforcing the fortress walls. These measures are essential for addressing vulnerabilities and ensuring the robustness of the digital defenses in safeguarding healthcare information.

Future of IoT Healthcare Security:

The IOT has a great job in revolutionizing the health care industry by providing immense care to patients, increasing operational efficiency, and overall health care outcomes. These devices also bring many security challenges as the privacy and integrity of health care data must be well protected to ensure patients privacy. As day by day the IOT in healthcare industries are growing in a rapid manner due to which healthcare security is a critical consideration.

Current State of IoT in Healthcare:

IOT devices in health care ranges from wearable connected devices to unconnected healthcare devices (devices that are connected through wi-fi, satellite, cellular, Bluetooth, Ethernet etc). These devices collect data, transmit and stores the sensitive details of patient's data, making them prime target for cybercriminals [11]. The cybersecurity landscapes in health care is very complex, along with challenges:

Diversity of devices: Hospitals often have thousands of IOT devices from different manufacturers in which each devices has its own security protocols and vulnerabilities

Data privacy concerns: Patient sensitive data and information's are highly valuable making it a prime target for hackers. Therefore, getting unauthorized access to such devices can lead to identity theft, financial fraud and many more.

5 Future Trends in IoT Healthcare Security

Enhanced Device Authentication

Future IOT healthcare security organisation going to focus on providing authentication to devices by integrating many devices like multi factor authentication, biometric recognition and block chain technology to ensure only authorized person can access and control devices [12].

Data Encryption and Blockchain

In healthcare data encryption protection is very important to secure the data transmitted between devices and storage systems. Integration of blockchain technology can ensure the integrity and confidentiality of patient data is well protected.

AI-Powered Security

AI and ML are going to play an important role in protecting IOT devices in healthcare industry. These technologies can detect abnormal behaviours in systems and can identify threats/risks.

Regulatory Compliance Automation

Different Automation tools will be developed to continuously monitor the healthcare systems and regulations [13, 14]. This process can reduce the heavy burden on health care workers and minimize the risks.

Secure Development Practices

Manufactures need to prioritize security in IOT healthcare systems while making the devices they need to secure coding practices, regular security updates etc. in order to minimize vulnerabilities.

6 Challenges and Considerations

There are several challenges that needed to be addressed like many hospitals have limited resources for cyber security, some faces financial problems and some hospitals needed experienced staff trainers. As day by day the advancement in IOT technologies are improving immensely. At the same time the cybersecurity risks and threats also increase. Therefore, by prioritizing authentication, AI driven security, secure development practices, proper trainers in medical industries, best quality of tools (like with very less vulnerabilities) and many can evolve landscape of IOT healthcare security, while ensuring patient safety and privacy and integrity of health care data must be well protected to ensure patients privacy. As day by day the IOT in healthcare industries are growing in a rapid manner due to which healthcare security is a critical consideration.

7 Conclusion

In conclusion, this Chapter explores the transformation of healthcare through smart devices like IoT. It highlights the positive impacts, such as remote patient monitoring, while also addressing the challenges, particularly the need to safeguard our health data. Moving forward, the emphasis is on enhancing security measures. From securing devices to employing smart technologies, the objective is to maintain the privacy of our health information. As we move towards the future, it is evident that technology will continue to redefine healthcare, and ensuring the safety of our information remains a top priority.

References

1. Prabha, R.: IoT based smart healthcare monitoring systems: a literature review (December 2020)
2. Islam, M.R, Kwak, D., Kabir, M.H., Hossain, M., Kwak, K-S.: The Internet of Things for health care: a comprehensive survey. IEEE (2015)
3. Dadkhah, M., Mehraeen, M., Rahimnia, F., Kimiafar, K.: Use of Internet of Things for Chronic Disease Management: An Overview. National Library of Medicine (2021)
4. Peyroteo, M., Ferreira, I.A., Brito, L., Elvas, L., Ferreira, J.C.: Remote monitoring systems for patients with chronic diseases in primary health care: systematic review (2021)
5. Sadek, I., Codjo, J., Ul Rehman, S., Abdulrazak, B.: Security and privacy in the internet of things healthcare systems: toward a robust solution in real-life. Computer Methods and Programs in Biomedicine Update (2022)
6. Easttom, C.: Computer security fundamentals. In: Pearson IT Cybersecurity Curriculum (ITCC). Pearson Education (2019)
7. Sadek, I., Rehman, S.U., Codjo, J., Abdulrazak, B.: Privacy and security of IoT based healthcare systems: concerns, solutions, and recommendations. In: Pagán, J., Mokhtari, M., Aloulou, H., Abdulrazak, B., Cabrera, M.F. (eds.) How AI Impacts Urban Living and Public Health. Springer International Publishing, Cham (2019)
8. Akkaoui, R.: Blockchain for the management of internet of things devices in the medical industry. IEEE Trans. Eng. Manage. (2021)
9. Blendon, C.D., Robert J.: Future health care challenges. Issues Sci. Technol. (2023)
10. Minoli, D., Sohraby, K., Occhiogrosso, B.: IoT security (IoTSec) mechanisms for e-Health and ambient assisted living applications. In: 2017 IEEE/ACM International Conference on Connected Health: Applications, Systems and Engineering Technologies, CHASE (2017)
11. Abdulrazak, B., de Marassé-Enouf, A., Mokhtari, M. (eds.): Participative Urban Health and Healthy Aging in the Age of AI. Springer International Publishing, Cham (2022)
12. Saravanan, M., Shubha, R., Marks, A.M., Iyer, V.: SMEAD: a secured mobile enabled assisting device for diabetics monitoring. In: 2017 IEEE International Conference on Advanced Networks and Telecommunications Systems, ANTS (2017)
13. Mohapatra, S., Yesubabu, M., Sahoo, A., Mohanty, S., Mohanty, S.N.: IoT-based Ubiquitous Healthcare System with Intelligent Approach to an Epidemic, Recent Patents on Engineering. https://doi.org/10.2174/0118722121240884230926092316
14. Mohapatra, A.G., Mohanty, A., Pradhan, N.R., Mohanty, S.N., Gupta, D., Alharbi, M., Alkhayyat, A.: An Industry 4.0 implementation of a condition monitoring system and IoT-enabled predictive maintenance scheme for diesel generators. Alexandria Eng. J. **76**, 525–541 (2023). https://doi.org/10.1016/j.aej.2023.06.026

Smart City: Challenges and Opportunities Detection and Identification of Autonomous Vehicles Using Sensor Synthesis

B. Ravi Chandra, J. Krishna Chaithanya, Ajay Roy, and C. Lokanath Reddy

Abstract Future ground transportation is predicted to be enhanced, transformed, and revolutionized by autonomous vehicles (AV). It's predicted that one-day intelligent automobiles will displace conventional ones, capable of autonomous decision-making and driving activities. Self-driving cars are outfitted with sensors to see and understand their immediate surroundings as well as the environment in the distance using more advanced communication technologies, such as 5G, to accomplish this goal. Local perception will still be a useful tool for short-range vehicle control in the interim, just like it is for humans. The car can reach its goal while maintaining a set of standards (security, energy conservation, reduction of congestion, leisure), but with the help of expanded perception, which enables anticipating of distant events. Despite significant advancements in sensor technologies over the past few years in the context of their usefulness to audiovisual systems and efficacy, it is not advised to rely solely on one sensor to perform any of the self-sufficient tasks related to driving because sensors continue to malfunction because of vibration, conditions outside, production flaws, or other variables. Several models of architecture have been put out as guides. To tackle this complexity, autonomous systems must be created, created, operated, and deployed. We provide a brief review of sensors and the fusion of sensors in self-driving cars in this study. We concentrated on the lens, radar detectors, and laser sensor integration from the main autonomous car sensors. The state-of-the-art in this field will be discussed, including occupancy grid mapping for navigating and location in changing circumstances, using both pictures and three-dimensional point cloud data, 3D object identification techniques are used, along with moving object tracking and tracking systems. It has been demonstrated that adding more sensors to a sensor fusion system improves both the performance and durability of the outcome.

B. Ravi Chandra (✉) · C. Lokanath Reddy
G. Pullaiah College of Engineering and Technology, Kurnool, Andhra Pradesh 518002, India
e-mail: Chandrabrc11@gmail.com

J. Krishna Chaithanya
Vardhaman College of Engineering and Technology, Hyderabad, India

A. Roy
Lovely Professional University, Jalandhar, Punjab, India

S. N. Mohanty et al. (eds.), *Explainable IoT Applications: A Demystification*, Information Systems Engineering and Management 21,
https://doi.org/10.1007/978-3-031-74885-1_9

Additionally, using camera data for localization and mapping, which are typically handled by sonar and laser data, enhances the perception of the world as a whole. One of the fastest-growing fields in the self-driving car arena is sensor fusion because it is essential to autonomous systems as a whole.

Keywords Self-driving automobiles · Self-driving cars · Fusion of sensors · Laser radar detectors · Cameras

1 Introduction

Autonomous vehicles, commonly referred to as self-driving cars, are acknowledged as an emerging field in development and research, with daily contributions from several important research organizations, automotive manufacturers, and academic institutions. The purpose of this work is to summarise current developments in sensor fusion and autonomous car sensors. Given that sensors are essential to self-driving cars, the integration of sensor data, accurate interpretation of that data and vehicle control constitute the core of autonomous driving.

According to Fig. 1, the autopilot system may be classified into four major types. Several various sensors put on the vehicle are used to sense the environment. These are pieces of gear that collect environmental data. The sensation block processes the sensor data by converting it through its components into meaningful information. The output from the perception block is used by the planning subsystem for both short- and long-range path planning as well as behavior planning. The control module issues orders to the vehicle and makes sure it travels along the course laid out by the planning subsystem.

In the latter half of the twentieth century, the initial effective efforts at building autonomous cars were made. In 1984, researchers at Mercedes-Benz, University of Munich [1] and the Carnegie Mellon University [2, 3], created the first completely autonomous automobiles. Since then, a large number of businesses and research institutions have created autonomous vehicle prototypes and are actively developing the complete autonomy of cars.

In the Grand Challenge competitions held by the Defense Advanced Research Projects Agency of the United States (DARPA) in 2004 and 2005 [4, 5] and the Urban Challenge competition held in 2007 [6], significant advancements in the area of self-driving cars were made. Six of the 11 self-driving automobiles in the 2007 DARPA Urban Competition finals managed to effectively navigate an urban setting to cross the line of victory, which is seen as a significant accomplishment in robotics.

Scene perception, localization, visualization, controlling the car, path optimization, and higher-level planning choices are the current barriers to the development of autonomous cars Two innovative trends in autonomous driving are comprehensive learning [7–9] and learning through reinforcement [10, 11].

Levels of Automation (0 to 5):

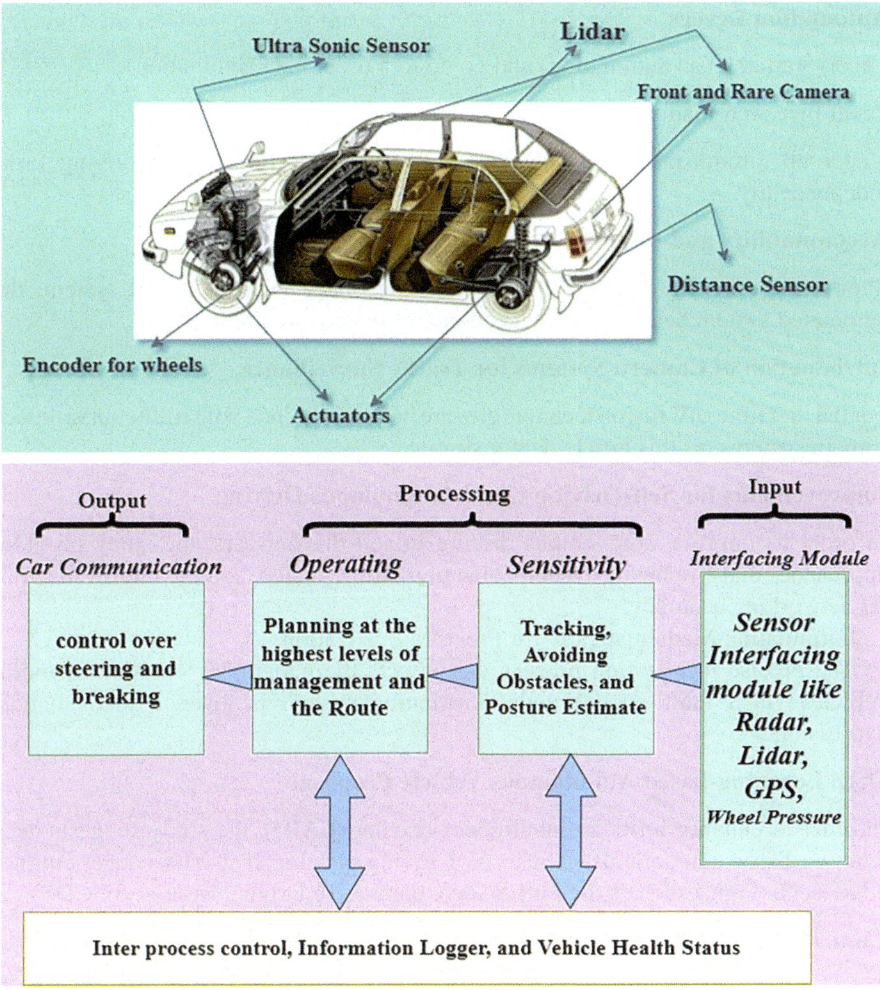

Fig. 1 The self-driving car system's block diagram

Level 0: Not using automation.
Level 5: Complete mechanization.

Fifth Level Automation Responsibilities: Under any circumstances, the vehicle can autonomously manage all driving operations thanks to the fifth level technology.

The automated system bears full responsibility and accountability in the event of malfunctions, failures, or accidents.

Important Points Format: Automated System Design Complexity:

A completely automated system, such as a self-driving car, is difficult to design.

Automation Levels:

Levels start at 0 (no automation) and go up to 5 (complete automation).

Level Five Automation:

Under all circumstances, a vehicle is capable of performing all driving tasks independently.

Accountability and Liability:

Should there be malfunctions, mistakes, or mishaps in a fifth-level system, the automated system bears complete accountability and culpability.

Introduction of Camera Systems for Traffic Surveillance:

For the first time, AV fusion technologies are being combined with traffic surveillance camera systems, as this article demonstrates.

Improvements for Self-Driving Cars' Autonomous Driving:

In order to improve autonomous driving in self-driving cars, the study provides approaches that emphasise exact localization, 4D detection, and the improvement of AI networking accuracy.

Formulating Mathematically for Precise Localization.

The precise localization process that affects the geographic location of motor vehicles via a multi-anchor node positioning system is given a mathematical expression.

Deep Learning-based Autonomous Vehicle Proposal

In order to enhance artificial intelligence driving (EAID), the study suggests deep learning-based autonomous vehicle (AV) driving systems. It also introduces cutting-edge technologies such as the Full-Scale Cooperative Driving System (FSCDS).

Contributions of Technology

This article presents breakthroughs in deep learning-based AV systems for EAID, accurate localization, autonomous driving, traffic monitoring, and other cutting-edge technologies that go beyond traditional methods. These are all shown in the following table.

Automation level	Driver engagement	Assistance operating mode
0 (Manual control)	Full responsibility on the driver for all driving tasks	No automated features in place
1 (Driver support)	Driver maintains control with potential assistance features	Not applicable
2 (A portion of automation)	Integration of automated functions (e.g., steering, acceleration) requiring consistent driver attention	Not relevant

(continued)

(continued)

Automation level	Driver engagement	Assistance operating mode
3 (Automata with conditions)	Driver not obligated for continuous monitoring but must be ready to take over	Not relevant
4 (Elevated automation)	Vehicle capable of handling all driving tasks in specific conditions; optional driver-assist systems available	Not relevant
5 (Total automation)	Car capable of autonomous driving in any scenario; optional driver-assist systems available	Not relevant

2 Hybrid Vehicles' Sensors

There are several viewpoints from which a complex system might be seen when it is represented using an architectural model, such as a collection of tangible elements, development phases, Logic operations, often known as function segments, are described in [12]. In the current research, autopilot architectures are examined from two perspectives:

(1) A technical perspective that focuses on software and hardware parts and their execution, and
(2) A practical strategy that contains an overview of the analysis phases that a self-driving automobile has to go through as the natural fundamental components of the complete system.
(i) **A Technical Perspective**

The two primary technological levels of a self-driving vehicle design are hardware and software, and every single layer contains components that reflect various system components. Numerous of those elements can be seen as distinct categories, although others operate as the backbone of their layer, setting the rules and giving structure for how each component interacts with the others. In Fig. 2, this description is shown.

These days, autonomous cars are huge, complicated systems with numerous sensors for internal and exterior monitoring, and they produce enormous volumes of data every day. Networking and computation units in cars are not anymore limited to a few Electronic Control Units (ECUs) connected via narrow band connections to handle all of that data, as they formerly were. The information collected by the sensors is gathered and analysed using more powerful tools like Graphical Processing Units (GPUs), heterogeneous computing systems with numerous cores, and Field Programmable Gate Arrays (FPGAs). External information is also accessible through the internet, other cars, or infrastructure along with data produced by the vehicle.

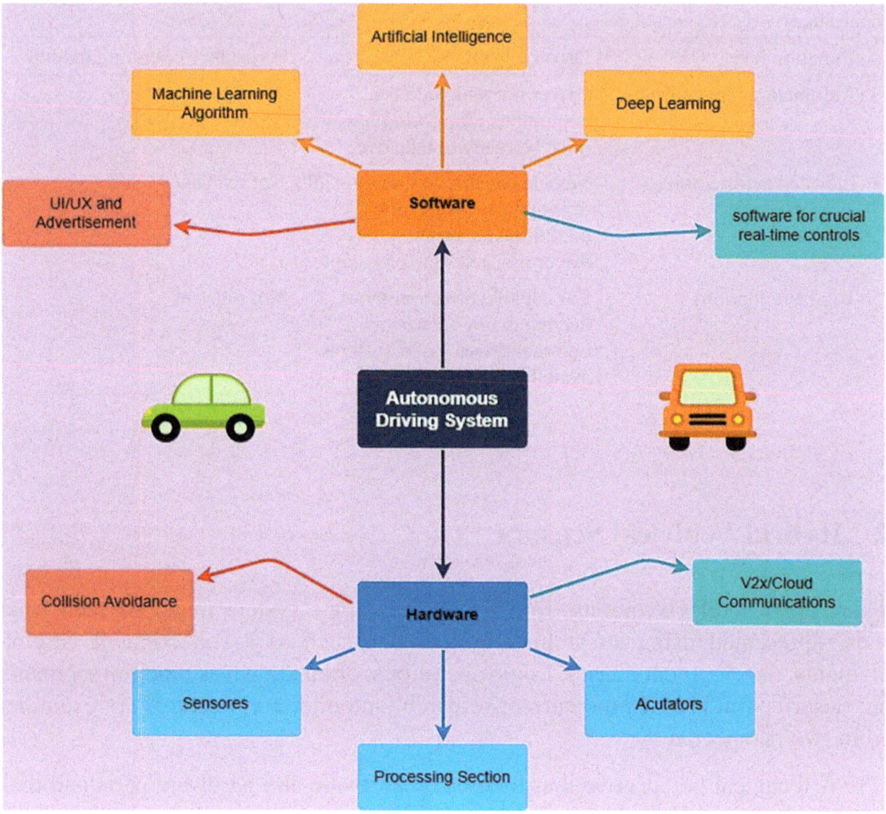

Fig. 2 The system's technical layout

This is known as vehicle-to-anything (V2X) communication. The physical part also includes the actual automobile, which serves as the movable system, as well as the actuating elements, which can vary in kind according to the application and the environment in which the system will be used.

Each subsystem's internal networking interfaces, such as high bandwidth connections like USB connection or Gigabit Ethernet that are used for transporting sensor data or low bandwidth interfaces like CAN and LIN networks, enable information exchange across the subsystems.

(ii) **Practical View**

From a different angle, autonomous cars are made up of conceptual or operational components that are established based on the information flow and processing steps carried out from data collecting through vehicle control, including internal system monitoring. This allows for the identification of four primary functional blocks that are shared by the majority of suggested designs and solutions in academic and

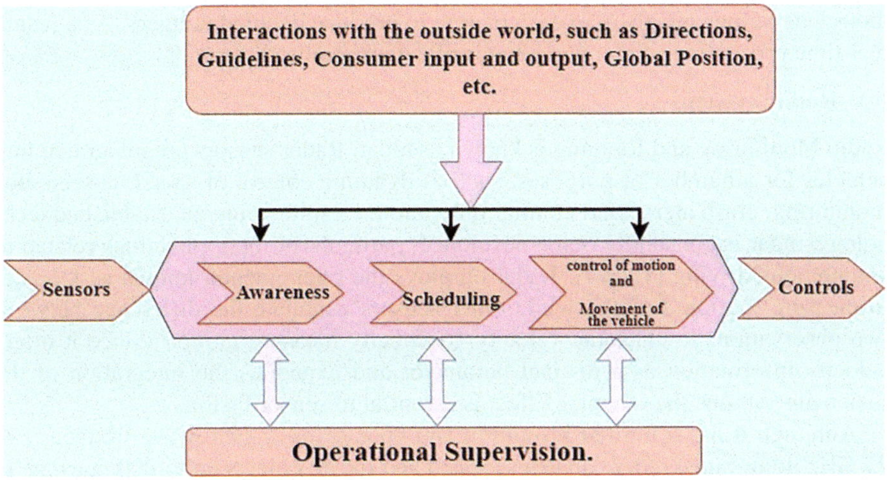

Fig. 3 A self-driving system's functional architecture

industrial literature: Vehicle vision, scheduling, making decisions, and motion, and computer monitoring. Figure 3 shows a representation of these blocks.

Gathering data from cameras along with additional reports, such as the sensor setup of the vehicle and map databases, is the main goal of the perception stage and create a model of the surrounding environment and an illustration of the vehicle's condition. Proprioceptive and tactile sets of sensors are separated on the basis of how well they accomplish these two functions. Proprioceptive sensors, such as Inertial Navigation Systems (INS), inertial measuring units (IMUs), Global Navigation Satellite Systems (GNSS), and Encoders, are those that sense the state of the vehicle. These are employed to obtain information on the platform's position, movement, and odometry.

Tactile sensors keep an eye on the area around the vehicles to gather information on the terrain, surroundings, and extraterrestrial objects. This group includes surveillance footage, lasers (light detection and ranging), radar, and ultrasonic detectors.

Following the gathering of all incoming sensor data, the perception stage performs two key tasks: localization and mapping and object recognition.

(a) **Digital Camera**

Cameras were one of the first kinds of detectors used in automated cars, and they are still the most popular choice among automakers, even though vision in self-driving cars is performed with several sensors and sensor systems. Numerous different cameras are fitted on novel vehicles. An autonomous car can see its surroundings directly thanks to cameras. They are less costly and more widely available than sonar or lidar, which is and they are especially good at recognizing and deciphering material. The camera's flaw is the amount of processing power needed to handle the data. Using thirty to sixty frames per second, the most current HD cameras can produce thousands of megapixels in in each picture to produce detailed imagery. As a result, real-time processing of many gigabytes of data is required.

(b) **Radar systems**

Radio Monitoring and Ranging is known as radar. Radar sensors are integrated into vehicles for a number of purposes, such as dynamic control of speed, unseen area monitoring, crash alerts, and conflict mitigation. Despite being an established technology, radar is constantly being developed, particularly for applications related to autonomous driving [13, 14]. Radar employs the phenomenon known as Doppler to measure motion directly, while other sensors calculate the difference between two observations to calculate velocity. to directly measure motion. Since it offers velocity information as a distinct parameter and expedites the integration of the fusion algorithms, the Doppler Effect is essential to sensor fusion.

Although it has a limited resolution and operates at a microwave frequency of 77 GHz, long-range radar can assess speed and locate objects up to 200 m away. In the 24 GHz and 76 GHz bands, short- and medium-range radar is an established and reasonably priced technology. This sensor can measure distance and velocity, but its precision is limited by broad beams and lengthy wavelengths, which can lead to complicated return signals.

Although cameras and lidar are more effective than radar in some circumstances, such as severe weather, sonar has lower angular precision and produces fewer images than lidar. In contrast to cameras, which must handle video feeds that are data-intensive, the radar must process data output at slower rates than lidar and cameras.

By creating sonar images of the surroundings, radar may be utilized for localization. It can look below other vehicles and identify structures and items that would be otherwise hidden. Sonar is hardly impacted by mist or rain of all the car's sensors, and it can see a wide area—roughly 150 degrees—and a distance of about 200 m. Radars have limited precision when compared to cameras and radars, particularly in the direction of the sky.

(iii) **Light Detection and Ranging (LiDAR)**

LiDAR is referred to as "lidar." Lidar measures the separation between an instrument and a nearby object using an infrared laser beam. Most current lidars use sunlight in an emission range, however, some employ wavelengths that are longer, which are more effective in mist and rain.

The laser beam is scanned over the field of vision by a revolving swivel in the most prevalent lidars. Lasers that shine are pulsed, and things reflect the pulses. A point cloud representing these objects is returned by these reflections. Because of the highly concentrated laser beam, the greater number of vertical scan layers, and a large number of lidar points per layer, lidar has a substantially better spatial resolution than radar. Since these kinds of lidars can't measure an object's velocity directly, they must instead compare the positions of objects among multiple scans. Weather and grime on the sensor have a greater impact on lidars.

Lidars, on the other hand, can scan laser beams thanks to MEMS (Micro-Electro-Mechanical System) vibrating micro mirrors. The laser beam can be moved using a method akin to split matrix sonar rather than mechanically. The phase connection between the waveguides may be changed to vary the laser beam's direction by splitting one beam of light into numerous waveguides.

Velocity may be measured directly using coherent lidars. Having a high resolution is useful for item identification. Lidars are capable of mapping a static scene as well as seeing and identifying moving objects like people, animals, and automobiles. Cost is the current barrier; however, technology is heading in the right way to reduce both the price and the dimensions of the sensor. References [15–19] represent creative development and research in the field of automotive lidar applications.

A novel solution in this field combines detection from radar and lidar ranges with a camera-based predictor after providing it important areas from a lidar cloud of points. The integration section, which is utilised to compile a list of moving objects, is where the tracking module gets its data from. By combining item classification from numerous sensor detections, the world as it appears is improved. Figure 4 depicts a block diagram of the various sensor awareness modules [20].

In sensor fusion from the device, radars, and laser data, low-level fusion is often applied to pre-processed radar and lidar data rather than running the data through an algorithm for obtaining attributes or object information. Then, an advanced fusing block that takes camera inputs into account includes this merged information. In this situation, high-level fusion produces detection, while low-level fusion addresses localization and mapping. One trend in the vision in autonomous cars may be the combination of minimal fusing as a stimulus to excellent fusion.

By employing visual category and shape information while selecting an object detection technique, the accuracy of information connection and activity classification may be improved. A system for tracking can transition between a point model and a 3D boxed representation depending on how close the item is to the vehicle [21]. This implies that footage from cameras is also required for location and tracking functions. Tracking technologies are going to be able to monitor more precisely in the future as a result of the analysis of pertinent information about urban traffic surroundings.

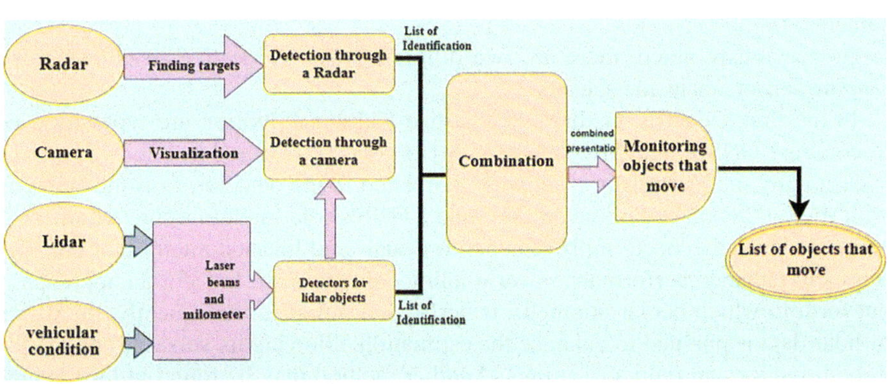

Fig. 4 A system for multiple sensing processing [20]

3 Perceptual Data Fusion

Data fusion, also known as multi-sensor combining data, sensor fusion, or sensor fusion, has been defined in many ways by various writers across the discipline [22–26], however, the following questions might help clarify what it is.

- How does data fusion work?

Homogenous or heterogeneous data can be combined, merged, or integrated.

- What purpose does data fusion serve?

How to implement data fusion? Improve the quality of the data, infer underlying information, and get a more accurate depiction of something or the environment.

Fusion of data is a complex activity that depends on the type of sensors being used, the environment, and the intended use.

Because information often travels from sensors to programmes, following filtering, enhancement of data, and detail extraction steps, multi-sensor data integration is a diverse area. As a result, expertise in a variety of domains is necessary, including artificial intelligence, machine learning, probability, and signal processing. [27].

By calibrating and integrating the sensors in the system, the initial stage in the fusion of multiple sensors is to gather and connect the information in both time and space. This phase is crucial since the effectiveness of the fusion stages depends on the consistency and alignment of the data from many sources with a single reference frame. The utilization of a multi-sensor integration system incorporating cameras, radar, and lidar in expansive off-road vehicles brings attention to challenges related to temporal synchronization. Some of the hurdles associated with aggregating these sensors are elaborated in reference [28]. Reference [29] also discusses a system with several sensors for self-driving cars, highlighting the difficulties in combining diverse sensor data and the advantages of following a model of software design based on clear elements and connections. You may find other instances of recent advancements in the synchronization and calibration of sensors (lidar, radar, and/or cameras) in [30–32].

As previously noted, there are two different ways to localize an object: using ranging sensors or inside sensors.

In the first scenario, methods that utilize Kalman Filtration are typically used to combine GNSS and IMU data [33, 34]. This is a tried-and-true technique, and occasionally the processing is already completed in the detector, as is the situation with some INSs that combine an IMU and a GNSS solution into a single unit.

There have also been improvements in visual- and lidar-odometry that use both senses to enhance performance over a unimodal approach. In [35], an approach is put forth in which ocular odometry is performed initially, subsequently, the fusion of lidar data is pursued to enhance the estimation, allowing its versatile application in both indoor and outdoor settings. Another method may be found in [36], which couples two reliable algorithms, LOAM for lidar odometry and VISO2 for visual odometry. The authors of [37] propose an alternative method in which they create virtual lidar data using a multi-camera system made up of four panoramic lenses and integrate it into the odometer procedure.

Various SLAM methods are found in the literature, employing different combinations of range sensors. Solutions range from using single cameras, groups of cameras, stereo cameras, depth cameras, to 2D and 3D lidars. As an illustration, Yang et al. [38] suggests a multi-camera SLAM system that generates a panoramic image by integrating input from 5 cameras. The system's performance with a single lens method and with 3, 4, or 5 cameras set up is shown. A review of multimodal localization strategies for mobile robots is provided in Ref. [39], which compares Visual Odometry, Place Recognition, and Visual SLAM, in light of the need to respond to changing environments and landscapes, including those that are unstructured, dirt roads, and outside. In [40], a separate review that was solely concerned with SLAM technology is offered.

For the sake of object identification and tracking, all of these range sensors have also been included. The use of Deep Learning and Neural network approaches is the foundation of the majority of current research in this field.

In Ref. [41], a lidar-camera interaction device using repeated neural systems is proposed. They evaluate both the combined approach and the specific efficiency of every sensor [42]. Additionally includes a Deep-Learning-based framework for combining radar and pictures for identifying objects applications.

In a recent investigation [43], a Camera-Radar synthesis approach is introduced for object recognition. To integrate 2D and 3D data, enhance your network system by incorporating Region Proposal Networks (RPN) as an additional layer. On the other hand, Ref. [44] describes a method for doing 3D tracking of objects based on the Kalman filter by combining INS, camera, and Lidar data [45]. The user offers a more thorough examination of multi-sensor fusion in autonomous driving, focusing on the combination of various data from GNSS, radar, camera, Lidar, IMU, ultrasonic's, and V2X communications.

4 Application of This Project

The detection and identification of autonomous vehicles using sensor synthesis have various applications across different domains. Here are some key applications:

1. **Traffic Management**:

Improved detection and identification of autonomous vehicles contribute to efficient traffic management. This includes better traffic flow, reduced congestion, and optimized signal control at intersections.

2. **Collision Avoidance**:

Sensor synthesis aids in detecting surrounding vehicles, pedestrians, and obstacles, enabling advanced collision avoidance systems. This is critical for enhancing safety in both urban and highway driving environments.

3. **Automated Parking**:

Autonomous vehicle detection and identification play a crucial role in automated parking systems. Sensors can help in accurately identifying available parking spaces and guiding vehicles for safe parking.

4. **Fleet Management**:
   - For businesses employing autonomous vehicles in their fleets, sensor synthesis supports effective fleet management. It allows real-time tracking, monitoring, and coordination of autonomous vehicles for logistics, delivery services, and transportation.
5. **Urban Planning and Infrastructure Design**:
   - Data collected through sensor synthesis can be utilized in urban planning to understand traffic patterns, optimize road infrastructure, and design autonomous vehicle-friendly city layouts.
6. **Security and Surveillance**:
   - Advanced sensor synthesis assists in identifying autonomous vehicles for security and surveillance purposes. This is valuable in applications such as border control, monitoring restricted areas, and ensuring compliance with traffic regulations.
7. **Smart Infrastructure Integration**:
   - Integration with smart city infrastructure, such as intelligent traffic lights and dynamic road signage, relies on accurate detection and identification of autonomous vehicles. This facilitates seamless interaction between vehicles and the surrounding infrastructure.
8. **Emergency Response Coordination**:
   - Quick and precise identification of autonomous vehicles is essential for emergency response services. It enables efficient coordination during emergencies, accidents, or medical situations involving autonomous vehicles.
9. **Public Transportation Enhancement**:
   - In public transportation systems, the detection and identification of autonomous vehicles contribute to the integration of autonomous shuttles or buses. This enhances the overall efficiency and reliability of public transportation services.
10. **Research and Development**:
    - Sensor synthesis applications also extend to research and development in the field of autonomous vehicles. It facilitates the testing and validation of new technologies, algorithms, and sensor configurations for improved autonomy and safety.

5 Conclusion

Two important aspects in the development of autonomous vehicles were the decrease in fatal collisions and the human factor as a contributing factor. The goal was to create an autonomous vehicle that could drive itself more reliably than a person. Vehicle performance must improve in addition to mimicking human conduct in order to accomplish this difficult goal. This process depends on the sensors' and the sensor fusion systems' dependability. For the cars to be completely autonomous, they must also contain sensors like radar and lidar that can identify obstacles and map their surroundings in addition to cameras that can mimic human vision.

The largest advancement in sensor technology is anticipated to come from lidar with MEMS micro mirror; it may generate a quick-density cloud of points at a fast speed while decreasing the size and cost of the sensor. Reaching this is essential since the primary obstacles to the widespread use of lidar in self-driving cars are the sensor's size and cost.

The determination of the capability of path planning, driving corridors, calculation of location with dignity to the regarding structure, landmark-based translation, obstacle avoidance, and the present challenges of integrating sensors in autonomous cars include the addition of historical knowledge regarding urban traffic conditions.

The features of vision, localization, and mapping tasks for autonomous cars were examined and critiqued in this study, with a focus on those that use deep learning algorithms for data-driven knowledge discovery as opposed to physics-based models. Summarizing possible study fields to advance the field of autonomous cars is the main goal of this section. In particular, deep learning techniques may be used to improve sensor fusion network performance and improve environmental perception, localization, and mapping.

6 Severe Weather Situations

One ongoing issue with self-driving cars is that they perform worse in bad weather—rain, snow, dust storms, or fog, for example—and are less maneuverable. Visibility distance can be impacted by several situations, which can also degrade range perception and eyesight. In such cases, the performance of the current sensors is severely impaired, potentially producing inaccurate results. Deep learning algorithms might be used to evaluate the risk of failure early on by leveraging learnt experiences and historical data. By doing this, the driver could be able to stop or turn off the autonomous system. Two possible strategies are proposed: investing in sensor hardware technologies, such as short-wave gated cameras and short-wave infrared LiDAR, and improving fusion algorithms using deep learning with current sensors.

7 Map-Matching of Landmarks

For autonomous car systems to achieve sub-decimeter accuracy, localization and mapping must be improved. A recently developed method compares the apparent position of recurring and unique landmarks (such as traffic signs and light poles) with an offline map. Previous work has employed traditional fusion techniques with inefficient detection methods; however, deep learning algorithms might replace them and speed up learning and dependability. Similar to how well-suited deep learning techniques are for object detection and recognition, their capacity for generalization can improve landmark matching as well. The identification of landmarks can be greatly enhanced by the development of 3D computer vision and shape-based techniques.

8 Problems to Be Solved: Repeatability, Reliability, and Cybersecurity

Although deep learning techniques have significantly enhanced several autonomous car perception and localization modules, they are dependent on substantial datasets produced over long periods of time. The dependability and applicability of the results are strongly impacted by the caliber and comprehensiveness of these datasets. The robustness and dependability of current methods can be improved by combining deep learning techniques with traditional model-based methods. Maintaining high efficiency is necessary to overcome problems in the certification and homologation of deep learning systems based on perception and localization. Data-driven techniques' susceptibility to interferences and disruptions in sensor data raises concerns that cybersecurity, repeatability, and reliability issues must be addressed. Tests have shown weaknesses, but they have also proven how resilient deep learning systems are to computational errors. It is essential to strike a balance between these factors for the safe and continuous development of autonomous car technologies.

References

1. Dickmanns, E.D., Zapp, A.: Autonomous high-speed road vehicle guidance by computer vision. IFAC Proc. Vol. **20**(5), 221–226 (1987)
2. Kanade, T.: Autonomous land vehicle project at CMU. In: CSC '86 Proceedings of the 1986 ACM Fourteenth Annual Conference on Computer Science (1986)
3. Wallace, R.: First results in robot road-following. In: JCAI'85 Proceedings of the 9th International Joint Conference on Artificial Intelligence (1985)
4. Thrun, S., et al.: Stanley: the robot that won the DARPA grand challenge. J. Robot. Syst. Special Issue DARPA Grand Challenge **23**(9), 661–692 (2006)
5. Montemerlo, M., et al.: Winning the DARPA grand challenge with an AI robot. In: Proceedings of the 21st National Conference on Artificial intelligence, pp. 982–987 (July 2006)

6. Buehler, M., Iagnemma, K., Singh, S.: The DARPA Urban Challenge: Autonomous Vehicles in City Traffic. Springer Tracts in Advanced Robotics (2009)
7. Bojarski, M., et al.: End-to-end learning for self-driving cars (2016). Available: https://arxiv.org/abs/1604.07316
8. Kocić, J., Jovičić, N., Drndarević, V.: Driver behavioral cloning using deep learning. In: 2018 17th International Symposium INFOTEH-JAHORINA (INFOTEH), East Sarajevo, pp. 1–5 (2018)
9. Kocić, J., Jovičić, N., Drndarević, V.: End-to-end autonomous driving using a depth-performance optimal deep neural network (submitted for publication) (2018)
10. Riedmiller, M., Montemerlo, M., Dahlkamp, H.: Learning to drive a real car in 20 minutes. In: 2007 Frontiers in the Convergence of Bioscience and Information Technologies, Jeju City, pp. 645–650 (2009)
11. Fridman, L., Jenik, B., Terwilliger, J.: DeepTraffic: driving fast through dense traffic with deep reinforcement learning. arXiv:1801.02805, [cs.NE], Jan 2018. Available at: https://arxiv.org/abs/1801.02805
12. Kruchten, P.: The 4+1 view model of architecture. IEEE Softw. **12**(6), 42–50 (1995)
13. Patole, S.M., Torlak, M., Wang, D., Ali, M.: Automotive radars: a review of signal processing techniques. IEEE Signal Process. Mag. **34**(2), 22–35 (2017)
14. Steinbeck, J., Steger, C., Holweg, G., Druml, N.: Next generation radar sensors in automotive sensor fusion systems. In: 2017 Sensor Data Fusion: Trends, Solutions, Applications (SDF), Bonn, pp. 1–6 (2017)
15. Han, J., Kim, D., Lee, M., Sunwoo, M.: Enhanced road boundary and obstacle detection using a downward-looking LIDAR sensor. IEEE Trans. Veh. Technol. **61**(3), 971–985 (2012)
16. Ogawa, T., Sakai, H., Suzuki, Y., Takagi, K., Morikawa, K.: Pedestrian detection and tracking using in-vehicle lidar for automotive application. In: 2011 IEEE Intelligent Vehicles Symposium (IV), Baden-Baden, pp. 734–739 (2011)
17. Thakur, R.: Scanning LIDAR in advanced driver assistance systems and beyond: building a road map for next-generation LIDAR technology. IEEE Cons. Electron. Mag. **5**(3), 48–54 (2016)
18. Kutila, M., Pyykönen, P., Ritter, W., Sawade, O., Schäufele, B.: Automotive LIDAR sensor development scenarios for harsh weather conditions. In: 2016 IEEE 19th International Conference on Intelligent Transportation Systems (ITSC), pp. 265–270 (2016)
19. Zermas, D., Izzat, I., Papanikolopoulos, N.: Fast segmentation of 3D point clouds: a paradigm on Lidar data for autonomous vehicle applications. In: 2017 IEEE International Conference on Robotics and Automation (ICRA), pp. 5067–5073 (2017)
20. Chavez-Garcia, R.O., Aycard, O.: Multiple sensor fusion and classification for moving object detection and tracking. IEEE Trans. Intell. Transp. Syst. **17**(2), 525–534 (2016)
21. Cho, H., Seo, Y., Kumar, B.V.K.V., Rajkumar, R.R.: A multi-sensor fusion system for moving object detection and tracking in urban driving environments. In: 2014 IEEE International Conference on Robotics and Automation (ICRA), 2014, pp. 1836–1843 (2014)
22. White, F.E.: Data Fusion Lexicon, The Data Fusion Subpanel of the Joint Directors of Laboratories, Technical Panel for C3 **15**(0704), 15 (1991)
23. Luo, R.: Multisensor fusion and integration: approaches, applications, and future research directions. IEEE Sens. J. **2**(2), 107–119 (2002)
24. Luo, R.C., Chang, C.C., Lai, C.C.: Multisensor fusion and integration: theories, applications, and its perspectives. IEEE Sens. J. **11**(12), 3122–3138 (2011)
25. Elmenreich, W.: A review on system architectures for sensor fusion applications. In: Obermaisser, R., Nah, Y., Puschner, P., Rammig, F.J. (eds.) Software Technologies for Embedded and Ubiquitous Systems, pp. 547–559. Springer, Santorini Islands, Greece (2007)
26. Bostrom, H., Andler, S., Brohede, M.: On the Definition of Information Fusion as a Field of Research. University of Skövde, Technical Report (2007)
27. Velasco-Hernandez, G.: Multisensor Architecture for an Intersection Management System. Universidad del Valle, Technical Report (2019)

28. Yeong, D.J., Barry, J., Walsh, J.: A review of multi-sensor fusion system for large heavy vehicles off road in industrial environments. In: ISSC (2020)
29. Giacalone, J.P., Bourgeois, L., Ancora, A.: Challenges in the aggregation of heterogeneous sensors for autonomous driving systems. In: SAS 2019—2019 IEEE Sensors Applications Symposium, Conference Proceedings, pp. 3–7 (2019)
30. Hu, H., Wu, J., Xiong, Z.: A soft time synchronization framework for multi-sensors in autonomous localization and navigation. In: IEEE/ASME International Conference on Advanced Intelligent Mechatronics, AIM, vol. 2018-July, pp. 694–699 (2018)
31. Domhof, J., Julian, K.F., Dariu, K.M.: An extrinsic calibration tool for radar, camera, and lidar. In: Proceedings—IEEE International Conference on Robotics and Automation, vol. 2019-May, pp. 8107–8113 (2019)
32. Yang, L., Wang, R.: HydraView: A Synchronized 360-View of Multiple Sensors for Autonomous Vehicles, pp. 53–61 (2020)
33. Panzieri, S., Pascucci, F., Ulivi, G.: An outdoor navigation system using GPS and inertial platform. IEEE/ASME Trans. Mechatron. **7**(2), 134–142 (2002)
34. Wahyudi, Listiyana, M.S., Sudjadi, Ngatelan.: Tracking object based on GPS and IMU sensor. In: 2018 5th International Conference on Information Technology, Computer, and Electrical Engineering (ICITACEE). IEEE, Sep 2018, pp. 214–218 (2018)
35. Zhang, J., Singh, S.: Visual-lidar odometry and mapping: lowdrift, robust, and fast. In: Proceedings—IEEE International Conference on Robotics and Automation, vol. 2015-June, no. June, pp. 2174–2181 (2015)
36. Yan, M., Wang, J., Li, J., Zhang, C.: Loose coupling visualizer odometry by combining VISO2 and LOAM. In: Chinese Control Conference, CCC, pp. 6841–6846 (2017)
37. Xiang, Z., Yu, J., Li, J., Su, J.: ViLiVO: virtual LiDAR-visual odometry for an autonomous vehicle with a multi-camera system. In: IEEE International Conference on Intelligent Robots and Systems, pp. 2486–2492 (2019)
38. Yang, Y., Tang, D., Wang, D., Song, W., Wang, J., Fu, M.: Multicamera visual SLAM for off-road navigation. Robot. Auton. Syst. **128**, 103505 (2020)
39. O'Mahony, N., Campbell, S., Carvalho, A., Harapanahalli, S., Velasco-Hernandez, G.A., Riordan, D., Walsh, J.: Adaptive multimodal localization techniques for mobile robots in unstructured environments: a review. In: IEEE 5th World Forum on Internet of Things, WF-IoT 2019—Conference Proceedings. IEEE, Apr 2019, pp. 799–804 (2019)
40. Singandhupe, A., La, H.: A review of SLAM techniques and security in autonomous driving. In: Proceedings—3rd IEEE International Conference on Robotic Computing, IRC 2019, no. 19, pp. 602–607 (2019)
41. Melotti, G., Premebida, C., Goncalves, N.: Multimodal deep-learning for object recognition combining camera and LIDAR data. In: 2020 IEEE International Conference on Autonomous Robot Systems and Competitions, ICARSC 2020, no. April, pp. 177–182 (2020)
42. Nobis, F., Geisslinger, M., Weber, M., Betz, J., Lienkamp, M.: A deep learning-based radar and camera sensor fusion architecture for object detection. In: 2019 Symposium on Sensor Data Fusion: Trends, Solutions, Applications, SDF 2019 (2019)
43. Li, Z.T., Yan, M., Jiang, W., Xu, P.: Vehicle object detection based on RGB-camera and radar sensor fusion. In: Proceedings—International Joint Conference on Information, Media, and Engineering, IJCIME 2019, pp. 164–169 (2019)
44. Asvadi, A., Girao, P., Peixoto, P., Nunes, U.: 3D object tracking using RGB and LIDAR data. In: IEEE Conference on Intelligent Transportation Systems, Proceedings, ITSC, pp. 1255–1260 (2016)
45. Wang, Z., Wu, Y., Niu, Q.: Multi-sensor fusion in automated driving: a survey. IEEE Access **8**, 2847–2868 (2020)

An IoT Based Real Time Traffic Monitoring System

Rama Devi Burri, Satish Reddy Nalamalapu, Musham Prashanthi, and Bussa Sathwik

Abstract The project focuses on developing a density-based dynamic traffic signal system to address traffic congestion issues. This paper compares the conventional fixed-time signal system with the utilization of infrared (IR) sensors for detecting traffic density at intersections. The Arduino Controller processes sensor data to dynamically adjust green and red-light durations based on real-time traffic conditions. This intelligent traffic control system aims to optimize signal timings by allocating longer green times to lanes with higher traffic density. The proposed framework leverages technology to enhance traffic flow efficiency and mitigate congestion, providing a more adaptive and responsive solution for urban traffic management.

Keywords Internet of Things (IoT) · Traffic monitoring · Network traffic analysis · Sensor data · Real-time monitoring

1 Introduction

Today's lifestyle, traffic congestion is a huge problem in day-to-day activities. Miscommunication will lead to a waste of time and both individual and group output in the workplace. Traffic congestion is caused by several factors, including an excessive number of trucks, poor road conditions, and an illogically placed signalling system. It contributes to the growth in pollution in a roundabout way since continually running engines waste a lot of natural resources like fuel and diesel. Therefore, novel approaches including the introduction of a sensor-based automated approach, in this area of traffic signalling are required to eliminate or greatly decrease the aforementioned problems.

The main goal is to reduce the potential for gridlock and long wait times at traffic lights, especially during times of low internet activity. It was planned for use around the locations of traffic signals to alleviate congestion at such junctions. The number

R. D. Burri · S. R. Nalamalapu (✉) · M. Prashanthi · B. Sathwik
Department of Information Technology, Institute of Aeronautical Engineering, Hyderabad, India
e-mail: satish.bannu45@gmail.com

S. N. Mohanty et al. (eds.), *Explainable IoT Applications: A Demystification*, Information Systems Engineering and Management 21,
https://doi.org/10.1007/978-3-031-74885-1_10

of trucks using each route is tracked, and the lights are timed accordingly. If there are more cars on the road, the light will be set for a longer wait. The fundamental objective of this effort is to bypass the other signal if it will not be used for online traffic. The system will move on to the next candidate if the signal is not present. This leads to the implementation of an intelligent traffic management system based on the Arduino Uno ATMega 328 that can automatically adjust its timing based on the density of web traffic. In this setup, traffic is detected using electronic infrared (IR) sensors, which also serve to identify vehicles by their unique emission patterns. Sensors set along the roadside to monitor and change the timing of traffic lights depending on the amount of internet traffic. The data from all of the infrared sensors is sent into an Arduino Uno. The system relies on light-emitting diodes (LEDs) to make its traffic signals, with two LEDs utilised for each lane.

2 Related Work

The Emergency Management Research Institute (EMRI) highlights an alarming increase in average ambulance response times, reaching 40 min due to heightened traffic density in both urban and suburban areas [1]. This delay in emergency services has severe consequences, with about 30% of road accident-related deaths attributed to late ambulance arrivals, as reported by the Times of India in 2016. The escalating frequency of accidents underscores the urgency of timely emergency response [2]. To address traffic congestion issues, previous approaches such as a system using ultrasonic and sound sensors aimed to detect traffic densities and adjust signals in urban cities. Although this system prioritized ambulance lanes [3]. It faced drawbacks, particularly in crowded ambulance routes, leading to potential delays. Additionally, the use of sound sensors proved ineffective in distinguishing emergency service sirens from other sources of sound [4]. Another proposed system relied on GSM and GPS for emergency service coordination and traffic signal control. However, the queuing transfer technique in GSM resulted in longer message transfer durations [5]. Various methodologies, including inductive sensors and visual cameras, have been explored for traffic congestion detection, but these approaches present challenges such as error estimation [6]. Existing works typically focus on single junctions and handle one emergency vehicle at a time, emphasizing the need for synchronized monitoring and road clearance systems across multiple junctions to address critical scenarios concurrently [7]. These short comings in previous methods prompt the development of a new approach to enhance emergency vehicle response and mitigate traffic congestion effectively. IoT will be the cornerstone of these Smart Cities. The regular daily chores like electricity consumption, pollution and health monitoring will be conducted by the smart devices. Designing one general framework for smart city will not be feasible due to the different types of sensors, total number of sensors, mode of connectivity and its security levels [8]. A comprehensive blueprint of developing a smart city using IoT, which is actually motivated and strongly demanded for improvement of the traffic control system [9]. Signifies the traffic light is being used improperly. In the future, appropriate sensing units may be used to improve it. Sensors are strategically in order to accurately determine the traffic volume [10]. Emphasizing the role of IoT in enhancing emergency response efficiency in order to ambulance system which deals to the system in order to signaling way [11]. The sensor is designed not only to count vehicles but also to classify them intelligently. Such a system could have implications for enhancing the efficiency of traffic surveillance by providing real-time data on vehicle flow and types [12].

3 Proposed System

This paper is set to revolutionize the current traffic signalling model by introducing a dynamic system driven by traffic density. In order to address the prevalent issue of traffic congestion in cities across the globe, the project aims to transition from traditional fixed-time traffic signals to an automated system with smart decision-making abilities [13]. Standard fixed-time systems may struggle when dealing with disparities in traffic flow across different lanes.

To solve this problem, we developed an intelligent driving management system that adjusts the timing of traffic lights according to the actual vehicle speed at the intersection. The Arduino controller acts as a decision center that analyzes data collected by purpose-designed sensors [8, 14, 15]. Arduino controller ensures best traffic management by carefully controlling the traffic speed. It has been suggested that our model can produce better results with higher accuracy than existing models in the field of buildings.

All the data collected by these devices goes here in Fig. 1. It's like a huge brain that analyzes the information, predicting where traffic might get heavy and suggesting smart solutions. The central control system, often referred to as the brain, communicates with the traffic lights situated on the roads. These traffic lights, distinguished by the smart capabilities, possess ability to adjust their timings in response to directives from the central control system. In contrast to traditional lights, these smart lights have the capability to adjust their schedules dynamically. For instance, if the system detects increased traffic on a specific road, it may allocate a longer green time to facilitate a smoother flow of vehicles.

Integration of automatic traffic control or traffic police can help manage the internet at many important intersections. But the traditional traffic management system is based on the idea of determining a certain time for both sides of the intersection, there is no way to adjust this time to match the traffic change. During rush hours, traffic lights need to be adjusted to accommodate many vehicles waiting on the same road, including VIP vehicles, ambulances and other emergency vehicles. It is recommended to use the traffic light as a thickness measurement, which involves analyzing the traffic volume at the intersection to make changes to the light.

The prototype, constructed utilizing IR sensor modules and Arduino technology, features custom programming for operational efficiency. The IR sensing devices play a crucial role in quantifying traffic density on specific routes, and the method

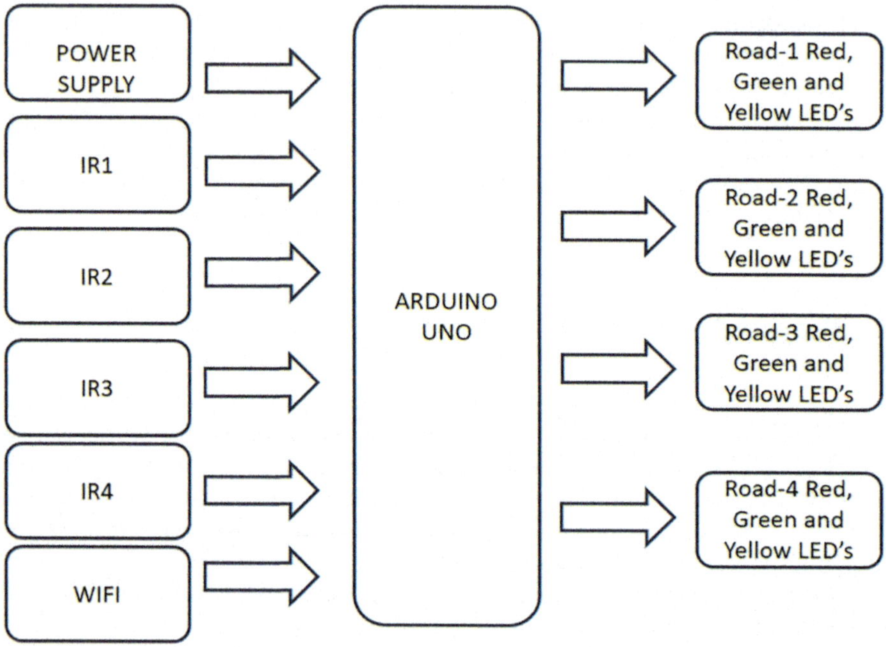

Fig. 1 Architecture diagram

is distinguished by its affordability and user-friendly design. It is possible that IR sensors can't function properly under regular lighting conditions.

This signifies the traffic light is being used improperly. In the future, appropriate sensing units may be used to improve it. IR sensors are strategically placed on each route in order to accurately determine the traffic volume; these sensors are constantly monitoring the road in question. The arduino serves as a hub to connect all of these sensors. The controller monitors and controls the whole web traffic system based on the data collected by these sensors.

The availability of trucks is used to determine how traffic is managed as shown in the Fig. 2. The concept supports the premise of adjusting the duration of traffic lights in response to the volume of vehicles passing through a certain section of route. In order to determine how many vehicles pass through a given location, four sensors are strategically positioned on each side of a four-lane road.

3.1 Arduino Uno

Arduino is an open source software and hardware company, project and community that designs and develops Arduino control boards and Arduino control packages to create digital devices and interactive devices that can be stored and managed

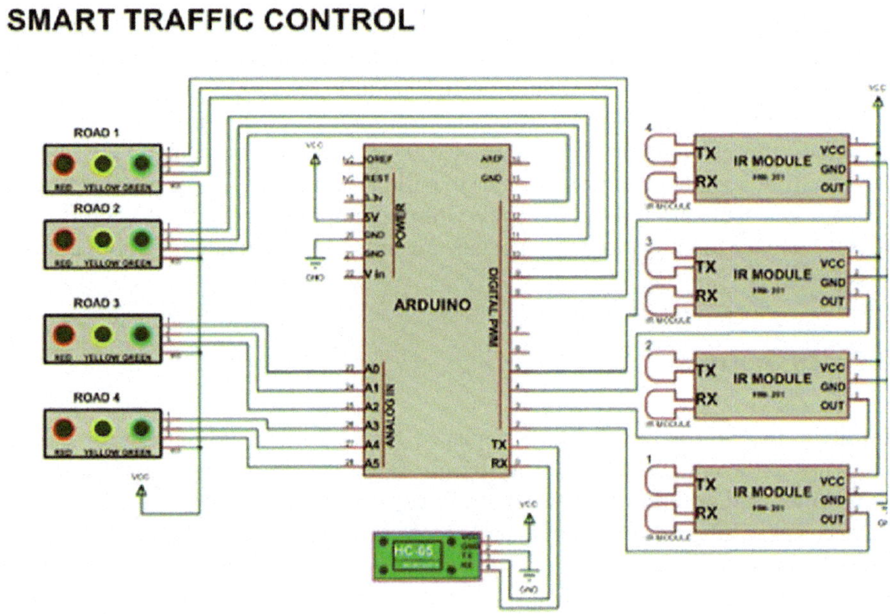

Fig. 2 Arduino circuit representation

electronically. The Arduino board and software are licensed under the GNU Lesser Public License (LGPL) and the GNU Public License (GPL).

Numerous kinds of microprocessors and controllers may be found on Arduino circuit boards. The boards have arrays of digital and analogue I/O pins for connecting to other circuitry, such as additional development boards, breadboards, or shields. The boards include serial interaction ports, such as USB on certain designs, that are also used to transfer software from personal computers as shown in Fig. 3.

Fig. 3 Arduino Uno

Creating an Arduino controller necessitates the use of a C-like, C++-like language. The Arduino Project offers in addition to traditional compiler toolchains a well integrated programming environment—IDE built on the Processing Language.

The Arduino project started as a class for students at the Institute support the Design Communication in Ivrea, Italy in 2003. It was meant to make it possible for anyone from beginners to professionals to build low cost and simple devices that use sensors and actuators to communicate with the environment around them. Simple robots, activity monitors and thermostats are some devices made for hobbyists.

3.2 IR Sensors

Infrared (IR) sensor devices have replaced the traditional traffic control system, allowing us to implement a density-based traffic signalling scheme. Both the IR transmitter and the IR receiver (photodiode) are included within the IR sensing device. These infrared transmitters and receivers will be stationed on opposite sides of the road, separated by a predetermined distance [16–18]. Infrared sensors placed in this area will detect the car and relay that information to an Arduino controller as it passes past.

The Arduino Controller will keep tally of the vehicles present, adjusting the LEDs' on time in response to traffic congestion. For lanes or roads with a larger thickness, the LEDs will shine for longer than usual, and vice versa. The traffic signals start off operating with a delay of 1000 ns, therefore the total delay is 1 ns longer than what is acceptable. This full embedded system is installed at that junction. LEDs and IR sensors are connected to an Arduino controller. There must be a total of four IR sensors and eight LEDs. That's why they're connected to any two of the Arduino's ports.

The product which needs a cautious use is the electro luminescent IR LED. IR LED's are developed from nanostructures with energy gap from 0.25 to 0.4 eV. Transmitter creates IR rays in planar wave. Although infrared rays radiate in all directions, it travel in a straight line in forward motion. IR rays are the one which produces the wavelets when it strikes with anything coming in its way. This feature is made use of in this case.

Infrared photo receiver is a two terminal P–N junction device as shown in Fig. 4 and operates under the reverse bias. It has a low transparent window, which allows light to impinge on the PN junction. A photodiode is a kind of photo detector that can convert light into current or voltage by its operation in one mode or another. Most photodiodes, however, will look like light emitting diodes [19–21]. There will be two leads, or wires, connected from the bottom. At its shorter end it is the cathode, and at its longer end it is the anode.

A photodiode is composed of PN junction or PIN construction. When a photon of specified energy interacts with a diode, electron is excited and becomes a mobile electron and a positive electron hole is created. When the carriers are swept out of the junction by the field of the depletion region, the absorption occurs in the junction's

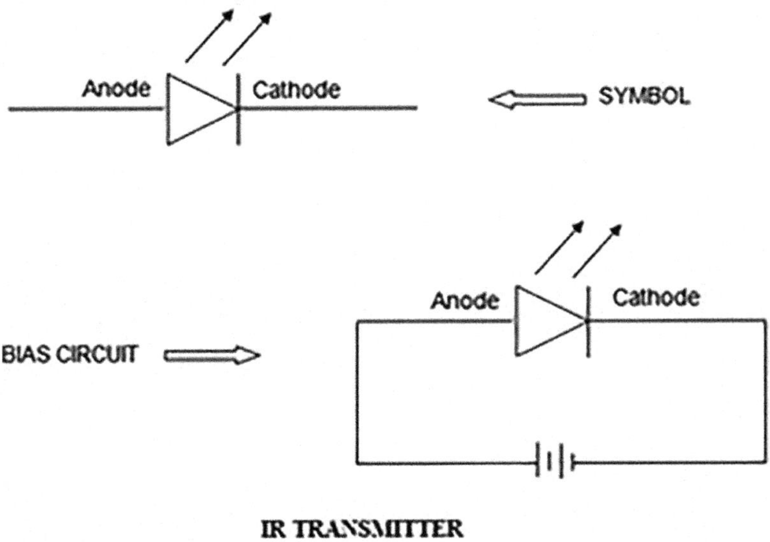

Fig. 4 IR transmitter

depletion region itself or approximately within one diffusion length away from it. So holes travel toward anode, but electrons toward the cathode, and a photocurrent is produced.

A built-in transmitter and receiver are equipped on the infrared sensor module. As shown in Fig. 5, density-based traffic signal control is the main function: this algorithm adjusts traffic signal timings to suit vehicle density and flow. Infrared detecting units are located at each junction, some predetermined distances from the signal box. The duration of time a traffic light will be on depends on how many cars are in transit at any particular time. For instance, infrared (IR) sensors count how many trucks are driving through. The Arduino Controller determines which thoroughfare gets priority at a traffic light and for what duration by referring to IR counts.

3.3 *Working of Infrared Communication*

A number of different types of software utilizing infrared technology are already available. Due to the embedding of transmitter and receiver elements in circuits, this one can't be used for other purposes but infrared-based ones. Our approach is universal as the infrared LED within our infrared interaction circuit architecture can be replaced by any other application type. This simplifies the development process of software using infrared technology. The circuit is split into two separate parts.

1. The Sender Part

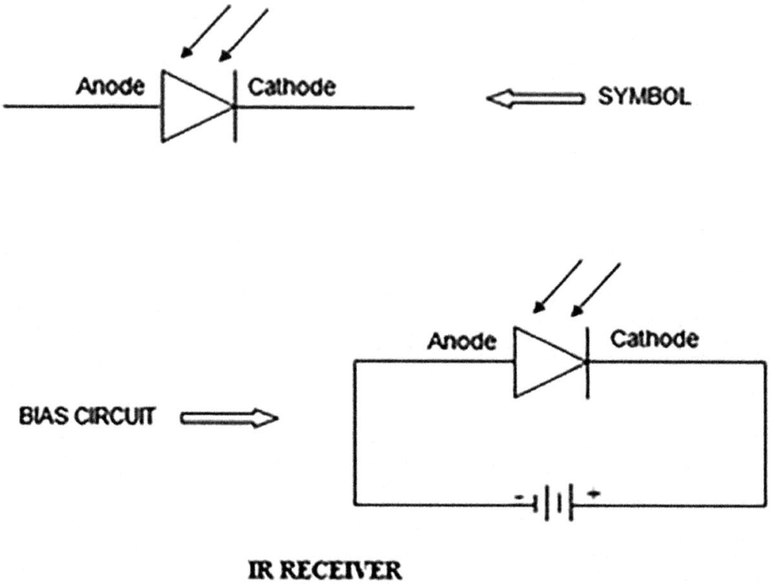

Fig. 5 IR receiver

2. The Receiver Part

First, the transmitter uses an integrated circuit 555 timer operating in as table mode. The wiring diagram looks like this. The output of the as table setting is routed into an infrared LED using a resistor, which limits the current flowing through the LED. The IR radiation emitted by the infrared LED in the transmitter is focussed into a tight beam using plastic lenses (optics).

The silicon phototransistor at the point of infrared radiation photo generating the current. It is sensitive to a transmitted fast-pulsating signal only and needs an exceptional effort to convert the available sunlight into infrared. An infrared detector, and an electronic screen, the main components, are what make up the receiver circuit. Once the signals are off then IR led turns on after a predefined time according to the value of RC pair.

Placing the lens in the path of the light beams emitted by the infrared emitter will improve the effective range of light between the emitter and the receiver. Following the IR sender to receiver circuit connections and providing the needed 6 V power is all one needs for the circuit to function. The adaptability of this circuit makes it perfect for use in the most diverse set of applications. For example, buzzers are used as feedback devices; they would ring when the signals are interrupted. Simply a dual set of circuit boards (e.g., the bread board or printed circuit board) would be required in which both the transmitter and the receiver are housed. The IR receiver must be placed at the back of the IR lead to eliminate any interference caused by infrared leakage. The IR lead gives an output when casted on a nearby moving object.

Fig. 6 Photodiode

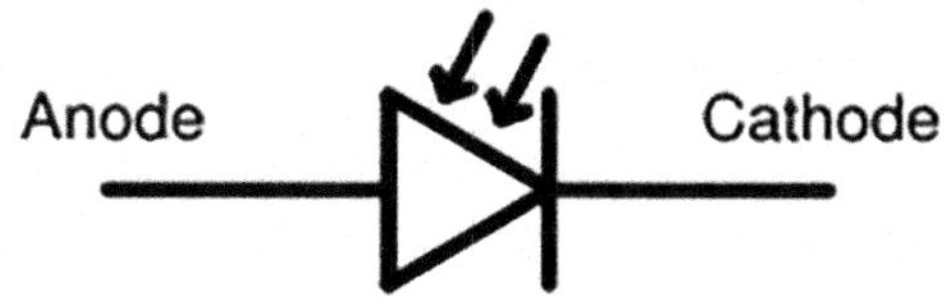

Diode cameras

A photodiode, which is a semiconductor image detector, is a device that converts light into photocurrent. This is achieved by patterning the light-sensing area such that it can be reached by light, manufacturers may use a window or optical fiber connections. It can be used on the inside without the window, as it will detect X or UV led light from a cleaner, an example is depicted in Fig. 6. It is nothing more than a bipolar transistor close to the base–collector junction in a transparent housing. Indicating that a phototransistor is also a photodiode because of the fact that photons are a base–collector junction create electrons, which penetrate the base, just like in a photodiode.

4 Result

The utilization of an IR sensor enhances the system's ability to perceive and respond to its environment in real-time. This dynamic approach ensures that the traffic control system is not only reactive but also proactive, adapting swiftly to changing conditions on the road.

A standout feature of this innovative system is its emphasis on prioritizing emergency vehicles. In emergency situations, time is of the essence, and seconds can make a crucial difference. The integration of IR sensors enables the system to detect the approach of emergency vehicles, triggering an immediate response to clear the traffic path and facilitate their swift passage. This priority-based mechanism significantly enhances the efficiency of emergency services, contributing to improved public safety and potentially saving lives.

The three-tiered priority system further highlights the sophistication of this traffic control solution. The highest priority assigned to emergency vehicles is complemented by a density-based priority level. IR sensors strategically placed on both sides of the road continuously monitor traffic density. This data is then processed by the Arduino controller, allowing the system to dynamically adjust signal timings. By responding to real-time traffic conditions, the system optimizes traffic flow and minimizes congestion, resulting in a more fluid and efficient transportation network.

The third priority level employs a timer-based system for regular traffic. While this may seem traditional, it acts as a reliable baseline, ensuring that the system maintains a semblance of order during normal traffic conditions. The timer-based mechanism provides a structured approach to traffic control, contributing to overall road safety and efficiency.

In comparison to existing traffic control systems, the accuracy achieved by this advanced technology is noteworthy. The continuous data collection from IR sensors and the intelligent decision-making capabilities of the Arduino create a system that is not only more accurate but also adaptable to the complexities of urban traffic. The system's ability to prioritize emergency vehicles, dynamically respond to traffic density, and maintain a baseline efficiency through timers collectively represents a leap forward in traffic management, showcasing the potential for future urban transportation systems to be smarter, safer, and more responsive to the needs of the community.

The implementation of the code is embedded into the Arduino Uno through a type B-cable which is inserted in the USB port as represented in the above Fig. 7. The working of the code is done in Arduino Uno, which will allow the manipulation of the traffic based on input of the lane. The working can be explained in a Four road junction here the traffic signal is adjusted according to the timer based, if any emergency vehicle is struck in traffic then according to the input the traffic signal will turn into green, if the input is "1" then the lane 1 will be green and all other lane signal will be in Red.

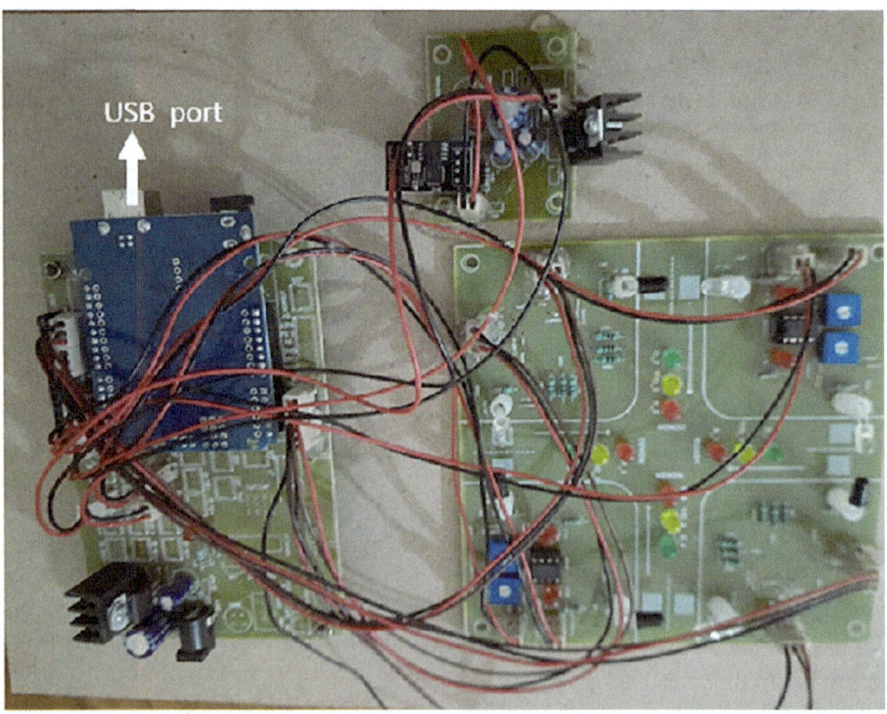

Fig. 7 Establishment of connection

```
if(inputString[2] == '1')
        {inputString="";
         digitalWrite(led1_green, LOW);digitalWrite(led1_yellow, HIGH);digitalWrite(led1_red, HIGH);
         digitalWrite(led2_green, HIGH);digitalWrite(led2_yellow, HIGH);digitalWrite(led2_red, LOW);
         digitalWrite(led3_green, HIGH);digitalWrite(led3_yellow, HIGH);digitalWrite(led3_red, LOW);
         digitalWrite(led4_green, HIGH);digitalWrite(led4_yellow, HIGH);digitalWrite(led4_red, LOW);
```

If the input the traffic signal will turn into green, if the input is "2" then the lane 2 will be green and all other lane signal will be in Red.

```
if(inputString[2] == '2')
        {inputString="";
         digitalWrite(led1_green, HIGH);digitalWrite(led1_yellow, HIGH);digitalWrite(led1_red, LOW);
         digitalWrite(led2_green, LOW);digitalWrite(led2_yellow, HIGH);digitalWrite(led2_red, HIGH);
         digitalWrite(led3_green, HIGH);digitalWrite(led3_yellow, HIGH);digitalWrite(led3_red, LOW);
         digitalWrite(led4_green, HIGH);digitalWrite(led4_yellow, HIGH);digitalWrite(led4_red, LOW);
```

If the input the traffic signal will turn into green, if the input is "3" then the lane 3 will be green and all other lane signal will be in Red.

```
if(inputString[2] == '3')
        { inputString="";
digitalWrite(led1_green, HIGH);digitalWrite(led1_yellow, HIGH);digitalWrite(led1_red, LOW);
digitalWrite(led2_green, HIGH);digitalWrite(led2_yellow, HIGH);digitalWrite(led2_red, LOW);
digitalWrite(led3_green, LOW);digitalWrite(led3_yellow, HIGH);digitalWrite(led3_red, HIGH);
digitalWrite(led4_green, HIGH);digitalWrite(led4_yellow, HIGH);digitalWrite(led4_red, LOW);
```

If the input the traffic signal will turn into green, if the input is "4" then the lane 4 will be green and all other lane signal will be in Red.

```
if(inputString[2] == '4')
        {inputString="";
         digitalWrite(led1_green, HIGH);digitalWrite(led1_yellow, HIGH);digitalWrite(led1_red, LOW);
         digitalWrite(led2_green, HIGH);digitalWrite(led2_yellow, HIGH);digitalWrite(led2_red, LOW);
         digitalWrite(led3_green, HIGH);digitalWrite(led3_yellow, HIGH);digitalWrite(led3_red, LOW);
         digitalWrite(led4_green, LOW);digitalWrite(led4_yellow, HIGH);digitalWrite(led4_red, HIGH);
```

The input of lane number is passed through the application where it goes to the IoT sensor then the sensor will be transfer to the Arduino Uno here the complete task will be done.

5 Conclusion

The traffic control recommended for emergency vehicles in this study is based on traffic situation monitoring using infrared sensors. Here the parties, ambulance and control room can see the traffic situation displayed in the generated application. With the help of this program, emergency vehicles can reach their destination as quickly as possible and without traffic jams. IR sensors on both sides of the road can be used to restrict traffic based on density. The design and development of the traffic light system ensured proper integration of hardware and software. The entire workflow of the road system is synchronized with this interface. This system can be configured to operate automatically.

Future Scope

This study is created using density and timer based approaches for emergency vehicles. The system can be integrated with artificial intelligence (AI) to improve its efficiency by automatically identifying emergency vehicles and traffic density. Artificial intelligence and machine learning combined will improve the system's capacity to identify intricate data patterns and adjust dynamically to shifting traffic situations. Furthermore, energy-efficient traffic management, taking into account the effects of transportation on the environment, and promoting sustainable behaviors will be prioritized in the future.

References

1. Rizvi, S.R., Olariu, S., Weigle, M.C., Rizvi, M.: A novel approach to reduce traffic chaos in emergency and evacuation scenarios. In: IEEE Vehicular Technology Conference (2007). https://doi.org/10.1109/vetecf.2007.407
2. Chandrakar, S., Thomas, A.: Combating man-made disaster using remote sensing. In: Second International Conference on Emerging Trends in Engineering and Technology (2009). https://doi.org/10.1109/icetet.2009.51
3. Rizvi, S. R., Olariu, S., Rizvi, M., Weigle, M.C.: A traffic chaos reduction approach for emergency scenarios. In: IEEE International Performance, Computing, and Communications Conference (2007). https://doi.org/10.1109/pccc.2007.358943
4. Stewart, D.H., Ivancic, V.W.D.: NASA Glenn Research Center Terry L. Bell, Lockheed Martin Global Telecommunications
5. Hsu, T., Sou, S., Lin, C.: Architecture and recipient selection of emergency messaging for ambulance traveling. In: IEEE Conference on Vehicular Technology (2014). https://doi.org/10.1109/vtcspring.2014.7022837
6. Derekenaris, G., Garofalakis, J., Makris, C., Prentzas, J., Sioutas, S., Tsakalidis, A.: An information system for the effective management of ambulances. IEEE Comm. Mag. (2002). https://doi.org/10.1109/cbms.2000.856910
7. Stewart, D., Ivancic, W.D., Bell, T., Kachmar, B.A., Shell, D., Leung, K.C.: Application of mobile router to military communications. In: 2001 MILCOM Proceedings Communications for Network-Centric Operations: Creating the Information Force (Cat. No. 01CH37277). (2002). https://doi.org/10.1109/milcom.2001.985825
8. Balid, W., Tafish, H., Refai, H.H.: Intelligent vehicle counting and classification sensor for real-time traffic surveillance. IEEE Trans. Intell. Transport. Syst. **19**(6), 1784–1794 (2018). https://doi.org/10.1109/TITS.2017.2741507
9. Mohamed, S.A.E., Al-Shalfan, K.: Intelligent traffic management system based on the Internet of Vehicles (IOV). J. Adv. Transp. **2021**, 1–23 (2021). https://doi.org/10.1155/2021/4037533
10. Jin, J., Gubbi, J., Marusic, S., Palaniswami, M.: An information framework for creating a smart city through internet of things. IEEE Internet Things J. **1**(2), 112–121 (2014). https://doi.org/10.1109/jiot.2013.2296516
11. Humagain, S., Sinha, R., Lai, E., Ranjitkar, P.: A systematic review of route optimisation and pre-emption methods for emergency vehicles. Transp. Rev. **40**(1), 35–53 (2019). https://doi.org/10.1080/01441647.2019.1649319
12. Bali, V., Mathur, S., Sharma, V., Gaur, D.: Smart traffic management system using IoT enabled technology. In: 2020 2nd International Conference on Advances in Computing, Communication

Control and Networking (ICACCCN) (2020). https://doi.org/10.1109/icacccn51052.2020.9362753
13. Balid, W., Tafish, H., Refai, H.H.: Intelligent vehicle counting and classification sensor for real-time traffic surveillance. IEEE Trans. Intell. Transp. Syst. **19**(6), 1784–1794 (2018). https://doi.org/10.1109/tits.2017.2741507
14. Balid, W.: Fully-Autonomous Self-Powered Intelligent Wireless Sensor For Real-Time Traffic Surveillance In Smart Cities, University of Oklahoma (2016)
15. Balid, W., Refai, H.H.: Real-time magnetic length-based vehicle classification: case study for inductive loops and wireless magnetometer sensors in Okla-homa state, Transport. Res. Rec. **2672**(19), 102–111 (2018). https://doi.org/10.1177/0361198118791612
16. Barbagli, B., Manes, G., Facchini, R., Marta, S., Manes, A.: Acoustic sensor network for vehicle traffic monitoring. In: VEHICULAR 2012: the First International Conference on Advances in Vehicular Systems, Technologies and Applications, 2012, pp. 1–6. May. http://www.thinkmind.org/index.php?viewarticle&articleidvehicular_2012_1_10_60034
17. Brindle, B.: How does Google maps predict traffic? (2020) https://electronics.howstuffworks.com/how-does-google-maps-predict-traffic.htm
18. Broccardo, L., Culasso, F., Mauro, S.G.: Smart city governance: exploring the institutional work of multiple actors towards collaboration. Int. J. Public Sect. Manag. **32**(4), 367–387 (2019). https://doi.org/10.1108/IJPSM-05-2018-0126
19. Camero, A., Alba, E.: Smart City and information technology: a review. Cities **93**(May), 84–94 (2019). https://doi.org/10.1016/j.cities.2019.04.014
20. Castelnovo, W., Misuraca, G., Savoldelli, A.: Smart cities governance: the need for a holistic approach to assessing urban participatory policy making. Soc. Sci. Comput. Rev. **34**(6), 724–739 (2016). https://doi.org/10.1177/0894439315611103
21. Cheung, S., Coleri, S., Dundar, B., Ganesh, S., Tan, C.-W., Varaiya, P.: Traffic measurement and vehicle classification with single magnetic sensor. Transport. Res. Rec.: J. Transp. Res. Board **1917**(January), 173–181 (2005). https://doi.org/10.3141/1917-19

Internet of Things Enabled Technological Devices Empowering Expertise in Improve Smart City Operations

Parimal Kumar Giri and Chandrakant Mallick

Abstract Innovative urban communities are constructed upon intricate socio-specialized frameworks that integrate human actors and technical apparatus. The widespread adoption of the Internet of Things (IoT) has facilitated the global expansion and technological advancement of smart city projects and associated efforts. This chapter looks at relevant experiences, urban activities, IoT-enabled amenities, and consequences to assess the current state of Internet of Things (IoT)-enabled Smart Cities. During this period, many smart city initiatives have been launched, primarily as proof-of-concept endeavours, but also progressively expanding into permanent, production-level implementations that enhance urban operations and enhance the well-being of individuals. The IoT and its breakthroughs in technology have greatly influenced the creation of smart cities, making them a vital source of inspiration in this chapter. The most important industrialized skills have been discussed, along with how they enhance living in the Smart City.

Keywords Smart city · ICT solutions · Industrialised skills · Internet of things · Machine learning · Socio-specialized · Computerized gadgets · Internet of everything

1 Introduction

In the current smart city landscape, the advancement and widespread adoption of internet of everything (IoE) and internet of things (IoT) technologies play a important role in advancing the notion of smart cities to a large-scale data size. In fact, 3.0 billion nodes will be connected by 2022, as demonstrated in [1] based on study literature. According to a 2019 Statista Exploration report, it is estimated that by 2025, there will be 75 billion interconnected IoT devices. This might result in an annual

P. K. Giri (✉) · C. Mallick
Department of Computer Science and Information Technology, GITA Autonomous College, Bhubaneswar, India
e-mail: Parimal.6789@gmail.com

S. N. Mohanty et al. (eds.), *Explainable IoT Applications: A Demystification*, Information Systems Engineering and Management 21,
https://doi.org/10.1007/978-3-031-74885-1_11

financial effect of 11 trillion USD [2]. These numbers so indicate that IoT would be among the most challenging advancements, presenting new opportunities, possible repercussions, and Challenges in developing exceptional services and apps. IoT is becoming more and more important in the creation of smart urban communities, as it addresses a primary concern for enhanced strategic progress and the establishment of a sustainable trajectory of events [3].

Smart metropolitan areas should be able to progress by adding new features and capabilities while requiring less human interaction overall. This is what creating and managing Internet of Things systems are all about. Focusing on the social difficulties that have been resolved and the cultural advantages that have resulted from acceptance of these advances is equally essential [4]. A number of the most challenging technological problems for contemporary IoT-enabled smart urban communities are addressed by main requirements; these encompass the provision of assistance to several diverse information providers, the management of a broad spectrum of conventions and information architectures, the facilitation of component sharing, and the assurance of adaptability and interoperability [5]. IoT is necessary for smart cities.

Enhancing Resource Administration

They maximize resources like water, electricity, and communication by utilizing technologies such as Machine Learning and IoTs. This may result in lower expenses, less waste, and more productivity.

A Higher Standard of Living

In numerous ways, they raise the standard of living for the populace. In order to improve public safety, education, and healthcare, for instance, they can leverage data and technology. Citizens may find it simpler to engage in civic life and obtain information thanks to them.

A Higher Level of Sustainability

Smart cities are made to be less wasteful, encourage the adoption of renewable energy sources and the reduction of overall energy consumption.. Cities can become more robust and the effects of climate change can be lessened.

Financial Progress

They stimulate economic expansion by attracting new customers and creating job opportunities. Moreover, they have the potential to enhance the efficiency of existing businesses, so positively impacting the local economy.

Urbanization is an endless process. It is anticipated that the global population would undergo relocation into cities at a rate of 3 million per week by 2040, accounting for a staggering 65% of all people. By 2050, population growth is predicted to bring this number to 6.3 billion. By 2030, the worldwide smart city market is expected to reach $6,061 billion in sales. In 2023, the smart utility segment held a 33% market share and encompassed various areas, including infrastructure management, water treatment, and electricity distribution [6].

It is crucial to avoid conducting additional responses, warehousing, and research when processing information in order to reduce employable costs and improve the sustainability of the urban environment. The following section presents a current and thorough overview of research literature on smart city regions, arrangements, and systems. It also includes information on significant advancements and the utilization and implementation of the IoTs that are integrated into smart city components.

2 Stages of IoT

The engineering of IoT encompasses a range of architectural structures that might vary depending on the conditions. However, generally, the design of IoT involves four main stages:

Networked Devices

The aforementioned devices are comprised of transducers, actuators, and sensors. These are the authentic devices that collect data and transmit it for processing. They can convert real amounts into electrical indicators that may be conveyed across an organization, making them suitable for receiving continuous information.

Data Aggregation

This stage is crucial because it involves turning the raw data collected by sensors into actionable insights that can be used to guide decisions. Furthermore, it integrates Information Procurement Frameworks and Web Interfaces. The transformation occurs from the basic indicators provided by sensors to more sophisticated signals.

Concluding Evaluation

In order to make information more efficient and execution-ready, this phase includes edge IT research and information management. It also involves managing and precisely locating all the devices.

Analysis of Cloud Computing

The most recent data is obtained and thoroughly analysed in data centres. They communicate and purify the data to free it from errors and omissions of any kind. Information is now ready to be returned and used to carry out tasks. Figure 1 represents the basic block diagram of IoT which comprises four phases.

The sensors, devices, actuators, and other components that collect data from the real climate, process it, and then transmit it throughout the company make up the Sensing Layer, the first stage of the Internet of Things. The Network Layer of the Internet of Things comprises information acquisition frameworks and organization entryways, constituting the second phase of this technological framework. The basic information obtained from sensors is transformed into advanced information by DAS. It also finds viruses and provides board information.

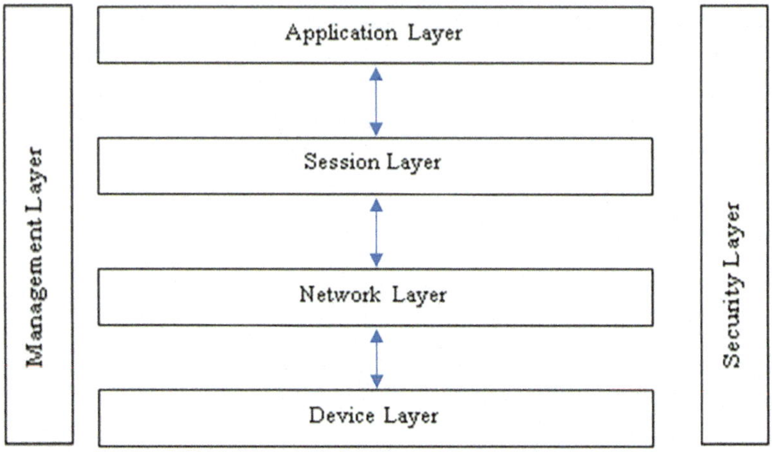

Fig. 1 IoT based smart city reference architecture

The third and primary stage of the Internet of Things is the Data Processing Layer. In this context, data is pre-processed and segregated as necessary. Subsequently, it is transported from server farms. Edge IT is employed in this context. The Application Layer, which is the fourth phase of the Internet of Things (IoT), consists of Cloud/ Server farms where data is managed and exploited by various applications such as agriculture, security, and healthcare.

3 Utilization of Smart City Technology

It's likely that you live in a city if you're like the great majority of people reading this post. Indeed, cities and metropolitan areas currently accommodate more than half of the world's population, and this percentage is projected to increase to 70% in the near future due to a growing number of individuals relocating to cities in pursuit of job opportunities [7]. However, this also means that in order for cities to be environmentally friendly, energy-efficient, and to provide a high standard of living, they must have improved infrastructure and planning. Nevertheless, this suggests that in order for cities to be environmentally friendly, energy-efficient, and to provide a high standard of living, they must have improved infrastructure and planning. Hence, let us examine the potential of civic planning and advanced technology in generating more intelligent and economically efficient cities [8]. The utilization of the Internet of Things (IoTs) presents numerous opportunities for enhancing municipal efficiency across various domains, such as smart buildings, traffic management, waste management, pollution control, disaster preparedness, and other areas [9]. Consequently, let us examine how the combination of civic planning and advanced technology might generate more intelligent and economically efficient communities.

Intelligent Governance

Smart governance oversees the integration of integrating ICT into responsible citizens administration. It shows how to promote more intelligent cooperation amongst social entertainers and partners, such as Public entities, municipal authorities, privately owned enterprises, and inhabitants, to accelerate regulatory and managerial systems and further develop dynamic interaction. E-government and citizen engagement in the process of making decisions are the most important components of smart governance [10]. The tools utilized to accomplish these include the following:

Utilizing ICT

It follows that digital tools like computers, the internet, and telecommunications are used for data collection, processing, sharing, and retrieval. Enhanced data transfer via satellite systems, cable, radio, and phone networks, among others. Transportation, finance, healthcare, energy, and security administrations use Geographic Information Systems (GIS) for texting, video conferencing, and mobility [11].

Online Consultation

People participation is the main attribute of wise governance. There must be effective communication between the public and the government. They must be granted the freedom to voice their thoughts and opinions regarding policies, plans, and other matters of government concern. Leaders, counsellors, city managers, or the local head should receive their input directly.

E-Data

Online accessibility to public data and information on government funding, investments, and expenditures is required. All information must be provided freely and publicly, with the exception of critical data pertaining to the security and safety of the public. Government transparency will increase as a result, and public participation in its operations will rise [12].

4 Models for Intelligent Governance

The following are the government models used for the software system as:

The G2C Model: Government to Citizen

This is a software system that makes connections easier between open organizations and citizens [13]. Examples include online interfaces for policy management, as well as portable apps and virtual entertainment channels that are used for communication and cooperation between residents and neighbouring states. Typically, these features are anticipated to provide the range of services offered by public entities, advice citizens about their personal information related to public administrations, and so on, thereby improving a significant amount of correspondence and communication between the public and legislative experts.

The G2B (Government to Business) Model

This paradigm respects information sharing amidst open groups and organizations. This strategy involves the use of e-obtainment arrangements, or computerized devices, to disseminate contests, tenders, projects, offices for the acquisition or disposal of commodities, and more general administrations from and for privately held enterprises to neighbourhood states. Advances in IoT are typically made in government-to-government (G2B) exercises, collaborating with neighbouring states and organizations to strengthen ties and provide inhabitants with private and public help [14].

The G2G Model: Government to Government

The approach focuses on the direct relationships between departments, agencies, and government organizations. The goal is to combine all governance channels to produce a more comprehensive and user-friendly solution. This will lead to more accountability, transparency, and efficient administration. Paperless, digital services will become the norm as a result of ICT use. Corrupt practices in government agencies will consequently decline, as will unnecessary clutter. The establishment of efficient two-way interaction between officials and residents, particularly at the urban and metropolitan levels, is necessary in order to enhance transparency as well as effectiveness in government operations [15].

G2E Model: Government to Employee

The aim of this model is to provide companies with government agencies, and employees internet tools and software for communication. It is best to keep each employee's bank account number, social security number, and other private data in a different account. Online forms are available for a variety of worker-related tasks, such as banking, clinical compensation, annuity plans, timely assets, and bank credits [16].

5 Smart Existence and Physical Structures

The term "smart living space" refers to all aspects of developing more brilliant city frameworks as well as managing and enhancing public administrations, such as social activities that are the tourism industry, and learning are key components that contribute to enhancing the overall quality of life for residents.

Intelligent Structures

IoT permits the quickly developing execution of numerous sorts of offices for brilliant structures, for instance, water waste, video reconnaissance, human action checking, security frameworks for monitoring confirmed entry to buildings, executive cooling, and occasion alarms, these include gas leaks and fires, as well as tools for determining if buildings are structurally upright [17] and other things. The home and systems space encompasses a range of IoTs improvements, which fluctuate based on their use case or environment [18].

Smart Homes

One item that could greatly simplify our lives is a smart house foundation. The limits of IoT for smart home applications end where our imagination does. A smart house can include anything we want to computerize or require to simplify our lives, including the framework for a smartphone. Currently, a smart residence typically serves as the foundation for a smart city. The smart city represents advancement in the concept of a smart house. At this point, state backing is also a critical component of a smart city, and we firmly believe that if states adopt this action, smart cities will fully integrate the Internet of Things.

Intelligent Indoor Thermostats

Temperature-detecting and temperature-controlling indoor regulators are a feature of smart houses. This manages the flow.

Localized Intelligent Devices

Intelligent devices can track you and swiftly communicate with many devices to function. Smart indoor regulators, for example, have the ability to track your location from a smart car and turn on the air conditioning before you arrive.

Voice-Activated Technology

These devices are capable of translating human speech into text that computers can understand. The basic task is then completed by machines. shrewd security architectures IOT-based security frameworks make use of features like iris scanners, facial recognition, and other security modes.

Recognizing Someone with a Face

One of the IoT's most exciting uses is this. Facial recognition models use the features of the face—such as the lips, noses, and jawlines—to predict an individual's outcome. The machine is also ready for advancement or sent for more preparation based on the accuracy.

Movement Location

Following the implementation of these modifications, the model is subsequently dispatched for further examination. Subsequently, these modifications in behaviour are exported from the model for further examination.

The Utilization of Fingerprint Authorization

Fingerprints have emerged as the essential foundation of safety mechanisms in any enterprise. They are readily controllable and devoid of difficulties. The professionals and personnel primarily record their fingers or iris marks on their faces, while the system records their attendance. This strategy provides both time and cost efficiencies.

Intelligent Living Systems

The utilization of Internet of Things (IoT) devices is prevalent across diverse domains and endeavours, facilitating the enhancement of personal fulfilment among educated individuals [19]. In order to efficiently manage and enhance the user experience of its collaborators, the travel industry is incorporating many technology breakthroughs, including smartphone apps, GIS-enabled and location-based offerings, multimedia infrastructure, virtual and augmented reality technologies, and entertainment via the internet [20].

6 Transportation and Intelligent Mobility

Intelligent transportation and urban traffic management systems are transforming the approach of urban communities towards adaptability and emergency response, while also reducing congestion on city highways. The practice and examination of transitioning from one location to another is an inherent aspect of human existence, both in the present and throughout history. Moving is an inherent aspect of human existence, encompassing various modes such as chariots, horses, carriages, automobiles, steam trains, and shuttles. Human civilization has achieved significant advancements in the use of ponies and camels for transportation between different locations [21]. Intelligent public transportation has been introduced as an emerging field stage of advancement with the emergence of automated public transportation and the rise of the Internet of Things (IoT). If the concept appears unclear or evokes mental imagery of autonomous vehicles with the ability to fly and high-speed tubes resembling hamsters, take a seat and unwind. This article aims to elucidate the concept of smart transportation, its operational mechanisms, and the numerous benefits it offers, along with an examination of some current operational models.

In addition, we will explore the many types of intelligent transportation systems currently being used. Vehicle demand management (TDM) plays a crucial role in the worldwide advancement of smart city initiatives and smart portability [22]. The role of innovation is expected to grow as ICTs continue to be integrated in urban areas in intelligent ways. Due to the rise of the sharing economy, innovative transportation methods, and other application-based mobility services, people now have more choices for transportation than ever before [23].

The issues confronting organizers and major organizations are upon enhancing productivity and facilitating employees' comprehension and implementation of these novel choices in a manner that is both secure and environmentally friendly [24]. While innovation is typically deemed indispensable for the advancement of smart mobility, it is imperative to cultivate and execute intelligent initiatives and tactics to bolster its implementation. Considering these aspects, it is necessary to explore a few emerging concepts related to intelligent mobility and intelligent transportation. The benefits of smart innovation and its advantages for mobility within a smart city are many. Some of them include the following.

More Secure

Intelligent machines (AI) and the internet of Things (also known as IoT) combination, and 5G technology in intelligent transportation systems (ITSs), encompassing both fixed-base and mobile infrastructure, has demonstrated the ability to mitigate the "human variable" in accidents. PCs do not become occupied, fatigued, or in close proximity to one's residence [24].

Well Managed

The gathering and organization of data are of the utmost significance in the efficient administration of infrastructure. Smart transportation offers comprehensive data on every aspect of the public transportation system, allowing managers to efficiently oversee operations, track repairs required, and pinpoint the underlying causes of difficulties that need to be resolved.

More Effective

Enhanced administration leads to increased efficiency in utilization. Accurate and reliable information can be important in identifying specific areas where production can be enhanced to a higher degree. Perhaps implementing a minor modification in train schedules could improve occupancy rates, or alternatively, transport routes could more effectively cater to the local region in case of unforeseen stop designations.

Cost-Effective

In light of the utilization of existing resources, intelligent transportation possesses the capacity to mitigate expenses by means of proactive maintenance, diminished energy usage, and lower allocation of resources towards accidents. Reduced expenses can be achieved by riders when economical public transportation demonstrates a level of efficiency that is comparable to that of private vehicle ownership.

Provides Rapid Insights

In order to enhance their ability to effectively collaborate with additional groups and emergency responders, city traffic enforcement communities (TMCs) can benefit from timely visibility and notifications pertaining to traffic jams or major problems that affect roadways in cities, public health, and responding to emergencies.

7 Intelligent Financial Plan

Smart Budget is a system that leverages modern information and communication technology (ICT) to establish connections between aboriginal and global markets. It offers e-commerce and e-business features that enhance the efficiency of delivery and throughput [25]. This domain also covers the concept of a distributed budget, whereby individuals, who are otherwise private enterprises, provide services by employing their personal assets and reliable marketplaces. Another alternative is the utilization of peer-to-peer labour facilities, where individuals and shareholders provide their services and engage in specific task assignments [26]. The application of computational intelligence (AI) and learning technologies has significantly improved both forecasting algorithms and systems for recommendations specifically tailored for online shopping as well as internet purchasing [27]. Wireless sensor technologies and NFC have made the payment and transaction processes stress-free. Smartphones and mobile phones are taking the place of cash and bank cards in everyday transactions and information access [28].

8 Intelligent Industry and Manufacturing

The concept of smart industry and business 4.0 is a transitional phase characterized by progress in the Internet of Things, digital physical structures (CPS), M2M technology for communication, and cloud-based production [29], which facilitate a more inventive and less reliant learning atmosphere. Regarding the mechanization of product supply chains, it is possible to effectively track their progression from the manufacturing system to targeted distribution through the utilization of sensor technologies such as RFID and NFC. In addition to the assessment of the items' quality and use, ongoing data can be collected and analysed for distribution in the future [30]. This includes the integration of advanced horticulture and cultivation techniques, addressing the challenge of establishing a sustainable food supply. Insightful horticultural frameworks often employ IoT devices to enhance the efficiency of water systems [31]. Dublin Air Terminal has implemented smart industry 4.0 to replace the luggage handling procedures at the 2nd Terminal [32].

9 Smart Power

Smart electrical networks aim to reduce energy consumption by efficiently administering and incorporating diverse and eco-friendly sources of energy. ICT and Network of Things, systems are employed in smart grids to enhance the management of energy generation and delivery. Examples of these technologies include prediction models based on aggregated consumption data and frequently provide self-restoration of the power grid supply [33]. Smart grids that adapt to consumption and availability facilitate smoother power distribution. These days, it's possible to predict future power consumption, switch to unconventional energy sources automatically, and determine the price and availability of power in this way.

Sensors such as light dependent resistors (LDRs), light beam brightness sensors, and energy and solar radiation usage sensors are among the several Internet of Things

(IoT) sensors utilized within the realm of smart energy [34]. Smart energy management has advanced through the scheduling of the consumption of power in both home and commercial environments by nations such as Nice, France, among others [35]. These devices are equipped with photometer sensors that accurately measure the strength of the emitted beam from the lamps. Additionally, these sensors verify the proper functioning of the bulbs [36]. Energy optimization platforms control household appliances and, when available, convert to solar and battery power. Smart grids facilitate energy conservation use in Helsinki by 15% [37]. The municipality has devised a solar-powered rooftop array with solar panels with a combined output of more than 10 MW. This can supply energy to about 40,000 people when combined with wind energy collection systems [38].

10 Intelligent Surroundings

The smart ecosystem encompasses the collection of ecological data, the surveillance and examination of pollution levels, the tracking of water quality and resources, and the surveillance of climatic and environmental occurrences for executives. Moreover, due to its consideration of diverse weather influences, effective waste disposal is widely recognized. The waste production processes is facilitated by intelligent waste containers equipped with sensors, which are specifically designed to provide ongoing monitoring of the available resources [39]. The strain for analysing the water's use rate is computed using a mixture of electrical and ultrasonic measurements [24]. The utilization of Wireless Sensor Networks (WSNs) in both quantity and quality of water monitoring systems has facilitated the development of more sophisticated water observation frameworks, enabling more sophisticated governance of the environment and continuous engagement.

Technologically proficient climate applications and services typically depend on external and artificial sensing devices to quantify real quantities that convey ecological parameters and situations such as temperature, pressure, humidity, and various pollutants. Advanced technologies such as satellites and LiDAR have proven to be valuable tools in the fields of land use and greenhouse gas (GHG) emissions [24]. The utilization of cutting-edge IoT technologies is employed to evaluate enhanced process efficiency and reduced greenhouse gas emissions [27]. The Green City Watch is an artificial intelligence station located in Amsterdam that employs computer-based intelligence algorithms and satellite imagery to continuously monitor urban green infrastructure [18]. In Stockholm, the implementation of sunlight-based regulated trash containers has been initiated. Busan, a smart city in South Korea, employs intelligent water management systems throughout its urban water cycle.

11 Concept of Smart Healthcare

During the ongoing Coronavirus pandemic, there has been widespread use of IoT developments and omnipresent computing to provide the deployment of adaptable healthcare solutions for remote monitoring, telemedicine, including and remote nursing [29]. These firms collaborate to integrate diverse information sources in order to protect patients' biometric and physiological data with IoT medical care submissions [2]. Brilliant clinics utilize IoT innovations to give aid for patients and clinical personnel through the identification and verification of patients in emergency clinics, as well as the management of clinical equipment that support dynamic cycles in clinics [13]. Guidelines such as High Level Seven (HL7) and PACS-DICOM in biological picture management [22] are clearly required by these application sectors and their accompanying needs.

Singapore has provided support for the Health Hub stage, which comprises clinical data about patients and residents as well as executive health records [3]. A continual finding framework (RTLS) was employed by the Helsinki College Medical Clinic to gather and disseminate anonymized area data regarding on-site advancements for monitoring specific areas during the coronavirus pandemic [15]. Furthermore, cloud administrations are provided to allow caretakers and medical professionals to interact with patients who have coronaviruses from a distance.

12 Smart Cities Enabled Technology

There are other issues that arise, including transportation congestion and the loss of energy and resources, and so on, are emerging worldwide as population growth and metropolitan concentration pick up speed. To address these problems, countries are turning to IoT-enabled smart cities. Nevertheless, it is evident that the progress of urban development is still incomplete and presents some challenges that must be addressed. Thus, the authors of this study have accumulated numerous experiences on smart urban areas by enhancing smart city administrations and infrastructures in accordance with global IoT standards [17]. Based on these experiences and the analysis conducted in this research, certain challenges faced by successful urban communities are deduced.

13 Utilisation of Silos

Most existing smart city programs were industrialized in silos and often combined robustly essential data resources and applications, particularly sensor devices. Due to necessity of each application being explicitly unified with every possible information source and any technological change, such an upgrade to the fundamental sensing apparatus network infrastructure [38], requires a technological adaption, it is challenging to exchange and reuse information as a result. It can be challenging enough to deploy a true Internet of Things infrastructure for a given application, but the establishment of a similar or equivalent configuration for every application that comes along. Infrastructure sharing can foster collaborations, and in many cases, it is the only way to enable infrastructure deployment. To do this, it is vital to separate the applications from the information sources.

One possible approach to achieve this objective is the implementation of an Internet of Things framework and the utilization of least in formativeness techniques [39]. Applications are sometimes regarded as being "possessed" by urban subdivisions or administrations, who seek to maintain authority and have made financial investments in the infrastructure. Furthermore, distribution is made difficult by the frequent strict allocation of financial plans. Therefore, it is imperative to have a global perspective, set high-level objectives, such as citywide ones, and collaborate across administrative units to accomplish these objectives—thereby repeatedly stifling personal rivalries.

Overview of Adaptable IoT Frameworks

Every smart city now has a unique appearance in terms of the applications that are available and the infrastructure that is in place, additionally, it is common for IoT systems to be utilized. This phenomenon is a hindrance to the growth of the smart city market and presents difficulties in reclaiming established elements from one city to another. Moreover, the cost feasibility of creating personalized platforms for each location is limited. Numerous merchants have initiated the provision of Internet of Things (IoT) infrastructures tailored for smart cities, typically drawing inspiration from the existing framework implemented in a pilot city, as previously outlined. The establishment of consensus over the connections as well as data modelling designated by OASC as the foundational interoperability approaches is of utmost importance.

Potential for Entrepreneurship

The digital revolution is creating new opportunities in addition to increasing the range of urban amenities that are typically provided to residents, typically overseen by local government officials in order to incorporate efficiency and sustainability [40]. The adoption of the smart city concept is crucial in fostering the development of novel business models through the facilitation of data and service exchange. Cities are increasingly serving as the arena for the sharing economy. The rapidity of money transfers and the widespread accessibility of specialized knowledge in urban areas enable the efficient and location-specific exchange of goods and services among citizens, businesses, and business-to-consumer (B2C) interactions.

By utilizing information as the new fundamental tool for conducting business in the recently created digital marketplaces, creative business concepts such as [41]:

(1) Promotes collaborative relationships—as opposed to the typically hierarchical structure and legally enforceable relationships.

(2) Establish new value allocation and management within the local community where profit is not the primary focus.
(3) Leverage a publicly accessible infrastructure that upholds privacy, resulting in a stock of business capital that is typically vulnerable and conducive to entry, replication, and imitation.

That being said, data distribution is no little matter, especially if it isn't covered by current agreements. As such, attention and taking this perspective into account are essential while making agreements, especially when revisiting exercises. The data should be adaptable, yet there is frequently a discrepancy between the data's apparent value and its actual value. An authentic value must be linked to adaptation; this is similar to the case with online advertising, where a financial value is directly linked to an expected level of usage.

Internet for Multiple Organizations

Urban areas do not exist in isolation; rather, they typically coexist with surrounding cities and are integral components of areas, jurisdictions, districts, and nations. Numerous applications necessitate operation within urban boundaries in order to sustain their operations. It is not necessary for individuals to revise their applications or request alternative ones when they relocate across city limits, as they may engage in employment in a single town, live in a different one, and allocate their leisure time in a third city. The app should possess the capability to assist users in locating a place to park by considering traffic conditions, regardless of their current location. Given the involvement of various stakeholders and the requirement for the federated system to operate across businesses, it is imperative to establish clear duties and distribute money. Therefore, technically, such configuration is feasible.

Strive for Transparency and Acknowledgement

The perspective of smart metropolitan networks has long been influenced by the sales pitches of developers and system integrators. In this context, the presence or absence of fundamental social distinctions worldwide holds significant importance in establishing platforms for individuals to express their perspectives in the ongoing debate surrounding IoT technology. In order for magnificent urban institutions to have the ability to genuinely influence residents, and can be observed by all of the collaborators and performers drawn in by the perplexing natural frameworks of metropolitan networks [39]. Moreover, it is of great significance to challenge the viewpoint about the decline in safety in order to establish a framework that can be addressed by various municipal resources, as predicted in light of the present circumstances and the data they generate.

Therefore, intelligent urban areas must provide skilled professionals who are knowledgeable about the stimulating forces and benefits of participating, ultimately ensuring the effectiveness of the entire system and the urban environment. It is crucial to promote a security-by-plan approach that enhances the transparency of all IoT asset management processes. These responses are in accordance with the requirements of the providers and the requests of the purchasers.

The Process of Reconciling Future Organizations

With the completion of new Smart metropolitan organizations, the number of devices associated with and integrated into the organization. Various IoT services should adhere to essential requirements, such as a substantial data transfer capacity and minimal inactivity [21]. The conventional support providers' distributed computing tactics necessitate a central Internet of Things (IoT) stage to effectively manage the diverse data generated by the devices. This has led to a rapid increase in the backhaul data traffic, resulting in a decrease in administration speeds [32].

Exemplary Figuring Models

For instance, various associations such as 3GPP and ETSI, as well as IoT normalization associations like M2M, acknowledge the importance of edge registration normalization. This recognition aids different organizations in establishing guidelines and enhancing similarity.

Guidelines Pertaining to Data in Diverse Nations

Due to the substantial amount of data collected and utilized by a smart city, there is undoubtedly a significant risk of personal data breaches. For example, the utilization of sensing equipment on IoT devices enables the collection of data within residential settings. It is also possible to remove sensors from a smart car to observe the driver's driving habits of data mining techniques using evidence that can elicit explicit people's IDs [42]. Europe implemented GDPR to regulate the protection and utilization of personal data [43]. Given that this law is applicable to all frameworks responsible for managing information, it is generally expected that information-based intelligent urban environments would comply with it. These claims should be replicated as system essentials and implanted into stage features [44]. This is why the entry approval component of the present stages needs to be expanded to include the ability to consciously obtain customer consent and organize such data inside a technologically advanced urban framework. Figure 2 displays the feature model that was produced for IoT-based smart city systems. A comprehensive elucidation of the distinct characteristics and the interconnectedness is presented in [6].

14 Conclusions

This chapter highlights the current availability of technology empowered agents, which now being distributed in metropolitan areas and integrated with certified public administrations. As we've shown, intelligent urban communities are made possible by significant advancements on the specialist side. Given the need for introduction projects, budgets may need to be adjusted. These factors could potentially be offset by future investment dollars, but not necessarily at the location where the initial investments were made. As we've shown, intelligent urban communities are made

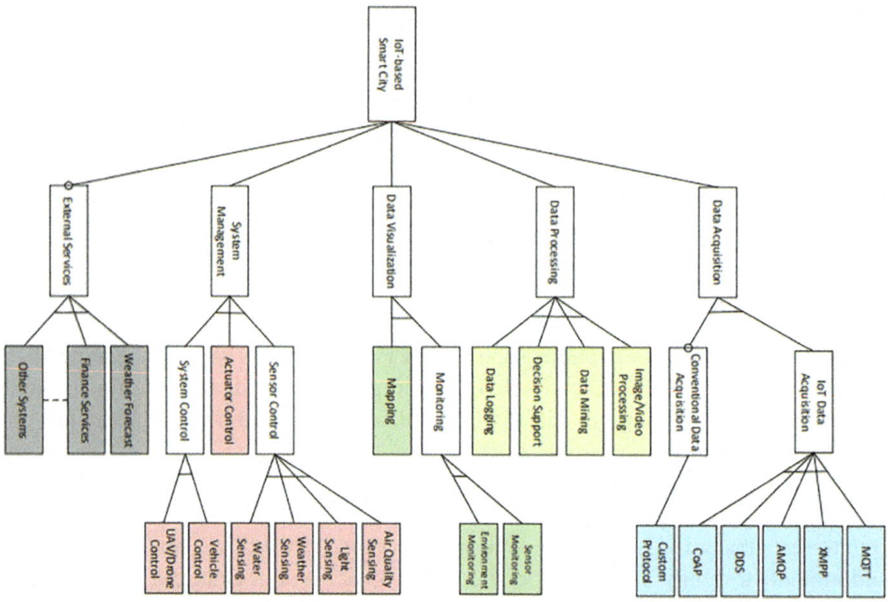

Fig. 2 Feature model for IoT-based smart cities [6]

possible by significant advancements on the specialist side. Given the need for introduction projects, budgets may need to be adjusted. If further pieces of information are obtained, it may be necessary to make future actions. Nevertheless, as their operations become more efficient and their citizens' levels of personal happiness rise, wise urban regions will provide significant benefits in the long run. Therefore, it is beneficial to address the obstacles, both at the technical and management levels, as well as within the hierarchical structure.

References

1. Ejaz, W., Anpalagan, A.: Internet of Things for Smart Cities: Technologies, Big Data and Security. Springer, Berlin/Heidelberg, Germany (2019)
2. Bauer, M., Sanchez, L., Song, J.S.: IoT-enabled smart cities: evolution and outlook. Sensors **21**, 4511 (2021)
3. Janani, R.P., Renuka, K., Aruna, A., Lakshmi Narayanan, K.: IoT in smart cities: a contemporary survey. Glob. Transit. Proc. **2**, 187–193 (2021)
4. Badii, C., Bellini, P., Difino, A., Nesi, P.: Sii-mobility: an IoT/IoE architecture to enhance smart city mobility and transportation services. Sensors **19**, 1 (2018)
5. Barcevicius, E., Cibaite, G., Codagnone, G., Gineikyte, V., Klimaviciute, L., Liva, G., Matulevic, L., Misuraca, G., Vanini, I.: Exploring Digital Government Transformation in the EU. Publications Office of the European Union, Luxembourg, JRC118857 (2019)
6. Tekinerdogan, B., Köksal, Ö., Çelik, T.: System architecture design of IoT-based smart cities. Appl. Sci. **13**, 4173 (2023)

7. Shi, Q., Zhang, Z., He, T.: Deep learning enabled smart mats as a scalable floor monitoring system. Nat. Commun. **11**, 4609 (2020)
8. Lombardi, M., Pascale, F., Santaniello, D.: Internet of things: a general overview between architectures, protocols and applications. Information **12**, 87 (2021)
9. Tang, S., Shelden, D.R., Eastman, C.M., Pishdad-Bozorghi, P., Gao, X.: A review of building information modeling (BIM) and the internet of things (IoT) devices integration: present status and future trends. Autom. Constr. **10**, 127–139 (2019)
10. Kezai, P.K., Fischer, S., Lados, M.: Smart economy and startup enterprises in the Visegrád Countries—a comparative analysis based on the crunch base database. Smart Cities **3**, 70 (2020)
11. Bellini, P., Nesi, P., Palesi, A.L.I., Pantaleo, G.: Fashion retail recommendation system by multiple clustering. In: Conference on Visualization and Visual Languages (DMSVIVA), p. 202 (2021)
12. Mallick, C.K., Giri, P.K., Paikaray, B.K., Mishra, S.N.: Machine learning approaches to sentiment analysis in social networks. Int. J. Work Innov. (IJWI) Inderscience **3**(4), 317–337 (2022)
13. Sethi, P., Sarangi, S.R.: Internet of things: architectures, protocols, and applications. J. Electr. Comput. Eng. 9324035 (2017)
14. Lopes, S.F., Pereira, R.M., Lopes, S.O., Coutinho, M., Malheiro, A., Fonte, V.: Yet a smarter irrigation system. Lect. Notes Inst. Comput. Sci. Soc. Inform. Telecommun. Eng. LNICST **323**, 337–346 (2020)
15. Six Real World Examples of Digital Transformation. https://www.smartindustry.com/blog/smart-industry-connect/six-real-world-examples-of-digital-transformation/
16. Shirazi, E., Jadid, S.: Autonomous self-healing in smart distribution grids using multi agent systems. IEEE Trans. Ind. Inform. **15**, 6291–6301 (2019)
17. Silva, B.N., Khan, M., Han, K.: Towards sustainable smart cities: a review of trends, architectures, components, and open challenges in smart cities. Sustain. Cities Soc. **38**, 697–713 (2018)
18. Zanella, A., Bui, N., Castellani, A., Vangelista, N., Zorzi, M.: Internet of things for smart cities. IEEE Internet Things J. **1**, 22–32 (2014)
19. Sankaran, V., Chopra, A.: Creating global sustainable smart cities. J. Phys. Conf. Ser. **1706**, 012141 (2020)
20. Belli, L., Cilfone, A., Davoli, L., Ferrari, G., Adorni, P., Di Nocera, F., Dall'Olio, A., Pellegrini, C., Mordacci, M., Bertolotti, E.: IoT-enabled smart sustainable cities: challenges and approaches. Smart Cities **3**, 52 (2020)
21. Ramirez-Moreno, M.A., Keshtkar, S., Padilla-Reyes, D.A., Ramos-Lopez, E., García-Martinez, M., Hernandez-Luna, M.C., Mogro, A.E., Mahlknecht, J., Huertas, J., Peimbert-Garcia, R.E.: Sensors for sustainable smart cities: a review. Appl. Sci. **11**, 8198 (2021)
22. Quadar, N., Chehri, A., Jeon, G., Ahmad, A.: Smart water distribution system based on IoT networks, a critical review. In: Human Centred Intelligent Systems. Springer, Berlin/Heidelberg, Germany, pp. 293–303 (2020)
23. Martinez, R., Vela, N., El Aatik, A., Murray, E., Roche, P., Navarro, J.M.: On the use of an IoT integrated system for water quality monitoring and management in wastewater treatment plants. Water **12**, 1096 (2020)
24. Huseien, G.F., Shah, K.W.: Potential applications of 5G network technology for climate change control: a scoping review of Singapore. Sustainability **13**, 9720 (2021)
25. Nitoslawski, S.A., Galle, N.J., Van Den Bosch, C.K., Steenberg, J.W.N.: Smarter ecosystems for smarter cities? A review of trends, technologies, and turning points for smart urban forestry. Sustain. Cities Soc. **51**, 101770 (2019)
26. Ghazal, T.M., Hasan, M.K., Alshurideh, M.T., Alzoubi, H.M., Ahmad, M., Akbar, S.S., Al Kurdi, B., Akour, I.A.: IoT for smart cities: machine learning approaches in smart healthcare—a review. Future Internet **13**, 218 (2021)
27. Asghari, P., Rahmani, A.M., Javadi, H.S.: Internet of things applications: a systematic review. Comput. Netw. **148**, 241–261 (2019)

28. Tian, S., Yang, W., Le Grange, J.M., Wang, P., Huang, W., Ye, Z.: Smart healthcare: making medical care more intelligent. Glob. Health J. **3**, 62–65 (2019)
29. Haak, D., Page, C.E., Deserno, T.M.: A survey of DICOM viewer software to integrate clinical research and medical imaging. J. Digit. Imaging **29**, 206–215 (2016)
30. Bauer, M., Sanchez, L., Song, J.S.: IoT-enabled smart cities: evolution and outlook. Sensors **21**, 4511 92021
31. Mallick, C.K., Giri, P.K., Mishra, S.N.: A multi-objective LGBBO algorithm for overlapping community detection in a social network analysis. Malays. J. Comput. Sci. **36**(2), 173–192 (2023)
32. Apanaviciene, R., Urbonas, R., Fokaides, P.A.: Smart building integration into a smart city: comparative study of real estate development. Sustainability **12**, 9376 (2020)
33. How Smart Hospitals Can Improve Healthcare. https://stlpartners.com/digital_health/how-smart-hospitals-can-improve-healthcare/
34. Sotres, P., Lanza, J., Sánchez, L., Santana, J.R., López, C., Muñoz, L.: Breaking vendors and city locks through a semantic-enabled global interoperable internet-of-things system: a smart parking case. Sensors **19**, 229 (2019)
35. Appio, F.P., Lima, M., Paroutis, S.: Understanding smart cities: innovation ecosystems, technological advancements, and societal challenges. Technol. Forecast. Soc. Chang. **142**, 1–14 (2019)
36. Giri, P.K., De, S.S., Dehuri, S., Cho, S.-B.: Biogeography based optimization for mining rules to assess credit risk. Intell. Syst. Account. Finan. Manage. **28**(1), 35–51 (2020)
37. Gutiérrez, V., Amaxilatis, D., Mylonas, G., Muñoz, L.: Empowering citizens toward the co-creation of sustainable cities. IEEE Internet Things J. **5**, 668–676 (2018)
38. Rao, S.K., Prasad, R.: Impact of 5G technologies on smart city implementation. Wirel. Pers. Commun. **100**, 161–176 (2018)
39. Badii, C., Bellini, P., Difino, A., Nesi, P.: Smart city IoT platform respecting GDPR privacy and security aspects. IEEE Access **8**, 23601–23623 (2020)
40. Ganayak, M., Mohanty, S.N., Jagadev, A.K.: Agricultural recommendation system for crops using different machine learning regression methods. Int. J. Agric. Environ. Inf. Syst. **12**(1), 1–20 (2021), ISSN: 1947–3192. https://doi.org/10.4018/IJAEIS
41. Liu, L., Guo, X., Lee, C.: Promoting smart cities into the 5G era with multi-field Internet of Things (IoT) applications powered with advanced mechanical energy harvesters. Nano Energy **88**, 106304 (2021)
42. Giri, P.K., De, S.S., Dehuri, S.: Adaptive neighbourhood for locally and globally tuned biogeography-based optimization algorithm. J. King Saud Univ. Comput. Inf. Sci. **33**(4), 453–467 (2018)
43. Giri, P.K., Choudhury, S., Roy, D.S., Paikaray, B.K.: Fog computing approaches for sustainable smart cities. Int. J. Reasoning Based Intell. Syst. (2024) (in Press) https://doi.org/10.1504/IJRIS.2024.10062540
44. Mallick, C.K., Mishra, S.N., Giri, P.K., Paikaray, B.K.: A meta heuristic optimization based deep learning model for fake news detection in online social networks. Int. J. Electron. Secur. Digit. Forensics Inderscience (2023) (In Press) https://doi.org/10.1504/IJESDF.2024.10057139

Enhancing Smart City Retail: An Innovative IoT Driven Smart Billing-Enabled Shopping Cart

Debasish Sahu, Swarna Prabha Jena, Sujit Mahapatra, Sujata Chakravarty, and Bijay Kumar Paikaray

Abstract The "Smart Billing enabled Shopping Cart" is an Internet of Things invention that will transform the shopping experience. It uses the ESP32 CAM, QR codes, and Arduino IDE. This project intends to improve and streamline the purchasing experience in a future where efficiency and convenience are critical. Traditional shopping carts no longer meet the modern consumer's needs, which require laborious human data entering. This paper aims to develop a shopping cart that minimises errors, streamlines data entry, and speeds up checkout. Users may scan products with a mobile app using the ESP32 Cam to affix QR codes to each item. It calculates the total cost and produces an itemised receipt. The work shows notable increases in user satisfaction and shopping efficiency. It is a viable option since it speeds up checkout, lowers error rates, and improves the shopping experience. In conclusion, the ESP32 CAM and QR code-powered Smart Shopping Cart provides a creative way to make shopping easier. It solves the drawbacks of conventional shopping carts, which eventually helps consumers. This whole work demonstrates how IoT technology can be used to optimise daily tasks in shopping malls.

Keywords Smart shopping · Remote billing · ESP32-CAM · IoT · QR-codes

D. Sahu · S. Mahapatra · S. Chakravarty
Department of Computer Science and Engineering, Centurion University of Technology and Management, Bhubaneswar, Khurda, Odisha, India

S. P. Jena (✉)
Department of Electronics and Communication Engineering, Centurion University of Technology and Management, Bhubaneswar, Khurda, Odisha, India
e-mail: prabha.jena@gmail.com

B. K. Paikaray
Siksha' O' Anusandhan (Deemed to Be) University, Bhubaneswar, Odisha, India

S. N. Mohanty et al. (eds.), *Explainable IoT Applications: A Demystification*, Information Systems Engineering and Management 21,
https://doi.org/10.1007/978-3-031-74885-1_12

1 Introduction

In the contemporary era, where time is considered the most precious commodity, the efficiency of daily activities has become a paramount concern. Shopping, an integral part of our lives, often entails an unwarranted expenditure of time, particularly during the billing process. The ubiquitous sight of long queues at checkout counters is not only time-consuming but also a source of frustration for shoppers, especially during peak seasons or festivals. Recognising this challenge, our project aims to revolutionise the traditional shopping experience by introducing a cutting-edge solution that significantly reduces the time spent at billing counters. The core objective is to empower customers to conduct their own billing seamlessly, thereby streamlining the entire shopping process.

With the integration of ESP32 CAM technology and QR code functionality, our intelligent shopping cart provides customers with a user-friendly and efficient way to estimate and settle their bills. The innovative approach not only enhances the overall shopping experience but also addresses the challenges retailers face during peak periods, such as festivals, where the influx of customers can lead to extended billing times. One of the key advantages of our smart shopping cart is its potential to reduce the reliance on manpower at billing counters. This not only optimises operational efficiency for shopping malls but also frees up valuable space that can be utilised to enhance product displays and improve the overall aesthetic appeal of the retail environment.

Beyond enhancing the billing process, our project extends its impact by incorporating a seamless payment mechanism. Customers can complete transactions by scanning a payment QR code from popular UPI apps such as PhonePe, Google Pay, and others. This integration not only adds a layer of convenience for shoppers but also aligns with the contemporary trend of digital payment preferences. In contributing to the ongoing evolution of the retail landscape, our SMARTER BILLING system strives to make the shopping experience more efficient, customer-centric, and technologically advanced. This work provides a new era of innovation in the retail sector through the amalgamation of self-service billing, digital payments, and a focus on enriching the overall shopping environment.

2 Literature Survey

A state-of-the-art initiative called Smart Shopping Cart seeks to transform the shopping experience. With cutting-edge technologies like RFID, IoT, and AI, this system is intended to improve buyers' overall enjoyment, convenience, and efficiency. This literature review has examined several important sources that provide insight into the ideas and technology behind the Smart Shopping Cart. The Smart Shopping Cart concept is centred upon a set of revolutionary technologies. The Internet of Things (IoT), artificial intelligence (AI), and radio-frequency identification (RFID) are some of the technologies that are used to develop a shopping cart that is more than just what it used to be. While IoT links the cart to a network and allows real-time data sharing, RFID tracks objects and their movements. Conversely, artificial intelligence (AI) infuses automation and intelligence into purchasing.

The RFID-based Smart Shopping Cart idea is introduced by [1]. This system uses RFID technology to monitor and identify things in the cart. With the cart, shoppers can easily keep track of the things they have chosen, and the shopping process can be more smoothly managed. This reference examines the benefits and distinctive features of this system. The system [2] offered in "ShopNDrop Using Smart Trolley" adds novel characteristics to the shopping trolley, contributing to the literature. The cited work presents new algorithms and developments that enhance the Smart Shopping Cart's functionality. To grasp the state-of-the-art in this subject, use this reference extensively. Karjol [3] provide details on an Internet of Things (IoT)-based Smart Shopping Cart that uses the Internet of Things capabilities to make shopping easy. IoT technology enables automation, individualised suggestions, and data sharing by establishing a connection between the cart and the larger retail environment. The reference provides insightful viewpoints on how the Internet of Things affects purchasing. It also presents a Smart Shopping trolley system [4] that combines RFID and IoT capabilities. This source describes how IoT provides connection and real-time data processing, while RFID technology helps with item tracking. Combining these technologies makes shopping more convenient for customers and increases process efficiency [5].

An inventive method [6] for creating smart shopping carts is presented about Intelligent shopping carts using Bolt ESP8266 based on IoT. This system shows how IoT technology may enable linked shopping carts to provide extra features and functionality by using the Bolt ESP8266 platform. ESP8266 integration creates new opportunities for intelligent shopping carts [7, 8].

One crucial area of research [9] is integrating artificial intelligence with smart shopping carts. A forward-thinking system that investigates the combined effect between AI and IoT. This source demonstrates how artificial intelligence (AI) may improve judgment, provide suggestions, and create a more thoughtful and tailored buying experience. The system described in "The IoT-Based Smart Shopping Trolley System" illustrates how smart trolley technology is developing with continuous improvements. This reference shows how the integration of IoT is continuing to alter shopping by providing insights into the latest advancements and trends [10].

Integrating machine learning and vision [1, 11] into these types of carts has also been investigated. The above reference delves into how these technologies facilitate item recognition, manage inventory, and enhance the shopping experience. Ryumin [12] examines the relationship between robotics and smart shopping carts. Offers a novel method for integrating technology into shopping by exploring the possibility of human–robot interaction to improve the capabilities of shopping carts.

In [13] Mobile apps are another Smart Shopping Cart initiative aspect. The reference discusses how mobile apps provide ease and smooth cart integration to improve the purchasing experience. RFID-based shopping cards are being developed [14–16].

Some work has also been done for real-time product tracking and inventory management using IoT [17, 18]. Also, an AI-powered shopping cart was developed for customer Behavior analysis [19], product recommendation [20, 21], recommendation behaviour analysis [22], and Fraud Detection [23, 24].

The literature review has illuminated several aspects of Smart Shopping Cart technology, ranging from AI, machine vision, and human–robot interaction to RFID and IoT integration. The development of the Smart Shopping Cart project is guided and inspired by the full awareness of current research and advances in the area that these references together give.

3 Proposed Methodology

The methodology employed in this research aims to comprehensively evaluate the effectiveness and practicality of the "Smart Billing enabled Shopping Cart" in revolutionising the shopping experience. Leveraging the Internet of Things (IoT) framework, the research investigates the integration of key components such as the ESP32 CAM, QR codes, and Arduino IDE to develop a sophisticated and user-friendly shopping cart system.

3.1 Research Design

This study adopts a multifaceted approach, encompassing quantitative and qualitative methods to understand the Smart Shopping Cart's functionality and user experience. The research design allows for exploring various facets, including technical performance, user satisfaction, and efficiency improvements. The study aims to generate comprehensive findings that inform technical enhancements and user-centric design improvements by combining quantitative data analysis with qualitative insights.

3.2 Data Collection

This study adopts a multifaceted approach, encompassing quantitative and qualitative methods to understand the Smart Shopping Cart's functionality and user experience. The research design allows for exploring various facets, including technical performance, user satisfaction, and efficiency improvements. The study aims to generate comprehensive findings that inform technical enhancements and user-centric design improvements by combining quantitative data analysis with qualitative insights.

1. **Quantitative Data**

In our data collection process, we meticulously gathered quantitative insights into the performance of the Smart Shopping Cart system. This involved a series of well-defined steps to capture objective metrics and conduct a thorough analysis.

(1) **Implementation in Real-World Scenarios**

We initiated the data collection phase by deploying the Smart Shopping Cart system in various real-world shopping environments. This encompassed supermarkets, retail stores, and convenience stores to ensure a diverse representation of user interactions.

(2) **Diverse Sampling**

We strategically selected various shopping scenarios to achieve comprehensive data coverage. Factors such as store layout, foot traffic patterns, and product assortment were considered to capture a representative sample of user experiences.

(3) **Sensors and Data Logging Mechanisms**

a. Sensor Integration: The Smart Shopping Cart system was outfitted with sensors capable of capturing key real-time performance metrics. These sensors included barcode scanners, weight sensors, and motion detectors to track item scanning speed, cart movement, and checkout duration.
b. Data Logging: We implemented robust data logging mechanisms within the Smart Shopping Cart system to ensure accurate data capture. This allowed for continuous recording of user interactions, enabling detailed analysis at a later stage.

(4) **Statistical Analysis and Comparison**

a. **Data Analysis**: Following the data collection phase, collected quantitative data underwent meticulous statistical analysis. This involved employing tools like Python's NumPy and Pandas libraries to compute summary statistics, frequency distributions, and correlation analyses.
b. **Benchmark Comparison**: To contextualise our findings, we compared the performance metrics obtained from our data collection efforts against established benchmarks within the retail industry. This facilitated assessing the Smart Shopping Cart system's performance relative to industry standards and user expectations.

2. **Qualitative Data**

In addition to quantitative measurements, we employed qualitative methodologies to gather nuanced insights into user perceptions and experiences with the Smart Shopping Cart system.

(1) **User Feedback Mechanisms**

a. Surveys: Participants were invited to complete structured surveys to assess various aspects of their interaction with the Smart Shopping Cart system. Survey questions covered overall satisfaction, ease of use, and perceived usefulness of the system.

b. Interviews: In-depth interviews were conducted with select users to delve deeper into specific aspects of their experiences. These interviews provided an opportunity to explore user preferences, pain points, and suggestions for improvement in greater detail.

(2) **Observational Studies**

a. Direct Observation: Researchers conducted observational studies in real-time shopping scenarios to observe and document user behaviours first-hand. This involved observing how users interacted with the Smart Shopping Cart system, including navigation patterns, decision-making processes, and interactions with store personnel.
b. Focus Groups: Focus group discussions were organised to facilitate group interactions and uncover collective insights into user perceptions and experiences. These discussions allowed participants to share their perspectives, exchange ideas, and contribute to a deeper understanding of user preferences and expectations.

Our comprehensive data collection approach, integrating both quantitative measurements and qualitative feedback, yielded a wealth of insights into the performance and user perceptions of the Smart Shopping Cart system. By meticulously executing each step of the data collection process and employing various methodologies, we gained valuable insights that inform the refinement and optimisation of smart shopping technologies for enhanced user experiences.

3.3 Implementation

The Smart Shopping Cart system is meticulously implemented, integrating the ESP32 CAM for product scanning, QR codes for seamless identification, and Arduino IDE for data processing. The system minimises errors, streamlines data entry processes, and expedites user checkout. Iterative prototyping and user testing are conducted to refine system functionality and user interface design iteratively. Continuous integration and testing practices ensure the stability and reliability of the Smart Shopping Cart system across diverse shopping environments.

3.4 Data Analysis

Quantitative data analysis employs statistical methods to analyse performance metrics and identify patterns in user interactions. Descriptive statistics, inferential statistics, and multivariate analysis techniques may be utilised to quantify the impact of the Smart Shopping Cart system on shopping efficiency and user satisfaction. Qualitative data analysis utilises thematic coding techniques to extract meaningful insights from user feedback and observations. Integrating both approaches facilitates a comprehensive evaluation of the Smart Shopping Cart's efficacy and identifies opportunities for refinement and optimisation.

3.5 *Ethical Consideration*

Ethical considerations are paramount throughout the research process to ensure the well-being and rights of research participants. Informed consent is obtained from all participants, and their privacy and confidentiality are safeguarded. Measures are implemented to ensure transparency, fairness, and integrity in data collection, analysis, and reporting. Ethical guidelines provided by institutional review boards (IRBs) or ethics committees are strictly adhered to, mitigating potential risks and ensuring the ethical conduct of the research.

3.6 *Limitations*

Acknowledging potential limitations is crucial for maintaining the integrity and validity of the research findings. Environmental constraints, participant biases, and logistical challenges may impact the research process and outcomes. Clear documentation of limitations enhances the credibility and reliability of the research findings, providing context for interpreting results and informing future research directions. Sensitivity analyses and robustness checks may be conducted to assess the robustness of research findings and mitigate the influence of potential confounding factors.

3.7 *Validations*

Peer review and expert consultation are sought to validate the research findings and ensure the robustness of the methodology. Feedback from stakeholders, domain experts, and interdisciplinary scholars enriches the interpretation of results and strengthens the validity of conclusions drawn from the study. Peer-reviewed publication and dissemination of research findings contribute to advancing knowledge in the field of IoT-enabled systems and user-centric design. Replication studies and follow-up research efforts further validate the generalizability and applicability of research findings in diverse contexts.

4 Hardware Requirements

We present a visual depiction of the Smart Shopping Cart prototype in operation, showcasing its functionality and practicality. It is an actual illustration of our goal for a smooth and effective buying experience.

4.1 ESP32-CAM Module

Description: The ESP32-CAM module served as the core processing unit of the Smart Shopping Cart system, providing wireless connectivity and image processing capabilities.

Key Features:

1. Integrated ESP32 microcontroller with Wi-Fi and Bluetooth connectivity.
2. The OV2640 camera sensor is for capturing images and QR codes.
3. GPIO pins for interfacing with external devices and sensors.

Purpose: The ESP32-CAM module facilitated real-time image recognition and processing, enabling the Smart Shopping Cart system to identify products and extract relevant information from QR codes.

4.2 USB to TTL Converter

Description: The USB to TTL converter was an essential interface for establishing communication between the ESP32-CAM module and external devices, such as computers or microcontrollers.

Key Features:

1. Use a USB Type-A connector to connect to the host computer.
2. TTL-level serial interface for communication with the ESP32-CAM module.
3. LED indicators for power and data transmission status.

Purpose: The USB to TTL converter facilitated the programming and debugging of the ESP32-CAM module, allowing for firmware updates and troubleshooting during the development process.

4.3 USB Cable

Description: The USB cable provided the physical connection between the USB to TTL converter and the host computer, enabling data transfer and power supply.

Key Features:

1. Type-A to Type-B USB connectors.
2. High-quality shielding for data integrity and reliability.
3. Various lengths and configurations are available to suit different deployment scenarios.

Purpose: The USB cable was the conduit for transmitting programming instructions, firmware updates, and debugging information between the ESP32-CAM module and the host computer.

4.4 Adapter

Description: The adapter supplied the necessary power to the ESP32-CAM module, ensuring reliable operation in standalone or embedded configurations.

Key Features:

1. Output voltage and current ratings compatible with the ESP32-CAM module's requirements.
2. Overcurrent and overvoltage protection mechanisms for safety and reliability.
3. Compact form factor suitable for integration into the Smart Shopping Cart system.

Purpose: The adapter provided the required electrical power to the ESP32-CAM module, allowing it to function autonomously without relying on external power sources.

Table 1 outlines the detailed specifications of each hardware component to facilitate replication and deployment of the Smart Shopping Cart system. Considerations of functionality, compatibility, and reliability guided the selection of hardware components for the Smart Shopping Cart system. By leveraging high-quality components such as the ESP32-CAM module, USB to TTL converter, USB cable, and adapter, we ensured the robustness and effectiveness of the system in real-world deployment scenarios.

Table 1 Components and its description

Component	Description	Specification	Quantity
ESP32-CAM module	ESP32-CAM module	ESP32 microcontroller with OV2640 camera sensor	1
USB to TTL converter	Interface converter	USB Type-A connector	1
TTL-level serial interface			1
USB cable	Data transfer	Type-A to Type-B connectors	1
Adapter	Power supply	Output voltage: 5 V	1

5 Software Requirements

5.1 Arduino IDE

Version 2.2.1 of the Arduino Integrated Development Environment (IDE) is essential to properly executing the Smart Shopping Cart project.

5.2 Python 3.0

The Smart Shopping Cart project harnesses the power of Python 3.0 as its primary programming language, supplemented by an array of essential libraries. This section delves into the pivotal role of Python 3.0 and these associated libraries, emphasising their collective significance in the project's successful execution.

The software components of the Smart Shopping Cart are based on Python 3.0. Many advantages support its adoption, including as readability, adaptability, and a robust library environment. Python is the primary programming language used in this project for its graphical user interface (GUI), data processing, and user-friendly interaction with the Smart Shopping Cart. Python is a great option for GUI development because of its readability and versatility.

The Smart Shopping Cart's interactive face, or GUI, offers users an easy-to-use platform for managing quantities, choosing products, and completing checkout quickly. Python-compatible libraries like as Tkinter and PyQt make GUI creation easier and guarantee that users can easily traverse and interact with the cart.

5.3 OpenCV

The Open Source Computer Vision Library, or OpenCV, is essential to Python's function in this project. By making it possible to recognise and interpret visual data specifically, QR codes used to identify products, it expands Python's capabilities. The Smart Shopping Cart can scan, analyse, and decode QR codes in real time thanks to OpenCV's powerful capabilities, guaranteeing the project's accuracy and efficiency.

To install OpenCV using Python command, run the following command in your terminal or command prompt: "pip install opencv-python".

5.4 Numpy

The project utilised Numpy, a numerical operations package, to enhance Python's capabilities. Its contribution is found in the areas of array management, data processing, and sophisticated mathematical computations. Because of its significance in data processing, Numpy is an essential part of real-time product information, quantity, and price management.

To install NumPy using the Python command, run the following command in your terminal or command prompt: "pip install numpy".

5.5 Urllib.Request

The Urllib.request library serves as the project's gateway to external data sources, ensuring secure data retrieval. Within the Smart Shopping Cart, Urllib.request plays a pivotal role in acquiring data essential for secure payments via UPI scan. Its function is paramount in enabling the project to access and process data securely, reinforcing the reliability of the payment process.

To install urllib.request using Python command, no additional installation is required as it is part of Python's standard library.

5.6 Pygame

Python's functionality is further enriched by the Pygame library, a versatile tool for graphical user interface (GUI) development. Pygame facilitates the creation of interactive and visually engaging interfaces. It is a cornerstone in crafting the project's user-friendly and visually appealing GUI, enriching the shopping experience and ensuring seamless customer interaction. To install Pygame using Python command, run the following command in your terminal or command prompt: "pip install pygame". The effectiveness, precision, and user-friendliness of the Smart Shopping Cart are supported by the synergy of Python 3.0 and its related libraries, such as OpenCV, Numpy, Urllib. request, and Pygame. Every part of the project contributes in a different but complementary way to safe payments, real-time data processing, and a captivating user experience.

5.7 Tkinter for GUI

Tkinter, a standard GUI library for Python, is specifically leveraged in the development of the graphical user interface (GUI) for the project. Although Tkinter is typically included with Python installations, some systems might require additional installation steps.

To ensure the Tkinter GUI library is installed, you can run the following command in your terminal or command prompt: "sudo apt-get install python3-tk" (for Debian/

Ubuntu based systems) or sudo yum install python3-tkinter (for Fedora/Red Hat-based systems). These commands will install Tkinter and its dependencies if they are not present.

5.8 *Jupyter Notebook*

Jupyter Notebook is essential to improving the effectiveness and usefulness of the Smart Shopping Cart project, especially when combined with Conda 4.10.1 and Tkinter for GUI. An in-depth discussion of Jupyter Notebook and related elements is given in this part, highlighting the importance of these elements to the project's successful completion.

The Jupyter Notebook, Conda 4.10.1, and Tkinter combination for GUI development enhance the functionality of the Smart Shopping Cart project. Conda simplifies environment management, Jupyter Notebook facilitates data analysis and collaboration, and Tkinter ensures the graphical user interface is responsive and easy to use. Tkinter, a standard GUI library for Python, is specifically leveraged in the development of the graphical user interface (GUI) for the project. Although Tkinter is typically included with Python installations, some systems might require additional installation steps.

6 Results

The work visually depicts the Smart Shopping Cart prototype in operation, showcasing its functionality and practicality. It is an actual illustration of our goal for a smooth and effective buying experience. Beyond its visual appeal, the prototype shown in Fig. 1 is evidence of the meticulous planning, collaborative teamwork, and technological prowess invested in the Smart Shopping Cart project. It encapsulates the essence of our goal, not merely to ideate but to materialise a real-world solution that addresses the evolving needs of modern consumers. As depicted in Fig. 2, we showcase demo products integrated with QR codes shown in Fig. 3, an essential Smart Shopping Cart system component. These demo products are designed to exemplify the seamless interaction between physical products and digital technology within the Smart Shopping Cart ecosystem. Figure 4, introduces the Graphical User Interface (GUI) for the Billing System, a crucial part of the Smart Shopping Cart project. Customers may easily evaluate their chosen goods, control quantity, and begin checkout with this interface. Figure 5 illustrates the Payment System, a vital point in the Smart Shopping Cart's operation. This picture illustrates the safe and user-friendly payment method that facilitates the execution of transactions.

Fig. 1 Top and front view of functional model

Fig. 2 Functional model with sample products

7 Discussion

Integrating ESP32 CAM, QR codes, Python GUI, and UPI payment in the Smart Shopping Cart project has significant implications for the retail industry. This discussion section explores key aspects that underscore the transformative potential of our innovative solution.

(1) **Impact on Customer Shopping Behaviours**

Fig. 3 Sample products

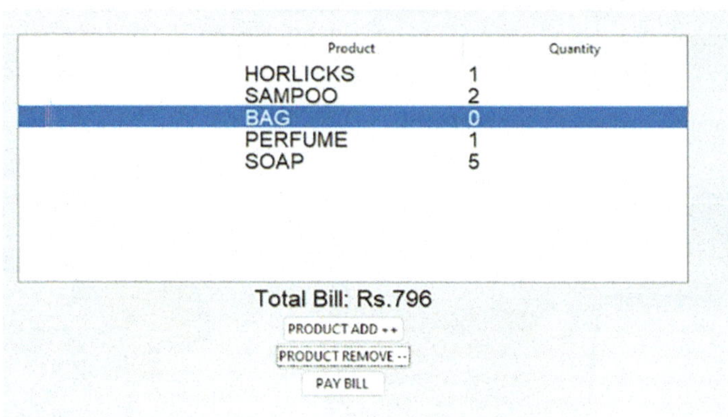

Fig. 4 Product billing system GUI

Fig. 5 Payment system

The infusion of technology into the shopping experience is poised to reshape customer behaviours. The Smart Shopping Cart makes the product identification, selection, and payment process more intuitive and efficient. Customers are likely to embrace this streamlined approach, leading to a shift in preferences towards technologically enhanced retail interactions.

(2) **Operational Efficiency and Reduced Checkout Times**

One of the primary benefits of the Smart Shopping Cart lies in its potential to enhance operational efficiency for both customers and businesses. The system significantly reduces checkout times by automating product scanning, quantity adjustments, and payment. This efficiency improvement saves customers time and contributes to a more streamlined and agile retail environment.

(3) **Secure and User-Friendly Payment Methods**

The integration of UPI payment in the Smart Shopping Cart addresses the crucial aspect of secure and user-friendly transactions. With the ease of scanning a QR code for payments, customers experience a seamless and secure payment process. adds convenience and fosters trust, a vital component in modern retail transactions.

(4) **Challenges and Considerations**

While the Smart Shopping Cart presents numerous advantages, it is essential to acknowledge potential challenges and considerations. User adoption, system scalability, and data security require careful attention. Balancing technological advancements with user-friendly interfaces will be key to overcoming these challenges and ensuring the widespread success of such innovative retail solutions.

8 Conclusion

The Smart Shopping Cart project marks a dramatic leap forward in retail experiences. Solving the persistent obstacle of checkout waits has shortened the shopping process and empowered consumers with a smooth, data-driven retail experience. Our idea utilises cutting-edge technology, anchored by the ESP32-CAM module, Python decoding, a dynamic graphical user interface, and USB-B connectors, to create an environment where convenience, efficiency, and user-friendliness merge. The following essential accomplishments emphasise the project's success:

(1) **Revolutionising Checkout**: We've redefined the shopping landscape by rendering checkout queues obsolete. With the Smart Shopping Cart, customers enjoy a checkout-free experience, reducing waiting times and enhancing overall convenience.
(2) **Data-Driven Shopping**: Our advanced technologies enable real-time data capture and processing. Product information, quantities, and pricing are at the customer's fingertips, promoting informed purchasing decisions.

(3) **User-Friendly Interaction**: The intuitive graphical user interface (GUI) ensures customers easily navigate shopping. Product selection, quantity adjustments, and secure payments via UPI scan are all accomplished effortlessly.
(4) **Secure and Transparent Payments**: Implementing secure UPI scanning for payments guarantees the safety and transparency of transactions. Real-time verification of payments underscores our commitment to a secure shopping environment.
(5) **Endless Innovation**: As we conclude this project, we stand at the threshold of a retail future brimming with innovation. The possibilities are boundless, and we are excited to explore new horizons in the retail industry.

This work with the Smart Shopping Cart project is not simply a climax but also a launchpad for the future of retail. We have tackled one of the most significant concerns in the business, paving the way for customer-centric, efficient, and data-driven shopping experiences. As we look forward, we are enthused about the unlimited potential for innovation in the retail industry. The Smart Shopping Cart is not simply a project; it's a vision for a better shopping future.

9 Future Scope

The Smart Shopping Cart project is a step forward, but its journey is far from complete. The project's success opens doors to several opportunities for future refinement and investigation. Here are some places where the project may continue to evolve:

(1) **Enhanced Data Analytics**: The project can delve deeper into data analytics to provide insights into customer behaviour, preferences, and shopping patterns. This information can guide personalised marketing and product recommendations.
(2) **Mobile Applications**: Developing mobile applications for the Smart Shopping Cart can further enhance the customer experience. Mobile apps can provide additional features like shopping lists, location-based promotions, and seamless account management.
(3) **Integration with Inventory Management**: Expanding the project to encompass real-time inventory management can benefit customers and retailers. Customers can receive accurate stock information, while retailers can optimise their supply chains.
(4) **AI and Machine Learning**: Integrating artificial intelligence and machine learning may give sophisticated capabilities like face recognition for easy entrance and personalised product suggestions based on prior purchases.
(5) **Expansion to Various Retail Sectors**: The project's framework can be applied across a spectrum of retail sectors, from supermarkets and convenience stores to apparel shops and pharmacies, amplifying its impact.

(6) **Internationalisation**: Adapting the project for worldwide markets by including multi-language support, local currency management, and compatibility with international payment methods may make it a global retail innovation.

The future of the Smart Shopping Cart is an unfolding narrative of innovation, optimisation, and expansion. We are excited to explore these possibilities, continuously striving to make the retail experience efficient and extraordinary.

References

1. Sawant, M.R., et al.: The RFID based smart shopping cart. Int. J. Eng. Res. Gen. Sci. **2**, 275–280 (2015)
2. Khan, A., et al.: ShopNDrop using smart trolley. SAMRIDDHI J. Phys. Sci. Eng. Technol. **12**, 354–356, SUP1(2020)
3. Karjol, S., et al.: An IOT based smart shopping cart for smart shopping. In: International Conference on Cognitive Computing and Information Processing, vol 1, pp. 373–385. Springer Singapore, Singapore (2017)
4. Devi, K.G., et al.: Smart shopping trolley using RFID based on IoT. Int. J. Innov. Res. Comput. Commun. Eng. **5**(3), 5392–5398 (2017)
5. Ambekar, K., et al.: Smart shopping trolley using RFID. Int. J. Adv. Res. Comput. Eng. Technol. (IJARCET) **4**(10), 3875–3877 (2015)
6. Lekhaa, T.R., et al.: Intelligent shopping cart using bolt esp8266 based on internet of things. In: 2019 5th International Conference On Advanced Computing & Communication Systems (ICACCS), vol. 1, pp. 758–761. IEEE (2019)
7. Tahiri, N., Mazoure, B., Makarenkov, V.: An intelligent shopping list based on the application of partitioning and machine learning algorithms. In: Proceedings of the 18th Python in Science Conference (SCIPY 2019), vol. 1, pp. 85–92 (2019)
8. Athauda, T., Marin, J.C.L., Lee, J., Karmakar, N.C.: Robust low-cost passive UHF RFID based smart shopping trolley. IEEE J. Radio Freq. Identif. **3**(2), 134–143 (2018)
9. Kumar, P.P., Sai, K.S., Soniya, B., Koutil, U., Sania, A., Sagar, K.: Advanced smart shopping cart with integration of AI and IoT. In: 2023 2nd International Conference on Edge Computing and Applications (ICECAA), vol. 1, pp. 1348–1353. IEEE (2023)
10. Maurya, S., Sahu, G., Yadav, A., Shukla, B., Agrawal, G., Kumar, N.: The IoT-based smart shopping trolley system. In: 2023 International Conference on IoT, Communication and Automation Technology (ICICAT), vol. 1, pp. 1–6. IEEE (2023)
11. Swain, S., Deepak, A., Pradhan, A.K., Urma, S.K., Jena, S.P., Chakravarty, S.: Real-time dog detection and alert system using tensorflow lite embedded on edge device. In: 2022 1st IEEE International Conference on Industrial Electronics: Developments & Applications (ICIDeA), vol. 2, pp. 238–241. IEEE (2022)
12. Ryumin, D., Ivanko, D., Axyonov, A., Kagirov, I., Karpov, A., Zelezny, M.: Human-robot interaction with smart shopping trolley using sign language: data collection. In: 2019 IEEE International Conference on Pervasive Computing and Communications Workshops (PerCom Workshops), vol. 1, pp. 949–954. IEEE (2019)
13. Kowshika, S., Varshini, G.M., Megha, V., Lakshmi, K.: IoT based smart shopping trolley with mobile cart application. In: 2021 7th International Conference on Advanced Computing and Communication Systems (ICACCS), vol. 1, pp. 1186–1189. IEEE (2021)
14. Chakraborty, M., Jena, S.P., Chakravarty, S.: IoT-based employee location tracking with google maps using STM32 and ESP microcontroller-A technological comparison. In: 2022 1st IEEE International Conference on Industrial Electronics: Developments & Applications (ICIDeA), vol. 2, pp. 100–105. IEEE (2022)

15. Yang, C., Chen, L., Guo, Y., Li, Z.: Smart shopping cart based on IoT and AI. Int. J. Distrib. Sens. Netw. **18**, 1–15 (2022)
16. Li, X., Gao, L., Li, T., Xu, G.: A smart shopping cart system based on RFID, IoT, and AI for smart shopping. IEEE Trans. Industr. Inf. **18**(2), 1152–1161 (2022)
17. Liu, C., Huang, G., Chen, H.: A smart shopping cart system with IoT-based real-time product tracking and inventory management. IEEE Trans. Autom. Sci. Eng. **19**(1), 198–209 (2022)
18. Zhang, J., Liu, Y., Wang, J.: A smart shopping cart system with IoT-based real-time product tracking and inventory management for smart grocery stores. IEEE Trans. Intell. Transp. Syst. **23**(12), 19398–19409 (2022)
19. Chen, Z., Huang, Q., Li, P.: A smart shopping cart system with AI-powered customer behavior analysis. IEEE Trans. Knowl. Data Eng. **34**(10), 2733–2746 (2022)
20. Li, Y., Wu, J., Li, X.: A smart shopping cart system with AI-powered product recommendation and navigation for visually impaired shoppers. ACM Trans. Accessible Comput. **15**(3), 1–18 (2022)
21. Zhou, Z., Huang, H., Zhang, Y.: A smart shopping cart system with AI-powered product recognition and navigation. Appl. Intell. **52**(10), 9323–9336 (2022)
22. Wang, J., Li, J., Zhang, D.: An AI-powered smart shopping cart system with personalised recommendations. IEEE Internet Things J. **9**(1), 457–467 (2022)
23. Zhang, X., Li, W., Wang, X.: A smart shopping cart system with AI-powered fraud detection and prevention. IEEE Trans. Syst. Man Cybern. Syst. **52**(10), 1093–1104 (2022)
24. Wang, X., Zhang, H., Li, C.: A smart shopping cart system with AI-powered fraud detection and prevention for smart shopping malls. IEEE Trans. Circuits Syst. Video Technol. **33**(1), 1–14 (2023)

Smart City: Challenges and Issues

Nidhi Agarwal, Sachi Nandan Mohanty, Bhawani Sankar Panigrahi, and Chinmaya Ranjan Patnaik

Abstract Smart City is one of the recent concepts which every city is craving for now a days. Research is going on extensively at both national and international levels in developing smarter and smartest cities for the upcoming era. This chapter is an effort to bring about underlying concepts, needs, provocations, exigencies behind the establishment of stylish cities. Complete flawless communication among cities poses a challenge for constructing completely automated niche and a new way to explore the unexplored domain. Contemporary findings reveal that for implementing stylish niche, in 2021 around 35 billion devices were utilized globally. Working flawlessly with such a humongous collection of devices posed as demanding issue, thus drawing major attention of investigators from varied domains like academic-based and industry-based towards the flawless construction of new-stylish niches called as "smart cities". A lot many concepts like smart grids, eHealth-cares, e Environmental-monitoring systems, smart-homes, smart-water purifiers, Smart-Air-Quality systems etc. are now put together to give a new concept of stylish city or rural niche. All this includes the latest 5G networks with high-end tools and technologies with full and flawless real time updates. The research work elaborated in this chapter includes a crisp study of the underlying technical confrontations lying in the path of style-oriented niches. We will also throw ample light on the implementation issues for the various other sub-domains as mentioned above.

N. Agarwal (✉)
School of Computing Science and Engineering, Galgotias University, Greater Noida, Uttar Pradesh, India
e-mail: nidhiagarwal82@gmail.com

S. N. Mohanty
School of Computer Science and Engineering, VIT-AP University, Amravati, Andhra Pradesh, India

B. S. Panigrahi
Department of Information Technology, Vardhaman College of Engineering, Hyderabad, Telangana, India

C. R. Patnaik
Department of Computer Science and Engineering, Ajay Binaya Institute of Technology, Cuttak, Odisha, India

S. N. Mohanty et al. (eds.), *Explainable IoT Applications: A Demystification*, Information Systems Engineering and Management 21,
https://doi.org/10.1007/978-3-031-74885-1_13

Keywords Smart city · Technological advancements · Smarter niche

1 Introduction

Today in India majority of cities are fighting to be upgraded as "smart cities" by incorporating new and advanced features. This upgradation is not so easy as they must compete a lot to achieve certain deliverables flawlessly. A lot many features and activities are to be taken care of to achieve this position. But these niche longing for "smarter niche" must deal with all the shortcomings lying in the various sections and sub-sections of the cities. They need to recover from all the problems and must overcome the challenges associated with the various flaws. It applies to all the governmental, non-governmental and semi-governmental domains. All the aspects related to these domains should be taken care of. The IT structure should be developed in such a way that it adds to the overall infrastructural development of the whole niche. All the latest advancements in terms of tools and technologies should be adopted. All the services should be delivered flawlessly. The data being floated among the various stakeholders, directly, indirectly should be managed and updated properly. The main backbone behind the smart development of cities is proper exhibition of IT services. All the hardware, software and other required components needed for the technological advances should be implemented without fail. The latest concepts of AI and ML should also be adopted as these are the basics of any new development. The fundamental issue is real time updating of all the associated elements. For this the internet-related services should also be imparted properly and flawlessly.

The plans for establishing "smart city" need to be developed quite judiciously. It takes into consideration the inhabitants, state offices, non-state offices, all private establishments and establishes a balance among all of them. A balance is a must among all these elements so that data moves in real time among all the deliverables. The IOT based devices must be functional properly with all real updates. The assurity of converting a normal city to smart niche by the city officials is rare as it solely depends upon the various constraints. First a draft is proposed which is approved by all the applicants. Then it is submitted to the concerned authorities for authenticity and approval. All the parts and sub-parts pertaining to the development cycle for a city are to be taken care of.

The developmental process for all the domains and sub-domains should be mentioned clearly in the draft. It should bear the data related to smart education, smart urbanization, smart medical devices, smart medical facilities, smart IOT, smart banking and various other domains to be smarter enough to upgrade any city to a smarter level. Even though the elevation of a city from normal to smarter niche is not certain, but if there are ample technological advances and fully efficient real time updates, then it can be accomplished. Also, that time is not far away when all the cities in India would raise their technological advancements and obtain the cadre of "smart niche" (Fig. 1).

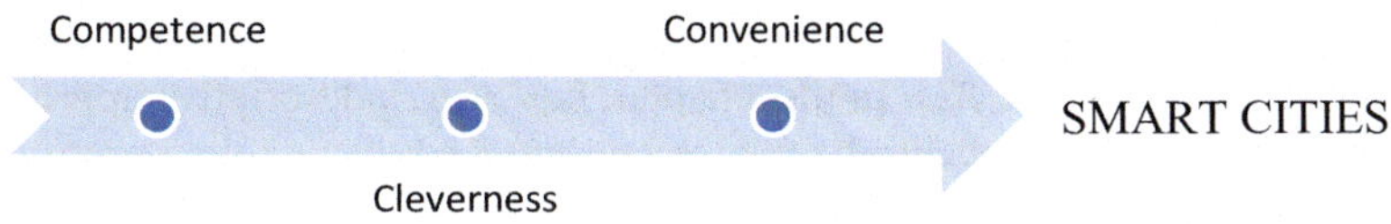

Fig. 1 Special features of smart cities

2 Literature Survey

A lot of work is going on in India now a days to implement the concept of "smart cities". But the assurity of converting a normal city to smart niche by the city officials is rare as it solely depends upon the various constraints. First a draft is proposed which is approved by all the applicants. Then it is submitted to the concerned authorities for authenticity and approval. All the parts and sub-parts pertaining to the development cycle for a city are to be taken care of. The developmental process for all the domains and sub-domains should be mentioned clearly in the draft. It should bear the data related to smart education, smart urbanization, smart medical devices, smart medical facilities, smart IOT, smart banking and various other domains to be smarter enough to upgrade any city to a smarter level.

The authors in [1–4] present current challenges of IoT and Blockchain while an analysis of the potential advantages of both has been evaluated. The literature survey in [5–8] identifies the components of the smart city to realize the concept. The authors discuss the IoT authentication issues in [9] providing a wide range of authentication protocols proposed in the literature. They analyze requirements for new domains and try to align elements within demand IOT tools and technologies. A systemic review of Internet of Things (IoT)-based smart cities and blockchain (BC) has been presented statically in [10]. The authors discuss the distributed nature of BC, which has been adopted by many businesses, posing challenges in IoT-based smart cities.

A comprehensive survey has been presented in [11, 12] on cyber-physical systems (CPSs) concerning applications, technologies, standards, and related security vulnerabilities, threats, and attacks. It further leads to identifying the key issues and challenges within this domain. Various security aspects, services, and best practices ensure resilient and secure CPS systems. Blockchain (BC) technology's evolution considering constituent technologies, consensus algorithms, and blockchain platforms have been presented in [13–15]. The authors discuss the security issues for smart cities and critically evaluate various smart applications enabled by blockchain-enabled solutions. The review in [16] presents an overview of layered architectures of IoT and associated attacks.

In their research work [17, 18], the authors have tried to find out the various basic technological issues pertaining to the flawless IoT based systems which are based on privacy, security, intelligent sensors/actuators design. The authors in [19, 20] discuss the characteristics of blockchain technology, focusing on the integration of distributed ledger technology in smart cities. This literature review is a continuation of a comprehensive review article that discussed the newly proposed solutions based on centralized and distributed block chain-based solutions for authentication IoT-enabled smart devices [21, 22]. In this chapter, however, a descriptive approach has been adopted to explore decentralized architectures and discuss the security issues in the authentication of these IoT-enabled smart devices.

3 Challenges and Issues Involved in Building Smart Cities

The various challenges which are being faced in the successful and fully authenticated smart cities and their full adoption by all the people in the society without any hesitation are discussed as below (Fig. 2).

3.1 Smart City IT Infrastructure Must Be Agile and Flexible to Scale

The basic infrastructural requirement for the establishment of a "smart city" is the scalability option for all the systems and sub-systems. It helps in growing and enlarging the capabilities and thus the standards of any city which aims to be a "future smart city". All the sub-systems should be capable of dealing with ever increasing data which should be updated in real time with fully efficient internet support. The sub-systems should also be capable of ever-increasing data and the vice versa should also be true i.e. the data should also be capable of getting acquainted with ever increasing support.

- ❖ Develop and then preserve open space
- ❖ Identification of the city
- ❖ Cost effective implementation
- ❖ Flawless services
- ❖ Implementing smart solutions
- ❖ Enhanced usage of resources
- ❖ Developing smart localities
- ❖ Improved transport facilities
- ❖ Public friendly governance
- ❖ Cost effective governance
- ❖ Improved communication services
- ❖ Improved e-commerce facilities

Fig. 2 Smart cities implementational demands

3.2 *Political Differences Can Be a Roadblock to Smart City Deployments*

The politized boundaries of the cities can also be a hindrance in the infrastructural development of the smart city initiative. As funding is the major requirement for the establishment of technological advances. Many a time, there are challenges related to the issue of timely funds. The funds need to be drawn from multiple resources like private, semi-private and non-private organizations. Many a times, these can be influenced by the government bodies and local political parties. It may result in delayed availability of funds or their unavailability also. Many a time, the concerned authorities have to intervene to make the funding process functional.

4 Smart City Architecture

The architecture of smart cities is divided into various layered methodologies in which operations are performed between the layers. A good amount of Internet speed and capacities required to exchange data among these layers. Dell data which passes from one layer to another needs to be highly secured and earlier below a particular layer passes the data to its upper layer. In this way the data is transmitted from lower layer to the higher layer as is usually done in the layered architecture system pertaining to other domains also. So, smart city architecture is majorly categorized as a division into 3 layers. The functionalities, challenges, issues and the weaknesses of these layered display architecture also need to be dealt with stringently to implement the smart city concept flawlessly and in the most effective way.

4.1 *Antagonist Smart-City Layered Architecture*

As all of us crave for the flawless implementation of smart cities, but it will also incur a lot of maintenance and replacement related functionalities. All this will obviously involve a lot of cost, but these things are imperative to enhance the efficiency of the system. The transfer of various data and information between the layers and the devices is the prime requirement for any smart city implementation. In such a scenario a lot of security and integrity with respect to the data transfer needs to be maintained by the implementors. Only then the people will turn up and accept the smart city implementation and can take the maximum benefit out of it. The system should be developed in such a manner that the unauthorized people should be first identified successfully and then restricted from accessing private data. As most of the data is on the Internet, the intrusion through the unauthenticated persons has more chances that is why security is a prime concern.

4.2 Application Layer Proneness

As application layer is the place where the communication takes place between the layered architecture and the users. So, it is the place which is most prone to attack by outside attackers. One must take utmost care and not leave this layer unattended and provide maximum security terms to this layer. A lot many sophisticated algorithms need to be applied on this layer so that it is most secure and as it is most vulnerable towards various types of attacks. The various types of attacks which can be they are on this layer are injection-based, cross-citing based, scripting-based, parametric tempering based, botnet based and buffer overflow-based attacks.

4.3 Transmission Layer Proneness

The transmission layer is responsible for the exchange of data between the layered architecture and the users. As this layer is prone to attackers from the outside place, it must be secured a lot. The attackers may target this layer to obstruct the network resources and by inserting fake data. This can pose quite a serious problem such as denial of services and various other type of attacks like Trojan horse attack, worm-based attack and various other man-in-the-middle type of attacks. The user is also many a times confused regarding which data is actual and which data is being faked. One must take utmost care in implementing this layer also.

5 Latest Elevations in Tools and Techniques and Upcoming Research

Here in this section, we will discuss the various advancements which have taken in the smart city implementation in the past and how the various shortcomings can be removed to have a fully secured and flawless smart city implementation for the future. Since we all know that for the best smart city implementation, we need to have the most secure system based on the Internet. Now since things are Internet based with the IoT enabled devices, we need to incorporate the security features largely by providing different logins and facilities to various people at various levels. It must be seen precisely which person should be given which type of rights so that privacy among people is not hindered. Here we will also discuss the various future works which can be done to remove the current shortcomings existing in the system and what can be the various ways to do so.

5.1 Newly Introduced Block Chain Related Techniques

As we all know that the block chain architecture adds an extra layer to the already layered smart city architecture. It must combine the IoT devices with the broken chain enabled devices using this additional layer. As the number of layers starts increasing one must take more care of the security issues related to it. The greater the number of layers the more are the implementation issues, and the security features. The implementors also need to see which type of rights are to be given to which person so that security breach is not there. The various block-chain related devices are joined as nodes in a decentralized system which needs heavy processing and a lot of high-speed computation overhead. All this incurs a lot of cost and complexity in the system. This is the current scenario which is being faced by the block chain architecture. In the future it is speculated that if the concept of development and deployment of smart cities is implemented successfully and people are not hesitant in adopting it, then money would not be a problem as far as increasing the layer is concerned. For the nodes, one can get some token sort of thing to enhance the security. This token would act as a crypto token which will be delivered only to those persons who have the various rights related to it and thus the keys. The various methodologies for encryption and decryption can be used at this stage. It would help in giving authentication only to those people who are authorized to access the data. In this way, the security can be implemented fully and with 100% satisfaction. If this can be implemented successfully, then more and more users will turn up for the concept which will no doubt increase the amount of fee which is being generated and thus more sophisticated and expensive security features can be used.

5.2 Latest Hashing Techniques

The latest hashing techniques based on secured hash algorithms can also be incorporated to implement the secured communication channel show that the data originality can be maintained, and it can move in a secured way from one place to another. Nowadays a lot many hashing algorithms are being introduced with high securities. Many organizations have started adopting the hashing techniques for various purposes but if it is adopted in on a wide range for the implementation of smart cities then the data integrity and security would also be maintained. The authentication process would be improved by the usage of hashing techniques and the smart city infrastructure would be improved by the introduction of a lot many public and private security keys in the distributed network.

5.3 IOT Enabled Upgradations

The various initiatives been taken to upgrade the firmware in actual time will help the smart city network to implement a secured collaboration between the IoT enabled devices with the blockchain. As data integrity and security are the key challenges which are imperative for the implementation of any secured smart city architecture, a lot many concerns to safeguard these aspects lie within the minds of the implementers. Various lightweight cryptography-based algorithms would also help in this way to maintain the resource-based methodology of these additional elements. As the data which is uploaded needs to be secured from the various intruders, participants, devices and from the cyber attackers.

5.4 Crypto Token Concept

We all know that the concept of crypto tokens is becoming very popular nowadays for the implementation which uses blockchain based smart elements. These tokens bear the rights of physically and digitally accessing the property based on its assets. Though these tokens come in the categorization of money, they can also be used for special purposes for keeping the tokenized assets. In the future they can be used for smart city authentication end to end with the various devices which would relate to these smart cities using the public key encryption system.

5.5 Anonymous Data Related Issues

If we talk about data integrity and data security by using a decentralized encryption scheme, then data anonymity also comes into picture. Coding of data is done so that it is not accessible by unauthorized persons. Anybody who wants to access the data needs to have the decoding key. If the implementation is done based on un-centralized proxy algorithms in which encryption can be applied again and again then it would enhance the data anonymity problem. Another issue which can play a major role in the successful implementation of this is trusting the data which is being sent by various people from various sources. Efficiency and to restrict the accessing power of the IoT enabled smart devices for the implementation of smart city architecture can only be achieved fully through the implementation of various data security and integrity related provisions so that the data can be accessed parallelly by more than one device. One should always try avoiding the overloading of data, especially during the peak hours for the smart city implementation.

6 Smart Cities Components from Indian Perspective

The word "smart city" has gained a lot of importance in India in the recent past when a lot many government schemes are being floated by the Prime Minister. India is aimed to develop 100 smart cities by 2023 according to the Prime Minister initiative. All these cities are expected to exhibit various features and facilities with respect to the smart niche. All these are supposed to work efficiently and fulfil all the motives of the smart city effectively. Here we are going to discuss the various features which are expected to be exhibited by "smart cities" in India and weather India is able to fulfil them or not or to what extent. We will also discuss that if the features are not fulfilled satisfactorily then what are the challenges which are still lying in the way of flawless implementation. We will also discuss the various schemes of government which are being proposed and/or developed for helping in the implementation of these smart niches.

6.1 Required Elements

The various elements which are required for the implementation of smart niche are those including the ample use of technology-based practices. These are adoption of technological usage practices, government policies, socio economic policies and various other elements related to their effective implementation. Effective usage of the available land area is also a must for the implementation of smarter niche. Proper utilization of available land resources, which includes non-wastage of the available land, is also to be taken care of. In such a scenario the area related to housing, common area for offices and schools, other areas in the city should also be taken care of so that the land area is utilized effectively. One also must see that while implementing such a concept the open areas, the areas where people can move and walk freely, do physical exercises and can use their non-fuel-based vehicles should also be left judiciously. This all will help in reducing the number of fuel-based vehicles and improve the usage of those vehicles which do not require any fuel. This will help reduce the amount of pollution. Another element for smart niche development is the establishment of open spaces where a lot many trees can be grown which can boost the health of the inhabitants. Reduced construction of skyscraper buildings and construction of more parks, playgrounds and clubs where various health boosting activities are done should also be accelerated. The availability of most of the Offices, schools, shopping complexes in the residential vicinity of the people is also encouraged so that transport facilities are used to their minimum. Development of various such smarter niche is a part of Smart City Mission of India. India is in the way of accomplishing all the elements pertaining to smarter niche and aims to accomplish 100% soon.

6.2 Decidability of Smart Niche

In today's competitive world, where every city wants to be upgraded as a smart city by the inclusion of various latest tools and technologies, there are certain evaluative criteria which categorize a city as "smart city". The various criteria are competing and challenging among themselves for all the cities and are specified as guidelines issued by "the smart city mission of India". It is also the responsibility of the government to check whether adequate funding has been given for the development of smart city if it is under the "smart city mission" of India. The various steps which are required to categorize the city as smart city are as below.

Step 1: Firstly, the various proposals which are sent by the concerned people for the extension of the city as "smart city" are checked and scanned for the inclusiveness and validity of the data. Then a unique number, which identifies that proposal, is assigned to that proposal. All this processing is done under the officials assigned by the Urban Development ministry of India.

Step 2: Secondly, all the submitted proposals are evaluated for genuineness and adequacy by the team of expert people who are assigned by the Government of India. Some foreigners are also added to the team so that the evaluation process is not biased.

Step 3: Thirdly, the decision whether the city will go for the smart city development process or not is taken by the officials. If "not", then the concerned people are informed about this. If "yes", then the various developmental schemes of the government, various policies, guidelines, fundings etc. are provided to the various sections and subsections of the city from time to time as specified by the government to help the city to raise its level to a "smarter niche". All the adequate funding must be provided from time to time to the concerned departments and their sub departments of the city so that the transition is there for a smarter niche.

6.3 Aims, Achievable, Policies and Schemes in India

The "smart city mission" in India was introduced in mid of year 2015 by the Prime Minister of India as a part of 5-year development program. It aims to accomplish the development mission in India by raising the standards of niches and bringing them to smarter level. The standards would be raised such that cities are able to compete in the urbanization and smarter cities movement. It's intended to provide those enhanced standards in terms of technologies and advancements. There are various schemes of government which aid in the accomplishment of this task. All these schemes deal with the advancements of various domains and sub-domains pertaining to various parts of a city. These schemes individually enhance the standards of various domains which improves the standard of the overall city, thus making it a "smart city". Some of those schemes are as below:

1. National Urban Sanitation Policy

2. Affordable Housing Policy
3. AMRUT Policy
4. TOD policy
5. Pan City Initiatives
6. SPV and ULB Policies

7 Conclusion

This chapter can suggest various theoretical and empirical ways for the successful development and implementation of smart cities. Various underlying problems are discussed and ways to resolve them are also elaborated in this research work. The work contains many suggestive tools and techniques, all supported empirically, to authenticate the suggestive processes. One of the approaches suggests an architecture which is not centralized and is developed for IOT based devices. The main work revolves around providing security to these IOT devices which is the basic and foremost important requirement of any successful smart city implementation. Though in prior studies, researchers have implemented security features but not completely, and it is that attribute which needs to be implemented with 100% perfection. This work is an effort for complete security implementation. Block chain and IOT based solutions are suggested to combat the situation and provide fully smart city architecture, an architecture which is fully satisfiable to all the stakeholders. The major problem which exists with the blockchain based systems is storing the humongous information which is being generated from the storage platforms which are not centralized. This technique will help to store large volumes of data so that the data storage capacity is not finished. Also, in this research paper it has been highlighted that the new techniques which people are adopting extensively nowadays are based on cryptographic methods which should be investigated and used to obtain enhanced security features with respect to data and attached devices integrity. This methodology in future will provide enhanced levels of security for real time-based applications because the IoT enabled devices have resource constraints. The decentralized methodologies should be explored in greater depth using various privately and publicly encrypted data and its storage methodologies. The various security issues pertaining to the problem of authentication for smart city architecture are discussed in this research work. Some suggestive architectures are also being supported with empirical evidence and some more theoretical architectures are being suggested through this research work.

References

1. Reyna, A., Martín, C., Chen, J., Soler, E., Díaz, M.: 'On blockchain and its integration with IoT. Challenges and opportunities.' Fut. Gener. Comput. Syst. **88**, 173–190 (2018). https://doi.org/10.1016/j.future.2018.05.046

2. Baker, S.B., Xiang, W., Atkinson, I.: Internet of things for smart healthcare: technologies, challenges, and opportunities. IEEE Access. **5**, 26521–26544 (2017). https://doi.org/10.1109/ACCESS.2017.2775180
3. Bhargava, K., Ivanov, S.: A fog computing approach for localization in WSN. In: Paper Presented at: 2017 IEEE 28th Annual International Symposium on Personal, Indoor, and Mobile Radio Communications (PIMRC), Montreal, Canada (2017)
4. Bhunia, S.S.: Sensor-cloud: enabling remote health-care services. In: Proceedings of the 2015 on MobiSys PhD Forum (PhDForum), Florence, Italy (2015)
5. Silva, B.N., Khan, M., Han, K.: Towards sustainable smart cities: a review of trends, architectures, components, and open challenges in smart cities. Sustain. Soc. **38**, 697–713 (2018). https://doi.org/10.1016/j.scs.2018.01.053
6. Bonomi, F., Milito, R., Zhu, J., Addepalli, S.: Fog computing and its role in the internet of things. In: Proceedings of the First Edition of the MCC Workshop on Mobile Cloud Computing (MCC), Helsinki, Finland (2012)
7. Cao, Y., Hou, P., Brown, D., Wang, J., Chen, S.: Distributed analytics and edge intelligence: pervasive health monitoring at the era of fog computing. In: Proceedings of the 2015 Workshop on Mobile Big Data (Mobidata), Hangzhou, China (2015)
8. Chung, K., Park, R.C.: Cloud based u-healthcare network with QoS guarantee for mobile health service. Cluster Comput. (2017). https://doi.org/10.1007/s10586-017-1120-0
9. El-Hajj, M., Fadlallah, A., Chamoun, M., Serhrouchni, A.: A survey of internet of things (IoT) authentication schemes. Sensors **19**(5), 1–43 (2019). https://doi.org/10.3390/S19051141
10. Lee, I., Lee, K.: The internet of things (IoT): applications, investments, and challenges for enterprises. Bus. Horiz. **58**(4), 431–440 (2015). https://doi.org/10.1016/j.bushor.2015.03.008
11. Jackson, K.R., Ramakrishnan, L., Muriki, K., et al.: Performance analysis of high performance computing applications on the Amazon web services cloud. In: Paper Presented at: 2010 IEEE Second International Conference on Cloud Computing Technology and Science, Indianapolis, IN, (2010)
12. Yaacoub, J.-P.-A., Salman, O., Noura, H.N., Kaaniche, N., Chehab, A., Malli, M.: Cyber-physical systems security: limitations, issues and future trends. Microprocess. Microsyst. **77**(103201) (2020). https://doi.org/10.1016/j.micpro.2020.103201
13. Liu, L., Yang, Y., Zhao, W., Du, Z.: Semi-automatic remote medicine monitoring system of miners. In: Adjunct Proceedings of the 2015 ACM International Joint Conference on Pervasive and Ubiquitous Computing and Proceedings of the 2015 ACM International Symposium on Wearable Computers (UbiComp/ISWC), Osaka, Japan (2015)
14. Maitra, A., Kuntagod, N.: A novel mobile application to assist maternal health workers in rural India. In: Paper Presented at: 2013 5th International Workshop on Software Engineering in Health Care (SEHC), San Francisco, CA (2013)
15. Khan, M.A., Salah, K.: IoT security: review, blockchain solutions, and open challenges. Fut. Gener. Comput. Syst. **82**, 395–411 (2018). https://doi.org/10.1016/j.future.2017.11.022
16. Burhan, M., Rehman, R.A.: IoT elements, layered architectures and security issues: a comprehensive survey. Sensors **18**(9), 1–37 (2018). https://doi.org/10.3390/S18092796
17. Miettinen, A.P., Nurminen, J.K.: Energy efficiency of mobile clients in cloud computing. In: Proceedings of the 2nd USENIX Conference on Hot Topics in Cloud Computing (HotCloud), Boston, MA (2010)
18. Rana, M.M., Bo, R.: IoT-based cyber-physical communication architecture: challenges and research directions. IET Cyber Phys. Syst. Theory Appl. **5**(1), 25–30 (2020). https://doi.org/10.1049/IETCPS.2019.0028
19. Hakak, S., Khan, W.Z., Gilkar, G.A., Imran, M., Guizani, N.: Securing smart cities through blockchain technology: architecture, requirements, and challenges. IEEE Netw. **34**(1), 8–14 (2020). https://doi.org/10.1109/MNET.001.1900178
20. Khalil, U., Malik, O.A., Uddin, M., Chen, C.-L.: A comparative analysis on blockchain and centralized authentication architectures for IoT-enabled smart devices in smart cities: a comprehensive review, recent advances, and future research directions. Sensors **22**(12), 1–52 (2022)

21. Yi, S., Hao, Z., Zhang, Q., Zhang, Q., Shi, W., Li, Q.: LAVEA: latency-aware video analytics on edge computing platform. In: Proceedings of the Second ACM/IEEE Symposium on Edge Computing (SEC), San Jose, CA (2017)
22. Yoon, J.: Leveraging sensor data content to assure sensor device trustworthiness in mobile edge computing. In: Paper Presented at: 2017 Second International Conference on Fog and Mobile Edge Computing (FMEC), Valencia, Spain (2017)

IoT Based Real-Time Ecological Monitoring System Deploying an Arduino Board and Cloud Computing

B. Ravi Chandra, G. Likhitha, K. Susmitha, K. Madhu Latha, and Amjan Shaik

Abstract Human participation in the gathering and processing of data has the potential to introduce inconsistencies or mistakes that can impair the quality of the data and delay updates. The research calls for using innovative technology to track local meteorological conditions and make the data obtainable from every corner of the world. This is made feasible by Internet of Things (IoT) technology, which can collect and communicate weather data in real-time, presenting customers with precise and up-to-date information. Ecological monitoring includes many but here we choose to monitor Weather. Because the weather reporting system makes it simple for everyone to access weather information online, a weather forecasting organization is not needed. The system monitors and controls factors in the environment like temperature, relative humidity, atmospheric pressure, intensity of light and CO levels using sensors. It then sends the data to a web page where the sensor data is plotted as graphical statistics. Data is constantly transmitted from the device to the microcontroller to be sent across wifi connections to internet web servers. An Android smartphone and cloud storage are sent the identified data. These outcomes are visible to users of cloud-based applications. The established system's updated data is available online from anywhere in the world. If there are any issues, we will send the mobile device unique notifications. The IoT app can be employed to change the dynamic triggers and send an email alert.

Keywords DHT sensor · LDR sensor · MQ135 sensor · ESP8266 module · Blynk app · BMP180 sensor · Graphical statistics

B. Ravi Chandra (✉) · G. Likhitha · K. Susmitha · K. Madhu Latha
G. Pullaiah College of Engineering and Technology, Kurnool, India
e-mail: chandrabrc11@gmail.com

K. Madhu Latha
e-mail: mudigetimadhulatha416@gmai.com

A. Shaik
St. Peters Engineering College, Maisammaguda, India

S. N. Mohanty et al. (eds.), *Explainable IoT Applications: A Demystification*,
Information Systems Engineering and Management 21,
https://doi.org/10.1007/978-3-031-74885-1_14

1 Introduction

Today's pollution monitoring systems focus on a few environmental parameters. We proposal an updated strategy to weather reporting that operate independently of traditional weather forecasting agencies. The Internet of Things (IoT) and cloud technologies have been integrated in this system to enable seamless data transmission from tool devices to end users over the internet. It also use graphics to Visually represent the parameter trend. This system enables it practicable to track weather conditions in real time from anywhere in the world and generates live updates that are accessible through an online server system [1, 2].

1.1 Introduction to Embedded Systems

A computing device that makes a specific, focused on duty is sometimes referred to as an embedded system. The air conditioner, VCD player, DVD player, printer, fax machine, cell phone, and other devices are examples of embedded systems. The term "firm ware" also refers to the embedded software. Embedded systems can only be programmed to perform one very specific task at a time. Particularly the memory, embedded systems' resources are extremely constrained. They typically lack secondary storage options like CDROMs and floppy disks. Systems that are embedded must meet certain deadlines. A certain task must be finished in a certain amount of time. Power consumption is restricted for embedded systems [3, 4].

1.2 Introduction to IOT (Internet of Things)

The Internet of Things, or IoT, is a breakthrough in technology that allows data to be collected and shared without human oversight via the connection of everyday commodities equipped with electronics, software, and sensors to the internet [5]. The term "Things" in the context of the Internet of Things refers to anything and everything that is easily accessible or connected online in daily life [6]. IoT is a state-of-the-art automation and analytics system that combines cloud messaging, artificial intelligence, networking, sensors, and electrical components to supply entire systems for the good or service [1, 7]. The Internet of Things system has increased performance, control, and transparency. We can associate everything around us because we have a platform, known as the cloud, that houses all of the data [8].

2 Literature Survey

2.1. H. Maenpaa, S. Varjonen, A. Hellas, S. Tarkoma and T. Mannisto [6], they are delivering papers for three years now, utilizing action research to educate Internet of Things device creation in a problem-based, real-world environment. They provide a comprehensive evaluation methodology, a sample course outline for planning IoT prototyping learning experiences, as well as suggestions for best practices for supporting customized learning in similar circumstances. The results demonstrate that generic evaluation criteria may bedeveloped, even in the face of the project outcomes of the students' varied complexity and adaptability.

2.2. G. Suciu et al. [7] Provided an example of how Internet of Things (IoT) devices Waspmote, Pycom, and Raspberry Pi all require very little electricity. These techniques improve efficiency, scalability, and structural optimization to get beyond well-known IoT flaws such resource and energy scarcity, delayed data processing, and restricted energy. In this study, waspmote devices and more Libelium technologies are shown. Pysense, LoPy and LoPy4 microcontrollers, Pytrack, SiPy, FiPy, GPy, WiPy, expansion boards, and five embedded sensors for temperature, humidity, altitude, pressure, and light are just a few of the many products that Pycom offers. The workings of the Adafruit cloud montage are also explained in this document. Pycom devices use long-range (LoRa) antennas that are specifically made for low-power wide-area networks.

2.3. Lee [9] researched for the purpose of estimating the disaster area and locationthey developed a multi-robot-multi-target path planning. The multi-robot-multi-target algorithm and heap optimization are used to improve the A* algorithm and decrease calculation time for multi-robot systems. Then, for precise location estimation for multiple Path planning the Kalman filter is utilized.

2.4. S. V. V. Srinivas [10], A. K. Singh, A. Raj, A. Shukla, R. Patel and A. Malaywere given the task of giving the concerned authorities a tool to help them gather information through reconnaissance using the rover and help them formulate a strategy for the rescue operation that is affordable, efficient, quick, as well as secure forthe rescue workers.

2.5. S. Sarkar [11], A. Patil, A. Hartalkar and A. Wasekar, comprehended the crucial necessity to provide a way to save lives and precious time during an earthquake as soon as possible. An earthquake causes significant damage and loss, and the buildup of debris makes searches and rescues challenging. In order to identify those stuck in a specific area, this article proposes the employment of rescue robot cars in such operations. Because they are small, these rescue robots have a high speed of action.

2.6. A. Ravendran [12], P. Ponpai, P. Yodvanich, W. Faichokchai and C. -H. Hsuresearched for rescue operations, a dependable and usable technical solution that can be changed in unknown circumstances with high limitations is essential. This research project focuses on creating low-cost mobile robots that can perform a variety of tasks and adapt to changing terrain and environmental circumstances. For various applications during rescue operations, a mobile robot with a gripper and a driller is designed and built. With the supplied power source, the devised system has a maximum lifting capacity of 2 kg and can drill various materials.

3 Problem Statement

The rapid growth of the Internet of Things (IoT) has led to an exponential increase in the volume of real-time data generated by various devices and sensors [1, 13]. To leverage this data effectively, existing IoT systems need to incorporate advanced data streaming and analysis capabilities. This report explores the integration of graph-based analysis in an IoT existing system to gain valuable insights from streaming data [14]. We present the architecture, key components, and benefits of this approach, along with a case study to demonstrate its effectiveness. The IoT has revolutionized the way devices interact, producing vast amounts of data that require real-time processing and analysis. Traditional batch processing is not sufficient to handle the continuous influx of data. Thus, integrating streaming analytics into an existing IoT system becomes crucial to enable proactive decision-making and improve operational efficiency [15].

4 Methodology

4.1 Proposed System

This report presents an innovative Internet of Things (IoT)-based proposed system that leverages streaming data analytics with graphs to enable real-time insights and decision-making [16, 17]. The proposed system aims to harness the potential of IoT devices and advanced data analytics techniques to enhance various applications, such as smart cities, industrial automation, healthcare [18], and more. This report outlines the system architecture, data flow, and the role of graphs in analyzing streaming data to gain valuable insights and optimize The Internet of Things (IoT) has revolutionized how devices and objects interact and communicate with each other. The increasing number of IoT devices generates massive amounts of streaming data that hold valuable information. To extract meaningful insights from this data deluge, a proposed system is designed to incorporate streaming data analytics with graphs, enabling efficient data processing and visualization processes (Fig. 1).

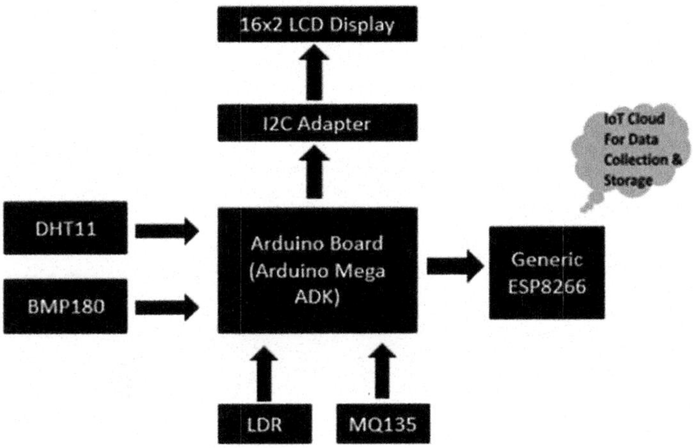

Fig. 1 Block diagram of weather monitoring

4.2 Working Process

Arduino Board (Arduino Mega ADK) is the central control unit for the project. The code is written for it to read data from sensors and control the display, WiFi module, and communicate with the Blynk app. Here about a DHT sensor, it's a type of sensor that measures temperature and humidity. BMP180 sensor is a pressure and temperature sensor which is connected to the Arduino board using the I2C adapter. LDR sensor is used to measure the intensity of light. MQ135 sensor is a gas sensor used to detect air quality, specially the presence of harmful gases like carbon monoxide. The code is then processed by the Arduino Mega ADK board, which then outputs it to a 16*2 LCD display. The ESP8266 module provides Wi-Fi connectivity to the Arduino Mega ADK. It acts as a bridge between Arduino and the internet, allowing Arduino to communicate with the Blynk cloud server and the Blynk app [2, 8] over Wi-Fi, where we can see the collected ecological data in the form of graphs and the data can be accessed remotely (Figs. 2, 3, 4, 5 and 6).

5 Results and Discussion

5.1 Software Installation

The necessary components are an Arduino-compatible microcontroller (anything from the article should work) and a computer running Windows, Mac OS X, or Linux [8, 15]. A USB A-to-B cable or an alternative suitable connection method for your Arduino-compatible microcontroller and PC.

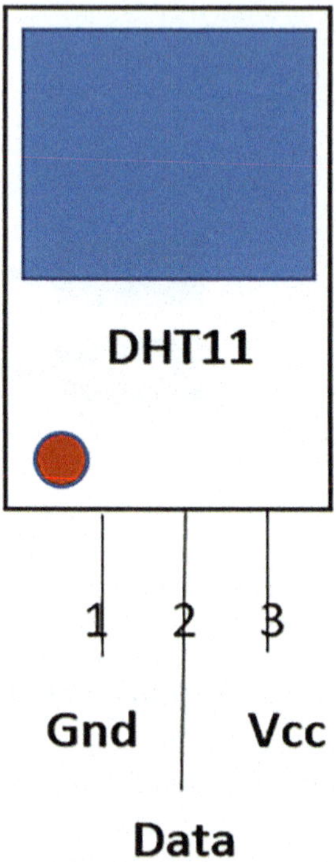

Fig. 2 DHT11 sensor

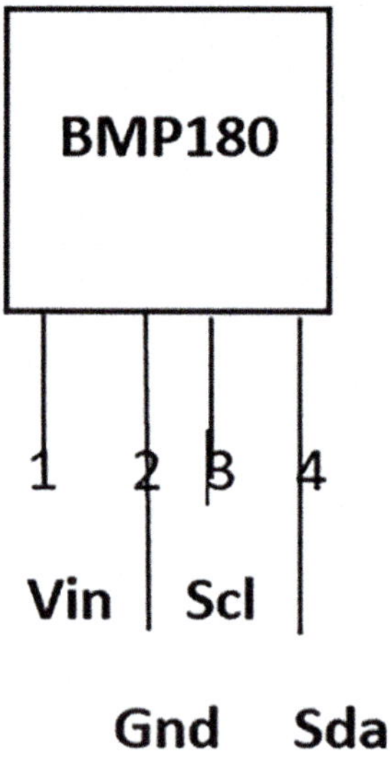

Fig. 3 BMP180 sensor

Fig. 4 MQ135 sensor

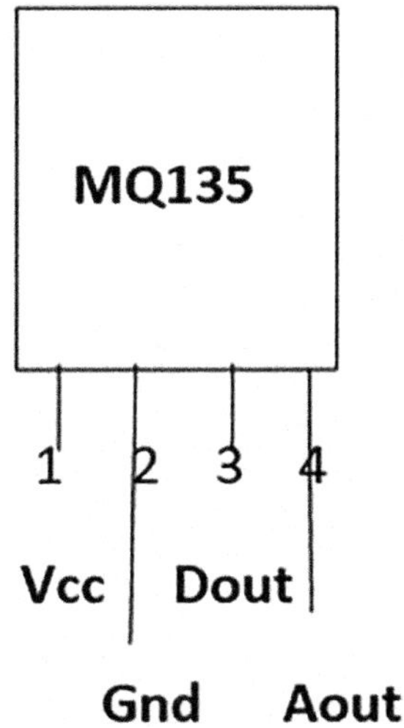

5.2 *Blynk App*

From the perspective of the Internet of Things developers, Blynk is a multifunctional mobile application. By offering a user-friendly interface to control and monitor connected devices using a range of widgets, it enables users to quickly construct unique IoT projects. Blynk is a useful tool for prototyping and delivering IoT solutions because to its compatibility with prominent hardware platforms and cloud integration, and its active community assures access to assistance and resources to build groundbreaking initiatives.

Make sure the internet and your Internet of Things (IoT) devices are linked to your microcontroller (Raspberry Pi, Arduino, etc.). For aid in completing this stage, Blynk offers libraries and code samples. Using the Blynk library or SDK for your selected hardware platform, create code for device that connects with the Blynk server [17]. The Blynk app can now send data to and receive commands from your hardware thanks to thecode. Utilising the Blynk app, you can now test your IoT project.

5.2.1 Quick Start: Arduino + Ethernet Shield

- From the Blynk app (Google Play, App Store), obtain the Auth Token.
- Bring this library into the Arduino Environment.
- To access Arduino_Ethernet in the Arduino IDE, go to File → Examples → Blynk → Boards_Ethernet.
- Adjust the Authorization Token of the sketch and submit it to Arduino.
- Utilizing the Ethernet shield, link your Arduino to the internet.

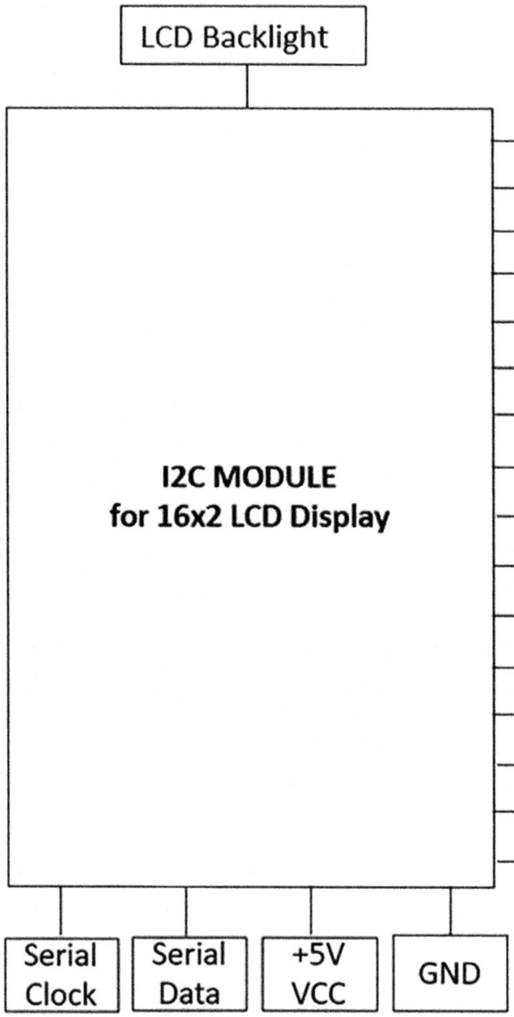

Fig. 5 I2C adapter

5.3 Result Analysis

See Figs. 7, 8, and 9; Table 1.

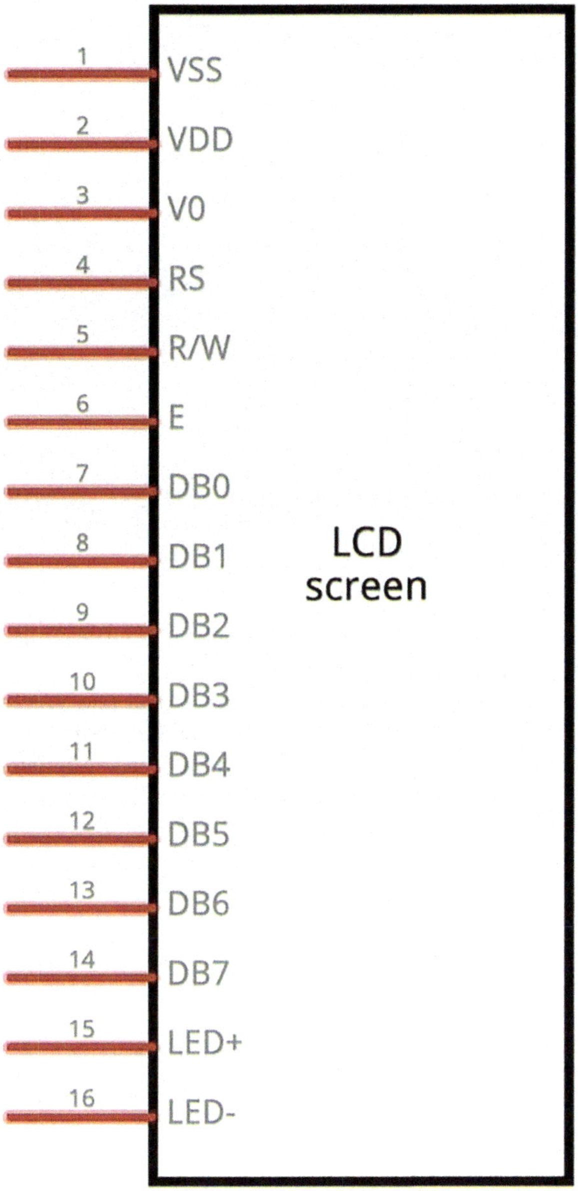

Fig. 6 LCD display

6 Applications

a. This data can be used for environmental research, urban planning, and pollution control measures [3].

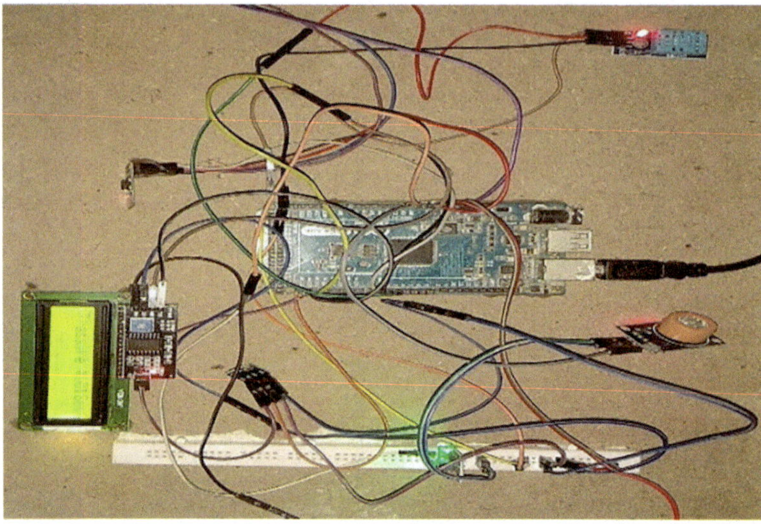

Fig. 7 Experiment kit

b. The monitoring system can be integrated into smart city initiatives [13] to monitor and manage various aspects of life.
c. Farmers can utilize the weather data and graphs to make informed decisions about irrigation, crop planting, and harvesting [14].
d. The project can provide real-time weather updates and graphical representations to assist pilots and air traffic controllers in making safe and efficient flight plans [19].
e. Travelers can access weather information and graphs to plan their trips, pack appropriate clothing, and make informed decisions about outdoor activities.
f. Construction companies can utilize weather data and graphs to schedule outdoor work, manage resources, and ensure worker safety.
g. The system can aid in monitoring and protecting wildlife habitats by tracking factors such as weather patterns.
h. The system can be used as an educational tool for students and researchers to learn about environmental monitoring, data analysis, and IoT technologies.

7 Advantages

a. You can continuously monitor ecological parameters with this technology. You can simply access and analyze the gathered data utilizing internet connectivity from anywhere by integrating the system with cloud computing. This makes it possible for researchers, decision-makers, and environmentalists to remotely monitor and visualize data.

Fig. 8 LCD results and graphs

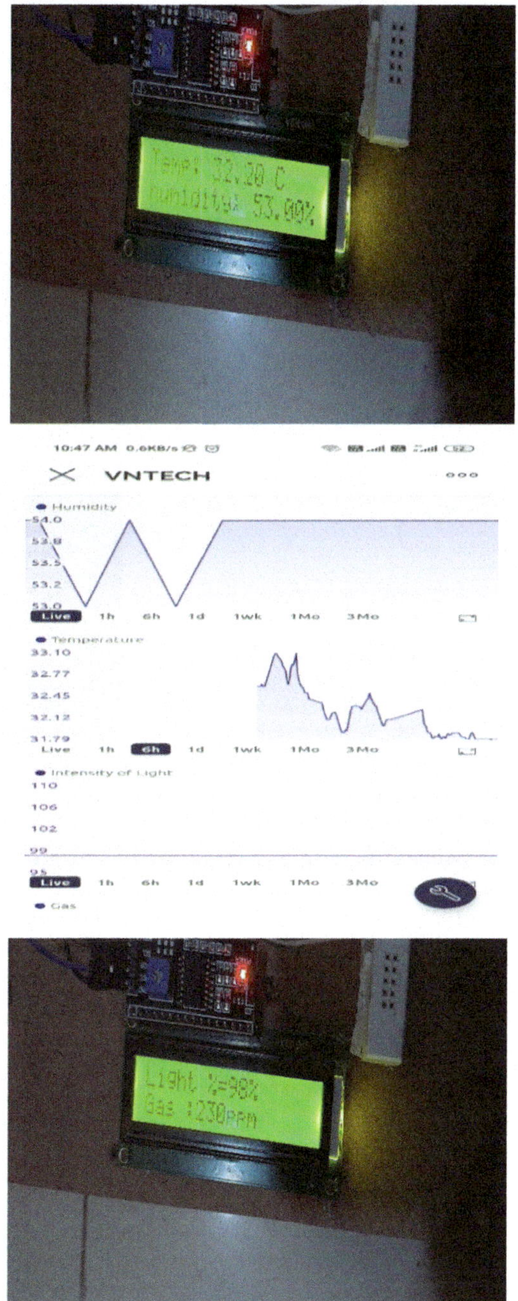

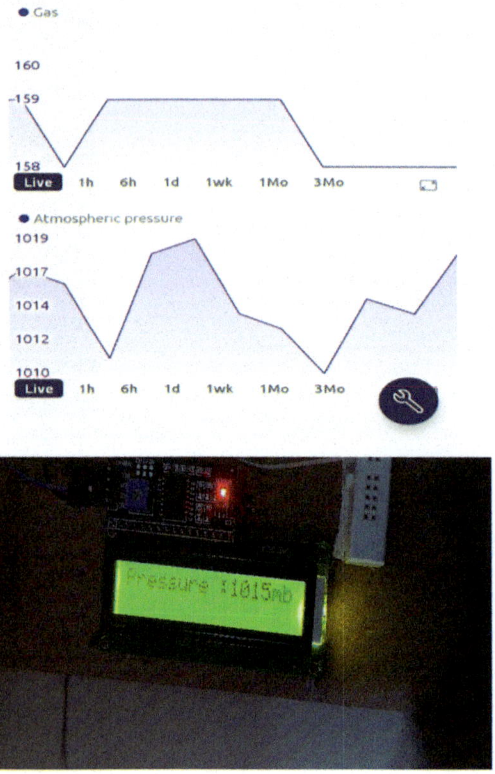

Fig. 8 (continued)

b. Scalability is made simple by the Arduino board and cloud computing combination. Without making any hardware changes, you can add more sensors or devices as needed to monitor various ecological parameters.
c. The increasing data load can be efficiently handled by the cloud-based infrastructure.
d. Collaboration and integration with various platforms and services are made easier by cloud-based technologies. Data sharing, teamwork, and integrating the monitoring system with existing environmental databases or apps are all options.

8 Conclusion

In conclusion, streaming with graphs using IoT is a groundbreaking approach to data analysis and decision-making. By harnessing real-time data streams and graph analytics, businesses can gain valuable insights, make data-driven decisions promptly, and improve overall efficiency and performance [16, 17]. However, challenges related to data privacy, complexity, and connectivity must be addressed to fully unlock the potential of streaming with graphs in IoT. With further developments, it

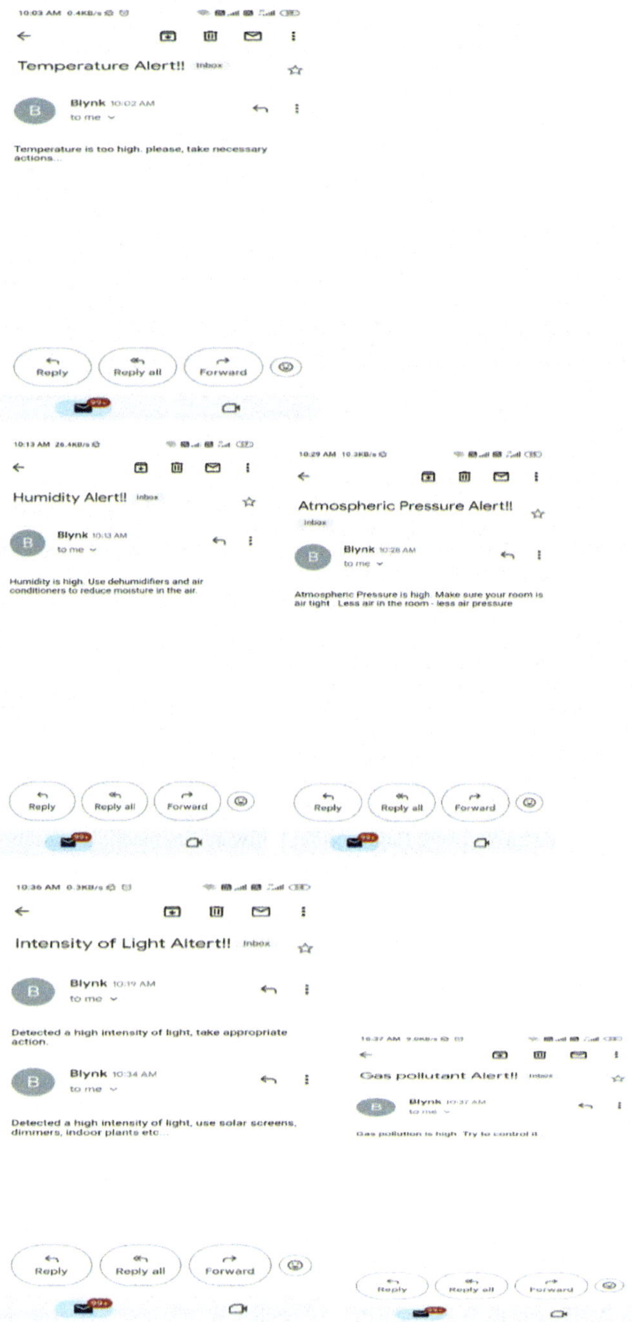

Fig. 9 Email alerts

Table 1 Real-time ecological monitoring system for different time slots

Parameters	Live	1 h ago	1 week ago	3 months ago
Temperature (C)	32.20	31.50	32.45	32.12
Humidity (%)	54	53	54	54
Intensity of light (%)	98	98	98	98
CO gas (ppm)	230	157	159	158
Atmospheric pressure (MB)	1015	1011	1016	1010

is obvious that this technology will be crucial in determining how data analytics develops in the future and how it affects decision-making procedures.

References

1. Rao, B.S., Rao, K.S., Ome, N.: Internet of things (iot) based weather monitoring system. Int. J. Adv. Res. Comput. Commun. Eng. **5**(9), 312–319 (2016)
2. Kulkarni, A.V., Satpute M.G.: Weather Reporting System Using FPGA: A Review, vol. 4, pp. 319–320 (2017)
3. International Journal of Advanced Research in Computer and Communication Engineering ISO 3297:2007 Certified Vol. 5, Issue 9, September 2016
4. International Journal of Engineering Trends and Technology (IJETT), Volume 32 Number 2, February 2016
5. Weber, R.H., Weber, R.: Internet of things, vol. 12. Springer (2010)
6. Maenpaa, H., Varjonen, S., Hellas, A., Tarkoma, S., Mannisto, T.: Assessing iot projects in university education-a framework for problembased learning. İn: 2017 IEEE/ACM 39th International Conference on Software Engineering: Software Engineering Education and Training Track (ICSE-SEET), pp. 37–46. IEEE (2017)
7. Suciu, G., Petrache, A.L., Badea, C., Buteau, T., Schlachet, D., Durand, L., Landez, M., Hussain, I.: Low-power iot devices for measuring environmental values. In: 2018 IEEE 24th International Symposium for Design and Technology in Electronic Packaging (SIITME), pp. 234–238. IEEE (2018)
8. Kodali, R.K., Sahu, A.: An IoT based weather information prototype using WeMos. In: Proceedings 2016 2nd International Conference Contemporary Computing Informatics, IC3I 2016, pp. 612–616 (2016)
9. Heo, S.N., Lu, S., Shin, J., Lee, H.: Multi-robot-multi-target path planning and position estimation for disaster area. In: International Conference on Information and Communication Technology Robotics, pp. 1–4 (2018)
10. Srinivas, S., Singh, A.K., Raj, A., Shukla, A., Patel, R., Malay, A.: Disaster relief and data gathering rover. In: 2018 3rd International Conference on Internet of Things: Smart Innovation and Usages (IoTSIU), pp. 1–4. IEEE (2018)
11. Sarkar, S., Patil, A., Hartalkar, A., Wasekar, A.: Earthquake rescue robot: a purview to life. In: 2017 Second International Conference on Electrical, Computer and Communication Technologies (ICECCT), pp. 1–7. IEEE (2017)
12. Ravendran, A., Ponpai, P., Yodvanich, P., Faichokchai, W., Hsu, C.-H.: Design and development of a low cost rescue robot with environmental adaptability. In: 2019 International Conference on System Science and Engineering (ICSSE), pp. 57–61. IEEE (2019)
13. Carlos, M., Jorge, B.P., Daniel, F., Pablo, S.: Design, development and implementation of a weather station prototype for renewable energy system journal. Energies **11**(1–13), 2234 (2018)

14. International Journal of Engineering Science and Computing, May 2017
15. Sagar, J.S.T., Balamurugan, M.S., Vivek, J.A.: A wireless framework for automotive monitoring systems. Indian J. Sci. Technol. **8**(19), IPL0146 (2015)
16. Basahel, S.B., Bajaba, S., Yamin, M., Mohanty, S.N., Lydia, E.L.: Teamwork optimization with deep learning based fall detection for IoT-enabled smart healthcare system. Comput. Mater. Continua **75**(1), 1353–1369 (2023). ISSN:1546-218. https://www.techscience.com/cmc/v75n1/51539
17. Potluri, S., Mohanty, S.N.: An efficient scheduling mechanism for IoT based home automation system. Int. J. Electron. Bus. **16**(2), 147–156 (2021). https://doi.org/10.1504/IJEB.2021.115719. ISSN:1470-6067
18. Swamy, K.C.T., Sarma Achanta, D., Supraja Reddy, A., Satya Srinivas, V., Somasekhar Rao, P.V.D.: Modelling of GPS signal scintillations with polynomial coefficients over the Indian region. Indian J. Radio Space Phys. **42**, 167–174 (2013)
19. Swamy, K.C.T., Venkata Ratnam, D., Suman, T., Ahmed, S.T.: Time-differenced double difference method for measurement of navigation with Indian Constellation (NavIC) receiver differential phase bias. Measurement **207**, 112385 (2023)

Iot Based Monitoring of Waste Management and Air Pollutants

Ravi Kumar Poluru, Madhuranjali Venigalla, R. Annie Richie, and Charani Madari

Abstract In order to raise the nation's cleaning standards, specific steps are now being implemented. An increasing number of people are taking proactive steps to keep their environment clean. Additionally, the government initiates a number of initiatives to improve sanitation. In order to remind the businesses to promptly empty the bin, we shall work to develop a mechanism. Using the Internet of Things (IoT) to monitor garbage collection systems at a reasonable cost has been the main focus of the majority of the literature's current work. While an IoT-based technology can monitor a waste collection system in real time, it cannot manage the overspill gasses that spread. Waste that is not properly disposed of results in harmful gasses, and radiation exposure has a negative impact on the environment, human health, and the greenhouse system. Given the significance of air pollutants, waste management and air pollution concentration monitoring and forecasting are highly necessary. Here, we describe an The Internet of Things (IoT) smart bin that forecasts air pollution in the vicinity of the bin and manages garbage disposal using an ESP 8266 model. For the purpose of generating alarm messages about bin condition and estimating the quantity of air pollutant carbon monoxide (CO) in the air at a given time, we experimented with a conventional model such as the ESP8266 and an ultrasound sensor. The generation and delivery of the alarm message to a sanitary worker was delayed by 4s as a result of the system. Together with messages from the warning mechanism, the system offered real-time garbage level monitoring. By using machine learning, the suggested works provide better accuracy than current solutions that rely on straightforward methods.

R. K. Poluru · M. Venigalla (✉) · R. Annie Richie · C. Madari
Institute of Aeronautical Engineering, Dundigal, Hyderabad, Telangana, India
e-mail: 20951A1237@iare.ac.in

R. K. Poluru
e-mail: p.ravikumar@iare.ac.in

R. Annie Richie
e-mail: 20951A1212@iare.ac.in

C. Madari
e-mail: 20951A1217@iare.ac.in

S. N. Mohanty et al. (eds.), *Explainable IoT Applications: A Demystification*,
Information Systems Engineering and Management 21,
https://doi.org/10.1007/978-3-031-74885-1_15

Keywords Google cloud server · Air monitoring · Forecasting · Air pollutant · Smart bin · Machine learning

1 Introduction

With growing population expansion and fast urbanization, solid waste management has become both necessary and challenging. The common dustbins that each city's respective Municipal Corporation has put in various areas might receive a lot of attention when it comes to waste collection and management. As the first step in actively gathering the garbage produced in society, it will ideally achieve the main goals of preserving social order, lowering environmental degradation, and managing a clean and hygienic environment, among other things. Through a smartphone application, this system tracks the amount of trash that is gathered in the trash cans and notifies users about it. The waste level is detected by the system using garbage sensors that are positioned over the bins, and it then compares that level with the depth of the bins. A buzzer, a Wi-Fi modem for data transmission, and an Arduino family microcontroller are used by the system. An adaptor that runs at 12 V powers the system. The level of trash collected in the bins is shown as of right now using the smartphone application. On the other hand, a web page is designed to display the status to the user who is keeping an eye on it. The software displays the amount of garbage collected in a graphical format by highlighting the collected waste in colour. The trash level status is displayed by the application. The dustbin rubbish collection system activates a siren when the collected level surpasses a certain threshold.

Hence, by delivering a graphical depiction of the bins via a mobile application, this system informs the user about the rubbish levels in the bins and maintains the city clean. Junk might comprise materials that are not wanted that are left behind from a variety of locations, including homes, businesses, public areas, colleges, and cities. "Internet of Things" (IOT) refers to the foundation of this initiative, which is associated with the "smart city." A smart lifestyle therefore requires cleanliness, and starting with the garbage can. With this project, the problem of disposing of trash will be lessened or resolved entirely. Not too far from now, commonplace items will be outfitted along with microcontrollers, digital communication transceivers, and suitable protocol stacks to allow them to communicate with people and one another, turning them into a crucial part of the Internet.

The Internet of Things, or IoT, is this. relatively new paradigm for communication. Our solution demonstrated better accuracy than conventional garbage collection systems in an innovative study we conducted on smart waste management systems using machine learning and an IoT-based methodology. Additionally, enough data is provided by the suggested system to monitor environmental air quality testing. The proposed method can allow accurate, real-time monitoring of waste levels in addition to alerts from an alert mechanism to municipal trash management. In smart cities with under-optimized garbage collection systems, it addresses the problems of managing contaminated waste. It offers continuous monitoring of various levels of harmful gases in the surrounding air. By predicting the next concentration in the air, air quality monitoring devices enable users to take prompt corrective action.

2 Related Work

Included in this document are An Internet of Things-based smart trashcan is suggested; it was constructed on an Arduino Uno board platform and linked to an ultrasonic sensor and GSM modem. The dustbin's lid held the sensor. Ten centimeters was designated as the maximum level. When the dustbin fills to the brim, the sensor detects it and triggers the GSM modem, alerting the relevant authority until the trash is removed. Ultimately, it was determined that the design of these smart bins addressed a number of concerns, including durability, affordability, and upkeep. In the process of developing a smart city, it also helped to create a clean and clutter-free atmosphere [1].

The rubbish management strategy is recommended by the researchers. A microcontroller-based system that included infrared wireless systems and a central system connected to a trashcan to display the amount of waste inside [2].

Using Wi-Fi, a mobile device was able to view the notifications. To cut costs, they just used weight-based sensors. Additionally, they just needed a Wi-Fi module on the sender side to send and receive data. The level of rubbish in the dustbin could not be measured by the sensor; it could only weigh the waste that was there. The author outlines a plan for allocating resources for the collection of waste in urban residential and business districts [3].

Using a GSM module, the ultrasonic sensor in this system measures the amount of trash in the trash can and transmits the data to the control room. Additionally, a MATLAB-based graphical user interface (GUI) was created to verify the garbage-related information for various locations. These two units were present in the system; the master unit was in the control room and the slave unit was in the garbage. It will be sent to the slave device once the sensor has determined the level of waste. Additional information is sent to the master unit, which alerts the authorities to clean the trash can. The Decision Support System that this paper proposes to be employed for waste collection in the cities [4].

Inefficient garbage collection in the parts of the city that were previously off-limits to this system. Cameras were positioned in the areas of the city that were most problematic. The first step in the process was to identify the waste collection companies. These companies should have trucks and the ability to assign drivers to pick up trash from different parts of the city and transport it to the city dumps or recycling organizations. Numerous inexpensive embedded devices were positioned across the city to track the amount of waste in the bins, along with other features [5].

Two distinct portions were suggested in this paper. The reception section is the other and the transmitter section is the first. To make it simpler to identify which bin is full and ready to be emptied, each bin was assigned a unique ID. The transmitter component has a microcontroller as well as sensors that are used to measure the amount of rubbish and employ radio frequency technology. In order to allow for the quickest possible emptying of the bin, transmitter data is fed into the system, where it is received by an RF receiver and forwarded to the relevant authorities [6].

In this study, a model for artificial systems-based cyber security systems with safe transactions is proposed [7] (Table 1).

Table 1 Literature review

S. no	Author name	Title	Characteristics
1	Monika KA, Rao N, Prapulla SB, Shobha G	Smart dustbin-an efficient garbage monitoring system	Smart dustbin built on Arduino Uno connected to GSM modem and ultrasonic sensor
2	Navghane SS, Killedar MS, Rohokale DV	IoT based smart garbage and waste collection bin	Microcontroller-based system with IR sensors, Wi-Fi for notifications
3	Kasliwal Manasi H, Suryawanshi Smithkumar B	A novel approach to garbage management using IoT	Proposal for IoT-based garbage management in smart cities
4	Medvedev A, Fedchenkov P, Zaslavsky A, Anagnostopoulos T, Khoruzhnikov S	Waste management as an IoT-enabled service in smart cities	System for waste management using IoT services
5	Schafer G	U.S. Patent No. 5,326,939	Patent related to waste management (not described in detail in provided text)
6	Anitha A, Paul G, Kumari S	A Cyber defence using Artificial Intelligence	Cyber defense model using artificial intelligence
7	Anitha A, Kalra S, Shrivastav	A Cyber defense using artificial home automation system using IoT	Cyber defense model using artificial intelligence for home automation via IoT
8	Memon SK, Shaikh FK, Mahoto NA, Memon AA	IoT based smart garbage monitoring and collection system using WeMos and Ultrasonic sensors	System for smart garbage monitoring and collection using WeMos and ultrasonic sensors
9	Siddique, MJ, Islam MA, Nur FN, Moon NN, Saifuzzaman M	BREATHE SAFE: a smart garbage collection system for Dhaka city	System for smart garbage collection in Dhaka city
10	Gollakota AR, Gautam S, Shu C-M	Inconsistencies of e-waste management in developing nations–facts and plausible solutions	Study on e-waste management issues in developing nations
11	Ayilara M, Olanrewaju O, Babalola O, Odeyemi O, Odeyemi O	Waste management through composting: challenges and potentials	Study on waste management through composting

3 Proposed System

The proverb goes beyond this statement, which states that cleanliness comes second only to godliness in human civilization, to emphasize the importance of maintaining a clean environment for health. That being said, managing rubbish in streets requires more than just reciting the proverb accurately. The aim is to decrease the duration needed for waste disposal by utilizing the real-time garbage level data, thus optimize the waste collection routes. (3) To reduce the need for human labor. (4) Let waste collectors arrange when to take up garbage every day or every week. (5) To mechanize the waste observation procedure.

Node MCU: In order to make it relatively simple to modify the Arduino IDE to support alternate tool chains and enable the compilation of Arduino C/C++ to these new processors, Arduino.cc had to modify the Arduino IDE as they started creating new MCU boards based on non-AVR processors, such as the ARM/SAM MCU and used in the Arduino Due. Introducing the Board Manager and SAM Core helped them achieve this. The set of software elements needed by the Arduino IDE and Board Manager to assemble an Arduino C/C++ source file down to the target MCU's machine language is called a "core". There is an Arduino core for the ESP8266 WiFiSoC accessible at GitHub, which was made by some inventive ESP8266 fans. 802.11b/g/n protocol, integrated TCP/IP protocol stack, soft-AP, Wi-Fi Direct (P2P). We can send the receiver side with the dustbin specifics thanks to the Wi-Fi Module.

Sound Sensing Device: This sensor tells you the exact amount of trash in the trash can and measures distance. The transmitter sends out ultrasonic waves, and the receiver picks up the waves that are reflected off of an object.

MQ135: A temperature, humidity, and sound sensor for monitoring noise pollution; a gas sensor for identifying hazardous substances. Every sensor is controlled by a microcontroller that transmits data from the smart bin via a LoRa linked communications module.

LCD Display: The underlying idea of liquid crystal displays (LCDs) is that molecules tend to untwist when an electrical current is supplied to them. This changes both the angle at which light enters the polarized glass molecule and the angle at which the top polarizing filter is located. As a consequence, a small portion of the LCD may see light through the polarized glass. In this instance, the air quality and bin level readings are displayed via LCD.

Schematic Diagram

We suggested installing a round-the-clock surveillance system to keep an eye on trash cans. Selective clearing is designed in this system with an intelligent and well-organized system. To determine how much rubbish is in the dumpster, an ultrasonic sensor is utilized. There is a signal from the dumpster if the containers are full. The corresponding dumpster can then be cleared by staff. With an Arduino esp8266 board, all of these sensors are connected. It may be utilized to regulate any mechanical setup according to the state of the current. Utilizing ultrasonic waves and an ultrasonic sensor, distances can be measured. The ultrasonic wave is sent by the sensor to the target, and it is returned with a reflection.

Schematic illustration for an intelligent waste management system. When the level of garbage is detected by US sensors, the message is sent to the database via Arduino and the internet, whereupon the web administrator receives a notification on the mobile app. The ultrasonic sensors in the bin will detect the current level of rubbish when the system is turned on. By producing sound waves that are over the range of human hearing, ultrasonic sensors are able to accomplish this. By receiving and releasing ultrasonic sound, the sensor's transducer doubles as a microphone. Ultrasonic sensors rely on a single transducer to fire a pulse and pick up the echo. The distance to a target is calculated by the sensor by timing the delivery and reception of an ultrasonic pulse. Operationally, the sensors are simple. When an object or impediment is detected, the 40 kHz ultrasonic pulse it emits through the air bounces back to the sensor. By squaring the travel time with the sound speed, one may approximate the distance. You can tell if the trash is wet or dry by using the moisture sensor. By utilizing additional factors such as electrical resistance, moisture content replacement, dielectric constant, and otherwise neutron interaction, the sensor assesses the volumetric water content of the material it is measuring. Following that, a serial communication will be used to send these values to the ESP8266 WiFi module. After obtaining those values, the WiFi module posts them online to the web server. Following the website's retrieval of the data from the web server, a graphical representation of the trash cans and the moisture content of the waste are displayed. In this application, an ultrasonic sensor is used to measure the amount of waste in a bin. Oriented towards the bin within its lower surface, the ultrasonic sensor is mounted atop the dustbin cap. Thus, the user will receive processed data that shows the amount of waste that has accumulated within the dustbin, which is continuously measured from the top lid position with the aid of a microcontroller. The dustbin has three levels: "Filled," "Half Filled," and "Empty." Three different levels of distance are available, depending on the size of the garbage. "Half Filled" refers to a level between 11 and 20 cm, "Empty" to a level between 21 and 30 cm, and "Filled" to a level between 1 and 10 cm. The code embodies this reasoning, which is also empirically confirmed. The level value is shown in the Android application next to the bin number when trash fills the bin for each of the threshold levels, and an alert message is presented on the webpage. The garbage collection vans will consume less time and fuel as a result of this. This website allows authorities to keep an eye on the condition of trash cans and whereabouts.

To determine the current location of each dust collector and to display the quickest route to the appropriate dustbins, an Android mobile application has been built.

Having a common database with bogus statistics was one of the biggest problems we ran into when creating this system. To target smart garbage collection and smart air monitoring, we could not find any real-world dataset to use in our program. Currently, there is just one air quality dataset available that shows the concentration of several airborne contaminants. 9358 records pertaining to 13 gas concentrations found in a contaminated environment are included in the dataset. In the contaminated urban setting of an Italian metropolis, multivariant sensors were utilized to gather this

Table 2 Database labeling with smart bin status

Test case	Distance level	Weight level	Bin status
Case 1	Low	Low	Unfilled
Case 2	Low	Medium	Unfilled
Case 3	Low	High	Half-filled
Case 4	Medium	Low	Unfilled
Case 5	Medium	Medium	Half-filled
Case 6	Medium	High	Filled
Case 7	High	Low	Half-filled
Case 8	High	Medium	Filled
Case 9	High	Low	Filled

dataset. A weight sensor, an odor sensor, an air monitoring module, and a distance sensor are all part of the smart bin we built. We utilized a sensor to find harmful gasses and air pollutants that could be harmful to people's wellbeing. CO, ethanol, and hydrogen atoms may all be found in the air using a mq135 sensors. As seen in Figure, a sensor was positioned at the side of the junk. CO has a little lower density than air at room temperature. There is a small difference in the concentration of CO and air. In addition to the type of waste inside the bin, incomplete burning of carbon-containing fuels such coal, oil, charcoal, wood, kerosene, natural gas, and propane can further raise the concentration of CO in the environment (Table 2).

4 Overall Design of the Research Object

Fig: Flow Chart

The suggested work's overall flow is depicted in Figure.

The following are the design philosophy and module design:

- A 12 V battery and solar panels are utilized as a power source (energy efficiency).
- The amount of rubbish is detected (working) by an ultrasonic sensor interfaced with a microcontroller; the status of the bin is displayed on an LCD interface, indicating whether it is filled with trash or contains hazardous gases.
- To control air quality, a mq135 sensor is utilized to track the presence of gases in the surrounding area.
- Smart bins are using NodeMCU to connect to IP servers like mobile apps. An IP server receives sensor data, evaluates it, and sends a notification to a sanitation worker near the bins.

5 Conclusion

Since the population has been growing so quickly over the past few decades, there have been more garbage piles. There is a higher concentration of harmful gases (CO) in the surroundings surrounding bins when a municipal corporation disposes of garbage carelessly. The health of people is seriously impacted by prolonged exposure to this gas. Establishing a system that forecasts air pollution and manages waste in order to prevent negative events in the future is essential to raising people's standard of living. The benefits and drawbacks of the available solutions were determined by a thorough review of the literature. The proposed work identified and addressed the limitations of the tradition system. The most effective machine learning classifier for classifying bin status as filled, half-filled, or un-filled was found after a thorough examination of several models using real-time garbage datasets. By obtaining five features as input, the machine learning algorithms were trained. Recall values in a real-time testing setting for the KNN model and logistic regression are 83% and 79%, respectively. The amount of air pollutants at a specific time slot was predicted using an LSTM-based model for sensor time-series data that took into account the prior entries. For predicting the future concentration of gases in the air, the modified LSTM and simple LSTM models have accuracy values of 90% and 88%, respectively. Along with notifications through an alert mechanism, the system offered real-time garbage level monitoring. Four sensors sent data to the Firebase database: one for weight, one for distance, one for odor, and one for air quality. Once the model was trained, a GCP server took the different features and labeled a specific bin. A double check was performed using posterior and prior probability to confirm the system's ambiguity. In contrast to current solutions based on straightforward methods, it was discovered that the suggested work offered an improved accuracy by utilizing machine learning. Deploying our system over a wider area and collecting data for an extended period of time are some of the next steps. For now, the fixed size of the bin makes it easy for the machine learning model to classify bin status. For bin status classification in the future, deep learning techniques may be applied. As things stand, the system is able to forecast a particular level of CO. A mathematical model that takes into account the variation of a single element's effect on various air pollutants found in the air can be developed in the future, and a relationship between various air pollutants can be investigated.

6 Result

The proposed garbage management system includes application of IoT and machine learning with use of ultrasonic sensors integrated near the waste bins as well as Arduino-based microcontroller that works to monitor real-time variation in amount of wastes. The bin level is shown at a mobile app, and web site gives an image view. While the garbage goes over a prescribed boundary, then the system sets off

Fig. 1 Summary of the SGCS system overall

buzzer. Its goal is to improve the city cleaner and create a less polluted environment. The literature review describes several smart dustbin systems highlighting the cost, maintenance and system effectiveness aspects. The use of NodeMCU, Ultrasonic sensors, MQ135 Gas Sensors and a display as an LCD component in order to monitor bin levels and air quality is Methodology. It is shown in that block diagram (Fig. 1) which by emphasized constant …

References

1. Prapulla, S.B., Monika, K.A., Rao, N., Shobha, G.: Smart dustbin: an effective garbage monitoring system science and technology. Int. J. **6**, 7113–7116 (2016)
2. Navghane, S.S., Killedar, M.S., Rohokale, D.V.: IoT based smart garbage and waste collection Bi. Int. J. Adv. Res. Electron. Commun. Eng. (IJARECE) **5**, 1576–1578 (2016)
3. Kasliwal Manasi, H., Suryawanshi Smithkumar, B.: A new method for waste management by utilizing internet of things in smart cities. Int. J. Curr. Trends Eng. Res. 348–353 (2016)
4. Medvedev, A., Fedchenkov, P., Zaslavsky, A., Anagnostopoulos, T., Khoruzhnikov, S.: Waste management as an internet of things-enabled service in smart cities. In: The Smart Spaces Conference Publications Springer International, pp. 104–115 (2015)
5. Schafer, G.: U.S. Patent 5,326,939. The US Patent and Trademark Office is located in Washington, DC, p. 1994
6. Anitha, A., Paul, G., Kumari, S.: A cyber defense using artificial intelligence. Appl. Pharmaceut. Sci. Int. J. **8** 25352–57 (2016)
7. Anitha, A., Kalra, S., Shrivastav.: Used artificial home automation systems with internet of things to fight cyberattacks. J. Pharm. Technol. Int. **8**, 25358–25364 (2016)

IOT Based Smart Dustbin Design and Implementation for Monitoring Under Uncertain Environments

Sayan Roy, Sandipan Jana, Anushka Sarkar, Jayanta Pratihar, and Arindam Dey

Abstract With the excessive increase in urbanization and population, waste generation also increases. One of the significant challenges we face is the handling, disposal, and management of solid waste. Since India is considered one of the most densely populated countries in the world, improper waste management and disposal is a significant issue in our country. That is why The Government of India launched the Swachh Bharat Abhiyan to make India clean, hygienic and healthy. Drawing inspiration from this noble mission, we have proposed the design of a smart dustbin monitoring under uncertain environments. The Dustbin is integrated to perform automatic opening actions on sensing a human motion in its proximity. Moreover, the Dustbin is well equipped with ultrasonic and gas sensors to properly monitor the garbage level and the amount of foul odour inside the dustbin. It sends the live Status of the Dustbin to an IoT server, which allows multiple users to remain notified whenever the Dustbin is complete or the foul smell inside the Dustbin is in excess amount. The entire system is integrated using the microcontroller NODEMCU. This would help in the no-touch disposal of indoor waste, preventing the spread of diseases and helping people maintain proper hygiene. At the same time, it also promises to provide systematic waste management that would prevent the overflowing of dustbins and the spread of foul smells that contain harmful, poisonous gases.

S. Roy (✉) · S. Jana · A. Sarkar · J. Pratihar
Department of Computer Science and Engineering, Budge Budge Institute of Technology, Nischintapur, Budge Budge, Kolkata, West Bengal 700137, India
e-mail: roysayan446@gmail.com

S. Jana
e-mail: sandy7sandipan@gmail.com

A. Sarkar
e-mail: anushkasarkar0807@gmail.com

J. Pratihar
e-mail: jpratihar7974@gmail.com

A. Dey
School of Computer Science and Engineering, VIT-AP University, Amaravati, Andhra Pradesh 522237, India
e-mail: arindam84nit@gmail.com

S. N. Mohanty et al. (eds.), *Explainable IoT Applications: A Demystification*, Information Systems Engineering and Management 21,
https://doi.org/10.1007/978-3-031-74885-1_16

Keywords Waste management · Smart dustbin · IoT · Sensors · Micro-controller

1 Introduction

The increasing population all over the world has made waste management an ever-growing problem worldwide. It has become a significant concern worldwide because garbage and waste materials are often overlooked, disregarded, neglected, and poorly dealt with. As a result, people are suffering from several infectious diseases, and the world is becoming an unhealthy place to live in. Recently, the sudden upsurge in the number of people affected by dengue and other contagious diseases is an alarming example of the consequences caused by improper waste disposal. Mr. Prakash Javadekar, former Union Minister for the Ministry of Environment, Forest and Climate Change of India, said to the Lok Sabha on November 22, 2019 that the amount of plastic waste generated in India is somewhat around 25,000 tonnes every day, out of which, approximately 40% is not collected and left littered in the open environment.

There are a lot of factors responsible for this global problem, the major one being the unavailability of proper dustbins or waste disposal options everywhere within the country. Thus, we can understand that modern techniques for better solid waste collection, management and disposal are the immediate need of the hour and, hence, as responsible citizens to contribute something to eradicate this national and global problem, we have proposed the design and implementation of an Iot Based Smart Mobile Dustbin in this paper. The Dustbin is integrated to open its lid automatically on sensing human motion in its proximity. It remains open for a long time for waste disposal and then closes its lid automatically. This means that people need not to worry about touching the dirty dustbin lid while disposing waste materials, instead the lid opens by itself as soon as the human proximity sensor senses a human approaching towards it. Here, we have used a microwave radar sensor to sense human motion and a servo motor to open and close the dustbin lid accordingly. This would ensure no contact disposal of solids and thus help maintain proper health and hygiene.

Moreover, internally, the Dustbin is provided with two sensors: one for level detection and another for detecting foul odour inside the Dustbin. The garbage level detection and monitoring inside the smart dustbin are done using an ultrasonic sensor. The sensor comes with two separate pins, namely, the Trigger pin and the Echo pin. Ultrasonic sound waves sent by the Trigger pin gets reflected from the garbage present inside the dustbin and received by the Echo pin which calculates the time taken by the sound waves to return. The time is then used to calculate the distance between the dustbin lid and the garbage, thus, letting the user or the concerned authorities know about the present condition of the dustbin.

The Gas sensor is responsible for the lousy odour detection inside the Dustbin. We have chosen the MQ series sensor which is widely used to check the air quality of an environment. It has been observed that the common gaseous ingredients inside dustbins are sulphur and nitrogen containing gases, and this sensor is calibrated to detect these gases specifically. This allows it to detect lousy odours in the dustbin more efficiently and notify the user or the concerned authorities, when necessary.

We have used the microcontroller NODEMCU to control these sensors and make the provisions for automation. NODEMCU has an in-built WiFi module, which sends the data from the ultrasonic and gas detector to the BLYNK IoT server. The BLYNK IoT server provides its dashboard for both computer and mobile devices, displaying the current condition of the Dustbin to the users or authorities as applicable. If the Dustbin gets filled to its maximum capacity or is filled with excessive foul smell, the BLYNK app notifies the concerned authorities/user about the information via an in-app notification and an email. This would help to remind when to empty or clean the Dustbin, thus maintaining proper hygiene. As a result, there would be no overflowing of waste materials, preventing the environment from getting contaminated.

2 Literature Survey

To tackle the issue of cleanliness, numerous countries have implemented diverse technologies to manage solid waste management systems. In [1], a Smart Dustbin is made that can be operated using a LAN server. The Arduino Uno microcontroller reads the garbage level inside the Dustbin. In [2], three sensor types are used. Dustbin levels monitoring is done using the ultrasonic sensor. The Mq series sensor detects smoke or foul odour, and the temperature and humidity sensor RHT03 is used to monitor the temperature of the surroundings. They have used NODEMCU to capture readings from several sensors. Arduino Mega is used to read all the sensor data sent by NODEMCU and send it to the IoT platform, GSM module SIM900A, to establish communication between the master station and IoT platform and ThingsSpeak IoT to provide statistical view and analysis of the data received from 3 types of sensors. Similarly, in [3], they have also used an ultrasonic sensor to track the dustbin level. A buzzer notifies the user when it's completely full. Arduino Uno controls all the components; communication with a server is established via an Arduino ethernet shield and a Yes Wimax router. Data is stored in UBIDOT cloud services, and the UBIDOTS Dashboard and Event Manager display the data and send alerts to authorized individuals when the dustbin is full. In [4], the smart dustbin utilizes a Raspberry Pi as the microcontroller and WiFi module to control and communicate with other devices. An ultrasonic sensor is employed to keep a check on the garbage level inside the dustbin, while three LEDs (red, orange, and green) indicate the status of the dustbin. RFID tags are assigned as unique identification markers for specific dustbins, and an RFID reader is used to gather data from these tags. A PHP server is responsible for collecting the data sent by the Raspberry Pi and presenting it to the relevant authorities. This system offers an efficient and automated approach to managing dustbins and providing real-time information to the appropriate personnel. These smart dustbins do not have any automatic opening and closing system.

The Arduino Uno microcontroller is the predominant choice among users, garnering substantial adoption within the authorship community. Several authors have explored alternative options, such as the NODEMCU microcontroller. In [5], they have used NodeMCU (ESP8266 Wi-Fi module) for connectivity, an ultrasonic sensor for measuring the distance and level of the dustbin, and IFTTT Webhooks for sending notifications and integrating with other web-based services. Jumper wires are used to establish connections between the components. The project also mentions a GSM modem in the literature survey, which alerts the relevant authority when the garbage exceeds a threshold. Still, it is not explicitly included in the proposed system. Intriguingly, specific authors have opted for a hybrid approach, incorporating Arduino Uno and NODEMCU within their projects. In [6], the main components used are Arduino, NODEMCU, Servo Motor, and Ultrasonic Sensors. The project aims to create a prototype of a dustbin that opens its lid upon detecting human hands and waste and sends notifications about the waste level inside the bin using LED indicators. The software component utilized is the Blynk application, which receives reports. This smart dustbin serves as a starting point for a Smart Waste Management System, enabling officials to efficiently manage waste collection based on real-time notifications rather than relying on individual reports. By incorporating these components, the project enhances waste disposal practices and promotes a cleaner and healthier environment.

Most smart dustbins come with an auto open/close feature or level detection feature, but very few deal with the problem of foul odour detection. In [7], the smart dustbin detects human motion using an IR sensor. If a human approaches the dustbin, it will automatically open the lid of the dustbin. If they move away, it will automatically close the lid with the help of a motor. It can also monitor the level of garbage using an ultrasonic sensor and the presence of foul odour inside the dustbin with the help of a gas sensor. If any of them reach the threshold, the GSM module will notify the concerned personnel via messaging. Reference [8] employs an ultrasonic sensor to monitor the garbage level and an mq-2 gas sensor to detect the presence of foul smell inside the dustbin. If either of them exceeds the threshold limit, the concerned authorities are notified in the form of a signal. They have implemented another gas sensor which is used to monitor the air quality around the dustbin. Thus, a hygienic environment is maintained. Transmitting signal and tracking location are implemented using the GSM module and GPS modem. Reference [9] utilizes various devices, including an Arduino Uno as the microcontroller, an I2C Module ADS1115 to extend the analogue signal pins, a servo motor for opening and closing the dustbin lid, an LCD for output display, a GSM module for sending status messages, a buck converter for voltage conversion, and several gas sensors such as MQ-136, MQ137 and MQ-2 for detecting harmful gasses. Additionally, two ultrasonic sensors were employed, one for dustbin lid automation and the other for garbage level monitoring. An IR sensor counted the successful trash drop, while an ESP8266 module facilitated WiFi communication with the BLYNK server. The BLYNK server and its user interface were utilized to display the user's dustbin's status and provide necessary notifications.

Some authors have used numerous components to make their projects, making the architecture complex and costly. Reference [10] utilizes various components for its implementation. The system is built around an Arduino microcontroller, which acts as a controller between the ultrasonic sensors, buzzer, LED, fan, and WiFi module. The project aims to address the issue of overflowing dustbins by providing an intelligent solution. When a person approaches the dustbin, the sensor detects their presence and automatically opens the cover. The level of garbage inside the dustbin can be monitored through an application called Blynk. Once the garbage reaches a certain level, a notification is sent to the cleaning department as a reminder. When the dustbin reaches its maximum capacity, the buzzer starts to beep, and an alert message is sent to the cleaning department. In [11], various devices were utilized. These devices include an Arduino Uno microcontroller as the central control unit, two ultrasonic sensors employed for dustbin lid automation and monitoring the dustbin's garbage level and a servo motor responsible for opening and closing the lid. A water sensor brick was also incorporated to distinguish between dry and wet waste, enabling automatic waste segregation. To establish a wireless internet connection between the Arduino Uno and the database, an ESP8266 module was utilized. The data transmitted by the Arduino Uno was stored and sent to the respective authorities using a MySQL database server. Reference [12] is an RFID-based Smart Dustbin. This system utilises RFID tags, ultrasonic sensors, Arduino, Raspberry Pi, and cloud-based monitoring to ensure effective waste management. It incorporates features such as solar power utilization, authentication mechanisms, distance measurement, lid control, motion detection, and integration with AWS IoT. It contains Django and Python for server-side programming and Aadhar API for user authentication. Reference [13] incorporates IoT technology, including sensors, detectors, and actuators, to create an intelligent system for waste management. The system utilizes dustbins equipped with Ultrasonic Sensors to measure the garbage level inside the dustbin and report the readings to the nearest corporation office through an Arduino board connected to an Ethernet module. Additionally, IR Sensors detect objects near the bins, triggering an alert buzzer to discourage improper disposal. Rain Sensors detect rainfall and prevent water from entering the bins. The system also includes relays and motors to automatically close the bin doors when they reach capacity. The hardware components used in the plan include Arduino UNO, Ultrasonic Sensors, IR Sensors, Rain Sensors, Ethernet modules, and alert buzzers. Software components such as Arduino IDE, HTML, and embedded C language are utilized for programming and data visualization.

3 Motivation

From the detailed literature survey, it can be concluded that we can automate dustbins and monitor their present status easily using IoT, which makes the disposal of garbage more accessible and convenient. Over the years, people have tried to utilize IoT for dustbin automation. As a result, several advancements have been made in this field to create a healthy environment. Some of these inventions include providing an automatic opening or closing option in dustbins to sense human motion in its proximity. This allows for no-contact waste disposal that prevents the spread of harmful diseases. Another such advancement is the detection and monitoring of the dustbin level and, at the same time, keeping the user notified about it through some messaging or other means. This helps to remind people to empty their dustbins whenever they are complete, thus preventing the overflow of bins.

Despite the advancements, there is still a lot to be done to make dustbins more convenient. One of the significant shortcomings we discovered from the above literature survey is that most of the models the respective authors proposed are based on complex architectures of many devices that are difficult to integrate. Not only that, such complex architectures also require a lot of financial resources, thus making those models unaffordable for the common public. To summarize, most of them aimed at creating a dustbin that would be maintained by the government and other similar organizations, not something to be used by every other household. Hence, we found an opportunity to work on this issue and create a model that can be of use for both indoor and outdoor environments.

Another major loophole of the above-mentioned papers is that very few addressed the issue of foul odour detection [2, 7–9]. However, still, none could provide a prominent solution to the problem. The most common gases responsible for the foul odour in garbage are nitrogen and sulphur-containing gases. Still, the sensors in the cited papers do not specifically detect them. In [2, 8], the authors have used an MQ-2 gas sensor to identify foul odour, which is basically a gas leak detection or smoke detection sensor detecting gases like Hydrogen, LPG, Methane, Carbon Monoxide, Alcohol, Smoke or Propane and is preferably used to detect gas leakage in households or in fire alarms. Detecting the presence of smoke or CNG inside the dustbin is not a proposal nor a solution to the problem.

Anilkumar et al. [7] proposed the use of a MQ-3 gas sensor for foul odour detection, which is basically an alcohol detection sensor preferably used in a breathalyzer for detecting alcohol concentration in a person's breath. This cannot be considered a perfect solution to the problem. Ahmed et al. [9] does address the situation in the best possible way up till now as they have used an integrated system of three different gas sensors, namely, MQ-2, MQ-136 and MQ-137 sensors. Although MQ-2 is not a good option, the other two are designed to detect Hydrogen Sulphide (H2S) and Ammonia (NH3). Although this solves the problem of foul odour detection to a greater extent, this system is way too costly, with each of MQ-136 and MQ137 costing around 2500/ – each, making the model unaffordable for ordinary people.

These are some of the drawbacks that we discovered in the existing systems and models where we found a scope of opportunity to work. With waste management and disposal emerging as one of the significant challenges worldwide, we are motivated to utilize this opportunity and address these shortcomings better.

4 Methodology

We have undertaken the following methodologies to achieve our objective. The NODEMCU microcontroller is used for controlling various components. It is compatible with Arduino Integrated Development Environment, which programs the microcontroller. RCWL-0516 microwave radar sensor is used for human motion detection. It works on the principle of the Doppler effect by emitting microwave radiation. It sends a digital input of 1 to the NODEMCU if it senses any human motion in its proximity; otherwise, it sends 0.

SG90 Servo Motor opens or closes the dustbin lid based on the radar sensor's input. If the radar sensor sends 1, the servo moves to the mentioned angle, and the dustbin is open if the radar sensor sends 0, then the servo does not move; thus, the dustbin remains closed.

HC-SR04 Ultrasonic Sensor is used for level detection. This sensor has two pins: a Trigger pin and an Echo pin. Ultrasonic sound waves emitted from the Trig pin are reflected from a nearby obstacle and are received by the Echo pin, which calculates the time the sound wave takes to return in microseconds. With the help of the time taken, we calculate the distance using the formula ($2d = V * t$) and convert it into a percentage.

MQ-135 Gas Sensor has been used to check the presence of foul odour inside the dustbin. It can detect gases like Ammonia, Sulphide, NOx, Alcohol, Benzene, Smoke, CO_2, etc. It has been observed that the common gaseous ingredients inside the dustbin that mostly contribute to the lousy smell inside it are sulphur and nitrogen containing gases. This makes this particular Gas Sensor the perfect candidate for our model. This sensor comes with two pins, a Digital Pin and an Analog Pin that helps it to precisely detect the presence of the harmful, foul odour producing gases. Here, we have used the Analog as it helps us understand the exact amount of those poisonous gases present inside the dustbin.

The Blynk application generates an Email Notification as well as an In-App Notification to alert the user when the garbage level of the dustbin crosses the safety threshold value or when the gas sensor value crosses the safety threshold value to help remind the user that it is time to clean the dustbin. The current live status of the dustbin can also be viewed in the Blynk application, where multiple users can view the percentage of dustbin getting filled and the amount of foul odour present inside it whenever they want from anywhere across the globe.

5 Design and Development

5.1 Hardware Components

NODEMCU: The microcontroller used in this system is capable of controlling all the other modules. It operates on an input voltage of 7–12 V which gives flexibility in powering the system through various sources like batteries or external power supplies, and it has an operating voltage of 3.3 V, which is crucial to ensure that the components connected to it are compatible with this voltage. Additionally, it is equipped with an in-built WiFi module ESP8266 which allows the NodeMCU to connect to the internet, making it suitable for IoT (Internet of Things) applications. The microcontroller also provides 11 Digital I/O pins, these pins are versatile and can be used to interface with various sensors, actuators, or other modules (Fig. 1).

Advantages Over Alternative Components: NodeMCU offers distinct advantages in the realm of microcontroller development boards. Its primary strengths lie in its integrated Wi-Fi functionality, offering seamless connectivity for IoT applications without requiring additional modules. The NodeMCU's compatibility with the Arduino IDE simplifies programming. Its low cost and availability make it an economical choice for various projects. The board's ample memory and processing power, based on the ESP8266 chip, enhance its capability to handle IoT tasks efficiently. Additionally, NodeMCU's onboard USB support streamlines interfacing with computers for programming and power supply.

RCWL-0516 Microwave Radar Sensor: The RCWL-0516 Microwave Radar Sensor is incorporated into this system. It is capable of detecting human motion in its proximity. Operating with an input voltage range of 4–28 V and consuming a maximum current of 3 mA, this sensor provides a versatile detection range of up to 7 m, which can be adjusted according to specific needs. Notably, it offers 360-degree detection, enabling it to sense human motion from all sides. Additionally, it can detect motion through various materials, including plastic, glass, and wood. When

Fig. 1 NODEMCU microcontroller

Fig. 2 RCWL-0516 microwave radar sensor

the sensor detects human motion, it sends a signal of 1 to the NODEMCU, while a signal of 0 is sent when no motion is detected (Fig. 2).

Advantages Over Alternative Components: The selection of the RCWL-0516 Microwave Radar Sensor for detecting human motion offers distinct advantages over alternative motion detection sensors. Unlike traditional infrared sensors, the microwave radar sensor is not affected by ambient temperature changes, ensuring consistent and reliable performance in diverse environmental conditions. Additionally, compared to ultrasonic sensors, the microwave radar sensor provides a more comprehensive detection range allowing for a wider and more flexible scope of motion sensing. Its capability to penetrate various materials sets it apart from passive infrared sensors that may be obstructed by obstacles.

SG90 Servo Motor: This servo motor performs the function of opening or closing the dustbin lid. Operating within an input voltage range of 4.8–6 V, it is synchronized with the radar sensor's output. When the radar value is 1, the servo motor rotates by the mentioned angle, thus, opening the dustbin lid. Conversely, when the radar value is 0, the servo motor remains at 0 degrees, ensuring that the dustbin lid remains closed. The SG90 Servo Motor is an integral component in facilitating the automatic open and close mechanism of the dustbin lid (Fig. 3).

Advantages Over Alternative Components: The SG90 Servo Motor stands out for its precise control, allowing specific angle rotations crucial for the delicate movement of the dustbin lid. Unlike other motors, the SG90's compact design and tailored movement make it an ideal choice for the smart dustbin mechanism.

HC-SR04 Ultrasonic Sensor: The HC-SR04 Ultrasonic Sensor is crucial for monitoring and quantifying the waste level within the dustbin. This sensor operates on a 5 V input supply and draws a current of 15 mA. It offers a detection range of 0.2-4 m, with a measuring angle of 15 degrees and works with a frequency of 40 Hz. Ultrasonic sound waves are emitted by the Trig Pin, while the reflected waves are received by the Echo pin after encountering an obstacle, enabling time calculation.

Fig. 3 SG90 servo motor

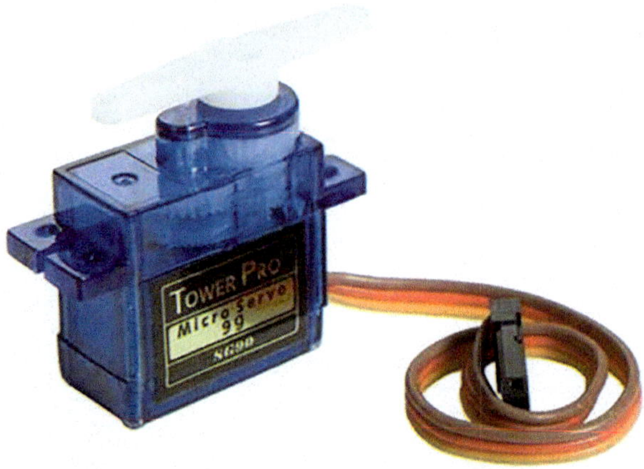

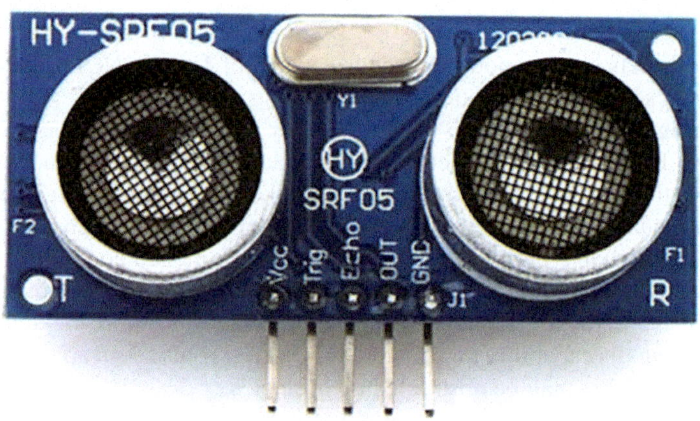

Fig. 4 HC-SR04 ultrasonic sensor

The distance is determined by the formula (2d = V * t), which is then converted into a percentage to ascertain the garbage level (Fig. 4).

Advantages Over Alternative Components: The HC-SR04 Ultrasonic Sensor emerges as the optimal choice for garbage level detection within the smart dustbin system, due to several key advantages. Its primary strengths include accuracy in distance measurement, offering reliable and precise readings in various applications. This sensor operates at a low cost while maintaining high performance. Its simple interface facilitates straightforward integration with microcontrollers like NodeMCU, Arduino, Raspberry Pi, and other popular platforms. Its non-contact measurement method, wide detection range and rapid response time ensures versatility across diverse environments and surfaces.

Fig. 5 MQ-135 gas sensor

MQ-135 Gas Sensor: The MQ-135 Gas Sensor is vital in assessing the air quality within the dustbin, contributing to effective waste management practices. This sensor operates on a 5 V input supply with analog and digital outputs. This system uses the analog output to obtain the precise concentration of gases in the dustbin, measured in parts per million (ppm) (Fig. 5).

Advantages Over Alternative Components: The inclusion of the MQ135 Gas Sensor in our smart dustbin project provides essential advantages over alternative gas sensors for evaluating air quality within the waste container, its cost-effectiveness positions it as an economical yet highly effective solution for our project. This sensor's key strength lies in its ability to detect a range of gases commonly found in dustbins especially sulphur and nitrogen containing gases.

5.2 *Software Components*

Arduino IDE: For our smart dustbin project, we employed version 2.0.2 of the Arduino IDE to develop and configure the NODEMCU microcontroller according to our project specifications. The Arduino IDE served as a comprehensive development environment, enabling us to seamlessly write, compile, and upload code to the NodeMCU. We utilised the platform's libraries to streamline the integration of sensors and other components, simplifying the management of code (Fig. 6).

Advantages Over Alternative Components: Arduino IDE stands out for its simplicity and user-friendly interface, ideal for beginners. It offers an intuitive platform to write, compile, and upload code. Supported by a vast community, it provides ample documentation, tutorials, and examples for troubleshooting and learning. It is compatible across multiple operating systems, thus, ensuring accessibility.

Blynk App: This application functions as a dynamic display, providing users with present and actual status of the dustbin. Furthermore, it enhances user engagement by sending notifications through email. This feature ensures that the users receives

Fig. 6 Arduino IDE version 2.0.2

Fig. 7 Blynk app

updates on time. The integration of the Blynk app not only provides a convenient means for users to stay connected with the dustbin but also enhances the overall user experience by offering a seamless and interactive way to interact with the smart system (Fig. 7).

Advantages Over Alternative Components: Blynk App provides effortless device connectivity, thus, providing a user-friendly platform for connecting IoT devices to the cloud. It simplifies the process of building IoT applications through its drag-and-drop interface, enabling quick and easy creation of interfaces for controlling and monitoring connected devices. It supports a wide range of hardware platforms, ensure secure data transmission, and is compatible with both iOS and Android platforms.

5.3 Circuit Diagram

See Fig. 8.

5.4 Block Diagram

See Fig. 9.

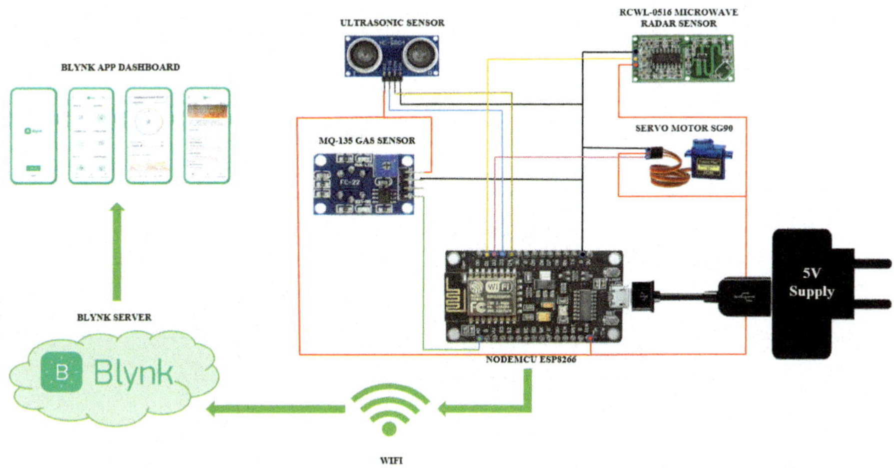

Fig. 8 Circuit diagram

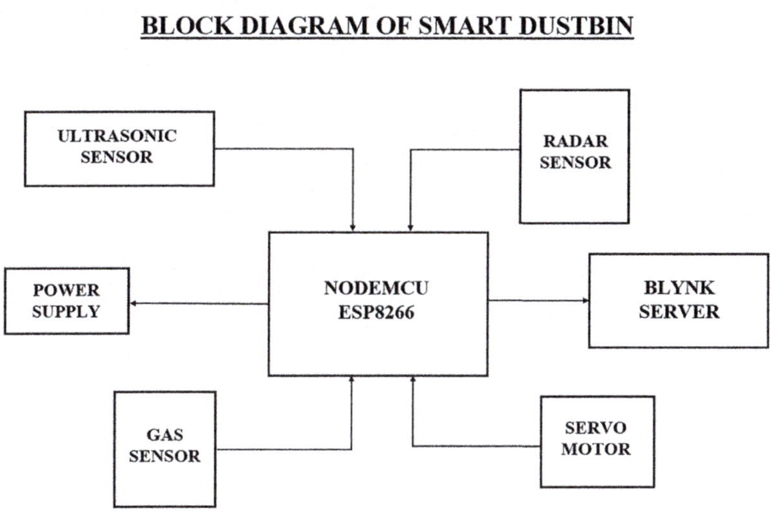

Fig. 9 Block diagram

5.5 Flowchart

See Figs. 10 and 11.

Fig. 10 Hardware flowchart

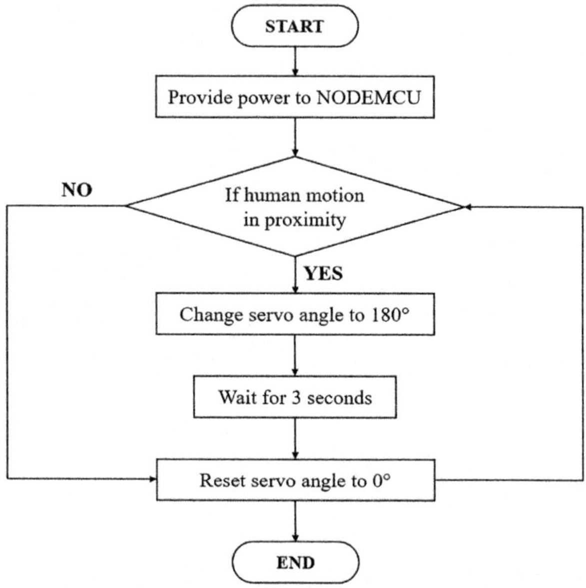

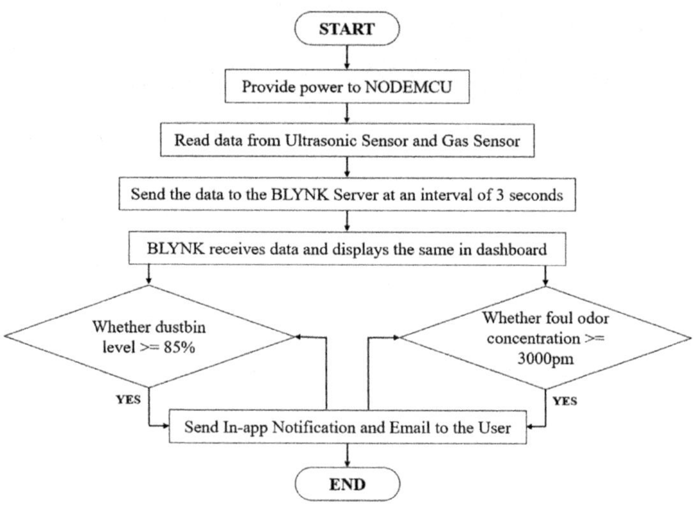

Fig. 11 Software flowchart

6 Result and Discussion

After implementing the connection and coding as mentioned above, we ended up with the following results:

Result 1: The dustbin lid automatically opens whenever someone approaches it in order to dispose garbage. It remains open for a mentioned period of time for garbage disposal and then, closes its lid automatically.

Result 2: The Blynk app generates an email as well as an in-app notification to alert the user when the dustbin is filled with garbage.

Result 3: The Blynk app sends an email as well as an in-app notification to notify the user when the garbage inside the dustbin begins to produce foul odours due to excess poisonous gases.

Result 4: The Current Status of the dustbin can be a viewed in the Blynk app or in the Blynk Dashboard by multiple users simultaneously from anywhere across the globe.

7 Conclusion and Future Scope

In conclusion and as a summary, it can be stated that our model is efficient, cost-effective, and offers solutions to numerous issues. Given our unwavering emphasis on preparing the model for various uncertain environments, this intelligent dustbin finds its utility in various settings, including residential homes, educational institutions, corporate offices, shopping complexes, dining establishments, roads, railway stations and more. Integrating this product into our daily routines can significantly diminish human exertion and expenses, optimize time management, and facilitate the organized disposal and upkeep of waste materials, thus, addressing one of the major global concerns. Ultimately, this endeavour would contribute to creating a tidier, eco-friendlier, and healthier society as a whole.

Although we have successfully solved some significant problems related to waste disposal in our project model, there is still scope for further improvement and modification. One potential future enhancement to improve waste disposal is the implementation of automatic segregation for biodegradable and non-biodegradable waste, enabling effective sorting and separate disposal. This would enhance efficiency and promote environment friendly waste management practices. Another area for improvement is integrating the smart dustbin with popular smart home devices like Alexa or Google Home, enabling users to control it through voice commands for enhanced convenience. Additionally, developing a proprietary IoT server instead of relying on third-party services like Blynk would improve cost-effectiveness and provide greater control and customization options, reducing dependency on external platforms. Overall, these measures would reduce labour and costs while fostering a more systematic and efficient approach to waste management, leading to a healthier and more sustainable environment.

References

1. Selvaraj, K., Chakrapani, A.: Smart dustbin monitoring system using lan server and arduino. Int. J. Adv. Comput. Electron. Eng. **2**(4), 20–23 (2017)
2. Awawdeh, M., Bashir, A., Faisal, T., Ahmad, I., Shahid, M.K.: Iot-based intelligent waste bin. In: 2019 Advances in Science and Engineering Technology International Conferences (ASET), pp. 1–6. IEEE (2019)
3. Soh, Z.H.C., Husa, M.A.A.H., Abdullah, S.A.C., Shafie, M.A.: Smart waste collection monitoring and alert system via iot. In: 2019 IEEE 9th Symposium on Computer Applications and Industrial Electronics (ISCAIE), pp. 50–54. IEEE (2019)
4. Kuppusamy, P.G., Prathyusha, K., Prasad, P., Rasagna, V., Sasi Eswar Reddy, N., Purushotham, N: Smart Dustbin using rfid Reader
5. Maddileti, T., Kurakula, H.: Iot based smart dustbin. Int. J. Sci. Technol. Res. **9**, 1297–1302 (2020)
6. Srinivasan, P., Thiyaneswaran, B., Jaya Priya, P., Dharani, B., Kiruthigaa, V: Iot based smart dustbin. Ann Rom Soc Cell Biol 7834–7840 (2021)
7. Anilkumar, C.S., Suhas, G., Sushma, S: A Smart Dustbin using Mobile Application (2019)
8. Rajapandian, B., Madhanamohan, K., Tamilselvi, T., Prithiga, R.: Smart dustbin. Int. J. Eng. Adv. Technol. (IJEAT) **8**(6), 4790–4795 (2019)
9. Ahmed, M., Shaha, R., Sarker, K., Mahi, R.B., Kashem, M.A.: Design and implementation of intelligent dustbin with garbage gas detection for hygienic environment based on iot. In: 2022 International Conference on Advancement in Electrical and Electronic Engineering (ICAEEE), pp. 1–7. IEEE (2022)
10. Alsayaydeh, J.A.J., Khang, W.A.Y., Indra, W.A., Shkarupylo, V., Jayasundar, J.: Development of smart dustbin by using apps. ARPN J. Eng. Appl. Sci. **14**(21), 3703–3711 (2019)
11. Rambhia, V., Valera, A., Punjabi, R., Chachra, S.D.: Smart dustbins-automatic segregation & efficient solid waste management using iot solutions for smart cities. Int. J. Eng. Res. Technol. **8**(12), 703–707 (2019)
12. Tripathi, A., Pandey, C., Narwal, A., Negi, D.: Cloud based smart dustbin system for metro station. In: 2018 3rd International Conference on Internet of Things: Smart Innovation and Usages (IoT-SIU), pp. 1–4. IEEE (2018)
13. Murugaanandam, S., Ganapathy, V., Balaji, R: Efficient iot based smart bin for clean environment. In: 2018 International Conference on Communication and Signal Processing (ICCSP), pp. 0715–0720. IEEE (2018)

Smart Garbage Monitoring System Using IOT for Commercial Purpose

B. Ravi Chandra, N. Rakesh, Boya Talari Nithin Kumar, Y. Praneeth, and Amjan Shaik

Abstract Making an Internet of Things-based Dustbin is the goal of this project guide. Using a cloud-based Blynk application, we will determine whether the trash can is full or empty. Every 5s, this trash can changes its percentage status, and when it reaches 70% or higher, it notifies you that it is almost full. This Internet of Things trashcan project was built using a Node MCU and an ultrasonic sensor. An exceedingly original method, the IoT Garbage Monitoring Technology proposal will help to maintain cities clean. This device monitors the garbage cans and reports data on the volume of rubbish that has been gathered in the cans via a Blynk cloud page. Through this Blynk application, all information is also transmitted to waste collection vans.

Keywords IOT (internet of things) · NodeMCUESP8266 · Smart bins · Ultrasonic sensors · Route optimization · Waste collection optimization

1 Introduction

Nowadays, the absence of environmental sanitation in reference to refuse management is one of the difficulties influencing the majority of localities and municipalities. This is a result of improper waste collection management. Due to this improper handling, waste spreads across the neighborhood, which in turn leads to hazardous conditions in the area [1–4]. Additionally, it detracts from the area's beauty and promotes a number of serious illnesses in the local population. A garbage monitoring system is created to stop improper trash management and improve societal cleanliness. The design of the Smart Trash Can System utilizing the Arduino ultrasonic sensor and lire detector will be investigated in this research article. As the

B. Ravi Chandra (✉) · N. Rakesh · B. T. N. Kumar · Y. Praneeth
G. Pullaiah College of Engineering and Technology, Kurnool, India
e-mail: chandrabrc11@gmail.com

A. Shaik
St. Peters Engineering College, Hyderabad, India

S. N. Mohanty et al. (eds.), *Explainable IoT Applications: A Demystification*, Information Systems Engineering and Management 21,
https://doi.org/10.1007/978-3-031-74885-1_17

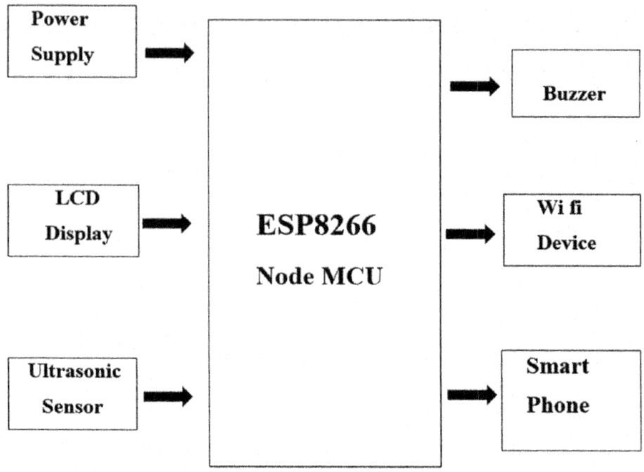

Fig. 1 Block diagram

population expands, so does the quantity of waste in and around metropolitan areas. Here, we offer an IOT-based and sensor-based smart receptacle that runs automatically to assist in resolving this problem [5–10]. Ordinary refuse cans need to be opened by pressing a foot against a lever before being filled with debris.

1.1 Block Diagram

See Fig. 1.

1.2 System Requirements

Hardware:

- ESP8266 Node MCU
- Flame Sensor
- Ultrasonic Sensor
- Gas Sensor MQ6
- LCD Display
- Buzzer
- Wifi module
- Jumper wires
- PCB

Software:

- Arduino IDE

- Embedded C
- Blynk Cloud

1.3 Working Process

The Node MCU, the controller of the system, is connected to two ultrasonic sensors and the Blynk program. The task of seeing an object at the dustbin's mouth falls to a single ultrasonic sensor. The MCU ESP8266 receives the object's distance calculation and transmission. The MCU informs the servo motor whether or not to rotate and open the lid after determining the object's distance. The amount of trash in the trash can is measured by a different ultrasonic sensor. It transmits the reading to MCU, which examines it and, if it rises beyond a certain value, alerts the user via the Blynk Application that the trash can is full and that he should empty it as soon as possible. As they can detect nearly all sorts of flickering flames, the three sensors—MQ6 for methane, Ultrasonic sensor for garbage level measurement, and flame detectors—are perfect for waste treatment facilities. This is crucial for the waste business since flames can come from a variety of materials, such as paper, plastic, and even metals, and they can intensify when the air is contaminated with garbage gas. An exceedingly original method, the IoT Garbage Monitoring Technology proposal will help to maintain cities clean. The MCU is constantly connected to the Blynk app during this time, and the level of the dustbin is constantly shown on the app's interface.

1.4 Node MCU Firmware

Based on the ESP8266 Api 0.9.5, Node MCU is a freely available Network of Things framework that uses the Python programming language. It contains free applications such as spiffs and decision. Originating from the ESP8266 release in December 2013, NodeMCU development began in October 2014, with the first firmware file committed by Hong. It expanded to open-hardware with Huang R's contribution of the gerber file for devkit 1.0. Tuan PM then ported the MQTT client library to ESP8266, enabling MQTT IoT support with Lua. A significant update in January 2015 incorporated u8glib, allowing Node MCU to drive various displays like LCD, OLED, and VGA screens.

1.5 Node MCU ESP8266

A short while after Node micro-controller unit (MCU) was created, the Epress if systems (ESP8266) was released. Beginning on December 30, 2013, Express if Systems began manufacturing the ESP8266. The ESP8266 is an Internet of Things

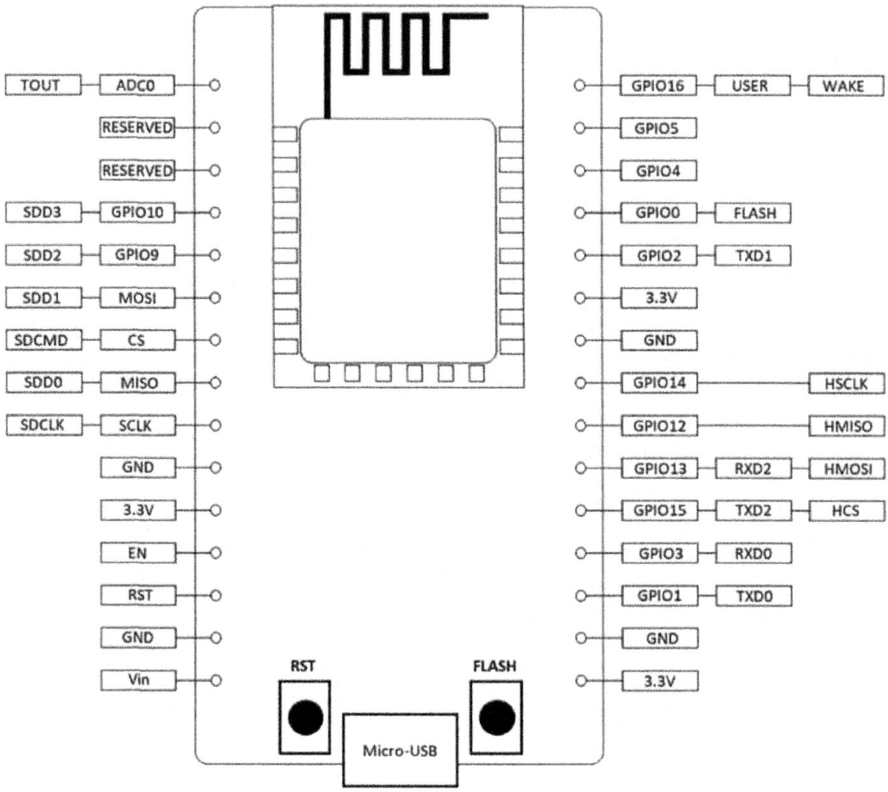

Fig. 2 Pins of node MCU

(IoT) application-specific integrated circuit (SoC) with a Tensilica Xtens LX106 core. The project was extended to incorporate an open-hardware platform after researcher Huang R contributed the Gerber file for the devkit v0.9 ESP8266 board two months after it was created. Soon after, Tuan PM modified the Contiki MQTT client library for the ESP8266 SoC platform in order to contribute to the Node MCU project. By using Lua to connect to the MQTT broker, Node MCU was able to build the MQTT Internet of Things protocol. Devsaurus published a key update for the Node MCU project on January 30, 2015, including u8glib and enabling Node MCU to easily control LCD, Screen, OLED, and even VGA displays (Fig. 2 and Table 1).

1.6 LCD Display

Liquid crystal displays, usually referred to as LCD (Liquid Crystal Display) panels, are currently the displays for electronics that are used the most frequently. If you look around, you should be able to see at least a few of them, perhaps on the landline

Table 1 General I/O pins

Input/Output index	ESPRESSIF systems (ESP8266) pin	Input/output index	ESPRESSIF systems (ESP8266) pin
0 [*]	General purpose I/O 16	7	General purpose I/O13
1	General purpose I/O5	8	General purpose I/O15
2	General purpose I/O4	9	General purpose I/O3
3	General purpose I/O0	10	General purpose I/O1
4	General purpose I/O2	11	General purpose I/O9
5	General purpose I/O 14	12	General purpose I/O10
6	GPIO12		

Fig. 3 LCD display 16 × 2

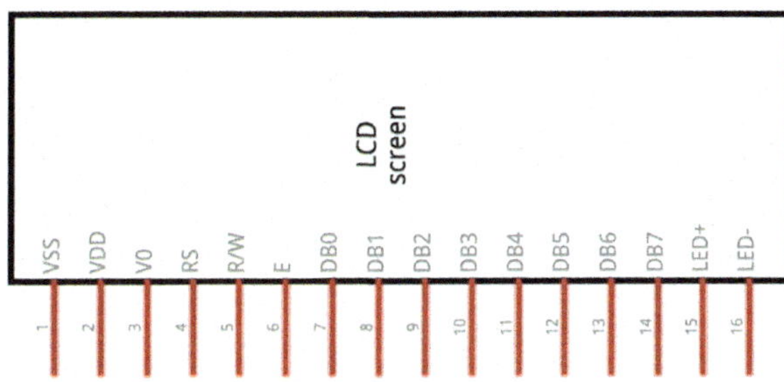

or microwave's caller ID display. Your LCD has a total of 16 pins. What each and every pin is meant to do is listed below: Pins 1 and 16 are where power and ground are situated. The LCD's brightness can be changed via Pin 3. Pins 4–6: These are used to operate the LCD. Pins 7 through 14 are the data lines. Use pins 15–16 to power the LCD's backlight (Fig. 3 and Table 2).

2 Literature Survey

Smart garbage monitoring systems using IoT technology are gaining traction for efficient waste management. Ultrasonic sensors and IoT devices are used to monitor waste levels in bins. Real-time data transmission via the Internet helps optimize collection routes and schedules. These systems aim to reduce operational costs and promote sustainability in smart cities. Research is ongoing to enhance accuracy and scalability, improving waste collection processes. IoT-based solutions promise cost-effective, eco-friendly waste management. Real-time data is crucial for waste collection planning and decision-making. This technology reduces fuel consumption, carbon emissions, and environmental impact. IoT is integral to creating smart, efficient, and sustainable waste management systems.

Table 2 Pins and functions

Pin no	Function	Name
1	Zero volts	Ground
2	5 V voltage supply (range: 4.7–5.3 V)	Vcc
3	Adjusting the contrast using a rheostat	VEE
4	When the command register is low, and when the data register is high	Register choosing
5	Low for register writing; high for register reading	Read/write
6	When a high to low pulse is provided, it transfers data to the data pins	Enable
7	Pins for 8-bit data	Data bit 0
8		Data bit 1
9		Data bit 2
10		Data bit 3
11		Data bit 4
12		Data bit 5
13		Data bit 6
14		Data bit 7
15	VCC for backlights (5 V)	Led+
16	(0 V) backlight ground	Led−

2.1 Proposed System

- An Arduino program is used to send a warning signal to the relevant authorities if the level of gas, flame, or dust exceeds the set point. One sensor employs an ultrasonic signal to monitor the quantity of rubbish in the dust bin and gases with flame.
- As a result, the environment is kept clean. Signal transmission and data tracking are done using the ESP8266 NodeMCU Wifi module (Fig. 4).

3 Results and Discussion

3.1 Hardware Deployment

Hardware deployment of a smart garbage monitoring system using IoT involves the placement of sensors in garbage bins, positioning microcontrollers nearby, and configuring IoT communication modules for data transmission. Power supply sources, such as batteries or solar panels, are essential. A network infrastructure,

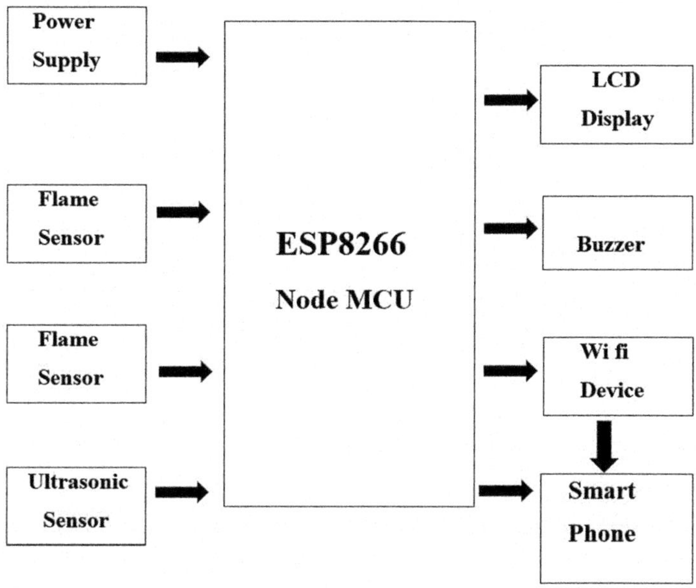

Fig. 4 Traditional versus smart dustbin

cloud server, and protective enclosures ensure data connectivity, storage, and hardware protection. Implement data validation, error handling, and user-friendly interfaces. Thorough testing, calibration, and scalability considerations are necessary. Deployment in target areas, regular maintenance, and data security measures are vital for a successful implementation.

3.2 Running Software

The A bootloader must be flashed via ST-Link (SWD) or USB to Serial. Read about flashing the bootloader. Note that before uploading a sketch, the board may need to be put into "perpetual bootloader" mode once the bootloader has been flashed. To do this, connect a resistor to pin PC14 and 3.3 V, then reset the board. After removing the resistor and restarting the board, you should now be able to flash a blank sketch. At that point, uploading new sketches ought to function as planned. You might need to modify the maple-upload script in the tools-folder if you discover that the IDE successfully resets your board but that dfu-util complains about no DFU-devices being present. Increase the value assigned to the line where it calls upload-reset.

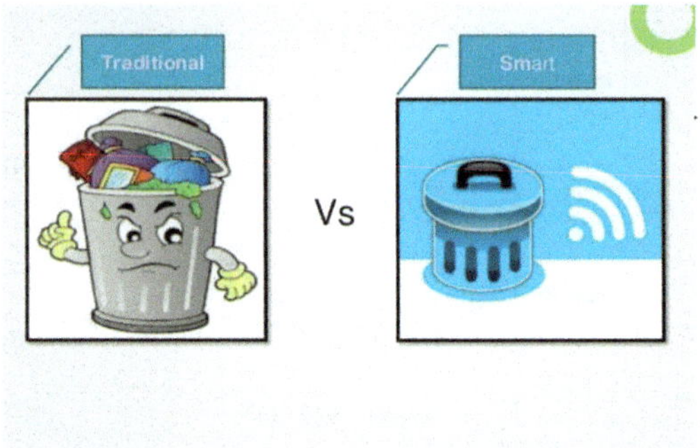

3.3 Application Blynk

Blynk was created with the World Wide Web of Things in heart Furthermore to many additional remarkable attributes, it can store and analyze data, present sensor data, and wirelessly operate devices. The structure of the platform consists of 3 primary components: **Blynk Library services, Blynk application, and Blynk Host**.

3.4 Features

Utilizes IoT-connected sensors (e.g., ultrasonic or infrared) to monitor garbage levels in real time.

Relies on IoT communication modules (e.g., Wi-Fi, LoRa) to send data to a central server.

Stores collected data in the cloud, enabling access from anywhere. Employs data analytics to optimize waste collection schedules, routes, and resource allocation. Sends notifications for bin status, facilitating efficient collection management. Offers a web or mobile app for user interaction, enabling real-time monitoring and historical data access May use low-power components or renewable energy sources to reduce environmental impact. Enhances operational efficiency, reducing collection costs and fuel consumption. Adapts easily to monitor additional bins or areas in a scalable manner. Implements robust data security and privacy measures to protect sensitive information and ensure compliance.

3.5 Result Analysis

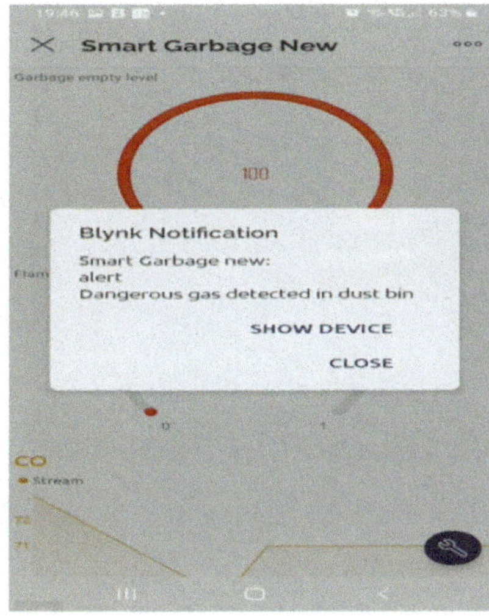

Scenario-1

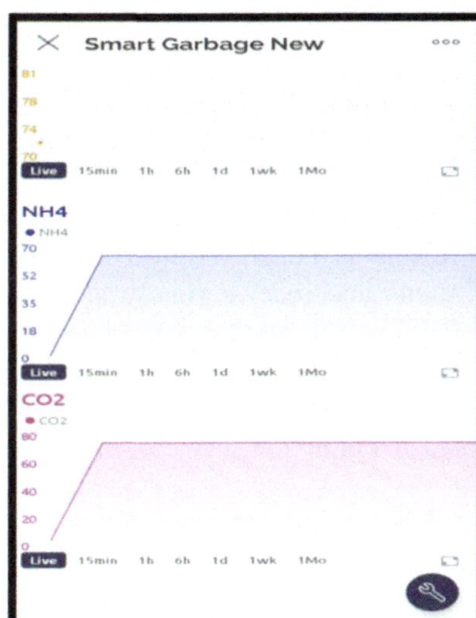

Scenario-2

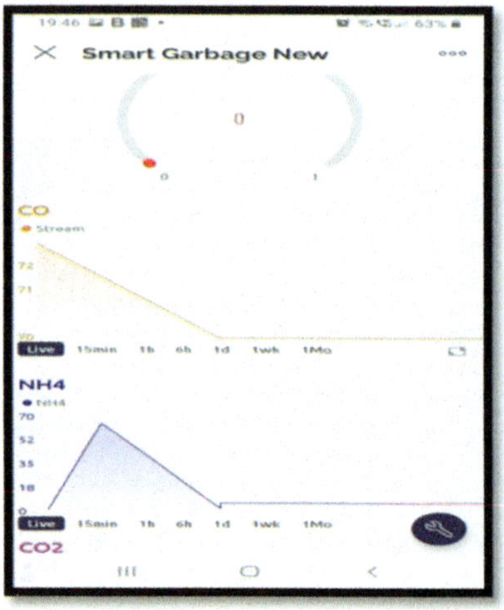

4 Utilization

1. It is characterized by its efficiency in waste collection and resource allocation.
2. By providing real-time data on garbage bin levels, it allows waste management authorities and service providers to optimize collection routes and schedules, reducing fuel consumption and operational costs.
3. This not only enhances the overall environmental sustainability of waste management but also contributes to the development of smarter and more sustainable cities.
4. Residents benefit from a user-friendly interface, enabling them to track bin status and receive timely alerts, which, in turn, minimizes the risk of overflowing bins and associated health and hygiene issues. Furthermore, this system fosters community engagement by raising awareness about responsible waste disposal practices and can be easily scaled to cover larger areas or additional waste collection services, making it adaptable to various urban environments.

4.1 Benefits

1. Optimizes waste collection routes, reducing fuel consumption and operational costs.
2. Minimizes carbon emissions by preventing unnecessary waste collection trips.
3. Enables efficient allocation of resources and personnel for waste management.
4. Provides real-time data for prompt maintenance and issue resolution.

5. Supports informed decision-making for waste management authorities and service providers.
6. Encourages public participation in responsible waste disposal practices while reducing health and hygiene concerns.

4.2 Negative Aspects

Smart garbage monitoring systems often outweigh the negative aspects, especially in terms of efficiency, cost savings, and environmental impact.

5 Conclusion

As real-time data systems in waste management become increasingly integral to urban infrastructure, their potential to revolutionize city living becomes even more pronounced. Beyond the immediate advantages of optimized collection routes and reduced operational costs, these systems lay the groundwork for a paradigm shift in how communities approach waste disposal. The efficient resource allocation facilitated by these technologies not only curtails fuel consumption and carbon emissions but also contributes to the broader global effort to combat climate change. While challenges such as initial implementation costs, data security, and maintenance are acknowledged, they serve as stepping stones toward refining and strengthening the long-term viability of these innovative solutions. One of the remarkable aspects of these systems is their ability to empower residents in the sustainability journey. By providing real-time insights into bin levels, individuals are not only informed but are actively encouraged to partake in responsible waste disposal practices. The prevention of overflowing bins, a direct outcome of this technological intervention, translates not only to cleaner and more organized neighborhoods but also to tangible improvements in public health. The future trajectory of waste management is undeniably intertwined with the continued development and widespread adoption of these cutting-edge systems. Looking ahead, the transformative impact of these technologies extends beyond mere operational efficiencies. They herald a cultural shift toward environmental consciousness and community engagement. As these systems evolve and become more ingrained in the urban fabric, they act as catalysts for a broader conversation on sustainability. The promise of cleaner, more efficient, and sustainable urban environments is not just a technological aspiration; it represents a fundamental shift in how cities conceptualize and manage their waste. The trajectory set by the adoption of these systems holds the potential to redefine the urban landscape, creating resilient and environmentally responsible communities for generations to come.

References

1. Kolhatkar, C., Joshi, B., Choudhari, P., Bhuva, D.: Smart E-dustbin. In: 2018 International Conference on Smart City and Emerging Technology (ICSCET), pp. 1–3 (2018). https://doi.org/10.1109/ICSCET.2018.8537245
2. Reddy, P.S.N., Naik, R.N., Kumar, A.A., Kishor, S.N.: Wireless dust bin monitoring and alert system using Arduino. In: 2017 Second International Conference on Electrical, Computer and Communication Technologies (ICECCT), pp. 1–5 (2017). https://doi.org/10.1109/ICECCT.2017.8117960
3. Parashar, S., Tomar, P.: Waste management by a robot- a smart and autonomous technique. J. Electron. Commun. Eng. **13**, 31–36 (2018). https://doi.org/10.9790/2834-1303023136
4. Fitzwatler G. Ang, Maria Karla Angel R. Gabriel, Jameson Sy, Jenny Jane O. Tan, Alexander C. Abad Department of Electronics and Communications Engineering De La Salle University, Man Ila2401 Taft Ave., Malate, Manila 1004, Philippines
5. Urlagunta, Nagaraju Smart Dustbin for Economic Growth (2017)
6. Sinha, A., Gupta, K., Jamshed, A., Singh, R.K.: Intelligent Dustbin: A Strategic Plan for Smart Cities
7. Rajapandian, B., Madhana, K., Tamilselvi, T.R.: Smart dustbin. Int. J. Eng. Adv. Technol. (IJEAT) **8**(6), 4790–4795 (2019). https://doi.org/10.35940/ijeat.F9359.088619
8. Omar, M.F.A.A.A., Zainal, D., Wahap, N.A., Ismail, N.M., Ahmad, N.: Implementation of spatial smart waste management system in Malaysia. In: IOP Conference Series: Earth and Environmental Science, vol. 37, p. 012059 (2016). https://doi.org/10.1088/1755-1315/37/1/012059
9. Gayanthika, W.A.L., Maduranga, G.K.C.D., Silva, A.I.S.: Smart dustbin for waste management. Int. J. Environ. Sci. Dev. **10**(4), 118–121 (2019). https://doi.org/10.18178/ijesd.2019.10.4.1159
10. Radhika, V., Rukkumani, V., Devasena, D., Ramya, R.: Smart waste management system using IoT. Int. J. Res. Arts Sci. **5**, 65–72 (2019). https://doi.org/10.9756/bp2019.1001/08

IoT Based Smart Home Systems

Akshet Patel, Shrey Sharma, and Princy Randhawa

Abstract The Internet of things is no longer the "Technology of the Future." We are living in a world of connected devices where every little device from smart watch to Wrist bands to Lamps are connected to the internet and sharing the data to the cloud platforms which makes them accessible from any part of the world via a Smartphone or a PC. This is the era of connected living; this is the era of the Internet of Things. IoT is the technology in which anything that can be controlled or monitored is connected to the cloud or Internet. Every individual device that is connected to the internet is having a unique address within the network. These IoT edge devices can communicate to other edge devices in the network and can be easily monitored or controlled with the help of a smartphone or PC. The Internet of things has numerous applications; however the most popular one is IoT based Smart Home Systems.

Keywords IoT · Internet of things · Smart homes · Smart systems · Smarden · Automation

1 Introduction

Turning on the lights when the motion is detected is not considered as an application of an IoT based smart home system. A Smart Home system consists of various parts of the home to be smart enough to control and maintain themselves without human presence. These various parts consist of Lighting, Heating, Cooling, Surveillance, and security. A system that is connected to the Internet and can share the data related to its state to the homeowner. Further the owner should be able to take actions or must be able to control the appliances and the actions in a smart home remotely via

A. Patel · P. Randhawa (✉)
Department of Mechatronics, Manipal University Jaipur, Jaipur, Rajasthan 303007, India
e-mail: princyrandhawa23@gmail.com

S. Sharma
Smarden, New Delhi 110034, India

S. N. Mohanty et al. (eds.), *Explainable IoT Applications: A Demystification*,
Information Systems Engineering and Management 21,
https://doi.org/10.1007/978-3-031-74885-1_18

a mobile application or dashboard, for instance predictive maintenance alert based on the power consumption parameters of the device or scheduled actions of multiple appliances or instant alert notifications when any unwanted motion is detected. All these along with all the other various use cases define a smart Home or an Intelligent home that is capable enough to alert its owner in case of emergencies and convenient enough that it can be controlled from any part of the world, at the same time saves energy, time, and money [1]. There was a time owning a Smart T.V. was luxury and now it is a necessity, owning a Smartphone was luxury and now a necessity and soon within a short span of time, owning a smart home will also become a necessity. This is all because our life is now dependent on DATA. Data has become an overly critical asset, and we all consume and compute massive amounts of data every day, our day to day life is dependent on data coming from multiple sources and similarly we need the data from our homes as well. We need this data to make life much more secure, convenient, and comfortable [2].

1.1 DIY Smart Home Design

It is possible to design a complete smart home environment for your home by yourself. However, there are a few things that anyone would need to understand before jumping straight into development and coding. There is an extremely popular saying when it comes to the technology of the Internet of things—"Just because you can do it, don't do it."

As fascinating as it sounds, the technology of the Internet of things opens a portal of security breaches if the proper security measures were not taken properly. Connecting anything to the cloud provides hackers with another gateway to get into your network and hack into your system. Even the door sensor that you are using to get the notifications from your main door could act as an edge device through which the hacker gains access to your home [3].

With proper security measures and proper architecture, one can design a complete smart home system on his own. To design a complete Smart Home system based on the technology of Internet of things, one must understand the architecture of IoT and must know how the devices communicate within an IoT network and how the data is transferred from one device to another or to the cloud and at the end, how we can interact with the data and perform operations on it. To get a closer look of the overall process, let us focus on the—Architecture of IoT [4].

1.2 Architecture of Internet of Things

For any technology, its architecture is the most important thing as it defines the overall working and the components of the technology. Without understanding the architecture, it is impossible to design products and solutions based on the technology.

Similarly, to design any product or solution based on the technology of Internet of things, one must understand its complete architecture [5].

One important thing to shift the focus before getting into the architecture—Internet of Things is a technology which is a stack of multiple things and a combination of both Hardware and Software with a combination of multiple technologies that helps us design IoT hardware and IoT software. One must focus on both aspects to design a complete solution.

Let us start with basic layers of architecture of IoT and we will dig deeper into each layer to understand the functionality and the technology.

There are 3 Basic layers that makes up the complete Architecture of IoT

- The Perception Layer
- The Network Layer
- The Application Layer

The Perception Layer

The perception layer is the basic layer of IoT, it is the first end point or the data collection layer, this layer consists of all the hardware that is connected in an IoT network. This layer consists of all the IoT edge devices, the edge devices are the IoT based hardware that are connected to the cloud or internet via some communication protocol [6].

Further the perception layer is divided into two parts—Sensors, Actuators, Relays and Micro-controllers.

Sensors, Actuators and Relays

These three are the basis constituents of any IoT based hardware, if we investigate the applications of IoT, the IoT devices are either used to monitor a certain asset or they are used to control the asset, and for both the applications—controlling and monitoring, we will be needing sensors, actuator, or relays [7].

- Sensors

A Sensor is an electronic component that is used to convert the physical data from the environment like temperature, humidity, motion, force, pressure, etc. in electrical or digital signals so that these electrical signals can be further fed into a microcontroller and one can perform on that data.

- Actuators

Actuators are those devices that are used to convert the electrical signals from the micro-controllers to some mechanical movement for e.g., stepper motors, servo motors, etc.

- Relays

Relays are used for the purpose of switching; it works on the signals from the microcontroller and helps in switching an electrical circuit.

Micro-controllers

Micro-controllers are the brain of any IoT based hardware that helps us to monitor or control anything. The edge devices are incomplete without any micro-controller, as it is the computational engine for any edge device and helps to process the data. The microcontroller is a single chip device that has a microprocessor, RAM, ROM, Clocks, Timers, GPIOs, etc. all together on the same circuit and can execute multiple tasks at same time.

It is impossible to design an IoT Edge device without using any microcontroller as the sensors, actuators and relays are connected to the microcontrollers and transfer the data to it so the microcontroller can perform operations on it. Similarly, the communication protocol device is also connected to the same microcontroller so that the data transmission can happen in the IoT network.

So if we take a look at the perception layer it is the layer that has all the hardware components in it, the sensors along with the micro-controller attached to any asset to control or monitor it makes it a part of the perception layer, In our case of smart homes the IoT edge devices could be anything like a temperature sensor, Motion sensor, Door Sensor, Smart Switches, smart bulbs, etc. and further we'll see how we can convert any existing thing to smart thing to monitor and control it from anywhere in the world with the help of a web-dashboard [8].

Network Layer

The network layer is the second layer in the IoT architecture, and it helps to collect the data from the perception layer and transfer it to the cloud or to the Internet. The edge devices need to get connected to the internet to make it accessible within the IoT network or from anywhere in the world with the help of a smartphone mobile app or a web dashboard.

Just like the perception layer has its multiple components like sensors and microcontrollers, the network layer is also divided into four distinct parts—the communication technology, gateways, IoT protocol and the cloud [9].

2 The Communication Technology

The edge devices must be connected to some external device that is responsible to transfer the data collected from the IoT device and needs to be transmitted to some other device or must be uploaded to the cloud [10].

Let us assume a situation where you need to connect a laptop/PC to the internet. What are those diverse ways by which you will be able to do so?

- Wi-Fi
- GSM
- LAN/Ethernet

These are the three commonly used technologies that connect any hardware to the internet but, in case of IoT edge devices, the device is designed to perform a specific task under some specific conditions and that's where we need multiple communication technologies like—Wi-Fi, Zigbee, Z-wave, LoRa, Sigfox, NRF, NB-IOT, RF, BLE, etc. came into picture.

There could be multiple factors before choosing any specific communication technology, here are a few things that one can keep in mind before selecting the technology.

- **Bandwidth**: Refers to the spectrum of frequencies contained within a specific range, crucial for transmitting signals. In the realm of computing, bandwidth represents the maximum data transfer rate along a designated pathway.
- **Latency**: It denotes the delay or the duration it takes for data to traverse from the sender to the receiver.
- **Power**: It is an integral to electronic devices for communication and signal transmission. However, the focus shifts towards energy, which encompasses power consumption over a defined period.
- **Service Level**: These are arrangements sought by industrial clients from their wireless service providers, outlining terms regarding connectivity, installation, and maintenance.
- **Topology**: These necessitates determining the configuration of devices, whether requiring point-to-point connections, server-client setups, or other network arrangements.
- **Range**: It signifies the furthest distance between a client and server within which signals can efficiently travel.
- **Value**: It pertains to the worth of a functional device to end-users or the price they are willing to pay for connectivity.
- **Product life**: Product Life spans in industrial settings typically exceed those of consumer devices, influencing the longevity of both the products and the wireless technology employed.
- **Form factor**: Form Factor considerations involve assessing device size, which can impact the selection of communication technologies.
- **Security**: It involve assessing device size, which can impact the selection of communication technologies. When selecting wireless technology, it is crucial to prioritize security. Some networks provide higher levels of security than others.

3 Gateways

As we have discussed, there are multiple other technologies that we can use in IoT networks while building an IoT edge device and not all of them provide direct access to the internet, many of them are meant for machine-to-machine communication as they act as the transceiver modules. Such communication technologies can be used in multiple areas depending on the application and numerous factors as we discussed above, however we eventually need their data on the cloud platform to control or monitor such devices and here the IoT gateways come into the picture. IoT Gateways are the devices that provide internet connectivity to those transceiver modules that can transfer and receive data from similar devices/modules. These IoT edge devices relate to the gateway and then gateway uses Wi-Fi/GSM/Ethernet or all of them to connect to the internet, this way the data from an IoT edge device based on Zigbee, BLE, Z-wave, LoRa etc. can be uploaded to the cloud via using an IoT Gateway [11].

Smart Application Layer

Once the data is uploaded from the IoT edge devices to the cloud, we need to perform operations on that data to create an application out of it. It is a service that provides users to take control over their IoT Edge devices. Smart Applications could be a Web Application or a mobile application or even a VUI application that helps the users to control and monitor the IoT edge devices [12].

Connecting Anything to the Internet

The Internet provides us the capabilities to control and monitor anything remotely via a mobile app or a web dashboard and by following the architecture of IoT we can connect anything to the internet.

Starting with the perception layer the first thing to do is to focus on the hardware part. We need to develop a smart home network and to start with that we need to define the structure or the architecture of the Home Automation System that we will be developing. Here are the results that we need.

- Smart Switches
- Smart Temperature and humidity Sensor
- Smart Motion Sensor
- Smart Door Sensor

To build these devices we also need to focus on the type of communication protocol that we need to accomplish the project. For a smart home system there are multiple communication protocols that we can use, like Wi-Fi, Zigbee, Z-wave, BLE, etc. but we will go ahead with the Wi-Fi as it matches our overall requirements, and it is very affordable.

The Microcontroller that we'll be using for this project is ESP8266-12E and we can also work with an open source MQTT broker—HIVE MQ which provides a free MQTT broker service or we can use any IoT Platform like—SmarDen.tech, Thingspeak, Ubidots, Adafruit, Etc. which provides the complete backend, cloud infrastructure, MQTT protocol and the front end mobile app or web dashboard to control our devices.

4 Data Transmission

The next most important part of building an IoT network is to choose the data transmission protocol, which can be HTTP, MQTT or CoAP. Mostly, MQTT is most widely used when it comes to IoT devices. Let us look at the MQTT protocol in detail and then we will use MQTT to transfer the data from our device to the local servers and vice-versa [12].

4.1 MQTT Protocol

MQTT, which stands for MQ Telemetry Transport, serves as a lightweight messaging protocol suited for scenarios with limited bandwidth. It employs a publish/subscribe mechanism, particularly useful for machine-to-machine (M2M) communication.

Developed to address the constraints of bandwidth and CPU resources in embedded environments, MQTT was designed with minimal overhead. It finds particular utility in wireless networks characterized by variable latency due to bandwidth limitations or unreliable connections. Industries such as automotive, energy, and telecommunications have adopted this protocol.

Originally conceived as a proprietary solution tailored for interfacing with SCADA systems within the oil and gas industry, MQTT has evolved into the primary open-source protocol facilitating connectivity among Internet of Things (IoT) and Industrial Internet of Things (IIoT) devices [13].

a. **MQTT Function**

The communication model of MQTT, known as publish/subscribe (pub/sub), optimizes available bandwidth in comparison to traditional client–server architectures. In pub/sub, the publisher operates independently of the subscribers. Brokers serve as intermediaries, connecting publishers with subscribers since direct communication between them is not established.

MQTT clients facilitate the publishing and subscribing of messages. The same MQTT client can handle both tasks. A device, acting as a client, transmits information to a server, or broker. Conversely, subscribing involves customers connecting to a broker and selecting the topics they wish to receive updates on.

In the event of a client losing connection with a broker, the broker retains messages and delivers them once the client reconnects. Similarly, if the connection between the publishing client and the broker is unexpectedly disrupted, the broker may dispatch cached messages to subscribers, including publisher directives [14].

b. **MQTT Broker**

An MQTT broker functions as an intermediary, facilitating the transmission of messages between clients and subscribers within the MQTT network. Conceptually akin to a post office, the broker serves as a centralized hub through which all communications must traverse before reaching their intended recipients.

When selecting an MQTT broker, organizations should carefully evaluate several factors to ensure it aligns with their requirements. Scalability is crucial, as the broker should be capable of handling increasing message loads as the network grows. Integration capabilities are also paramount, as the broker must seamlessly interface with existing systems and technologies within the organization's infrastructure.

An effective monitoring tools should be in place to track the performance and health of the MQTT broker, enabling administrators to promptly address any issues that may arise. Lastly, the broker should exhibit robust failure resistance mechanisms to mitigate the impact of potential disruptions, ensuring uninterrupted communication within the MQTT network.

In essence, the MQTT broker plays a vital role in facilitating efficient message exchange within the MQTT network, and organizations should prioritize considerations such as scalability, integration, monitoring, and failure resistance when selecting a suitable broker for their needs [15].

c. **MQTT-Message Types**

An MQTT session encompasses several stages: connection, authentication, communication, and termination. To initiate a session, a client establishes a connection with the broker via TCP/IP, utilizing either a standard or customized port designated by the broker's administrators. It's important to note that when establishing a connection, the server may restart an existing session if the client provides recurring identification.

For non-encrypted communication, the standard ports are 1883, while 8883 is used for SSL/TLS (TLS) encrypted communication. During the SSL/TLS handshake, the client verifies the server's certificate, and optionally, the client may present its own certificate to authenticate its identity to the broker. Despite not being part of the MQTT protocol, brokers often support SSL/TLS client-side certificates.

During the conversation phase, clients can perform various actions such as publishing, subscribing, unsubscribing, and pinging. Publishing involves disseminating the publisher's content to a specific topic. MQTT allows for the transmission of BLOBs up to 256 MB, accommodating application-specific content formats.

Subscription and unsubscription to topics are achieved using SUBSCRIBE/SUBACK and UNSUBSCRIBE/UNSUBACK messages. Topics are organized in a hierarchical structure using forward slashes (/), enabling the creation of a natural topic tree. Wildcard characters, including the plus (+) and hash (#), allow clients to subscribe and unsubscribe from topic branches dynamically. The dollar sign ($) is reserved for server- or system-specific communications, precluding subjects from wildcard root subscriptions.

Clients can ping the broker server using PINGREQ/PINGRESP messages to maintain a live connection, ensuring that TCP connections remain active and preventing interruptions from gateways or routers.

To terminate an MQTT session, clients can send a DISCONNECT message to the broker. A graceful shutdown enables clients to easily rejoin the session by providing their client identification, seamlessly resuming where they left off. If a publisher disconnects abruptly without sending a DISCONNECT message, the broker may furnish subscribers with a cached message from the publisher, offering guidance in case of sudden publisher termination [16].

d. **MQTT Packets**

CONNECT, PUBLISH, and SUBSCRIBE are MQTT messages.

e. **MQTT's Advantages**

MQTT's lightweight characteristics and minimal overhead allow seamless data transmission with little bandwidth and decrease CPU and RAM burden. MQTT's benefits versus competitors include.

Lightweight protocol provides effective data transmission and rapid implementation, low network utilization owing to decreased data packets, efficient data distribution, successful remote sensing and control, fast, efficient message delivery, utilizes tiny quantities of electricity, and optimizes network capacity.

f. **MQTT's Drawbacks**

- MQTT's send cycles are slower than CAP's (CoAP).
- CoAP employs a stable resource discovery approach against MQTT's flexible topic subscription.
- Unencrypted MQTT. It employs TLS/SSL for security encryption.
- It is hard to scale MQTT internationally.
- Security, interoperability, and authentication are MQTT problems. MQTT is utilized in secure back-end networks for application-specific reasons because it was not designed for security. MQTT's topic structure may form a large tree, and there is no obvious way to divide it into smaller federal domains. The challenge of building a globally scalable MQTT network increases as the topic tree grows.

Another disadvantage of MQTT is its lack of compatibility. Because binary message payloads lack encoding information, problems may arise, especially in open systems where programmers from multiple manufacturers must work seamlessly together.

MQTT has basic authentication. Any secure MQTT usage must include SSL/TLS, which is not a lightweight protocol [17].

4.2 IoT Uses MQTT

MQTT clients are designed to be lightweight, enabling them to operate efficiently on small-scale microcontrollers. The compact size of MQTT headers helps optimize network bandwidth usage. MQTT.org emphasizes that MQTT has the capability to interconnect millions of Internet of Things (IoT) devices, underscoring its scalability and versatility.

In the realm of IoT and Industrial Internet of Things (IIoT), MQTT stands out as one of the most prevalent protocols. Its widespread adoption facilitates seamless communication of data between utility providers, consumers, and various connected devices. This capability plays a vital role in enhancing the interoperability and efficiency of IoT and IIoT infrastructure.

4.3 MQTT Examples in IoT or IIoT Infrastructure

- Smart-metering-MQTT guarantees message delivery to offer real-time accurate meter readings. This improves billing.
- Sensor data collection—MQTT is an excellent match for IoT sensor build outs that require low-priority data delivery.
- Machine data. Ably, a pub/sub messaging platform, says a wind turbine needs machine health data sent to local teams before it reaches a data center.
- Invoicing. MQTT reduces redundant or missed billing message packets.

MQTT allows resource-constrained IoT devices to communicate or publish topic-specific information to a server that operates as a MQTT message broker. The broker sends the information to customers who already enrolled.

Topics appear like file paths to humans. Clients may subscribe to a certain level or use a wildcard to subscribe to many levels. The client might be an IoT sensor in the field or a data center application.

Carriots, Evrythng, and ThingWorx support MQTT.

4.4 Creating a Smart Device with ESP-12 Micro-controller

NodeMCU is an open-source firmware based Wi-Fi microcontroller that is used to develop IoT based prototypes. It is based on Espressif's ESP12 SoC Chip and widely used as a development board for rapid prototyping [18]. The ESP 12 chip and its pin diagram is shown in Fig. 1.

GPIO (General Purpose Input Output) Pins

As depicted in the pin diagram provided, NodeMCU incorporates versatile general-purpose input and output pins on its board. These pins can be configured to output digital signals, allowing for control over various components such as LEDs or power switches. Additionally, the GPIO pins support the generation of Pulse Width Modulation (PWM) signals.

ADC (Analog to Digital Converter) Channel (A0)

The NodeMCU board incorporates a single ADC channel/pin, enabling analog sensing functionality through the microcontroller's built-in Analog to Digital Converter (ADC) circuit. These analog pins are specifically designated for data reading purposes. They are capable of detecting voltage levels within a range of

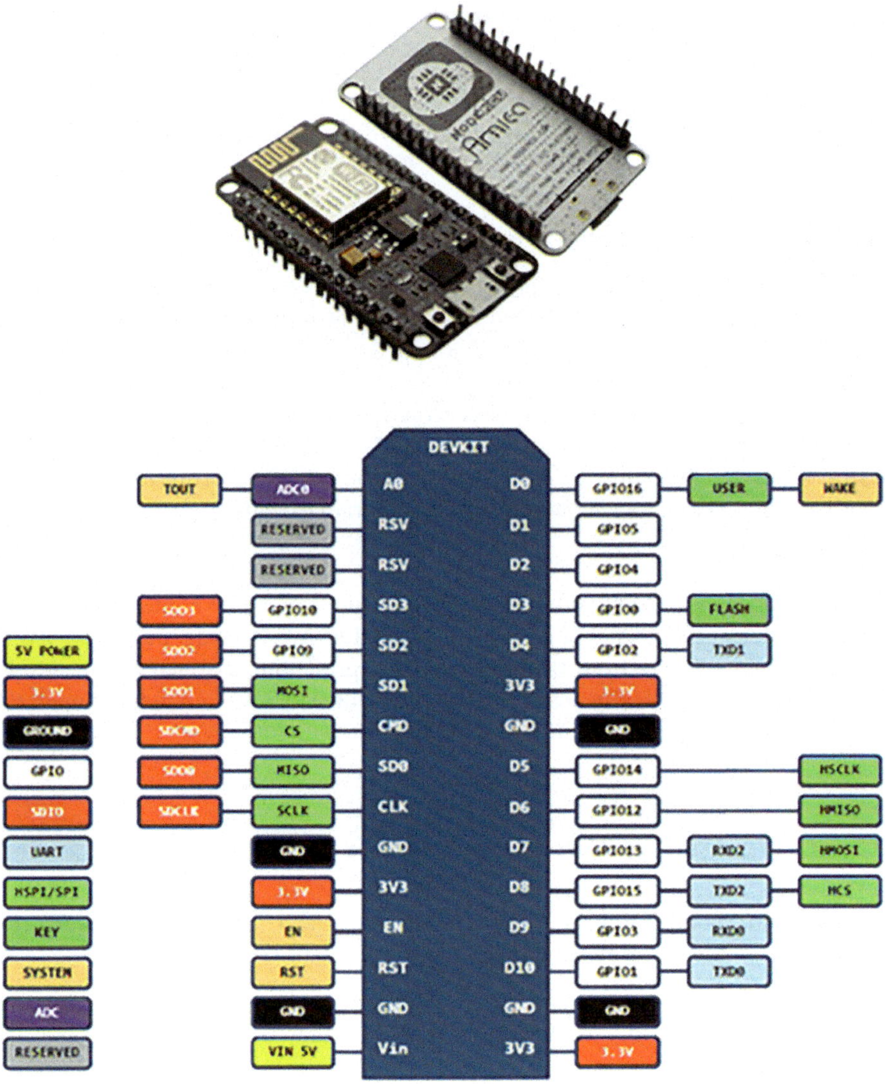

Fig. 1 ESP 12 chip and its pin diagram [18]

0–5 V or 0–3.3 V, depending on the microcontroller's specifications. Once voltage levels are detected, the analog signals are converted into digital format using the integrated ADC, and the resulting values, typically ranging from 0 to 255, are forwarded to the microcontroller's processing unit for further computation.

SPI (Serial Peripheral Interface) Pins

The NodeMCU, powered by ESP8266, features Hardware SPI (HSPI) capability, supported by four dedicated SPI communication pins. Additionally, it includes SPI pins designed for Quad-SPI communication. Utilizing this SPI interface, users can connect various SPI-enabled devices to the NodeMCU board for seamless communication.

I2C (Inter-integrated Circuit) Pins

NodeMCU, based on ESP8266, provides support for I2C functionality on its GPIO pins. However, not all GPIOs on the ESP-12E can be utilized for I2C due to internal functions. Therefore, it's advisable to conduct specific tests before assigning any GPIO for I2C applications.

UART (Universal Asynchronous Receiver Transmitter) Pins.

The ESP8266 NodeMCU features two UART ports, namely UART0 and UART1. As UART0 (RXD0 & TXD0) is essential for uploading firmware/codes to the board, it cannot be utilized concurrently for program execution.

PWM Pins

PWM (Pulse Width Modulation) harnesses the microcontroller's integrated digital-to-analog converter (DAC). Primarily utilized for data writing purposes, PWM outputs values ranging from 0 to 1023, each representing distinct voltage levels. The voltage range typically spans from 0 to 5 V or 0 to 3.3 V, contingent upon the specifications of the microcontroller in use.

4.5 Using Arduino IDE to Program the NodeMCU

To configure your Arduino Integrated Development Environment (IDE) for NodeMCU, follow these steps:

- Launch the Arduino IDE and go to File -> Preferences. In the Preferences dialog box, locate the "Additional Boards Manager URLs" field and enter the following link:

  http://arduino.esp8266.com/stable/package_esp8266com_index.json

- Close the Preferences window and navigate to Tools -> Board -> Boards Manager.
- In the Boards Manager window, type "esp" in the search bar; the "esp8266" option will appear below. Select the latest version of the board and click the "install" button.
- Once the board installation is complete, go to Tools -> Board, and choose "NodeMCU 1.0 (ESP-12E Module) [19]".
- Your Arduino IDE is now configured for NodeMCU development.

- Next, proceed to build your IoT device based on NodeMCU. This device can serve various purposes such as a smart switch, smart thermostat, or smart motion detector, depending on your requirements.

Since NodeMCU utilizes Wi-Fi for communication, your device will be directly connected to the internet. After establishing a successful Wi-Fi connection, the next step is to connect the device to an MQTT broker. The MQTT broker serves as the intermediary for all communication between the device and the frontend dashboard.

5 IoT Platforms

There are multiple IoT Platforms available for IoT prototyping that will help you to connect to the IoT edge device and will provide you a front end to control and monitor the devices. These IoT platforms help in rapid prototyping and developing IoT based applications by providing the protocol support, backend server database and a frontend to control and monitor those devices.

SmarDen.tech is one such IoT platform that helps in developing IoT based projects and products. It provides the users with complete backend support and a front end in the form of Web dashboard and mobile application along with the complete MQTT protocol.

Once we have the NodeMCU based device ready with us, all we need to do is to pair it with any such IoT platforms and then control it at our ease [20].

5.1 Smart Homes—Real Life Applications

The smart home market is increasing exponentially and is expected to reach USD 138.9 Billion by the end of 2026. The number of connected devices per user is also growing day by day. Smart homes were considered as luxury before but now they have become an integral part of our lifestyle, from controlling the temperature of the air conditioner to monitoring the door status, smart home devices are providing a secure, convenient, and comfortable lifestyle to their users.

One such example of a complete smart home network can be seen at Smarden Automation Pvt. Ltd. Smarden is an Indian IoT company that works on the technology of the Internet of things and one of their major domains is Wi-Fi based Retrofit Home Automation Solutions.

Retrofit Home Automation Systems are now in great demand as they do not require any special kind of wiring and the user does not need to replace any of their existing appliances. The purpose of retrofit smart home automation is to convert the existing things to smart things.

Figure 2 shows a retrofit smart home device by Smarden that converts the existing switches to smart switches which can be controlled and monitored from anywhere in

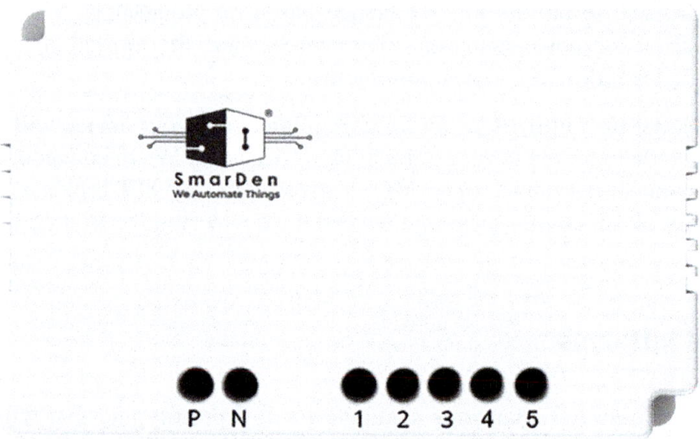

Fig. 2 SmarDen's 5 node smart module to convert five normal switches to smart

the world with the Smarden mobile application. There are multiple modules available for multiple switches in a switchboard ranging from one switch to six switches.

After installing those devices behind the existing switchboard, all the appliances can be controlled via SmarDen's Mobile application from anywhere in the world. Figure 3 shows the smart module installation behind the existing switch boards.

Fig. 3 Smart module installation

Fig. 4 Smarden's smart plug

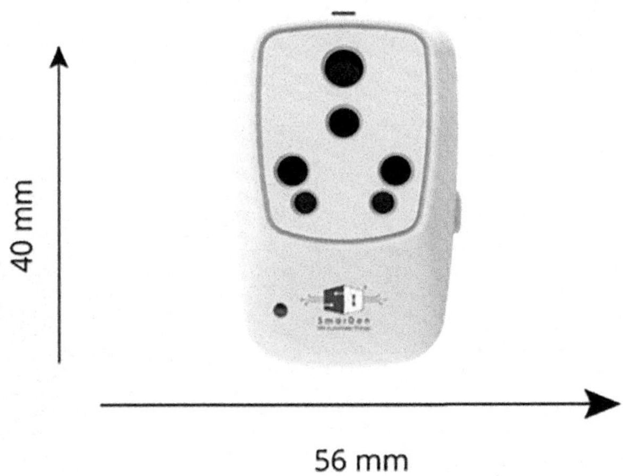

Along with the smart modules, there are multiple such IoT based Smart Home products that convert the existing things to smart things. Here is a list of a few of these devices.

- **Smart Plug**: This smart in Fig. 4. Plug converts any AC/Geyser/microwave/refrigerator to a smart appliance. Users only need to plug the socket of the appliances into the smart plug and then the plug can be controlled via the Smarden app.

Smart IR Blaster: The IR Blaster in Fig. 5 is an all-in-one remote system. It can control all your IR based appliances like TV, AC, Set top box, Home theatre systems, projectors, etc. at once via your phone. You will not have to keep ten different remotes anymore, just use your universal remote with the Smarden mobile app.

With the availability of devices like these, converting any home to a smart home becomes extremely easy. You can monitor the energy consumption, appliances, and control everything with the click of a button.

Fig. 5 Smarden's IR blaster

References

1. Naik, K., Patel, S.: An open source smart home management system based on IOT. Wireless Netw. **29**(3), 989–995 (2023)
2. Ma, Q., Tan, H., Zhou, T.: Mutual authentication scheme for smart devices in IoT-enabled smart home systems. Comput. Stand. Interfaces **86**, 103743 (2023)
3. Abdallah, R., Xu, L., Shi, W.: Lessons and experiences of a DIY smart home. In: Proceedings of the Workshop on Smart Internet of Things, pp. 1–6 (2017)
4. Hu, Y., Tilke, D., Adams, T., Crandall, A.S., Cook, D.J., Schmitter-Edgecombe, M.: Smart home in a box: usability study for a large-scale self-installation of smart home technologies. J. Rel. Intell. Environ. **2**(2), 93–106 (2016)
5. Gokhale, P., Bhat, O., Bhat, S.: Introduction to IOT. Int. Adv. Res. J. Sci. Eng. Technol. **5**(1), 41–44 (2018)
6. Khattak, H.A., Shah, M.A., Khan, S., Ali, I., Imran, M.: Perception layer security in Internet of Things. Futur. Gener. Comput. Syst. **100**, 144–164 (2019)
7. Yang, L., Motohisa, J., Takeda, J., Fukui, T.: Sensors Actuators (2007)
8. Gridling, G., & Weiss, B.: Introduction to microcontrollers. Vienna University of Technology Institute of Computer Engineering Embedded Computing Systems Group (2007)
9. Perlman, R.J.: Network layer protocols with byzantine robustness. Doctoral dissertation, Massachusetts Institute of Technology (1988)
10. Beniwal, G., Singhrova, A.: A systematic literature review on IoT gateways. J. King Saud Univ. Comput. Inf. Sci. (2021)
11. Karagiannis, V., Chatzimisios, P., Vazquez-Gallego, F., Alonso-Zarate, J.: A survey on application layer protocols for the internet of things. Trans. IoT Cloud Comput. **3**(1), 11–17 (2015)
12. Kim, D.Y., Jung, M.: Data transmission and network architecture in long range low power sensor networks for IoT. Wireless Pers. Commun. **93**(1), 119–129 (2017)
13. Yassein, M.B., Shatnawi, M.Q., Aljwarneh, S., Al-Hatmi, R.: Internet of things: survey and open issues of MQTT protocol. In: 2017 International Conference on Engineering & MIS (ICEMIS), pp. 1–6. IEEE (2017)
14. Atmoko, R.A., Riantini, R., Hasin, M.K.: IoT Real Time Data Acquisition Using MQTT Protocol (J. Phys. Conf. Ser.), vol. 853, No. 1, p. 012003. IOP Publishing (2017)

15. Soni, D., Makwana, A.: A survey on mqtt: a protocol of internet of things (iot). In: International Conference on Telecommunication, Power Analysis and Computing Techniques (ICTPACT-2017), vol. 20, pp. 173–177 (2017)
16. Singh, M., Rajan, M.A., Shivraj, V.L., Balamuralidhar, P.: Secure mqtt for internet of things (iot). In: 2015 Fifth International Conference on Communication Systems and Network Technologies, pp. 746–751. IEEE (2015)
17. Bansal, M.: Performance comparison of MQTT and CoAP protocols in different simulation environments. Invent. Commun. Computat. Technol. 549–560 (2021)
18. Schmitt, A., Carlier, F., Renault, V.: Dynamic bridge generation for IoT data exchange via the MQTT protocol. Procedia Comput. Sci. **130**, 90–97 (2018)
19. Son, B.J., Lee, D.W., Seo, D.H., Kim, M.S., Jo, J.I., Choi, B.Y.: Implementation of WIFI robot car using NodeMCU ESP-12 board. In: Proceedings of the Korean Institute of Information and Communication Sciences Conference, pp. 475–477. The Korea Institute of Information and Communication Engineering (2017)
20. Amudha, G., Priya, P.P., Dinesh, V., Akash, A., Akash, A.: Node MCU based DoS attack evaluation in IoT device. In: AIP Conference Proceedings, vol. 2519, no. 1, p. 30084. AIP Publishing LLC (2022)

A Survey on Various Secure Access Control and Authentication in a Block Chain-Enable Cloud IoT

V. Sahiti Yellanki and Basant Sah

Abstract The Internet of Things (IoT) is a hot topic these days, empowers everyday devices with identification and communication capabilities, spanning diverse domains like smart appliances, smart cities, wearables, and Electronic-health data. This expansive connectivity is anticipated to link tens to hundreds of billions of devices, each equipped with smart functionalities for autonomous data collection, analysis, and decision-making. Both academia and industry are determined to make progress in improving usability, maintainability, and security by establishing standards and best practices. Security is particularly important because it currently poses a major obstacle to wider adoption of the Internet of Things. There are various areas of research in the security field, including cryptography, network security, and identity management. In this paper, we focus on the application layer of the IoT environment, specifically in the areas of authentication, and authorization. This survey conducts an extensive comparative examination of contemporary IoT authorization and Authentication schemes, aiming to illuminate both their strengths and weaknesses. It subsequently delineates crucial requirements, elucidates authorization threats, and underscores weaknesses affecting IoT authorization. The survey concludes by outlining persisting open challenges in authorization and offering recommendations for prospective research.

Keywords IoT access control · Authentication · Block chain · Security threat · Security attacks

V. Sahiti Yellanki (✉) · B. Sah
Department of Computer Science and Engineering, Koneru Lakshmaiah Education Foundation, Guntur, Andhra Pradesh 522502, India
e-mail: sahithivellanki@gmail.com

V. Sahiti Yellanki
Department of CSE-AIML&IoT, VNR Vignana Jyothi Institute of Engineering and Technology, Bachupally, Hyderabad, India

S. N. Mohanty et al. (eds.), *Explainable IoT Applications: A Demystification*, Information Systems Engineering and Management 21,
https://doi.org/10.1007/978-3-031-74885-1_19

1 Introduction

The combination of block chain technology, Internet of Things and cloud Security has opened the door a new era of interconnected systems, presenting exceptional opportunities for innovation and efficiency. However, this merging also presents numerous security challenges, particularly in establishing strong access control and authentication mechanisms in block chain-enabled cloud-IoT environments. Given the growing number of interrelated devices and the distributed nature of block chain networks, ensuring secure access to resources and verifying the identities of users and devices becomes crucial to safeguard against unapproved access, data breaches, and further security risks. This survey aims to offer a wide-ranging overview of the different techniques for secure access control and authentication that are tailored for block chain-enabled cloud-IoT deployments. It sheds light on the contemporary solutions, emerging trends, and key considerations in designing secure and resilient systems in this complex and evolving landscape. By evaluating the strengths, limitations, and real-world applications of various approaches, this survey aims to provide valuable insights to researchers, practitioners, and stakeholders involved in securing block chain-enabled cloud-Internet of Things ecosystems.

The cool thing about Block chain technology teaming up with Cloud-IoT is that it brings some well-designed solutions to the counter when it comes to secure access control and authentication. Some recent research trends have really highlighted how important it is to use mechanisms powered by block chain to stop cyber-attacks and make sure data integrity and privacy are maintained in the Cloud-IoT world [1]. The integration of block chain technology with Cloud-IoT has led to significant advancements in secure access control and authentication mechanisms. Block chain, conversely, employs sophisticated cryptographic methods to establish an immutable and secure record of every transaction. Its decentralized nature ensures that no singular entity has control over it, enhancing its resistance to both failures and malicious attacks [2]. The development of these devices raises concerns regarding privacy and security, which for a method of controlling them [3]. Access control has been identified as the most important mechanism for managing unauthorized users and their actions [4, 5]. Many access control Techniques have been proposed to regulate resource access, but the reliance on a integrated entity creates a vulnerability as it becomes a single point of failure [6]. This centralized server can be compromised by adversaries in the network, allowing them to manipulate access policies in their favor. Furthermore, compromising the centralized entity exposes sensitive data, such as personal information and surveillance data of IoT users, to malicious actors [7].

To address the restrictions of integrated architectures and cater to large-scale scenarios, block chain technology is employed to make available decentralized access control in the Internet of Thing network [8, 9]. While Multi authority-based access control has been implemented in various mechanisms to enhance secure data sharing, it also gives rise to significant privacy concerns [10–12]. Present Techniques integrate the block chain by storing access policies in block chain nodes, which are then used to make access decisions [13]. These nodes act as delegates and handle the complete resource necessities of Internet of Things entities [14].

Now, the attentions on the protected access control & authentication aspects of block chain-enabled cloud-IoT systems. These systems combine the power of block chain knowledge, cloud Security, and IoT devices to create a protected and decentralized environment for data sharing, communication, and data. It's clear that we need some fresh approaches these models are built on a block chain-based authentication framework that really puts the de centralization of IoT devices and consensus methods front and center. This helps to keep data storage secure and lets us have better control over the system using encrypted blocks and smart contracts [5]. The massive total of devices connected in the IoT ecosystem. Traditional security measures primarily focus on securing a limited number of end points, such as computers or servers. In contrast, IoT security encompasses a wide-ranging of interconnected devices, including smart appliances, wearable devices and industrial machinery. This expansive network increases the potential attack surface, making it challenging to monitor and protect all devices effectively. Additionally, IoT security requires a multi-layered approach due to the variety of sensors and protocols involved compared to traditional security measures that primarily rely on firewalls and antivirus software. As a result, the complexity and scale of IoT security necessitate advanced threat detection and response mechanisms to ensure the safety and secrecy of users' information.

To create a trusted network for Internet of Things applications, organizations including processing systems, service providers, and sensor nodes must authenticate each other's nodes. Not only should the authentication protocol be resilient to malevolent attacks, but it would also be suitably "lightweight" in order to be implemented on underperforming edge devices. Security strategies such as multi-layered security, signatures, and key agreement-based location privacy model are used to overcome the security issues. In industrial applications enabled by the IoT, authentication is also a critical security component. A standard authentication system called Open authentication (OAuth) enables the user to safely access resources. When a user wants to enter an Internet of Things network, the security manager uses the authentication mechanism to verify the identity. Key Management for Block chain-Powered Cloud and Internet of Things Access control techniques are necessary to guaranteeing data safety and privacy in cloud-Internet of Things systems with block chain enabled. Traditional access control mechanisms used in centralized systems are not suitable for the decentralized and distributed nature of block chain-enabled Cloud-IoT. In this context, the use of block chain technology for access control in cloud-Internet of Things systems offers several advantages.

The incorporation of block chain technology with cloud and IoT systems has gained significant attention in modern Technology. Researchers have recognized the importance of secure access control and authentication mechanisms in this context. However, existing work in this area has focused on specific applications such as smart grids, public safety systems, and biometric imaging data processing. To address this gap, this survey paper goals to make available a comprehensive overview of various secure access control and authentication techniques in block chain—enabled cloud-IoT systems and their potential benefits and challenges. This paper attentions on the popularization of cloud computing as well as sightsees various preferences and practices in block chain-oriented security scenarios in IoT environments [8]. Here we highlighted the benefits and practices to be monitored in block chain-oriented security scenarios in IoT environments [15]. Proposed an efficient scheme using public key cryptography for IoT and cloud-based security. The incorporation of block chain technology with cloud and IoT systems has added important responsiveness in recent years. This integration aims to provide decentralized security mechanisms and address the security issues in IoT systems. Some key challenges in ensuring secure access control and authentication in block chain-enabled cloud-IoT systems include the mitigation of DDoS attacks, Sybil attacks, privacy protection, scalability, and interoperability of different block chain platforms [1, 14].

The left-over Paper is systematized as keep an eye on: Section 1 gives the Introduction, Sect. 2 highlights the various access control and authentication techniques in Cloud-IoT, Sect. 3 Principal Distinctions between IoT security and traditional security measures. Comparative investigation of current security techniques for cloud-based IoT environment is described in Sect. 4. To end with, we accomplish the article in Sect. 5 with future directions.

2 Various Access Control and Authentication Techniques

In this section gives overview on current techniques, which employ various strategies to carry out secure access control and authentication techniques, are compiled, the assumed categorization of the authentication and authorization methods is shown in Fig. 1.

Role-Based Access Control (RBAC): A common access control paradigm that regulates resources access based on the responsibilities of certain individuals inside an organization is known as role-based access control. By allocating rights based on roles rather than specific persons, RBAC streamlines access control [8]. It encourages scalability, security, and consistency in access control procedures.

Attribute-Based Access Control (ABAC)

To determine access permissions, ABAC dynamically assesses the user's, device's, or resource's properties. User roles, device characteristics, location, access time, and other contextual data are examples of attributes [2]. The access control approach known as Attribute-Based Access Control (ABAC) allows or prohibits access to properties based on characteristics related to persons, resources, and environmental factors.

All things considered, ABAC provides an adaptable and dynamic method of access control, enabling businesses to create access policies based on a variety of characteristics and circumstances. It enables fine-grained control over access to resources and supports dynamic adaptation to changing security requirements and environmental conditions.

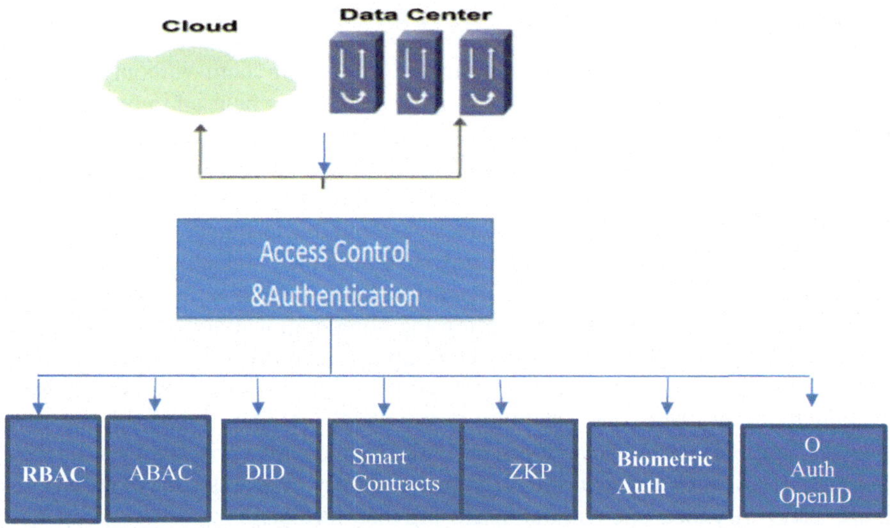

Fig. 1 Various access control and authentication techniques

Decentralized Identity (DID)

Decentralized Identity (DID) is a concept and set of standards that enable individuals, organizations, and devices to generate and switch their digital identities without relying on central authorities [2]. DID standards enable the management of decentralized identities on the block chain every user or devices are assigned a distinctive identifier (DID) stored on the block chain. DID be presented by the World Wide Web Consortium (W3C) and the Decentralized Identity Foundation (DIF), define specifications for creating, resolving, and interacting with DIDs.

Smart Contracts

Smart contracts can encode access control rules and execute them autonomously, Access permissions, such as granting or revoking access to resources, can be enforced directly through smart contracts [7]. Smart Contracts are contracts the execute themselves based on the terms written directly into code. Smart contracts revolutionize traditional contract execution by providing a vulnerable, transparent, and automated way to conduct transactions and enforce agreements in a decentralized manner. They are a fundamental aspect of block chain technology and are driving innovation in various industries. IoT devices can interact with smart contracts to request access permissions, ensuring transparent and auditable access control.

Zero-Knowledge Proofs (ZKPs)

ZKPs allow one party to demonstrate the veracity of a claim without disclosing any further details. In access control scenarios, ZKPs can be used to prove possession of certain attributes or credentials without disclosing sensitive data [5]. Zero-Knowledge Proofs, or ZKPs, are cryptographic techniques that allow a prover to convince a verifier that a claim is true without providing any further information. ZKPs make it possible to verify a statement without disclosing any private information that would raise questions. This property makes them valuable for applications where privacy is paramount, such as identity verification and authentication.

ZKPs also pose challenges, including the computational complexity of generating and verifying proofs, the potential for implementation errors leading to security vulnerabilities, the need for efficient and user-friendly protocols. A strong cryptographic tool that incursions a stability in the middle of security and privacy, zero-knowledge proofs enable privacy-preserving authentication, verification, and computation in a variety of applications. To improve privacy in access control procedures, a user can verify their age without disclosing their birthdate.

Biometric Authentication

Biometric data, such as fingerprints or facial recognition patterns, can be securely stored on the block chain. IoT devices can authenticate users based on their biometric characteristics, providing a convenient and secure access control mechanism. Biometric authentication relies on unique biological traits or behavioral patterns that are specific to each individual. These may consist of hand geometry, voiceprints, iris patterns, fingerprints, facial traits, and even behavioral biometrics like typing rhythm or stride.

Biometric traits are difficult to replicate or share, making them inherently more secure than traditional passwords or tokens. Since biometric characteristics are unique to each person and generally cannot be easily stolen or forged, they provide a higher level of confidence in identity verification on Non-Transferable.

Biometric authentication is widely used across various industries and applications, including: Smartphones and tablets often use fingerprint or facial recognition for unlocking and user authentication. Physical access control is used in Biometric scanners for secure entry to buildings, rooms, and sensitive areas. Biometric authentication is employed for secure login to banking apps, ATM withdrawals, and payment authorization in financial services. Biometrics is used for patient identification, access to electronic medical records, and prescription verification in Healthcare.

Multimodal Biometrics: Some systems combine multiple biometric modalities (e.g., fingerprint and facial recognition) to improve accuracy and security. Multimodal biometrics can provide a more robust authentication solution by mitigating the limitations of individual biometric traits. Biometric authentication offers a convenient, secure, and user-friendly method of verifying identity across a wide range of applications, improving security and reducing the reliance on easily compromised passwords or tokens. However, careful consideration must be given to privacy and security implications when implementing biometric authentication systems. Privacy concerns should be addressed by ensuring that biometric data is encrypted and accessed only when necessary.

Multi-signature Transactions

Multi-signature transactions (MST) require authorization from many parties' earlier execution. In access control scenarios, multi-signature schemes can be used to ensure that critical operations, such as data sharing or device management, require consent from multiple authorized entities. This enhances security by reducing the risk of unauthorized actions by a single party.

Multi-signature transactions, or multi-sig transactions, are a cryptographic mechanism requiring multiple authorized signatures to execute a transaction. This approach enhances security and trust in digital transactions, as it prevents any single party from unilaterally authorizing or executing a transaction. Multi-signature transactions are widely used in block chain and cryptocurrency systems, particularly in multi-signature wallets where funds are held by multiple parties, requiring a predefined number of signatures to spend or transfer assets. They also find applications in escrow services, corporate governance, and other scenarios where consensus among multiple parties is required for transaction authorization. With the flexibility to set the required number of signatures and the ability to implement them using smart contracts or multi-signature scripts, multi-signature transactions offer a robust and versatile solution for secure and collaborative decision-making in digital transactions.

OAuth and OpenID Connect

OAuth (Open Authorization) and OpenID Connect are widely adopted authentication and authorization protocols used to be responsible for protected access to web and cloud-based resources [16]. Through the use of delegated authorization, OAuth allows applications from outside the user to access resources on their behalf without disclosing their login credentials. To acquire and utilize access tokens, communication takes place in the middle of the Source and client application, resource server and authorization server. OpenID Connect builds upon OAuth, providing identity layer authentication on top of the authorization framework. It presents a standardized way of authenticating users across different applications and services by enabling clients to confirm end users' identities based on authentication carried out by an authorization server. Together, OAuth and OpenID Connect play crucial roles in enabling secure, scalable, and interoperable authentication and authorization mechanisms for modern web and cloud-based environments.

The vigorous and multifaceted nature of the Cloud-IoT environment necessitates a multifaceted approach to access control and authentication techniques. The integration of diverse techniques such as Role-based access control (RBAC), attribute-based access control (ABAC), and multi-factor authentication (MFA) enhances the overall security posture by providing a layered defense against unapproved access. Additionally, the use of biometric authentication, tokenization, and encryption further strengthens the authentication process, ensuring the confidentiality and integrity of sensitive data in the Cloud-IoT ecosystem. As the landscape continues to evolve, adaptive access control mechanisms become imperative, allowing for real-time adjustments based on contextual information. Striking a balance between usability and security is vital to foster the seamless process of Cloud-Internet of Things systems while safeguarding against potential threats and unauthorized intrusions. Regular updates, continuous monitoring, and a proactive security stance are essential to mitigate emerging risks in this dynamic and interconnected environment.

By employing these secure access control and authentication techniques, block chain-enabled cloud-IoT systems can successfully bring about access to resources, guard sensitive data, and ensure compliance with security policies.

3 Principal Distinctions Between IoT Security and Traditional Security Measures

There are some key contrast between the IoT and regular remote systems when it comes to managing security and privacy issues. For instance, the sending of IoT sensed data is pretty interesting compared to that of the typical internet. The IoT devices use LLNs, which are really stressed by dynamism, memory, and power [7]. These angles are not considered for the standard Internet. LLNs encounter incredible information misfortunes due to hub pantomime.

The incorporation of cloud computing and the IoT introduces a multitude of security challenges stemming from the convergence of distributed IoT devices and centralized cloud services. Some key challenges include data privacy and confidentiality concerns as sensitive IoT data is transmitted and stored in the cloud, requiring vigorous encryption and access control measures to guard against unapproved access. Additionally, the sheer scale and heterogeneity of IoT deployments exacerbate security risks, with diverse devices running different operating systems and firmware versions potentially introducing Legal responsibility and compatibility issues. Network security is paramount, with threats such as distributed denial-of-service (DDoS) attacks targeting IoT devices and cloud infrastructure. Identity and access management (IAM) become complex in cloud-Internet of Things environments, necessitating secure authentication and authorization mechanisms to make sure that only authorized users and devices can access resources. Ensuring the integrity and availability of data and services, implementing secure communication protocols, and addressing regulatory compliance requirements are also critical aspects of cloud-IoT security. Addressing these challenges requires a holistic approach that combines robust security architectures, regular security audits, ongoing monitoring, and collaboration among stakeholders to safeguard cloud-IoT deployments against evolving threats.

The security highlights of both the IoT and traditional network gadgets are likewise unique [4, 10]. One of the principal distinctions between IoT security and traditional security measures lies in the nature of the devices and systems being protected. Traditional security measures typically focus on protecting centralized systems, such as servers or computer networks, from external threats. Though, with the growth of the IoT, security measures need to adapt to the unique challenges posed by interconnected devices. IoT security must address a vast network of diverse and often resource-constrained devices, ranging from smart appliances to industrial machinery. Furthermore, IoT security also needs to consider threats from both external attackers and compromised devices with in the network itself. Therefore, the primary distinction between IoT security and traditional security measures is the scale, diversity, and complexity of the devices and systems that need protection, necessitating new strategies and approaches to make sure the security and integrity of the IoT ecosystem. IoT security faces the challenge of protecting a diverse range of devices, as well as smart appliances, wearable devices, medical equipment, and industrial machinery. Strong security measures may be challenging to install on these devices because to their potential limitations in terms of processor speed, memory, or battery life. Additionally, the sheer number of IoT devices and the complexity of their interactions require a scalable and adaptable security framework. Unlike traditional security measures that primarily deal with external threats, IoT security must also address internal threats posed by compromised devices within the network. As IoT devices transfer data with each other, a compromised device can become a gateway for attackers to gain access to the entire network, leading to potential data breaches or malicious activities.

To address these challenges, IoT security requires new strategies and approaches. These may include encryption and authentication mechanisms tailored for resource-constrained devices, secure protocols for communication, network segmentation to contain potential threats, continuous monitoring and vulnerability management, and strong access control measures. In the IoT observation layer, sensor hubs have limited computational power, and low stockpiling limit which make the recurrence jumping correspondence application and open key encryption to anchor the Internet of Things appliances pretty much impossible. The Internet of Things makes use of light weight cryptographic calculations, which include light weight encryption technology. There are security concerns associated with Internet of Things, including man-in-the-middle attacks and fake attacks.

There are two attacks that can exploit the system and send counterfeit data to the conveying hubs in the system [11]. Authenticity confirmation and information privacy components are utilized to forestall unapproved hubs.

At the application layer, data sharing is the fundamental component, which makes security issues in data protection, gets to control, and revelation of information [12]. The security necessities for the application layer incorporate authentication, key session, and assurance of client protection crosswise over heterogeneous systems.

Numerous research works have investigated the application of block chain technology in access control systems [6]. Suggest role-based access control (RBAC) methods that concentrate on data protection and knowledge management systems. References [1, 17] extend this work to network resource sharing and distributed attribute-based access control (ABAC) systems, respectively. These studies collectively highlight the potential of Block chain in enhancing security, transparency, and trust in access control mechanisms. Functions of the Access Control Model, ABAC [17] Real identities are used as a set of attributes that express access control policies in a fine-grained manner in attribute-based access control, or ABAC. RBAC [6]

“Roles” are the mechanism that Role-based Access Control, or RBAC, uses to assign permissions. Prior to the assignment of permissions, users are allocated an associated role. ACL Access control lists (ACLs) will automatically provide permission to users of that particular resource.

The ordinary security mechanisms are not appropriate for communication among machines. Although the security vulnerabilities in the two systems may be similar, each system’s security vulnerabilities are addressed using different approaches and techniques [13]. The attacker can successfully obtain the plaintext M by calculating the random secret number if they are able to determine the key Km using publicly available information [18].

The principal goal of this learning is to address the escalating vulnerabilities connected with conventional authentication, access control, and revocation mechanisms in cloud environments.

4 Comparative Investigation of Current Security Techniques for Cloud-Based IoT Environment

The Following are some observations that have been made in light of the literature review. This gives a comparative analysis of various security related mechanisms for cloud IoT environment, along with objective, Methods used and limitations.

References	Schemes	Objective	Algorithms/ methods used	Limitations
1	Access Control	Dynamic secure access control using the block chain (DSA Block)	Elliptic curve cryptography (ECC) and the practical Byzantine fault tolerance (PBFT) algorithm	Outstanding to the possibility of a delegator node single point of failure, hierarchical access control is required
2		Decentralized, flexible, and fine-grained authorization for IoT devices	Elliptic curve cryptography (ECC)	Complexity is high due to key size is large
3		Removes the single point of failure	Smart contracts, simulation using OAI, utilization of Ethereum network,	Need to address issues related to storage consumption
4		Trusted auditing of access attempts can be provided by a distributed ABAC system built on a block chain	Raft and Kafka as CFT ordering services, and the modular structure of hyper ledger fabric for component plug ability	Inability of the system to scale after 25 clients

(continued)

(continued)

References	Schemes	Objective	Algorithms/ methods used	Limitations
5		Combining zero knowledge token based access control (BZBAC) with block chain technology	Token based access control (BZBAC) with zero knowledge, Ethereum, smart contracts, off-chain computing, and block chain	Poor batch optimization on the requester side
6		Decentralized Internet of T hings network with smart contracts as its core element	Block chain technology, ABAC, smart contracts, and the trust and reputation system (TRS)	De-anonymization in the block chain
7		Smart contracts and the stack for things frame-work	RBAC, universally unique identifiers (UUIDs), and smart contracts	Lack of automate the monitoring of the smart contracts
8		Attribute-based access control in IoT	Consortium block chain, elliptic curve cryptography (ECC)	High complexity as a result of big key sizes
9		Revocation using attribute based encryption	Polynomial-time series algorithm, Diffehellman cryptography	Data integrity is low
10	Authentication	Secure and effectively resist all kinds of known attacks	BAN-logic verification, informal security, and automated security verification with ProVerif	Limited to unknown attacks
11		A block chain-based decentralized frame-work	Development of a blockchain-based decentralized framework, employing registration, certification, and revocation processes, and evaluating the performance	Lack of comprehensive solutions for securing communication in WSNs-enabled IoT

(continued)

(continued)

References	Schemes	Objective	Algorithms/ methods used	Limitations
12		ECC-based secure mutual authentication protocol	HTTP cookies are used in ECC-based secured mutual authentication protocol to enable secure communication between embedded devices and cloud servers	Inadequate evaluation of performance metrics
13		Internet of things dynamic security authentication administration	Attribute-based security authentication that combines block chain technology and access control, built on the hyper ledger fabric architecture	Lack of scalability in the proposed IoT-chain system
14		Reciprocal authentication between communication entities, secure and effective key creation and administration are required	Distributed internet of things architecture based on block chains that uses hash chains for safe key management	Context of IoT, there is a need for a reliable and effective access control system that protects user and data privacy without depending on third par- ties and avoiding a centralized design
15		Authentication protocols are prone to single point failure	Block chain based multi-WSN authentication scheme for IoT	Need for further exploration of the practical implementation and scalability of block chain-based authentication in IoT environments
16		Block Auth technique gives a added secure, strong fault tolerance and reliable decentralized novel authentication with high-level security in IoT environment	Block chain-based distributed verification modelling Technique, smart contract, and implementation of a block chain-based authentication platform for evaluation	One-side failure security risk or failure caused by outside attacks or inside duplicitous

5 Conclusion and Future Scope

In This Section gives exploration of various protected access control and authentication mechanisms within a block chain-enabled cloud-IoT ecosystem reveals a diverse landscape of solutions. The analysis underscores the importance of robust security measures in guaranteeing the integrity and privacy of data in this interconnected environment. While different access control and authentication approaches demonstrate strengths in certain aspects, they also exhibit limits as well as vulnerabilities. Future investigation should focus on addressing these challenges to enhance the overall security posture of block chain-enabled cloud-IoT systems. Additionally, continuous advancements and innovations in secure access control and authentication protocols are crucial to staying ahead of evolving cybersecurity threats in this dynamic and interconnected technological landscape.

In Future work we proposed an innovative AI-Hybrid Chain framework integrating cutting-edge methodologies—picturized authentication, Deep Reinforcement Learning (DRL) for access control, and a secure two-fold revocation system—to fortify cloud computing security.

References

1. Alshehri, S., Bamasaq, Alghazzawi, Jamjoom, A.: Dynamic secure access control and data sharing through trusted delegation and revocation in a blockchain-enabled cloud-IoT environment. In: IEEE Internet Things J. **10**(5), 4239–4256 (2023). https://ieeexplore.ieee.org/document/9930865
2. Zhang, Y., Li, B., Liu, B., Wu, J., Wang, Y., Yang, X.: An attribute-based collaborative access control scheme using blockchain for IoT devices. Electronics **9**, 285 (2020). https://doi.org/10.3390/electronics9020285
3. Ghaffari, F., Bertin, E., Crespi, N.: A novel approach for network resource sharing via block chain. In: SIGCOMM '21 Demos and Posters, August 23–27, 2021. Virtual Event, USA (2021). https://doi.org/10.1145/3472716.3472867
4. Bhatt, S., Pham, T.K., Gupta, M., Benson, J., Park, J., Sandhu, R.: Attribute-based access control for AWS Internet of Things and secure industries of the future. IEEE Access **9**. 107200–107223,%202021
5. Lin, X., Zhang, Y., Huang, C., Xing, B., Chen, L., Hu, D., Chen, Y.: An access control system based on block chain with zero-knowledge rollups in high-traffic IoT environments. Sensors **23**, 3443 (2023). https://doi.org/10.3390s23073443
6. Nyame, G., Qin, Z.: Precursors of role-based access control design in KMS: a conceptual framework. Information **11**, 334 (2020). https://doi.org/10.3390/info11060334
7. Zaidi, S.Y.A., Shah, M.A., Khattak, H.A., Maple, C., Rauf, H.T., El-Sherbeeny, A.M., El-Meligy, M.A.: An attribute-based access control for IoT using block chain and smart contracts. Sustainability **13** (2021). https://doi.org/10.3390/su131910556
8. Ding, S., Cao, J., Li, C., Fan, K., Li, H.: A novel attribute-based access control scheme using block chain for IoT. IEEE Access **7**, 38431–38441 (2019)
9. Yang, Q., Zhang, M., Zhou, Y., Wang, T., Xia, Z., Yang, B.: A non-interactive attribute-based access control scheme by block chain for IoT. Electronics **10**(15), 1855 (2021) [Online]. Available: https://www.mdpi.com/2079-9292/10/15/1855
10. Kousalya, A., Sakthidasan, K., Latha, A.: Reliable service availability and access control method for cloud assisted IoT communications. Wirel. Netw. **27**, 881–892 (2021)

11. Doghramachi, D.F., Ameen, S.Y.: IoT threats and solutions with blockchain and context-aware security design: a review. In: 2021 International Conference of Modern Trends in Information and Communication Technology Industry (MTICTI), pp. 1–8. Sana'a, Yemen (2021). https://doi.org/10.1109/MTICTI53925.2021.9664784
12. Ma, M., Shi, G., Li, F.: Privacy-oriented block chain-based distributed key management architecture for hierarchical access control in the IoT scenario. IEEE Access **7**, 34045–34059 (2019)
13. Li, Z., Hao, J., Liu, J., Wang, H., Xian, M.: An IoT-applicable access control model under double-layer blockchain. IEEE Trans. Circ. Syst. II Exp. Briefs **68**(6), 2102–2106 (2021)
14. Khalid, U., Asim, M., Baker, T., Hung, P.C.K., Tariq, M.A., Rafferty, L.: A decentralized lightweight blockchain-based authentication mechanism for IoT systems. Clust. Comput. **23**, 2067–2087 (2020)
15. Shivaramakrishna, D., Nagaratna, M.: A novel hybrid cryptographic framework for secure data storage in cloud computing: integrating AES-OTP and RSA with adaptive key management and time-limited access control. Alexandria Eng. J. **84**, 275–284 (2023) ISSN 1110–0168. https://doi.org/10.1016/j.aej.2023.10.054
16. Vivekanandan, M., Sastry, V.N., Srinivasulu Reddy, U.: Blockchain based privacy preserving user authentication protocol for distributed mobile cloud environment. Peer-to-Peer Netw. Appl. **14**, 1572–1595 (2021)
17. Rouhani, S., Belchior, R., Cruz, R.S., Deters, R.: Distributed attribute-based access control system using permissioned block chain. World Wide Web **24**, 1617–1644 (2021). https://doi.org/10.1007/s11280-021-00874-7
18. Panda, S.S., Jena, D., Mohanta, B.K., Ramasubbareddy, S., Daneshmand, M., Gandomi, A.H.: Authentication and key management in distributed IoT using block chain technology. IEEE Internet Things J. **8**(16), 12947–12954 (2021)

Uncovering the Truth: A Machine Learning Approach to Detect Fake Product Reviews and Analyze Sentiment

Dasari Kavitha, Gampa Srujankumar, Chigurupati Akhil, and Penna Sumanth

Abstract The proliferation of fake product reviews in the E-commerce industry has emerged as a significant challenge, undermining consumer trust and integrity in online platforms. This project addresses this pressing issue by developing an advanced system for detecting and eliminating fake reviews using cutting-edge methods for machine learning (ML) and natural language processing (NLP). The system integrates models such as Random Forest, Naive Bayes, MLP, and a Voting Classifier, trained on the "Amazon Yelp dataset" to ensure scalability and adaptability for large-scale applications. By providing a real-time assessment of review authenticity, the system empowers platform owners to take informed actions against spurious content, thereby preserving the integrity of online shopping experiences. The project highlights the importance of employing advanced NLP and ML techniques in combating fake reviews and contributes significantly to enhancing user trust in the online marketplace. The project aims to address the critical need for effective fake review detection and elimination systems in the E-commerce industry. The escalating prevalence of fake reviews on prominent platforms like Flipkart and Amazon undermines consumer trust, highlighting the urgency for a robust solution. By leveraging the "Amazon Yelp dataset" for model training, the study emphasizes scalability and adaptability for large-scale applications. Recognizing the escalating impact of fake reviews on user trust, this research addresses the critical need for platforms to combat spammers and keep alive the integrity of online shopping experiences. The proposed model not only provides a real-time assessment of review authenticity but also offers a foundation for website owners to take informed actions against spurious

D. Kavitha · G. Srujankumar (✉) · C. Akhil · P. Sumanth
Institute of Aeronautical Engineering, Hyderabad, India
e-mail: gsrujan9181@gmail.com

D. Kavitha
e-mail: d.kavitha@iare.ac.in

C. Akhil
e-mail: 20951A3304@iare.ac.in

P. Sumanth
e-mail: 20951A3354@iare.ac.in

S. N. Mohanty et al. (eds.), *Explainable IoT Applications: A Demystification*, Information Systems Engineering and Management 21,
https://doi.org/10.1007/978-3-031-74885-1_20

content. With potential applications for platforms of varying sizes, this sophisticated model demonstrates its efficacy in detecting spam reviews, contributing to the enhancement of user trust in the online marketplace.

Keywords Fake reviews · Machine learning (ML) models · Random forest · Naive bayes · MLP (multi-layer perceptron) · Voting classifier · Amazon yelp dataset · E-commerce industry

1 Introduction

In the rapidly expanding digital realm, The prevalence of fraudulent reviews presents a significant risk to the legitimacy of digital platforms, especially in the ever-changing E-commerce sector. This article delves into the imperative need for robust fake review detection and elimination, employing Machine Learning (ML) models. Through the integration of Random Forest, Naive Bayes, MLP, and a Voting Classifier (MLP + RF + NB), our proposed model strives to evaluate the authenticity of reviews within a given dataset. The surge in fake reviews on prominent platforms such as Flipkart and Amazon not only undermines consumer trust but underscores the pressing requirement for an effective solution. Leveraging the "Amazon Yelp dataset" for model training, this study emphasizes scalability and adaptability for large-scale applications. Acknowledging the escalating impact of fake reviews on user trust, our research addresses the critical need for platforms to combat spammers and uphold the integrity of online shopping experiences. The proposed model not only furnishes a real-time assessment of review authenticity but also serves as a foundation for website owners to take informed actions against spurious content. With its potential applications across platforms of varying sizes, this sophisticated model showcases its efficacy in detecting spam reviews, thereby contributing significantly to the augmentation of user trust in the online marketplace. For prospective buyers in particular, reading product reviews before to making a purchase becomes second nature. They typically examine user reviews of the current product before deciding to purchase it. When reviews are overwhelmingly positive, consumers are more likely to purchase the product; when they are primarily negative, they are more likely to purchase other goods. Positive customer reviews can bring in a lot of money for a company, but they can also be used as a basis for choices about the services that are offered to clients and the design of new products. Everyone is able to openly and anonymously voice their beliefs without worrying about repercussions Social media and internet publishing have made it easier than ever to post boldly and publicly. These ways of view have advantages and disadvantages. When they are employed to persuade others or give the right advice to the right person who may assist in solving the issue, they can be quite beneficial. These opinions are regarded as valuable. Because of this, it is easier for people with malicious intent To manipulate online conversations, some individuals post anonymous comments praising their own products or criticizing those of their competitors. These individuals are known as opinion spammers, and the actions they engage in are referred to as opinion spamming. In addition to undermining customer confidence, the rise in fraudulent reviews on well-known websites like Flipkart and Amazon highlights the urgent need for a workable solution. Using the "Amazon Yelp dataset" to train the model, this study highlights flexibility and scalability for large-scale uses. Our research emphasizes the important need for platforms to battle spammers and protect the integrity of online purchasing experiences, acknowledging the growing influence of bogus reviews on customer confidence. In addition to providing a real-time evaluation of review authenticity, the suggested methodology lays the groundwork for website owners to take well-informed action against fraudulent content. This advanced algorithm demonstrates its effectiveness in identifying spam reviews, which can be used to platforms of different sizes. This can lead to a notable increase in consumer confidence in the online marketplace.

2 Literature Survey

Mohawesh et al. [1] provides a comprehensive overview of the techniques and methodologies employed in detecting fake reviews. With a focus on the increasing prevalence of fake reviews in online platforms, the survey delves into various approaches ranging from linguistic analysis to machine learning algorithms. It outlines the challenges associated with fake review detection, including the evolving nature of fraudulent activities and the need for robust detection mechanisms. The paper discusses key features utilized in fake review identification, reviewer behavior analysis, and domain-specific characteristics. Furthermore, it highlights the importance of dataset quality and the role of deep learning techniques in improving detection accuracy.

Barbosa and Feng [2] proposed a robust sentiment detection method for Twitter, addressing the challenges of biased and noisy data. Their work emphasizes the importance of handling the unique characteristics of social media content to improve sentiment classification accuracy.

In a related study, Davidov and Tsur [3] introduced improved sentiment analysis with smileys and hashtags on Twitter. They leveraged additional contextual information to improve sentiment analysis accuracy, showcasing the significance of incorporating metadata for better understanding sentiments expressed on social media platforms.

Go et al. [4] explored distant supervision for sentiment classification on Twitter. They used distant supervision to automatically label a large dataset, demonstrating the feasibility of leveraging auxiliary information for training sentiment classifiers.

Moving beyond Twitter, Patel et al. [5] focused on fake review detection using opinion mining. Their work highlights the importance of sentiment analysis in identifying fake reviews, contributing to the broader field of opinion-based deception detection.

Ravi [6] conducted a comprehensive survey about sentiment analysis and opinion mining, providing insights into various tasks, approaches, and applications. The survey serves as a foundational resource for understanding the evolution of sentiment analysis techniques.

Khan et al. [7] delved into extracting viewpoint elements from unstructured reviews, emphasizing the extraction of specific components contributing to overall opinions. This work provides a nuanced perspective on opinion mining, considering the multifaceted nature of user reviews.

Dyar Wahyuni and Djunaidy [8] proposed a adapted framework for iterative computation technique for fake review detection from product reviews. Their approach demonstrates the importance of refining existing methodologies to enhance the accuracy of deception detection.

Saumya and Singh [9] focused on the identifying spam reviews using a sentiment analysis approach. By employing sentiment analysis, they aimed to spot fraudulent reviews, contributing to the broader efforts of maintaining the integrity of online reviews.

Xie et al. [10] explored review spam detection through temporal pattern discovery. Their work emphasized the temporal aspects of review patterns, revealing the importance of considering the evolution of sentiments over time.

Mukherjee et al. [11] conducted a categorization and evaluation of authentic and fraudulent reviews, contributing to the understanding of characteristics that distinguish genuine reviews from fake ones. Their work provides valuable insights into the nuances of fake review detection.

Heydari et al. [12] addressed fake opinion detection using time series analysis. By considering temporal patterns, they aimed to improve the accuracy of detecting fake opinions over time, recognizing the dynamic nature of deceptive practices.

Ren and Ji [13] investigated the use of neural networks for detecting false opinion spam. Their empirical research showcases the effectiveness of deep learning techniques in identifying deceptive opinions, highlighting the evolving landscape of sentiment analysis.

In addition to sentiment analysis and fake review detection, McCallum [14] provided insights into graphical models, particularly Bayesian Network Representation, offering a broader perspective on probabilistic models that can be applied to sentiment analysis tasks.

Joseph [15] conducted a an examination of data mining techniques for systems with intelligent computing, emphasizing the role of advanced algorithms in the broader context of intelligent systems. This work provides a comprehensive overview of data mining techniques, setting the stage for their potential application in sentiment analysis.

In conclusion, the literature surveyed highlights the diverse approaches and methodologies employed in sentiment analysis and fake review detection. From addressing the challenges of social media data to leveraging advanced techniques such as neural networks, the field continues to evolve, providing valuable insights for researchers and practitioners alike.

3 Methodology

3.1 Modules

Import necessary libraries for data analysis and manipulation tasks.
Load the dataset into the working environment for analysis.
Conduct exploratory data analysis using TextBlob, Vader Sentiment, and visualization.
Cleanse the text data by removing noise and irrelevant information.
Utilize Tfidf Vectorizer for transforming text data into numerical features.
Splitting data into train and test: Data from this module will be split into train and test categories.
Model generation: Model building—Random Forest, Naive Bayes, MLP, Voting Classifier (MLP + RF + NB)
User signup and login: Utilizing this module will result in login and registration.
User input: This module's use will provide input for Prediction.
Prediction: final prediction shown.

See Fig. 1.

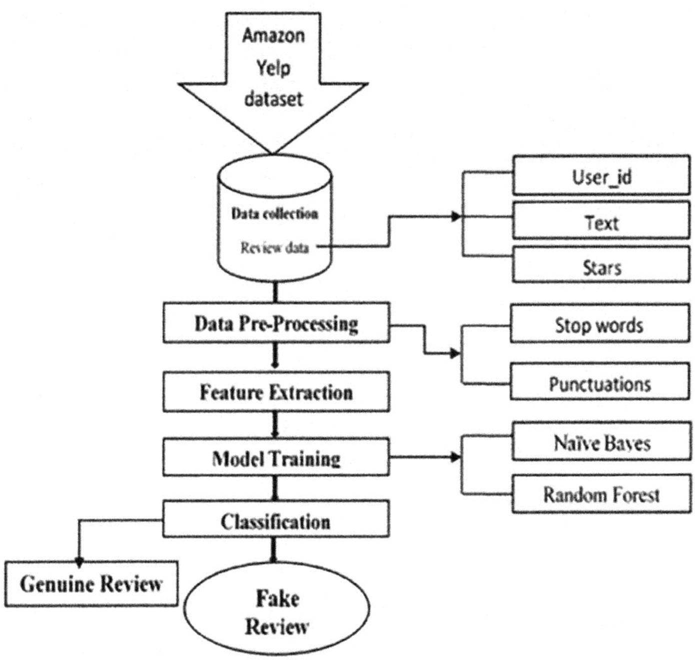

Fig. 1 System architecture

System Architecture We propose an advanced system for combating deceptive reviews in the E-commerce industry, integrating cutting-edge models for machine learning (ML). Our solution employs a combination of Random Forest, Naive Bayes, MLP, and a Voting Classifier, trained on the "Amazon Yelp dataset" for scalability. This system ensures real-time assessment of review authenticity, empowering platform owners to combat spurious content and preserve the integrity of online shopping experiences. Versatile and effective across platforms of varying sizes, our model contributes significantly to mitigating the threat of deceptive reviews, enhancing user trust and confidence in the expanding digital marketplace.

Dataset Collection The "amazon academic review" dataset is employed, and it includes numerous attributes such as user IDs, ratings, reviews, and helpful votes. For feature engineering, the relevant parameters are obtained. Thousands of real and fraudulent reviews are combined in this dataset to make it simple to evaluate the model's accuracy when it comes to implementation. There is data for 11,537 businesses in the Yelp dataset that was made available for the academic challenge. According to www.yelp.com/dataset, this dataset includes 229,907 reviews, 43,873 users, and 8282 check-in sets for these businesses. The dataset presents a challenge because it includes a wide range of reviews and parameters that can be used to train any algorithm.

Pre-processing To start data analysis and manipulation tasks, essential libraries like pandas for data manipulation, matplotlib and seaborn for visualization, TextBlob and VaderSentiment for sentiment analysis, and scikit-learn for text preprocessing will be imported. The dataset will be loaded using pandas read csv method, and sentiment information will be extracted using TextBlob and VaderSentiment for exploratory data analysis. Matplotlib and seaborn will aid in visualizing data distribution and patterns. Text data will be cleansed by removing noise and irrelevant information using regular expressions and TextBlob functions to ensure a cleaner dataset. Text data will be transformed into numerical features using Tfidf Vectorizer from scikit-learn for machine learning tasks. The process involves converting text into a matrix of TF-IDF features, indicating the importance of each word in the dataset. Overall, the data processing pipeline includes importing libraries, loading the dataset, exploratory data analysis with TextBlob and Vader Sentiment, cleansing text data, and transforming it into numerical features with Tfidf Vectorizer, preparing the data for further analysis and modeling.

Training and Testing The training and testing process involves several key steps to implement the described modules. First, the data needs to be split into training and testing sets using a module such as scikit-learn's train test split. This guarantees that the model is tested on a different subset of data and trained on a different one in order to precisely assess its performance. Next, the model generation phase involves building different classifiers, including Random Forest, Naive Bayes, MLP (Multi-Layer Perceptron), and a Voting Classifier. This can be achieved using libraries like scikit-learn for model creation. For user signup and login, a module handling user authentication and registration is necessary, employing techniques like user database management and secure authentication protocols. The user input module is responsible for collecting input features for prediction. This can be implemented through a user interface or an application form. Finally, the prediction module uses the trained models to make predictions on the user input, and the results are displayed to the user. Care should be taken to handle user interactions seamlessly and provide clear and understandable predictions. In summary, the training and testing process encompasses data preparation, model generation, user authentication, input collection, and prediction, ensuring a comprehensive and functional system for user interaction and accurate predictions.

Algorithms Random Forest: A Random Forest is a widely employed ensemble learning technique in machine learning that excels in both regression and classification tasks. During training, it constructs a large number of decision trees, which are then combined to produce the mean prediction (regression) or the most common class label (classification). The randomness incorporated into feature selection and dataset sampling helps to prevent overfitting and improves the generalization capacity of the model. This robustness makes Random Forest particularly well-suited for complex and high-dimensional datasets, where it often outperforms individual decision trees in terms of accuracy.

Naive Bayes: The Naive Bayes probabilistic classification algorithm is built upon the Bayes theorem, simplifying computations by assuming features are conditionally independent of the class label. Despite its simplicity, Naive Bayes demonstrates impressive performance in text classification, spam filtering, and other tasks. Its computational efficiency and low requirement for training data are advantages. However, in certain real-world scenarios, the independence assumption may not hold true, leading to potential inaccuracies.

MLP (Multilayer Perceptron): An artificial neural network called a multilayer perceptron is made up of several layers of connected nodes, or neurons. MLPs are trained using back propagation and can learn complex patterns and representations. They are versatile and applicable to various tasks, including image recognition, speech processing, and more. However, training large MLPs can be computationally intensive and may require substantial amounts of labeled data to prevent overfitting.

Voting Classifier: A Voting Classifier combines the predictions of multiple base classifiers to enhance overall performance. Each model contributes its prediction, and the final output is determined by majority voting. This ensemble approach leverages the strengths of individual models, leading to improved accuracy, robustness, and generalization. The diversity in model architectures ensures that the ensemble is effective across different types of data and underlying patterns, making it a powerful tool in machine learning ensemble methods (Figs. 2, 3, 4 and 5).

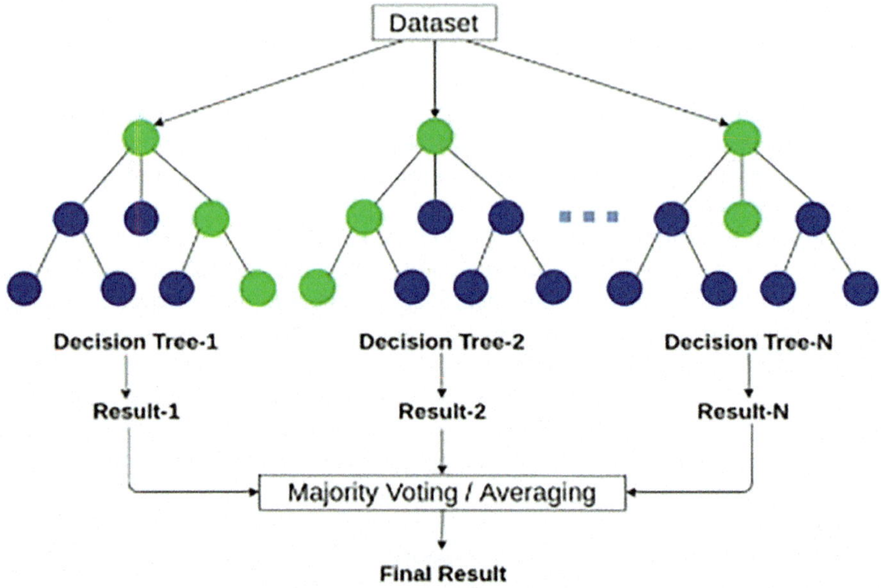

Fig. 2 Random forest

Fig. 3 Naive bayes

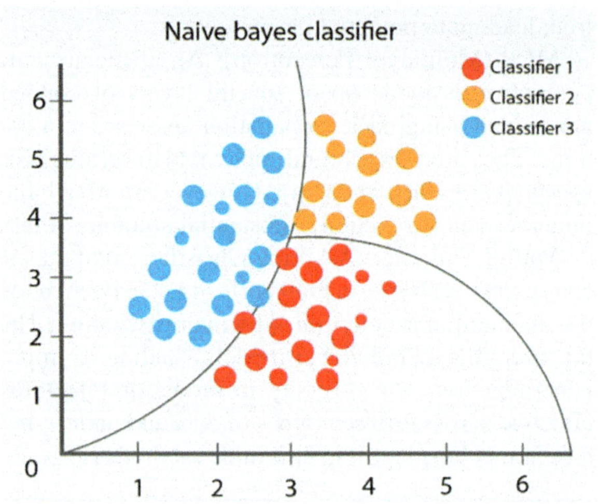

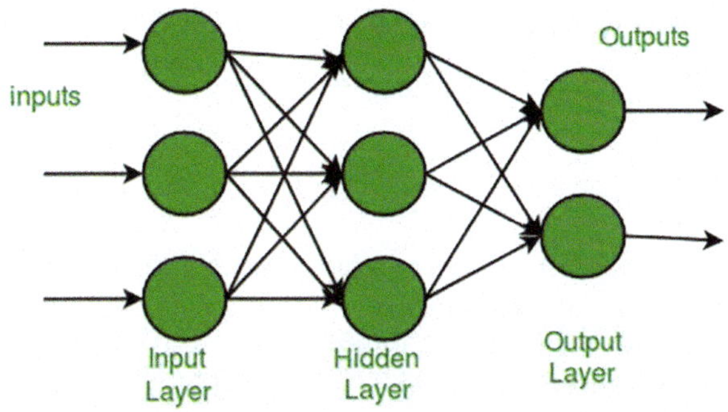

Fig. 4 Multilayer perceptron

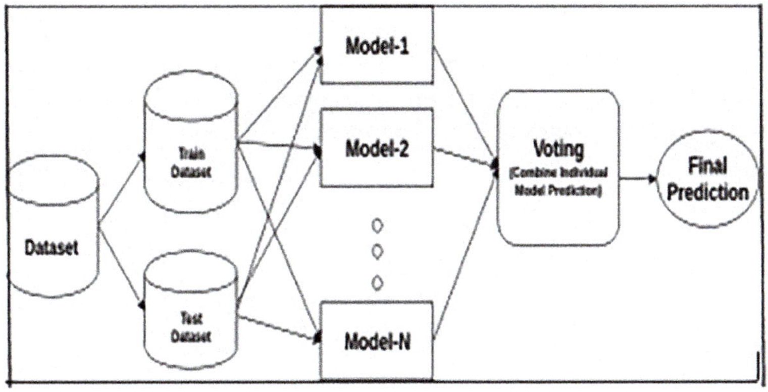

Fig. 5 Voting classifier

4 Experimental Results

4.1 Comparison Graphs → Accuracy, Precision, Recall, f1 Score

Accuracy: A test's accuracy is defined as its ability to recognize debilitated and solid examples precisely. To quantify a test's exactness, we should register the negligible part of genuine positive and genuine adverse outcomes in completely examined cases. This might be communicated numerically as:

$$\text{Accuracy} = \text{TP} + \text{TN}\ \text{TP} + \text{TN} + \text{FP} + \text{FN}.$$

$$\text{Accuracy} = \frac{\text{TP} + \text{TN}}{\text{TP} + \text{TN} + \text{FP} + \text{FN}}$$

Precision: Precision quantifies the percentage of correctly classified samples or occurrences among the positives. As a result, the accuracy may be calculated using the following formula:

$$\text{Precision} = \text{True positives}/(\text{True positives} + \text{False positives}) = \text{TP}/(\text{TP} + \text{FP})$$

Recall: Recall is a machine learning metric that surveys a model's capacity to recognize all pertinent examples of a particular class. It is the proportion of appropriately anticipated positive perceptions to add up to real up-sides, which gives data about a model's capacity to catch instances of a specific class.

$$\text{Recall} = \frac{\text{TP}}{\text{TP} + \text{FN}}$$

F1-Score: The F1 score is a machine learning evaluation measurement that evaluates the precision of a model. It consolidates a model's precision and review scores. The precision measurement computes how often a model anticipated accurately over the full dataset.

$$\text{F1 Score} = \frac{2}{\left(\frac{1}{\text{Precision}} + \frac{1}{\text{Recall}}\right)}$$

$$\text{F1 Score} = \frac{2 \times \text{Precision} \times \text{Recall}}{\text{Precision} + \text{Recall}}$$

4.2 Performance Evaluation Graph

See Fig. 6.

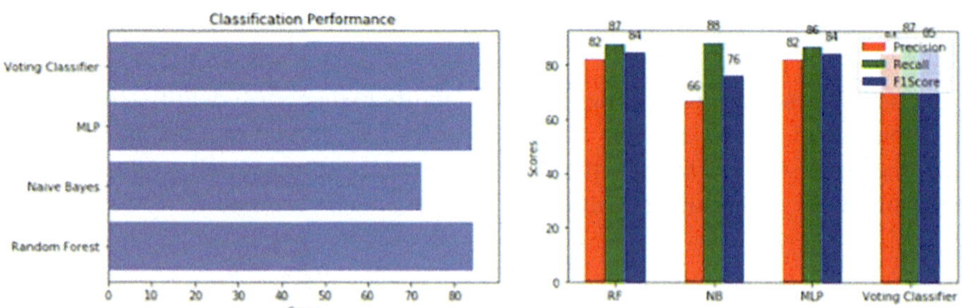

Fig. 6 Comparison graphs

4.3 Screens

See Figs. 7, 8, 9, 10, 11, 12, 13 and 14.

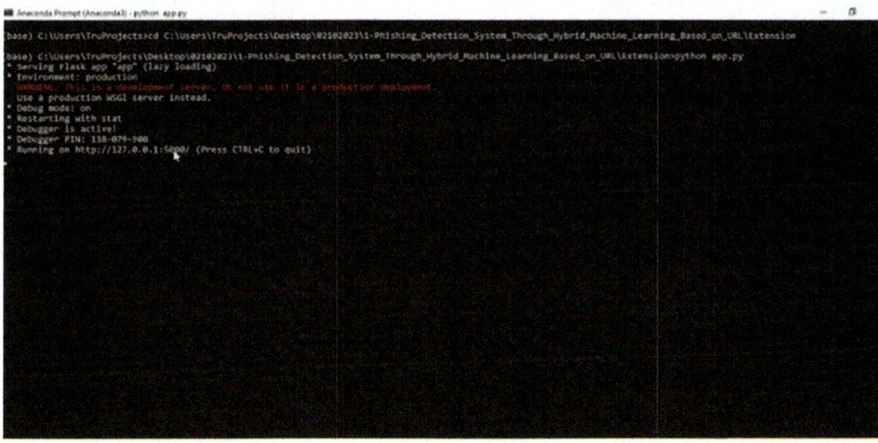

Fig. 7 Url link to web page

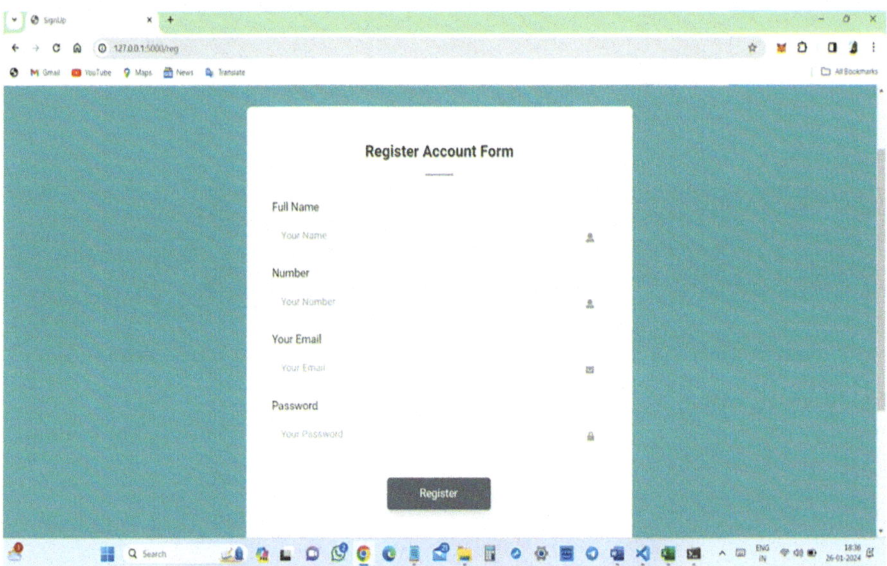

Fig. 8 Register user

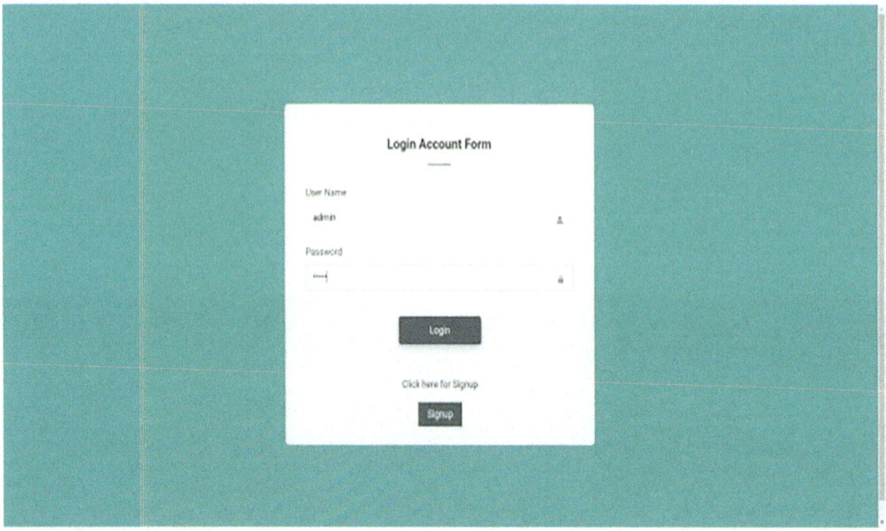

Fig. 9 User sign in page

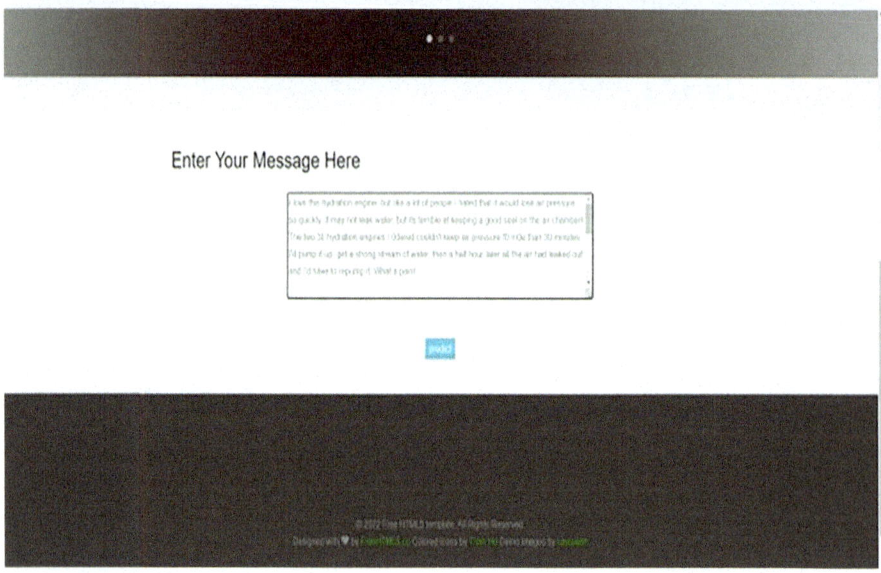

Fig. 10 Entered data

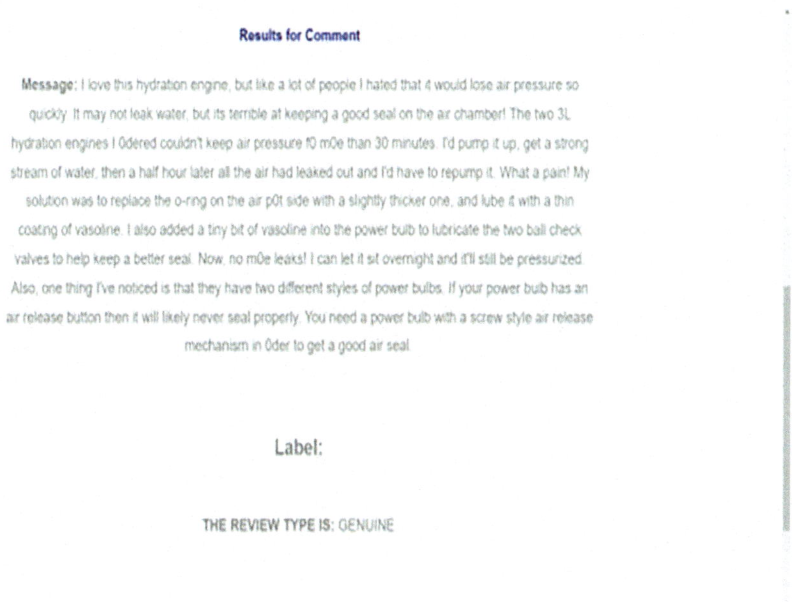

Fig. 11 Genuine

Fig. 12 Fake

Fig. 13 Entered other data

Fig. 14 72.0% positive

5 Conclusion

In conclusion, the escalating prevalence of deceptive reviews in the E-commerce industry necessitates a proactive and robust solution, and our proposed model, integrating Machine Learning (ML) models, stands as a promising step forward. The amalgamation of Random Forest, Naive Bayes, MLP, and a Voting Classifier demonstrates a comprehensive approach to evaluating the authenticity of reviews within the dynamic digital landscape. By leveraging the "Amazon Yelp dataset" for training, our study emphasizes scalability and adaptability, addressing the imperative need for large-scale applications. The detrimental impact of fake reviews on consumer trust, especially on renowned platforms like Flipkart and Amazon, underscores the urgency for effective countermeasures. Our model not only provides a real-time assessment of review authenticity but also empowers platform owners to take informed actions against spurious content, contributing to the preservation of the integrity of online shopping experiences. The project presents a robust and effective solution for combating fake product reviews in the E-commerce industry. By leveraging advanced NLP techniques and ML models such as Random Forest, Naive Bayes, MLP, and a Voting Classifier, the system demonstrates a high level of accuracy in assessing the authenticity of reviews. The integration of these models with the "Amazon Yelp dataset" for training ensures scalability and adaptability for large-scale applications. The system not only provides real-time assessment of review authenticity but also empowers platform owners to take informed actions against spurious content, thereby preserving the integrity of online shopping experiences. As the digital environment keeps changing, adopting these advanced solutions is crucial for the survival and expansion of the E-commerce sector. In conclusion, the project presents a robust and effective solution for combating fake product reviews in the E-commerce industry. By leveraging advanced NLP techniques and ML models such as Random Forest, Naive Bayes, MLP, and a Voting Classifier, the system demonstrates a high level of accuracy in assessing the authenticity of reviews. The integration of these models with the "Amazon Yelp dataset" for training ensures scalability and adaptability for large-scale applications. The system not only provides real-time assessment of review authenticity but also empowers platform owners to take informed actions against spurious content, thereby preserving the integrity of online shopping experiences. As the digital landscape continues to evolve, the implementation of such advanced solutions becomes imperative for the sustenance and growth of the E-commerce industry.

6 Future Scope

Furthermore, the versatility of our sophisticated model positions it as an effective tool across platforms of varying sizes, showcasing its efficacy in detecting spam reviews. In this way, our research makes a significant stride towards mitigating the threat of deceptive reviews, ultimately bolstering user trust and confidence in the online marketplace. As the digital realm continues to expand, the implementation of such advanced solutions becomes imperative for the sustenance and growth of the E-commerce industry. The future scope of the project includes several potential avenues for enhancement and expansion. One direction could involve further refining the NLP and ML models to improve their accuracy and efficiency in detecting fake reviews. This could involve exploring more advanced techniques in NLP, such as transformer models like BERT or GPT, which have shown promising results in natural language understanding tasks. Additionally, integrating user feedback and interaction into the model could help in continuously improving its performance and adaptability to evolving patterns of fake reviews. The system's scalability is another factor to take into account because it could need to handle larger datasets and a higher volume of reviews over time. This could involve optimizing the existing architecture for better performance and exploring distributed computing solutions to handle the increased computational load. Furthermore, there is potential for collaboration with E-commerce platforms and regulatory bodies to implement the model as a tool for monitoring and removing fake reviews, thereby enhancing the credibility and trustworthiness of online reviews for consumers.

References

1. Mohawesh, R., et al.: Fake reviews detection: a survey. IEEE Access **9**, 65771–65802 (2021)
2. Barbosa, L., Feng, J.: Robust sentiment detection on twitter from biased and noisy data. In: Coling 2010—23rd International Conference on Computational Linguistics, Vol. 2, pp. 36–44 (2010)
3. Davidov, D., Tsur, O.: Enhanced Sentiment Learning Using Twitter Hashtags and Smileys. ICNC/2, Institute of Computer Science, The Hebrew University (2010)
4. Go, A., Bhayani, R., Huang, L.: Twitter sentiment classification using distant supervision. Processing **150** (2009)
5. Patel, D., Kapoor, A., Sonawane, S.: Fake review detection using opinion mining. Int. Res. J. Eng. Technol. (IRJET) **5**(12) (2018)
6. Ravi, K., Ravi, A.: Survey on opinion mining and sentiment analysis: tasks, approaches and applications. Knowl. Based Syst. **89**, 14–46 (2015)
7. Khan, K., et al.: Mining opinion components from unstructured reviews: a review. J. King Saud Univ. Comput. Inf. Sci. (2014). https://doi.org/10.1016/j.jksuci.2014.03.009
8. Wahyuni, E.D., Djunaidy, A.: Fake review detection from product review using modified method of iterative computation framework. MATEC Web Conf. 58.03003 (2016) (BISSTECH 2015)
9. Saumya, S., Singh, J.P.: Detection of spam reviews: a sentiment analysis approach. CSIT **6**, 137–148 (2018). https://doi.org/10.1007/s40012-018-0193-0
10. Xie, S., Wang, G., Lin, S., Yu, P.S.: Review spam detection via temporal pattern discovery. In: Proceedings of the 18th ACM SIGKDD International Conference on Knowledge Discovery and Data Mining, pp. 823–831. ACM (2012)
11. Mukherjee, A., Venkataraman, V., Liu, B., Glance, N.: Fake Review Detection: Classification and Analysis of Real and Pseudo Reviews. Technical Report UIC-CS2013–03, University of Illinois at Chicago (2013)
12. Heydari, A., Tavakoli, M., Salim, N.: Detection of fake opinions using time series. Expert Syst. Appl. **58**, 83–92 (2016)
13. Ren, Y., Ji, D.: Neural networks for deceptive opinion spam detection: anempirical study. InfSci **385**, 213–224 (2017)
14. McCallum, A.: Graphical Models, Lecture2: Bayesian Network Represention (PDF). Retrieved 22 Oct 2019
15. Joseph, S.I.T.: Survey of data mining algorithm's for intelligent computing system. J. Trends Comput. Sci. Smart Technol. (TCSST) **1**(01), 14–24 (2019)

Real Time Fall Detection Monitoring on Elderly Using IoT and Deep Learning

Wanraplang Nongbri, Nissi Paul, Pushpanjalee Konwar, Abhijit Bora, and Mriganka Gogoi

Abstract Research has shown that there is an increase in number of elderly people living alone. Taking this into consideration, a smart detection system for elderly people has been proposed in the current research paper to detect fall of elderly people so as to prevent any serious injuries to the elderly people who stay at home by themselves. Fall has been identified as one of the leading causes of serious injuries and even fatalities. This system can assist caretakers and family members in monitoring elderly individuals, enabling them to take timely actions in any signs of a fall being detected. An alert notification is sent to the caretaker to provide medical care as early as possible to avoid any serious injuries. The work presented in the paper involves an approach to process the images and videos isolating the background and foreground and using optical flow and head speed movement for enhancing the accuracy of detection. Four deep learning models-VGG16, ResNet50, InceptionV3 and EfficientNetB7 were used for implementing the detection of falls. Two types of images/frames were considered, one with proper illumination and another with low light from www.falldataset.com and a created dataset. 1050 images with proper

W. Nongbri · N. Paul · A. Bora
Department of Computer Application, Assam Don Bosco University, Azara 781017, India
e-mail: theo.zenette@gmail.com

N. Paul
e-mail: nissi.paul@dbuniversity.ac.in

A. Bora
e-mail: abhijit.bora0099@gmail.com

W. Nongbri
Faculty of Computer Technology, Assam Down Town University, Panikhaiti 781026, India

P. Konwar
Department of Electrical and Electronics Engineering, Assam Don Bosco University, Azara 781017, India
e-mail: konwarpushpanjalee@gmail.com

M. Gogoi (✉)
Department of Electronics and Communication Engineering, Assam Don Bosco University, Azara 781017, India
e-mail: mrig.gogoi@gmail.com

S. N. Mohanty et al. (eds.), *Explainable IoT Applications: A Demystification*, Information Systems Engineering and Management 21,
https://doi.org/10.1007/978-3-031-74885-1_21

illumination and low light were used for training out of them, 150 images with proper light and 95 low light were used for testing. In both the categories of images, VGG16 accuracies reported 96% for images with proper illumination and 92.63% low light images while EfficientNetB7 reported 97.33% and 93.68% respectively. In both the categories VGG16 and EfficientNetB7 have outperformed better than the other two in terms accuracies. Also successful alert notification was demonstrated via email and phone number with hardware platform using Arduino Uno and GSM module for sending alert messages to registered phone number and Raspberry Pi 3 for sending to the registered email Id.

Keywords Deep learning · Optical flow · IoT · Raspberry Pi · Arduino

1 Introduction

Falls pose a significant public health risk, particularly for the elderly, prompting the development of various technologies for fall detection. These include non-wearable, wearable, and vision-based devices. Smart homes enhance the independence, comfort, and safety of elderly individuals, especially those with dementia. Despite device development, forgetfulness to wear them remains an issue, making intelligent systems valuable. Installing a smart fall detection system enables easy monitoring, notifying emergency contacts swiftly in case of a fall, preventing serious injuries or life-threatening situations for the elderly [1, 2]. The primary focus of this study is to highlight the importance of spatial information in categorizing activities. The proposed approach utilizes a two-stream spatial CNN for identifying human behavior in extensive video databases [3]. Another paper introduces an original OD segmentation model based on attention gates, employing a deep convolutional neural network [4]. The utilization of a two-stream inflated CNN with GRU layers for raw frames and optical flow analysis is proposed [5]. Additionally, a simplified end-to-end method for Finger-Knuckle recognition is presented using CNNs [6]. An innovative method for evaluating the reliability of the software-as-a-service deployment of the prototype for a coronavirus disease data analysis system is presented, employing a distinctive cycle of evaluation models [7]. In the modern era, intelligent systems drive advancements in self-driving cars, artificial pancreas, the Internet of Things, M2M-enabled manufacturing robots, and wearable health monitors [8]. Various fall detection methods have been proposed, ranging from wearable devices using IoT [9], vision-based approaches with convolutional neural network(CNN) models like VGG16 [10]. In [11], the system is based on computer vision which uses camera and deep learning for elderly and sick people systems. In [12], detection of human fall is assisted using IoT and mobile application. In [13], K Nearest Neighbors algorithm is used to classify as fall or not fall. In [14], fall detection is carried out using combination of CNN and recurrent neural network models. Systems utilizing depth images and 3D images with SVM models are expressed in [15–17]. The article reviews methodologies for pose estimation, tracking, human detection, and walking motion analysis [18]. Image registration methods are explored, emphasizing reduced potential matches for increased speed [19, 20].

2 Method and Implementation

2.1 *Proposed Method*

This study employs an IoT device equipped with a camera to capture live feed data. A deep learning model, specifically utilizing pre-trained CNN models such as VGG15, InceptionV3, EfficientNetB7, and ResNet50, is integrated to detect falls among the elderly. The dataset can either be prepared or sourced from existing ones. The camera is strategically placed for comprehensive room coverage to identify falls. Additional considerations include analyzing head movement and optical flow. The deep learning model is then embedded in a Raspberry Pi to create the hardware system. To send notifications, the same Raspberry Pi, along with Arduino and IoT devices featuring GSM modules, is utilized. Upon detecting a fall, the system promptly sends notifications to emergency contacts, aiming to prevent serious injuries or fatalities.

2.2 *Implementation*

The dataset used is an existing one which is collected from www.falldataset.com [21] that contains over 10,000 images. The images are divided into five different postures, they are: lying down, bending, crawling, sitting and standing. The images are labelled before training or building the model into two class, that is, Fall and Not Fall. In the fall class it will contain the lying down and crawling whereas in not fall it will contain standing and bending. Each image feature are extracted according to the pre-trained model type, which are VGG16, InceptionV3, EfficientNetB7, Resnet50 as portrayed in Figs. 1, 2, 3 and 4 respectively.

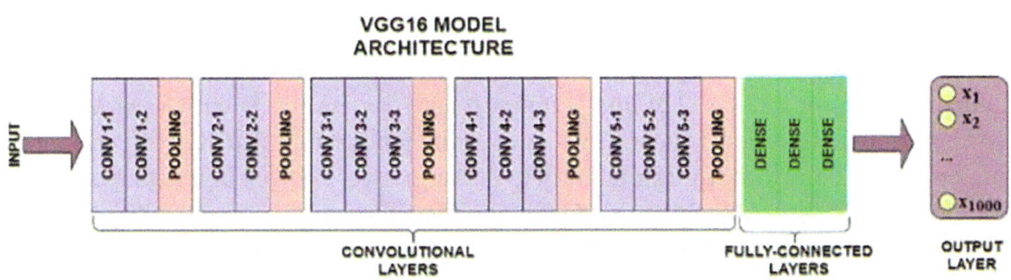

Fig. 1 VGG16 architecture

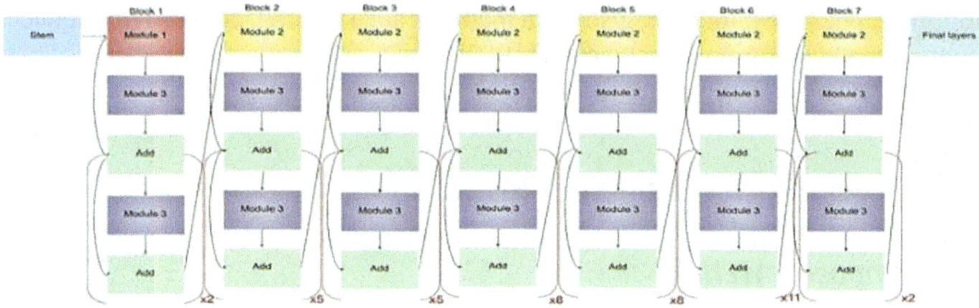

Fig. 2 InceptionV3 architecture

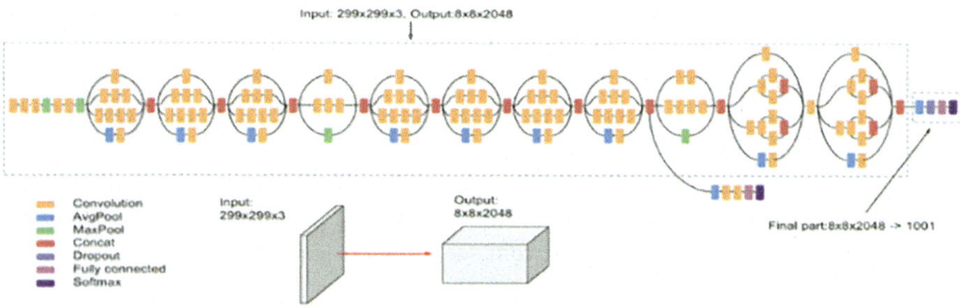

Fig. 3 EfficientNetB7 architecture

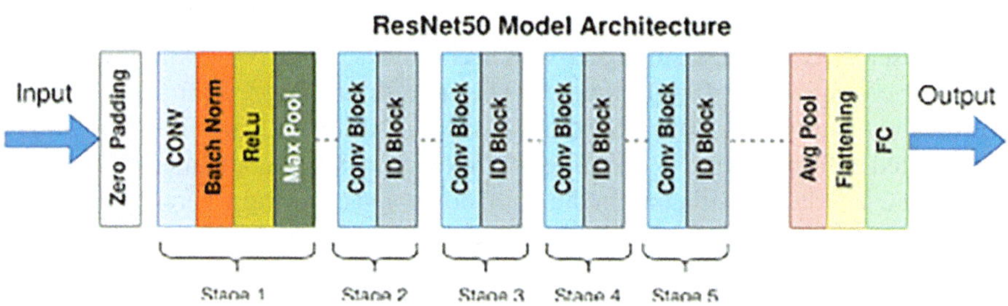

Fig. 4 ResNet50 model architecture

VGG16 comprises 16 layers, with 13 convolutional layers and 3 fully connected layers, processing a 224 × 224 image input. It uses 3 × 3 kernels with a stride of one, employing the Rectified Linear Unit Activation function (ReLu) exclusively in its hidden layers, contributing significantly to performance. Pooling layers play a crucial role in reducing dimensionality. VGG Net includes three fully connected layers, with the first two having adjustable 4096 channels, and the third layer comprising 1000 channels for each class. In contrast, EfficientNetB7 has over 813 layers, covering convolutional, pooling, and fully connected layers. The Stem Convolutional Layer extracts low-level features, and EfficientNet-B7 utilizes repeated blocks, including

Depth-wise Convolution, Swish activation, and a Squeeze-and-Excitation (SE) Block for channel-wise recalibration. An output convolution adjusts the channel count. Global average pooling follows, reducing spatial dimensions. The resulting features pass through fully connected layers, with the final layer's unit count determined by the task's class count. Typically, a soft-max activation function concludes, generating class probabilities for multi-class classification. After several tests that has been tested with different models the VGG16 and EfficientNetB7 are giving good results. Also for the images on low light it is detecting accurately.

Figure 5 illustrates various human postures, including Crawling, Lying down or sleeping, Bending, Standing, and Sitting. Additionally, the model has been trained to enhance detection performance in low-light environments, as depicted in Fig. 6, showcasing images of human standing postures in such conditions. In Fig. 7, flow chart of the proposed fall detection system is highlighted. In our model, Background-SubtractorMOG2 is specifically utilized for background subtraction, while head speed motion is considered a criterion for distinguishing between fake and actual falls.

To calculate motion or head speed, we employ the Optical Flow method, namely Lucas-Kanade, which describes image motion by identifying differences in a video over a short period. RGB images are converted to grayscale frames, and corners are reshaped to 3D. The comparison between current and old frames yields values like p1, status, and error, where the Lucas-Kanade equation is applied to determine the flow of points, distinguishing good points in the current and previous frames. For all points $(k, l) \in W$, where W is a window or a neighborhood like a small patch in an image.

The pseudo code for the suggested framework is,

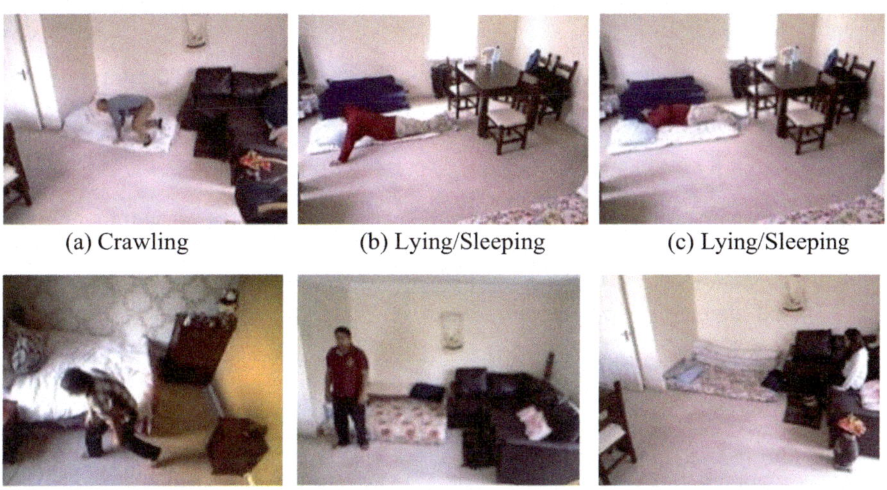

(a) Crawling (b) Lying/Sleeping (c) Lying/Sleeping

(d) Bending (e) Standing (f) Sitting

Fig. 5 Images of each different pose

Fig. 6 Low light images of a person standing

```
# camera = initialize_camera()
# threshold_value = set_threshold_value()
# background = get_background_subtraction()
# while True:
  #frame = capture_frame(camera)
  #subtracted_frame = apply_background_subtraction(frame, background)
  #binary_mask = apply_thresholding(subtracted_frame)
  #contours = find_contours(binary_mask)
  #frame_with_contours = draw_contours(frame, contours)
  # If contours exist
 if len(contours) > 0:
    # optical_flow = calculate_optical_flow(frame, frame_with_contours)
    # good_points = select_good_points(optical_flow)
#motion_movement = calculate_motion_movement(good_points)
#head_speed_movement = calculate_head_speed_movement(motion_movement)
    # if head_speed_movement >= threshold_value:
      # fall_detected = detect_fall(frame)
        # if fall_detected:
       # display_or_notify(frame)
  # display_frame(frame_with_contours)
  # If 'q' is pressed, exit the loop
  if key_pressed('q'):
```

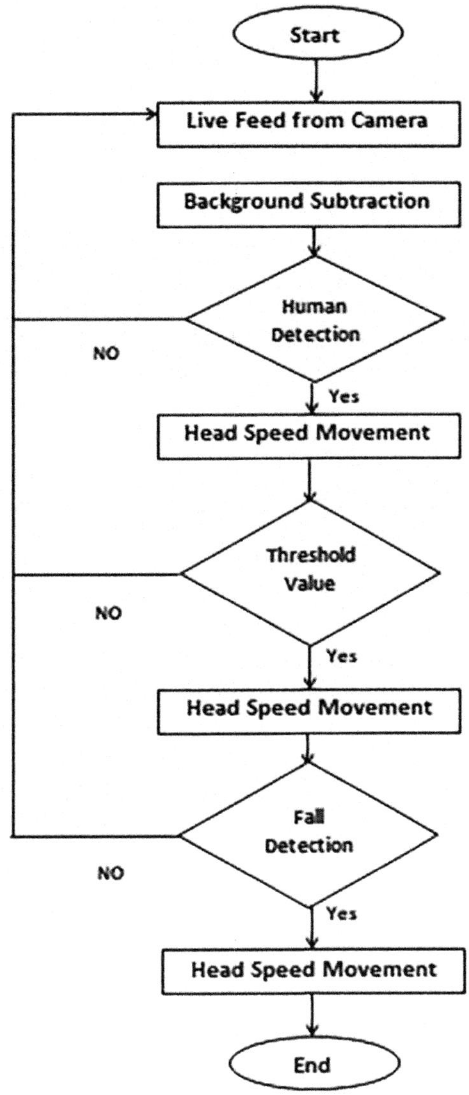

Fig. 7 Flow chart of the system

    break

release_camera(camera)

The model has been also trained to detect better in low light rooms. Some of the images of human standing postures in low light are shown in Fig. 6.

In Fig 7, flow chart of the proposed fall detection system is highlighted. In our model, BackgroundSubtractorMOG2 is used for background subtraction. Head speed motion is another criterion considered for detection of fake and actual fall of a person. For calculation of the motion or head speed in this process the Optical Flow methods namely, Lucas Kanade is used. Optical flow technique describes the image motion, which is applied to a series of images to find any differences of a video in a small time [22].

The RGB images/frames are converted to gray scale frames after that the corners are reshaped to 3D. A comparison between the current and old frame to obtain three values namely p1, status, and error. p1 are array points in the next frame, the status is like a Boolean flag which uses the value 0 and 1, whenever there is flow of point in the frame it will indicate as 1. So, we store those good points where status is 1 in a variable 'good_new' for the points of the current frame whereas the 'good_old' is for the old or previous frame. The mathematical equation for the Lucas Kanade can be written in this form:

For all points $(k, l) \in W$, where W is a window or a neighborhood like a small patch in an image.

We have the constraint equation as

$$I_x(k,l) * u + I_y(k,l) * v + I_t(k,l) = 0 \tag{1}$$

where (u, v) is the flow vector.

From the equation, the derivatives of the intensity in the x direction $\{I_x(k,l)\}$ times u, the derivatives of the intensity in the y direction $\{I_y(k,l)\}$ times v, and the derivatives of the intensity in time t $\{I_t(k,l)\}$ are obtained. In matrix form the equation can be stated as described in Eq. (2).

$$\begin{bmatrix} I_x(1,1) & I_y(1,1) \\ I_x(k,l) & I_y(k,l) \\ \vdots & \vdots \\ I_x(n,n) & I_y(n,n) \end{bmatrix} \times \begin{bmatrix} u \\ v \end{bmatrix} = \begin{bmatrix} -I_t(1,1) \\ -I_t(k,l) \\ \vdots \\ -I_t(n,n) \end{bmatrix} \tag{2}$$

The above equations can be further formulated as shown in Eqs. (3), (4) and (5), where A is the matrix comprised of derivatives of the intensity in x and y directions respectively, U is the flow vector and B is the matrix with derivatives of intensity in time. A^{T} is the transpose matrix of A.

$$A * U = B \tag{3}$$

$$A^{\mathrm{T}} * A * u = A^{\mathrm{T}} * B \tag{4}$$

In matrix form:

$$\begin{bmatrix} \sum_W I_x I_x & \sum_W I_x I_y \\ \sum_W I_x I_y & \sum_W I_y I_y \end{bmatrix} \times \begin{bmatrix} u \\ v \end{bmatrix} = \begin{bmatrix} -\sum_W I_x I_t \\ -\sum_W I_y I_t \end{bmatrix} \tag{5}$$

Equation (4) can be represented in as Eq. (6)

$$U = \left(A^{\mathrm{T}} * A\right)^{-1} * A^{\mathrm{T}} * B \tag{6}$$

For the optical flow to work the $A^{\mathrm{T}}A$ needs to be invertible, that is determinant ($A^{\mathrm{T}}A$) is not equal to zero. If the derivates are small from "W" or a patch it is regarded as bad region, in other words a region where the gradients are small and the visual are unclear it is considered as bad region. A good region is the one where there is rich texture, have clear visual and the points computed are large and reliable.

Once the motion detection calculation is determined, the subsequent step involves computing motion movement by isolating suitable points from p0 and p1. Here, p0 represents the 'good_old' points, while p1 denotes the 'good_new' points. Employing the Euclidean Distance, the algorithm measures the disparity between the 'good_old' and 'good_new' points. This computation enables the derivation of motion movement. To ascertain the speed of the detected motion, the sum of the motion movements is divided by the count of 'good_new' points, thereby yielding the head speed movement.

3 Results

All the four models VGG16, ResnNet50, InceptionV3 and EfficientNetB7 were trained using 1050 images including low light images. 150 images with proper light and 95 low light images used for testing. Consideration of Head speed moment enhances the accuracy of the models. Fake Pos represents the count of images where a human presence was inaccurately identified as fallen despite not being so. Fake Neg signifies the number of images wherein a fallen human presence was erroneously detected as not fallen. Real Pos is the number of images containing fallen humans which were correctly identified. Lastly, Real Neg denotes the count of images correctly recognized as not having a fallen human presence. Tables 1 and 2 describes the accuracy details of the models for proper light and low light images respectively without considering the head speed movement. Figure 8 gives a graphical comparison of accuracies of various models before considering head movement criteria.

Table 4 describes the accuracy details of the models when tested with low light images. After inclusion of Head Speed Moment as another criteria for determining

Table 1 Accuracy result for each model tested with 150 images with proper light

Model name	Accuracy (%)	Fake Pos	Real Pos	Fake Neg	Real Neg
VGG16	79.33	19	74	12	45
ResNet50	70.66	24	69	20	37
InceptionV3	74.00	22	71	17	40
EfficientNetB7	82.66	17	76	9	48

Table 2 Accuracy result for each model tested with 95 images with low light

Model name	Accuracy (%)	Fake Pos	Real Pos	Fake Neg	Real Neg
VGG16	71.57	14	38	13	30
ResNet50	66.31	17	35	15	28
InceptionV3	68.42	16	36	14	29
EfficientNetB7	75.78	12	40	11	32

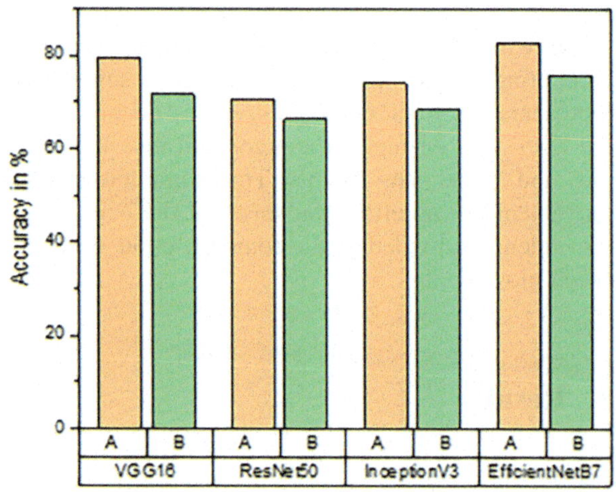

Fig. 8 Comparative analysis of all the four models in proper light (**a**) and low light (**b**) before considering head speed movement

of fallen human the accruacy of the models were enhanced. Tables 3 and 4 describes the accuracies for this case. Figure 9 gives a graphical comparison of accuracies of various models after considering head movement criteria. Out of all the models that has been trained the EfficientNetB7 and VGG16 have higher accuracies in detecting fallen person in both proper light and low light frames.

A GUI interface is design where an image, video or live feed can be uploaded. Figure 10 shows the correctly detection of not fallen person from an uploaded image using the four models. On the other hand Fig. 11 depicts the detection of a fallen person correctly. And for the image lying down which classified as fallen, and when the fall is detected it will send a message to the given email to notify. For sending

Table 3 Accuracy result for each model tested with 150 images with proper light and head speed moment feature

Model name	Accuracy (%)	Fake Pos	Real Pos	Fake Neg	Real Neg
VGG16	96	3	90	3	54
ResNet50	85.33	9	84	13	44
InceptionV3	90.66	5	88	9	48
EfficientNetB7	97.33	2	90	2	56

Table 4 Accuracy result for each model tested with 95 images with low light and head speed moment feature

Model name	Accuracy (%)	Fake Pos	Real Pos	Fake Neg	Real Neg
VGG16	92.63	4	48	3	40
ResNet50	82.10	10	42	7	36
InceptionV3	87.3	7	45	5	38
EfficientNetB7	93.68	4	48	2	41

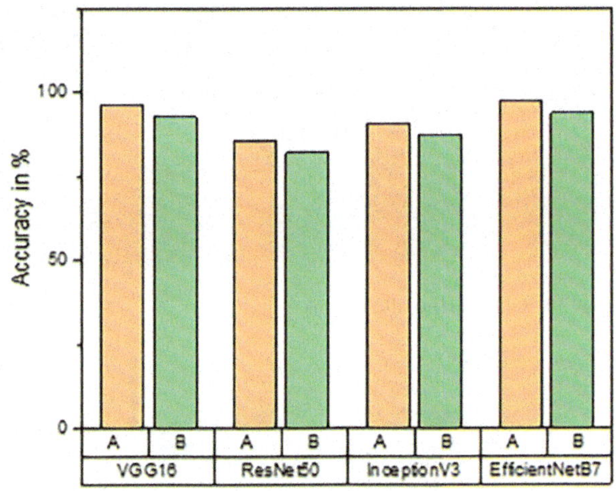

Fig. 9 Comparative analysis of all the four models in proper light (**a**) and low light (**b**) after considering head speed movement

the information to the right person hardware platforms are developed using both Arduino uno and Raspberry Pi 3. Using Arduino message is successfully send to a specified phone number integrated with a GSM module. While email is send to the destined email ID using the SMTP protocol and "*smtplib*" library with the support of Raspberry Pi. Figures 12 and 13 shows the Arduino with GSM module and Rasberry Pi modules used for sending the information on detection of fallen human along with fall detection messages.

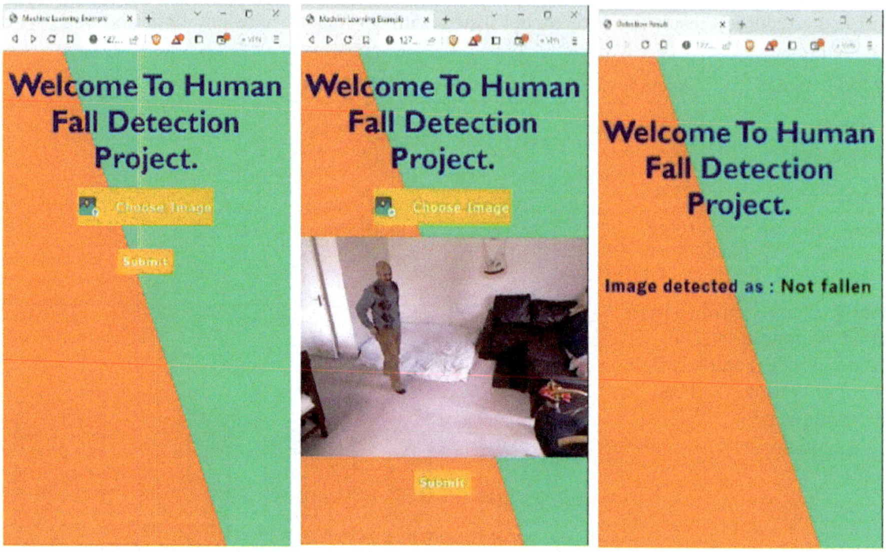

Fig. 10 GUI Testing the model

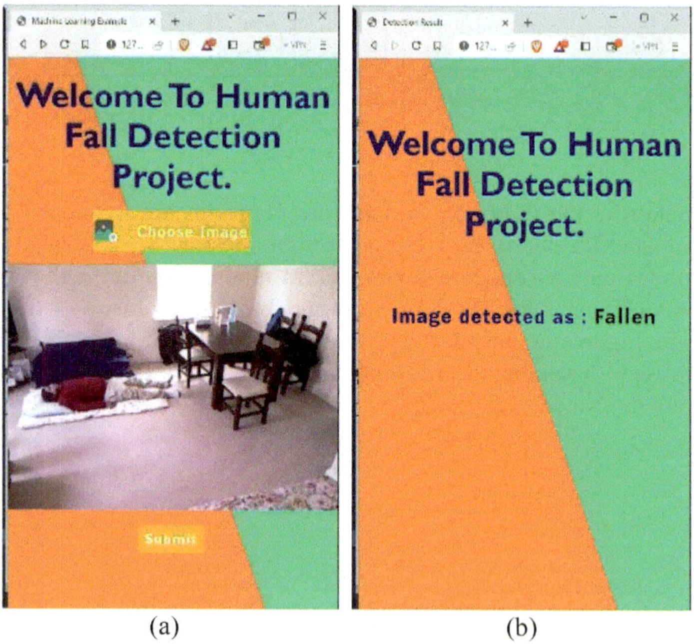

Fig. 11 **a** GUI Testing the model. **b** GUI model displaying the result of human fall detection

(a) (b)

Fig. 12 **a** GSM Module device was used. **b** Using Arduino board device

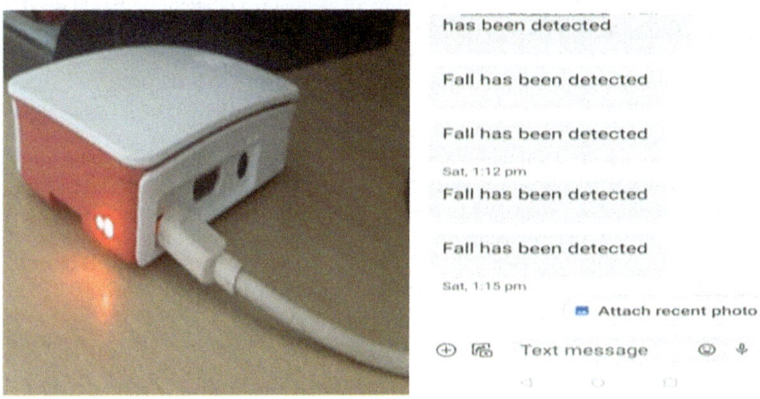

(a) (b)

Fig. 13 **a** Raspberry Pi3 module. **b** Fall detection message obtained on the phone

4 Conclusion

This research paper proposes a system for human fall detection for elderly people or a patient using computer vision or a camera base system. This system uses camera to get a live feed and run through an algorithm that uses deep learning models namely VGG16, ResNet50, InceptionV3 and EfficientNetB7 to detect human fall. Along with the models, background subtraction, head speed motion features are added to improve the detection accuracy. Experiment for detection was performed considering proper light and low light frames. Out of the four deep learning models VGG16 and EfficientNetB7 accuracies are better and found to be 96 and 97.33% for proper light frames while 92.63 and 93.68% for low light frames. To share the information of fall detection hardware platform is created using Arduino Uno and GSM module as well as Raspberry Pi 3. The work may be extended for more categories of images where multiple real-world locations (indoors and outdoors) are available. This can also be carried for further improvement for detecting fall in real time when there are more than one people around.

References

1. Brunete, A., Selmes, M., Selmes, J.: Can smart homes extend people with Alzheimer's disease stay at home? J. Enabling Technol. **11**, 6–12 (2017)
2. Lord, S.R., Sherrington, C., Menz, H.B.: Falls in Older People: Risk Factors and Strategies for Prevention. Cambridge University Press, Australia (2001)
3. Reddy, P.S., Chella, S.: Multimodal spatiotemporal feature map for dynamic gesture recognition from real time video sequences. Int. J. Eng. Trans. B **36**, 1440–1448 (2023)
4. Abaei Kashan, A., Maghsoudi, A., Shoeibi, N., Heidarzadeh, M., Mirnia, K.: An automatic optic disk segmentation approach from retina of neonates via attention based deep network. Int. J. Eng. Trans. A **35**, 715–724 (2022)
5. Savadi Hosseini, M., Ghaderi, F.: A hybrid deep learning architecture using 3d cnns and grus for human action recognition. Int. J. Eng. Trans. B **33**, 959–965 (2020)
6. Zohrevand, A., Imani, Z., Ezoji, M.: Deep convolutional neural network for finger-knuckle-print recognition. Int. J. Eng. Trans. A **34**, 1684–1693 (2021)
7. Bora, A., Bezboruah, T.: Evaluating the reliability of PwCOV: a loosely coupled software as a service for COVID-19 data processing system. Int. J. Eng. **33**, 2496–2502 (2020)
8. Harum, N., Abidin, Z.Z., Shah, W.M., Hassan, A.: Implementation of smart monitoring system with fall dectector for elderly using IoT technology. Int. J. Comput. **17**, 243–249 (2018)
9. Al-Kababji, A., Amira, A., Bensaali, F., Jarouf, A., Shidqi, L., Djelouat, H.: An IoT-based framework for remote fall monitoring. Biomed. Signal Process. Control **67**, 102532 (2021)
10. Chhetri, S., Alsadoon, A., Al-Dala'in, T., Prasad, P.W.C., Rashid, T.A., Maag, A.: Deep learning for vision-based fall detection system: enhanced optical dynamic flow. Comput. Intell. **37**, 578–595 (2021)
11. Soni, P.K.S., Choudhary, A.: A framework for fall activity detection and classification using deep learning method. INFOCOMP J. Comput. Sci. **20**, 56–65 (2021)
12. Vaiyapuri, T., Lydia, E.L., Sikkandar, M.Y., Díaz, V.G., Pustokhina, I.V., Pustokhin, D.A.: Internet of things and deep learning enabled elderly fall detection model for smart homecare. IEEE Access **9**, 113879–113888 (2021)
13. De Miguel, K., Brunete, A., Hernando, M., Gambao, E.: Home camera-based fall detection system for the elderly. Sensors **17**, 2864 (2017)
14. Ali, S.F., Muaz, M., Fatima, A., Idrees, F., Nazar, N.: Human fall detection. In: IEEE International Conference on Multi Topic, pp. 101–105. IEEE Press, Lahore (2013)
15. Soni, P.K., Choudhary, A.: Automated fall detection from a camera using support vector machine. In: 2nd International Conference on Advanced Computational and Communication Paradigms (ICACCP), pp. 1–6, IEEE Press, Gangtok (2019)
16. Bian, Z.P., Hou, J., Chau, L.P., Magnenat-Thalmann, N.: Fall detection based on body part tracking using a depth camera. IEEE J. Biomed. Health Inform. **19**, 430–439 (2014)
17. Paul, S.N., Jayanta Singh, Y.: Survey on video analysis of human walking motion. Int. J. Signal Process. Image Process. Pattern Recogn. **7**, 99–122 (2014)
18. Kwolek, B., Kepski, M.: Human fall detection on embedded platform using depth maps and wireless accelerometer. Comput. Methods Programs Biomed. **117**, 489–501 (2014)
19. Lucas, B.D., Kanade, T.: An iterative image registration technique with an application to stereo vision. In: 7th International Joint Conference on Artificial İntelligence, vol. 2, pp. 674–679 (1981)

20. Lucas, B.D.: Generalized Image Matching by the Method of Differences. Carnegie Mellon University, USA (1985)
21. Bouguet, J.Y.: Pyramidal implementation of the affine Lucas Kanade feature tracker description of the algorithm. Intel Corporation **5**, 1–10 (2001)
22. Basahel, S.B., Bajaba, S., Yamin, M., Mohanty, S.N., Lydia, E.L.: Teamwork optimization with deep learning based fall detection for IoT-enabled smart healthcare system. Comput. Mater. Continua. **75**(1), 1353–1369 (2023). ISSN: 1546-218. https://www.techscience.com/cmc/v75n1/51539

CNN's Augmented with IoT for Traffic Optimization and Signal Regulation

Kiran Sree Pokkuluri and N. SSSN Usha Devi

Abstract Real-time monitoring, analysis, and control of traffic systems is made efficient by the integration of IoT and Convolutional Neural Networks (CNNs) for traffic management, which completely transforms urban mobility. Large volumes of data are gathered about traffic flow, vehicle kinds, and ambient conditions by cameras and sensors. Wireless networks are used to transfer this data to cloud, where deep learning models in particular, CNNs analyse it to find incidents, detect vehicles, and estimate traffic density. The knowledge acquired helps to optimise traffic flow, provide dynamic traffic signal regulation, and improve safety by promptly responding to incidents. The real-time capabilities of this integrated approach significantly improve the overall effectiveness of traffic management systems. The accuracy, precision, latency and scalability of the proposed system was found promising.

Keywords Convolutional neural networks (CNNs) · Internet of Things (IoT) · Traffic management · Deep learning

1 Introduction

Travel times and pollution have increased as a result of the quick urbanization and exponential rise in the number of vehicles on the road. More intelligent and adaptable solutions have been required because traditional traffic management systems frequently are unable to dynamically adjust to real-time traffic situations. Road safety, travel efficiency, and congestion reduction all depend on traffic optimization

K. S. Pokkuluri (✉)
Department of CSE, Shri Vishnu Engineering College for Women, Bhimavaram, Andhra Pradesh 534202, India
e-mail: drkiransree@gmail.com

N. SSSN Usha Devi
Department of Computer Science and Engineering, University College of Engineering, JNTU Kakinada, Kakinada, India

S. N. Mohanty et al. (eds.), *Explainable IoT Applications: A Demystification*, Information Systems Engineering and Management 21,
https://doi.org/10.1007/978-3-031-74885-1_22

and signal regulation. These activities may lead to the reduction of fuel usage, emission thus leading to environmental sustainability [1, 2]. This indirectly improves the overall quality of life as the smooth and speed traffic grantees the flow which directly improves the urban mobility. A novel mechanism where IoT is integrated with CNN can be used for signal regulation and traffic optimization. This new phenomenon adds the benefits of strong data analysis and pattern recognition features of CNN and live real time data collected by IoT to generate an effective traffic control system. All the urban roads are furnished with traffic cameras [3], sensors of vehicle, environmental monitors that collect data pertaining to number of vehicles, speed, traffic flow and weather condition in real time. These sensors are capable of collecting the live data which can be trusted and utilized to monitor and control traffic. This real time feature of IoT enables the system to react swiftly to the events occurred, shifting the traffic and other possible impacts on the traffic flow [4].

CNN are useful for processing image and video efficiently. The strong ability of object detection, classification and segmentation are very successful by using CNNs. These are used to identify the accident prone places, density of the traffic, unlawful parking and categorize the cars based on the videos feed from the life cameras [5]. CNNs have potential capability to learn from large datasets and improves continue sly hence if handles the urban traffic dynamically. Set of procedure are adopted to augment IoT and CNN. IoT sensors and devices first gather unprocessed data i.e. readings of sensors, footage, unprocessed raw data, which is stored in cloud to analyze and respond [6].

By processing data locally, close to the source, edge computing is frequently utilized to lower latency and bandwidth consumption. Before being fed into CNN models, the data is pre-processed to improve quality and extract pertinent features. The pre-processed data is then analyzed using CNN models, such as SSD (Single Shot Multibox Detector), Faster R-CNN, and YOLO (You Only Look Once). In real time, these models spot accidents, estimate traffic flow, and detect and classify vehicles.

2 Literature Survey

Scholars outline diverse IoT frameworks for intelligent urban areas, underscoring the significance of IoT in the instantaneous monitoring and regulation of traffic [7]. Many scientists emphasize the significance of effective systems for data collecting and transmission in Internet of Things-enabled traffic management, with particular attention to edge computing and 5G wireless communication technologies [8]. When compared to conventional methods, the accuracy of vehicle detection and classification using CNNs is significantly improved, as demonstrated by researchers. A few scientists conducted a survey covering different CNN-based methods for traffic incident detection from video feeds, emphasizing how resilient they are under a range of circumstances [9].

Urban traffic systems are becoming more complicated, and controlling them requires traffic optimization and signal regulation [10]. There are several ways to enhance traffic flow and safety, including conventional techniques, adaptive control systems, and contemporary technologies like deep learning and the Internet of Things. Cities may create traffic management systems that are more adaptable, effective, and long-lasting by combining various technologies. Sustained investigation and advancement are necessary to tackle current obstacles and optimize the capabilities of these sophisticated approaches [11].

Few authors proposed how CNNs and IoT can be integrated for real-time traffic monitoring, utilizing edge computing to guarantee prompt replies and lower latency. A few academics have developed an integrated system that employs CNNs to evaluate traffic data gathered from IoT sensors and dynamically modify traffic signal timings to maximize flow [12]. A case study by a number of authors details the installation of an IoT-CNN system at a busy crossroads, which significantly reduced traffic jams and enhanced flow of traffic [13]. Researchers deployed an IoT and CNN-based traffic management system throughout a city in a thorough investigation, showing improved traffic regulation and shortened travel times. CNNs with IoT integration for traffic optimization and signal management marks a major technological breakthrough in smart city applications [14]. These technologies can significantly enhance urban traffic management by utilizing real-time data and sophisticated analytics, providing a route towards more effective, safe, and sustainable cities. To overcome present obstacles and fully realize the promise of this promising technology, more study and development are required [15, 16].

3 Architecture of CNN-IoT for Traffic Optimization and Signal Regulation

In order to gather, process, analyse, and act upon traffic data in real-time, CNN architectures enhanced with IoT for traffic optimisation and signal regulation often contain multiple interconnected components. An outline of the architecture is shown in Fig. 1.

Data Acquisition

Sensors installed at intersections and along roads collect data. This data could include Vehicle count, Vehicle type (car, truck, etc.),Speed, Lane occupancy, Pedestrian presence, Traffic light status (red, yellow, green).

Data Pre-processing

The raw data from the IoT sensors is cleaned and formatted for the CNN. This involves removing outliers, scaling numerical data, encoding categorical data (e.g., converting traffic light status to numerical values).

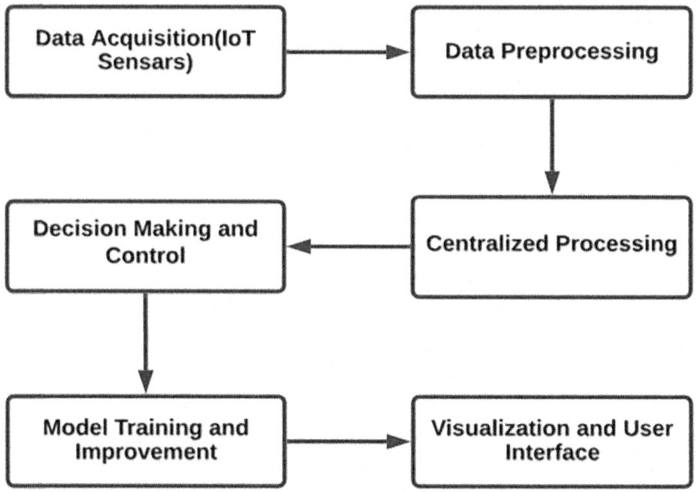

Fig. 1 Architecture of the proposed method

Centralized Processing

Data aggregation is the practice of gathering pre-processed information from several edge devices for additional study. To detect cars, estimate traffic flow, and identify incidents, Convolutional Neural Networks (CNNs) evaluate video feeds and sensor data. To optimise signal timings and traffic flow, algorithms analyse CNN results as well as other traffic data.

Decision Making and Control

Traffic signal control is the process of optimising traffic flow by modifying signal timings in response to real-time traffic data and optimisation algorithms. Rapid identification of events, such as traffic jams or accidents, and prompt action, such as diverting traffic or calling for emergency assistance, are known as incident detection and response.

Model Training and Improvement

To confirm the efficacy of optimisation algorithms and signal control modifications, feedback data is continuously gathered from Internet of Things devices and traffic sensors. Training and refining deep learning models and optimisation algorithms with gathered feedback data to achieve higher performance over time.

Visualization and User Interface

Dashboards for traffic: In-the-moment displays of traffic conditions, levels of congestion, timings of traffic signals, and incident alerts. User Interface: Interfaces via which operators of traffic management can monitor data, communicate with the system, and manually overrule control choices when needed.

4 Experimental Results and Discussion

Urban traffic difficulties can be greatly improved by integrating Convolutional Neural Networks (CNNs) with Internet of Things (IoT) technology for traffic optimisation and signal regulation. The results of recent research are presented in this section together with their implications for urban traffic management.

Real-time traffic analysis and monitoring is one of the main uses for CNNs enhanced with IoT. Research has indicated that CNNs can reliably identify incidents, estimate traffic density, and detect vehicles from video feeds and sensor data collected by Internet of Things devices as shown in Fig. 2.

Car Detection and Classification from traffic camera feeds, CNN-based models have successfully identified and categorised cars with a high degree of accuracy. This makes it possible to monitor traffic flow and vehicle movement patterns in real time. Traffic Density Estimation: CNNs can determine the amount of traffic density and congestion in various parts of the city by examining data from environmental monitors and vehicle sensors. CNNs are useful for detecting traffic occurrences from video feeds and sensor data, including accidents, breakdowns, and road closures. This makes it possible to respond quickly and lessen traffic interruptions as shown in Fig. 3.

Convolutional Neural Networks (CNNs) are highly precise in detecting automobiles because of their capacity to extract complex characteristics from photos. This makes CNNs connected with IoT devices exceptionally accurate. However, because CNNs require a lot of resources, implementing them on Internet of Things devices could be difficult. When it comes to vehicle recognition and classification, Support Vector Machines (SVMs) provide competitive accuracy and precision. SVMs deliver accurate results by efficiently dividing classes in high-dimensional space. Despite this, SVM accuracy may suffer when dealing with complex data distributions. Models

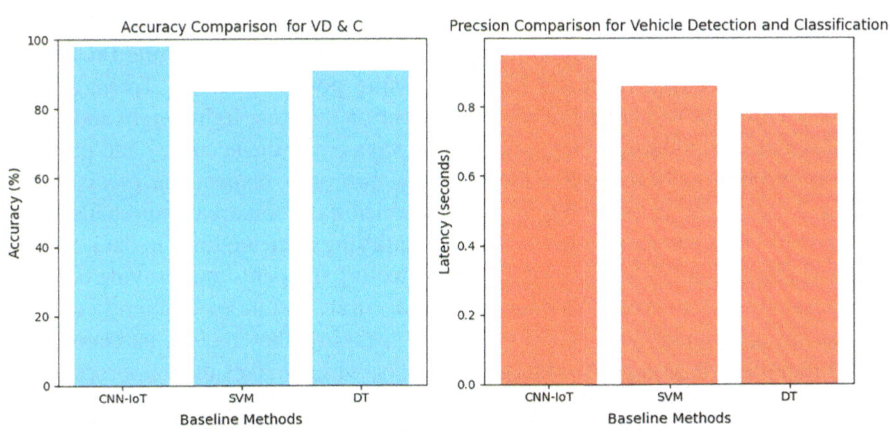

Fig. 2 Comparison of vehicle detection and classification with baseline methods

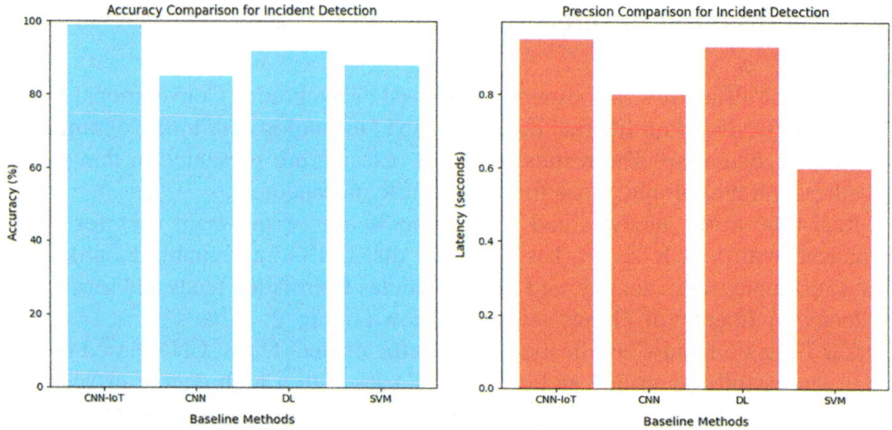

Fig. 3 Comparison of incident detection with baseline methods

for vehicle recognition and classification that are more straightforward but yet functional include decision trees (DTs). Especially in situations with interpretable decision rules, DTs provide respectable accuracy and precision. However, when handling complex picture characteristics, they might not have the same discriminative ability as CNNs and SVMs as shown in Fig. 2.

CNN + IoT, high accuracy and precision can be achieved by integrating Convolutional Neural Networks (CNNs) with Internet of Things devices. CNNs are excellent at identifying complex patterns in incident data because they use deep learning to facilitate feature extraction and classification. However, because of computational constraints, CNN deployment on resource-constrained IoT devices may be difficult and affect real-time performance. CNN, individual CNNs are well known for their precision and accuracy when it comes to incident detection. CNNs can achieve high accuracy and precision because of their capacity to automatically learn complicated features from data. They may not be as suitable for places with limited resources, though, as they usually demand high processing power. Decision Trees, or DTs, provide an alternative method of incident detection. DTs are lightweight and simple to understand, even though they might not always equal the accuracy and precision of deep learning techniques like CNNs. They can offer respectable precision and accuracy, especially in situations where the decision criteria are comprehensible. In contrast to CNNs, they could have trouble identifying intricate links in data. Support vector machines, or SVMs, are efficient at detecting incidents and provide competitive precision and accuracy. SVMs can yield accurate results by efficiently dividing classes in high-dimensional space. Generally speaking, they require less computing power than deep learning techniques like CNNs, which makes them appropriate for applications with constrained computing resources as shown in Fig. 3.

For real-time traffic management systems to guarantee prompt responses to shifting traffic circumstances, latency minimization is essential. This is how one manages latency, Data preparation and preliminary analysis can be done locally by placing edge computing devices close to IoT sensors, which lowers the latency between data gathering and processing. This makes it possible for incident detection and traffic signal modifications to happen more quickly. Algorithms that have been optimized for efficiency include CNN models and traffic optimization algorithms, which cut down on processing time. Further reducing delay in traffic data analysis include parallel processing techniques and streamlined data pipelines.

Particularly in big cities, scalability is crucial to managing the growing amount of traffic data provided by IoT devices. Scalability is handled as follows, By adding additional edge computing nodes or cloud resources as needed, the system's distributed architecture enables horizontal scaling. This guarantees that increasing traffic volumes and data volumes won't cause performance degradation for the system. To increase scalability and parallelize processing, traffic data processing activities are divided and dispersed over several computing nodes. As a result, the system can handle big datasets in real time with efficiency. Depending on traffic demand, computer resources are distributed using dynamic resource allocation algorithms. This guarantees that system performance increases in line with the volume of traffic.

5 Conclusion

Urban traffic management has advanced significantly with the use of Convolutional Neural Networks (CNNs) and the Internet of Things (IoT) for traffic optimization and signal regulation. CNNs can precisely recognize and categories vehicles, estimate traffic density, and identify events by utilizing real-time data from IoT devices. This approach enhances traffic flow, reduces congestion, and improves safety, while also contributing to environmental sustainability. We have achieved an accuracy, precision of 98.6, 98.3 for vehicle detection and classification and accuracy, precision of 99.1, 98.9 for incident detection.

References

1. Khalifa, A.A., Alayed, W.M., Elbadawy, H.M., Sadek, R.A.: Real-time navigation roads: lightweight and efficient convolutional neural network (LE-CNN) for Arabic traffic sign recognition in intelligent transportation systems (ITS). Appl. Sci. **14**(9), 3903 (2024)
2. Saillaja, V., Pasha, M.R., Krishnaveni, S., Ravinder, B., Srinivasan, S., Nandagopal, V.: IoT-embedded traffic cones with CNN-based object detection to roadwork safety. In: 2024 2nd International Conference on Intelligent Data Communication Technologies and Internet of Things (IDCIoT), pp. 120–125. IEEE (2024)
3. Khan, M.A., Park, H., Chae, J.: A lightweight convolutional neural network (CNN) architecture for traffic sign recognition in urban road networks. Electronics **12**(8), 1802 (2023)
4. Sirohi, D., Kumar, N., Rana, P.S.: Convolutional neural networks for 5G-enabled intelligent transportation system: a systematic review. Comput. Commun. **153**, 459–498 (2020)

5. Wang, M., Xiong, X., Kan, Y., Xu, C., Pun, M.-O.: Unitsa: a universal reinforcement learning framework for v2x traffic signal control. IEEE Trans. Veh. Technol. (2024)
6. Raju, P.J.R.S., Kiran, K.V.D., Sree, P.K.: Digital image watermarking based on hybrid FRT-HD-DWT domain and flamingo search optimisation. Int. J. Computat. Vis. Rob. **13**(6), 573–598 (2023)
7. Jilani, U., Asif, M., Rashid, M., Siddique, A.A., Talha, S.M.U., Aamir, M.: Traffic congestion classification using GAN-Based synthetic data augmentation and a novel 5-layer convolutional neural network model. Electronics **11**(15), 2290 (2022)
8. Pokkuluri, K.S., SSSN Usha Devi, N.: A novel cellular automata classifier for covid-19 prediction. J. Health Sci. **10**(1), 34–38 (2020)
9. Khan, S., Teng, Y., Cui, J.: Pedestrian traffic lights classification using transfer learning in smart city application. In: 2021 13th International Conference on Communication Software and Networks (ICCSN), pp. 352–356. IEEE (2021)
10. Pokkuluri, K.S., SSSN Usha Devi, N., Devi, U.: Crop disease prediction with convolution neural network (CNN) augmented with cellular automata. Int. Arab J. Inf. Technol. **19**(5), 765–773 (2022)
11. Ahmed, M., Masood, S., Ahmad, M., Abd El-Latif, A.A.: Intelligent driver drowsiness detection for traffic safety based on multi CNN deep model and facial subsampling. IEEE Trans. Intell. Transp. Syst. **23**(10), 19743–19752 (2021)
12. Trinh, H.D., Gambin, A.F., Giupponi, L., Rossi, M., Dini, P.: Mobile traffic classification through physical control channel fingerprinting: a deep learning approach. IEEE Trans. Netw. Serv. Manag. **18**(2), 1946–1961 (2020)
13. Reddy, D.R., Chella, C., Bala Ravi Teja, K., Baby, H.R., Kodali, P.: Autonomous vehicle based on deep q-learning and yolov3 with data augmentation. In: 2021 International Conference on Communication, Control and Information Sciences (ICCISc), vol. 1, pp. 1–7. IEEE (2021)
14. Pokkuluri, K.S., Usha, D.N.: A secure cellular automata integrated deep learning mechanism for health informatics. Int. Arab. J. Inf. Technol. **18**(6), 782–788 (2021)
15. Louati, A.: A hybridization of deep learning techniques to predict and control traffic disturbances. Artif. Intell. Rev. **53**(8), 5675–5704 (2020)
16. Potluri, S., Mohanty, S.N.: An efficient scheduling mechanism for IoT based home automation system. Int. J. Electron. Bus.

CVLSTMLW-CNN: A IoT-Enabled Hybrid CNN Model for Heart Disease Prediction

Shikha Singh, Archana Singh, Sanjay Singh, and Rachna Khurana

Abstract Cardiovascular diseases pose significant health concerns, particularly especially in locations where medical facilities are limited. Smart wearable devices have emerged as a viable solution to tackle this challenge. Internet of Things (IoT) based devices serve as monitoring systems empowered by machine learning algorithms. They collect cardiovascular data, including heart rate, blood pressure, and electrocardiogram (ECG) signals. Subsequently, this data is transmitted to a centralized server for preprocessing and analysis. This paper employs a CVLSTMLW-CNN hybrid model, which combines LSTM and LWCNN. The LSTM layer analyses electrocardiogram (ECG) signals, representing the heart's electrical activity over time, and learns patterns from this sequential data. Meanwhile, the LWCNN layer extracts spatial features from medical images. By enabling early detection and intervention, the system has the potential to increase survival rates. The algorithm's efficacy is measured using metrics such as F1-score, recall, accuracy, and precision. It reliably predicts the presence or absence of cardiovascular diseases with a 99.0% accuracy across both Kaggle and UCI datasets.

Keywords Wearable sensors · Cardiovascular diseases · Machine learning · Depth-wise separable convolutional neural network · Cloud data · Evaluation metrics

S. Singh (✉) · A. Singh · R. Khurana
Anand Engineering College, Agra, India
e-mail: shikhasinghjyoti@gmail.com

A. Singh
e-mail: archisingh.15@gmail.com

S. Singh
Hindustan College of Science & Technology, Mathura, India
e-mail: sanjaysanju1001@gmail.com

S. N. Mohanty et al. (eds.), *Explainable IoT Applications: A Demystification*,
Information Systems Engineering and Management 21,
https://doi.org/10.1007/978-3-031-74885-1_23

1 Introduction

Cardiovascular diseases are the most common disease and the primary cause of death, claiming approximately 17.9 million lives in 2019, which amounts to 32% of all worldwide deaths, which represents 32% of all fatalities worldwide. Heart attacks and strokes together accounted for 85% of these fatalities. Moreover, among the 17 million premature deaths occurring in the same year, with individuals under 70 years old, CVDs were responsible for 38% of them [1]. To start treatment early, it is essential to recognize cardiovascular disease [2].

Cardiovascular diseases include hypertension, arrhythmia, stroke, and myocardial infarction. Only highly skilled medical experts have the necessary training and experience to identify and treat these problems.

Healthcare institutions face the challenge of delivering high-quality treatment at affordable costs, as incorrect diagnoses and wrong treatment may result in unfavourable outcomes. IoT-based Cardiovascular disease Decision systems can help patients in a timely manner and at a minimal cost [3]. Here, a database will be maintained where the patient's health condition will be stored and accessed anywhere [4]. The broad adoption of the Internet of Things (IoT) presents solutions to problems facing the healthcare industry. Integrating IoT technologies into healthcare practices makes it possible to enhance patient care, improve efficiency, and optimize resource allocation. IoT devices can collect real-time patient data, facilitating ongoing observation of vital signs, compliance with medication, and disease progression [5, 6]. Such information can be saved for further study or seen in real-time, allowing healthcare providers to detect early warning signs of cardiovascular diseases and intervene promptly. An AI system would immediately prescribe the medication, incorporating multiple hospitals and doctors. Overall, the integration of IoT in healthcare practices holds excellent potential for revolutionizing the prevention, diagnosis, and management of cardiovascular diseases [7].

Machine learning algorithms have significantly advanced healthcare by streamlining disease diagnosis [8]. These algorithms are particularly beneficial for people residing in rural or distant areas and senior citizens [9]. Using machine learning techniques, healthcare providers can track and monitor physical health by recording parameters such as heartbeats and blood pressure. Furthermore, it has been demonstrated that deep learning and machine learning methods are efficient in offering trusted disease diagnosis. These algorithms can predict disease accurately and can provide timely intervention and treatment. Healthcare organizations are utilizing machine learning techniques that have revolutionized diagnostic capabilities, improved patient outcomes and enhancing healthcare delivery [10]. Disease detection can be easily handled by techniques for deep learning like convolutional neural networks [11, 12] CNN models can be trained using the original colossal dataset, and due to multiple layers, they increase the accuracy of the predictions and have the ability to reduce the number of parameters [13, 14]. This reduction in parameters allows CNNs to efficiently handle larger datasets, enabling the model to tackle more complex tasks such as disease prediction [15]. The primary finding of this paper is as follows:

In this paper, a hybrid model, CVLSTMLW-CNN, is developed for the prediction of long- and short-term memory (LSTM) and light-weight CNN (LWCNN)-based on heart disease. The LSTM component focuses on understanding patterns in the data over time, whereas the LWCNN component captures spatial features from an electrocardiogram (ECG) image. The construction and optimization of hybrid CNN structures utilizing Depth-wise Separable Convolution (DSC) units are applied for the identification of cardiovascular diseases using IoT sensors, with an enhanced scope based on routine sensor data. Kaggle datasets are used for both training and testing the model. Performance is measured by performance evaluation metrics such as Accuracy, F1-score etc.

The basic structure of the paper is as follows: Sect. 1 provides a brief introduction to the subject. Section 2 describes the many methods for detecting diseases. Section 3 shows the suggested CVLSTMLW-CNN approach for predicting heart disease. Section 4 shows the experimental findings from the suggested model employing various disease datasets. Section 5 concludes the paper.

2 Related Work

Numerous efforts have been dedicated to this trajectory with regard to the literature. Author proposed a framework for heart disease diagnosis that encompasses an IoMT, MSSO, and ANFIS to increase prediction accuracy among the frameworks tested, the MSSO-ANFIS framework showed the highest accuracy in predicting cardiac disease [4]. In this study author developed a Cloud-based IOT model; this m-healthcare model predicts disease using IOT and Cloud data [3].

The concept of computational sciences data is utilized to define key terminologies that enable the generation of User-centric health measurements, thereby facilitating a better understanding and application of health-related data. The main goal of this paper is to provide a method for health diagnostics. That gathers sensor data and produces user diagnosis results (UDR) through computational means.

This paper compares various classifiers to detect the best classifier for specific diseases. Data segmentation is employed for data analysis. Various classification techniques are validated for predicting diseases depending on the F-measure, sensitivity, specificity, and accuracy. In this study author created a Tri-logical IoT-fog-cloud (TIFC) model for gathering and evaluating data about Acute Encephalitis Syndrome (AES), which may be utilized in the healthcare system to monitor encephalitis. This viral infection affects the human immune system. The spatiotemporal-based Temporal-Recurrent Neural Network (T-RNN) prediction model, in conjunction with the Self-Organized Mapping (SOM) technique, is employed for early disease identification and geographical visualization of disease spreads. A fuzzy C-Means classifier is employed here to examine patients' health-related data parameters, facilitating the classification of patients according to their health condition and the severity thereof. The suggested system is validated using metrics like accuracy, f-measure, and reliability, and the result indicates its efficacy in handling and mitigating infectious viruses such as encephalitis [16].

A secure healthcare application based on blockchain was presented in this study to predict cardiovascular and diabetes diseases using fog computing. Patient health data is collected from fog nodes and safely kept on the blockchain for disease prediction. The patient health records are grouped using a particular rule-based clustering method, and disease prediction is done using an adaptive neuro-fuzzy inference system (FS-ANFIS) that uses feature selection. The proposed study accurately forecasts disease outcomes, achieving an accuracy rate exceeding 81%, surpassing other neural network algorithms [17].

An intelligent healthcare approach that integrates electronic medical records (EMR) with sensor data through a feature fusion technique to generate healthcare data. They employed an information gain technique to streamline computational processes and prioritize relevant features. Their deep learning system achieved an impressive accuracy of 98.5% in predicting heart disease [18]. In this study, the author developed a healthcare system specially designed for war soldiers, with the ability to monitor patients in remote locations. The system enables accurate –time tracking of soldiers' location and health monitoring using GPS modules and wireless body area sensor networks (WBASNsl) [19]. ZigBee modules are used to transport the collected data among allies wirelessly. The data can also be transferred to the cloud for further analysis and predictions using the K-Means Clustering technique.

3 Proposed Methodology

The primary goal of this work is to create a hybrid model for predicting cardiovascular disease called the Cardiovascular Long Short-Term Memory Light Weight Convolutional Neural Network (CVLSTMLW-CNN), built on LSTM and LWCNN.

Combining LSTM and LWCNN architectures enables the utilization of spatial and temporal information for heart disease detection. The LSTM component analyses temporal data, while the LWCNN component extracts spatial features from medical imaging data. Lightweight CNN can improve efficiency and handle computational resources, which is suitable for deployment on wearable devices. Fusing these two approaches can enhance the accuracy and effectiveness of heart disease detection systems (Fig. 1).

The IoT sensor is placed on the wearable device to collect patient data. This data is then saved to a cloud database for access at any time. The data is transferred from the cloud to the healthcare application database, where the patient's health report is stored. The proposed heart disease detection model uses a CVLSTMLW-CNN to predict the patient's heart condition. For this, the model goes through training and testing phases. UCI and Kaggle data repositories are provided to the model for training and testing. This paper proposed a model combining Convolutional LSTM (Long Short-Term Memory) and a lightweight CNN model for predicting

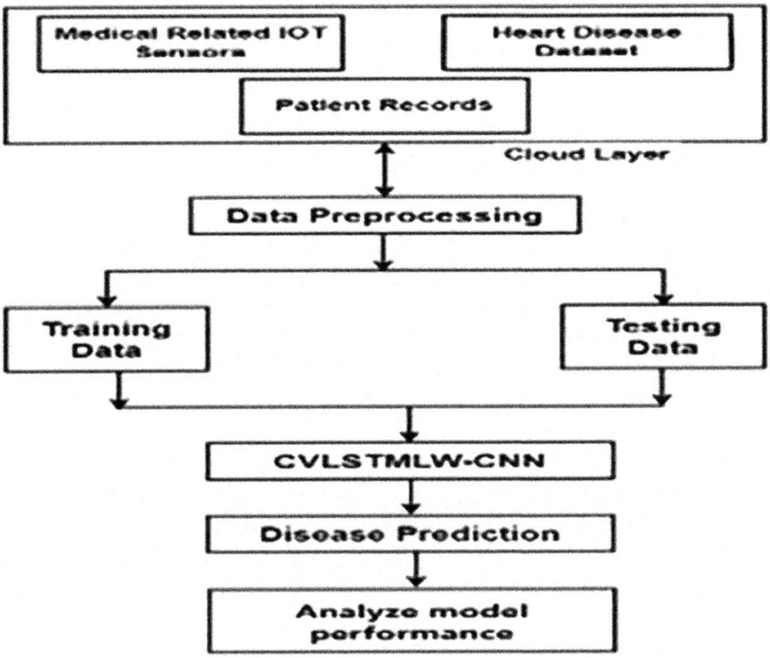

Fig. 1 Architecture of the proposed model

heart disease using IoT devices, which can be a powerful approach. In this section, the proposed model is interpreted.

3.1 Data Collection

The model requires physiological data from IoT devices, such as heart rate, blood pressure, ECG level, and activity level. Hospitals collect patient health records, including hypertension, diabetes, and blood pressure history. This study collected data from the UCI and Kaggle data repositories.

3.2 Data Preprocessing

The preprocessing stage is performed to ensure data quality, reliability, and usability for further analysis. This includes a meticulous process of finding the missing value, identifying and handling the anomalies, and rectifying any incorrect and mislabeled value in medical data, ensuring a comprehensive data cleaning process. Normalizing medical data involves standardizing numerical features, encoding categorical variables, and extracting relevant features. Data integration is performed to ensure compatibility and consistency across various data formats.

3.3 *Proposed Healthcare Model*

Figure 2 depicts the proposed CVLSTMLW-CNN model, which employs LSTM and LWCNN. In this paper, we use two datasets from Kaggle and UCI. Electrocardiogram (ECG) signals are an electrical representation of the heart's activity over time, and the LSTM layer analyzes them. LSTM learns the patterns from these sequential data. The LWCNN layer extracts spatial features from medical images.

The convolutional layer extracts abnormalities from the electrocardiogram (ECG) image. Multiple depth-wise separable convolutional units are utilized to obtain enough information from the input. The DSC unit receives the electrocardiogram (ECG) signal from the LSTM unit and then performs depth-wise convolutions across input channels, independently capturing each channel's features. To improve stability and performance, normalization of the previous layer is performed. After normalization, ReLU activation is applied to adopt linear 1 * 1 Convolutional. Finally, the Max Pooling and Sigmoid functions activate the fully connected layer for classification. Figure 2 represents the DSC unit.

3.3.1 DSC Unit

We replace the standard convolutional layer with a depth-wise separable convolutional unit. Depth-wise separable convolutional layer $f(d)$: $R^{n\times \text{cin}} \rightarrow R^{n\times \text{cout}}$ consisting of depth-wise and pointwise convolution:

The depth-wise convolution $f(d)$: $R^{n\times \text{cin}} \rightarrow R^{n\times \text{cin}}$ with kernel x^d is defined as

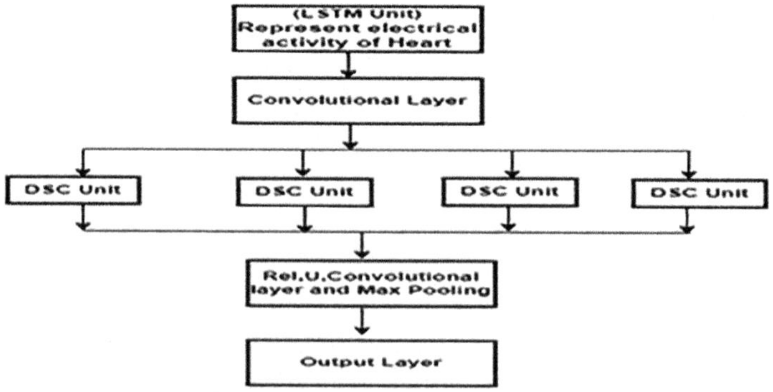

Fig. 2 The suggested model's architectural design

$$f(d)(x) = \sum_{i=1}^{n} xd_i * v_{1:n} + b_i \tag{1}$$

The pointwise convolution $f(p)$: $R^{n\times \text{cin}} \rightarrow R^{n\times \text{cout}}$ with kernel x_p is defined as

$$f(p)(x) = x_p * v_{1:n} \tag{2}$$

4 Performance Evaluation Metrics

This section provides various performance evaluation metrics. The accurate classification rate for the positive class is a true positive (TP). False positives (FP) are predictions of positive outcomes that are not true negatives (TN). An inaccurately predicted negative is known as a false negative.

Receiver operating characteristic curve: This curve exhibits the effectiveness of the proposed model across all classifying criteria. A graph compares the TPR and FPR, i.e., the real positive rate and the false positive rate.

$$\text{TPR} = \frac{\text{TP}}{\text{TP} + \text{FN}} \tag{3}$$

$$\text{FPR} = \frac{\text{FP}}{\text{FP} + \text{TN}} \tag{4}$$

Area under the ROC Curve: It calculates all the categorization thresholds that may be used. The AUC scale, a simple yet powerful tool, ranges from 0 to 1. The AUC value for a 100% correct categorized version is 1.0, whereas it is 0.0 for a 100% incorrect classification.

Accuracy: Accuracy measures the dataset that was adequately categorized.

$$\text{Accuracy} = \frac{\text{TP} + \text{TN}}{\text{TP} + \text{TN} + \text{FP} + \text{FN}} \tag{5}$$

Specificity: This refers to the accuracy with which negative items are detected.

$$\text{Specificity} = \frac{\text{TN}}{\text{TN} + \text{FN}} \tag{6}$$

Sensitivity: Positive entries accurately identified in sensitivity.

$$\text{Sensitivity} = \frac{\text{TP}}{\text{TN} + \text{FN}} \tag{7}$$

Recall: It determines how many true positives the model has recorded and labels them as the same.

$$\text{Recall} = \frac{\text{TP}}{\text{TP} + \text{FN}} \tag{8}$$

Precision verifies the model's accuracy by separating the true positives must be distinguished from false positives.

$$\text{Precision} = \frac{\text{TP}}{\text{TP} + \text{FP}} \tag{9}$$

F1-Score: This is the recall and precision function essential while trying to strike a balance between the two.

$$\text{F1 - Score} = \frac{\text{Recall} * \text{Precision}}{\text{Recall} + \text{Precision}} \tag{10}$$

4.1 *Performance Analysis*

The accuracy of the model is shown in Fig. 3. The accuracy defines the correctness of the model. The proposed Hybrid model exposed a high accuracy rate of 99.0 on the UCI and Kaggle datasets. In this paper, we compared the proposed model with CNN-Bi-LSTM, LWAMCNet, and MSSO-ANFIS, and higher accuracy was observed with the proposed model. The F1-Score is shown in Fig. 4. 98% was observed.

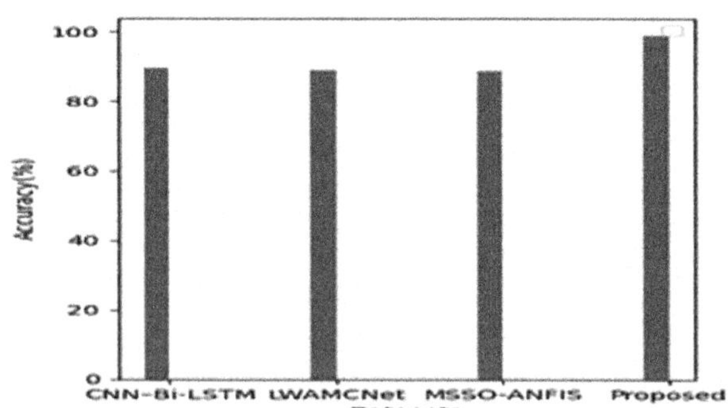

Fig. 3 Comparison of accuracy score

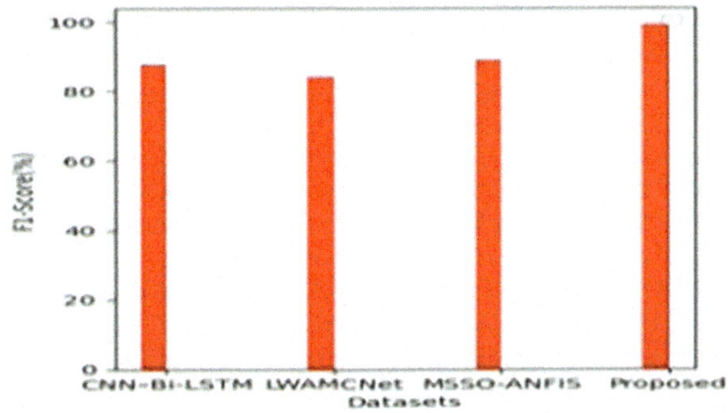

Fig. 4 Comparison of F1-score

Table 1 Dataset

Dataset	Reference	Instances	Attributes
Heart disease dataset (HDD)	[20]	303	76
Cardiovascular disease dataset (CDD)	[21]	1000	12

4.2 Dataset

To validate the algorithm's results, datasets from HDD and CDD are utilized. The dataset is presented in Table 1.

5 Conclusion

The CVLSTMLW-CNN model, implemented in Python, predicts cardiovascular diseases in real-time. Kaggle validates the entire process. And UCI datasets. Comparison with CNN-Bi-LSTM and LWAMCNet MSSO-ANFIS methods indicates superior performance, with accuracy and F1-score reaching 99.0% and 98.0%, respectively. The approach effectively predicts cardiovascular disorders like heart disease because it demonstrates the best accuracy and F1-score values. In spite of its optimal performance, future efforts will focus on testing the model on different datasets based on images and reducing the execution time. The aim is to extend its applicability to predicting various other diseases using health datasets.

References

1. WHO, Accessed 18 March 2024. https://www.who.int/news-room/fact-sheets/detail/cardiovascular-diseases-(cvds)

2. Dutta, A., Batabyal, T., Basu, M., Acton, S.T.: An efficient convolutional neural network for coronary heart disease prediction. Expert Syst. Appl. **159** (2020)
3. Verma, P., Sood, S.K.: Cloud-centric IoT based disease diagnosis healthcare framework. J. Parallel Distrib. Comput. **116**, 27–38 (2018)
4. Khan, M.A., Algarni, F.: A healthcare monitoring system for the diagnosis of heart disease in the IoMT cloud environment using MSSO-ANFIS. IEEE Access **8**, 122259–122269 (2020)
5. Khan, M.A.: An IoT framework for heart disease prediction based on MDCNN classifier. IEEE Access **8**, 34717–34727 (2020)
6. Khatun, M.A., Yousuf, M.A., Ahmed, S., Uddin, M.Z., Alyami, S.A., Al-Ashhab, S., Akhdar, H.F., Khan, A., Azad, A., Moni, M.: A Deep CNN-LSTM with self-attention model for human activity recognition using wearable sensor. IEEE J. Transl. Eng. Health Med. **10** (2022)
7. Ganesan, M., Sivakumar, N.: IoT based heart disease prediction and diagnosis model for healthcare using machine learning models.
8. Chowdhury, N.K., Kabir, M.A., Rahman, M.M., Islam, S.M.S.: Machine learning for detecting COVID-19 from cough sounds: an ensemble-based MCDM method. Comput. Biol. Med. **145** (2022)
9. Khatun, M.A., Memon, S.F., Eising, C., Dhirani, L.: L: Machine learning for healthcare-IoT security: A review and risk mitigation. IEEE Access **11**, 145869–145896 (2023)
10. Khan, M.F., Ghazal, T.M., Said, R.A., Fatima, A., Abbas, S., Khan, M.A., Issa, G.F., Ahmad, M., Khan, M.A.: An iomt-enabled smart healthcare model to monitor elderly people using machine learning technique. Computat. Intell. Neurosci. (2021)
11. Liu, J., Li, M., Luo, Y., Yang, S., Li, W., Bi, Y.: Alzheimer's disease detection using depth wise separable convolutional neural networks. Comput. Methods Prog. Biomed. **203** (2021)
12. Moses, D.A.: Deep learning applied to automatic disease detection using chest X-rays. J. Med. Imaging Radiat. Oncol. **65**(5), 498–517 (2021)
13. Khan, Z.Y., Niu, Z.: CNN with depthwise separable convolutions and combined kernels for rating prediction. Expert Syst. Appl. **170** (2021)
14. Koushik, J.: Understanding convolutional neural (2016)
15. Bassi, P.R.A.S., Attux, R.: A deep convolutional neural network for COVID-19 detection using chest X-rays (n.d.)
16. Bhatia, M., Kumari, S.: A novel IoT-fog-cloud-based healthcare system for monitoring and preventing encephalitis. Cogn. Comput. **14**(5), 1609–1626 (2022)
17. Shynu, P.G., Menon, V.G., Kumar, R.L., Kadry, S., Nam, Y.: Blockchain-based secure healthcare application for diabetic-cardio disease prediction in fog computing. IEEE Access **9**, 45706–45720 (2021)
18. Ali, F., El-Sappagh, S., Islam, S.M.R., Kwak, D., Ali, A., Imran, M., Kwak, K.S.: A smart healthcare monitoring system for heart disease prediction based on ensemble deep learning and feature fusion. Inform. Fusion **63**, 208–222 (2020)
19. Gondalia, A., Dixit, D., Parashar, S., Raghava, V., Sengupta, A., Sarobin, V.R.: IoT-based healthcare monitoring system for war soldiers using machine learning. Proced. Comput. Sci. **133**, 1005–1013 (2018)
20. Janosi, A., Steinbrunn, W., Pfisterer, M., & Detrano, R.: Heart disease. UCI machine learning repository. UCI Machine Learning Repository (1988)
21. Doppala, B.P., NagaMallik Raj, S., Stephen Neal Joshua, E., Thirupathi Rao, N.: Automatic determination of harassment in social network using machine learning. In: Smart Technologies in Data Science and Communication: Proceedings of SMART-DSC 2021, pp. 245–253 (2021)

Navigating IoT Security, Risks and Complexities

Advancements in Security Technologies for Smart Cities: A Comprehensive Overview

Lokesh Singh, Deepshikha, and Megha Agarwal

Abstract The emergence of Smart Cities, facilitated by advancements in technology, heralds a new era of urban development characterized by interconnectedness, efficiency, and sustainability. The integration of many cutting-edge technologies, such as blockchain, Internet of Things (IoT), and artificial intelligence (AI), is essential to achieving the potential of Smart Cities. These technological advancements allow Smart Cities to better manage their resources, provide better public services, and raise the standard of living for its citizens. Nonetheless, the swift expansion of interconnected systems presents intricate obstacles in the domains of cybersecurity and privacy preservation. To solve these issues, practitioners and researchers are actively investigating blockchain solutions, AI-driven security measures, and cryptography techniques. Furthermore, AI is used in Smart Cities not just for security but also for predictive analytics, self-governing systems, and intelligent infrastructure supervision. Though these technologies hold great potential, worries about algorithmic bias, data privacy, and the moral implications of using AI in cities still exist. As Smart Cities continue to evolve, interdisciplinary collaboration and innovative solutions are essential to harness the full potential of technology while safeguarding the rights and well-being of citizens.

Keywords Smart cities · Cybersecurity · Internet of Things (IoT) · Security overview · Privacy protection

L. Singh (✉) · M. Agarwal
Department of Electronics and Communication Engineering, BPIT, Delhi, India
e-mail: search.familiar001@aleeas.com

M. Agarwal
e-mail: meghaagarwal@bpitindia.com

Deepshikha
Department of Electronics and Communication Engineering, MSIT, Delhi, India
e-mail: deepshikha@msit.in

S. N. Mohanty et al. (eds.), *Explainable IoT Applications: A Demystification*,
Information Systems Engineering and Management 21,
https://doi.org/10.1007/978-3-031-74885-1_24

1 Introduction

In the 21st-century, urban-landscapes are undergoing a radical-transformation propelled by the integration of advanced technologies, giving rise to the concept of Smart Cities. These intelligent urban environments leverage cutting-edge technologies, such as the Internet of Things (IoT), data analytics, and artificial intelligence, to enhance the quality of life for residents and streamline city operations. While the benefits of Smart Cities are immense from efficient resource management to improved citizen services, this digital metamorphosis also introduces unprecedented challenges, with security emerging as a paramount concern. As Smart Cities evolve into interconnected hubs of innovation, the multitude of systems operating within this complex ecosystem becomes vulnerable to diverse security threats. This chapter aims to delve into the critical need for security measures across various systems within Smart Cities, examining potential risks and presenting examples of security solutions implemented in real-world scenarios. The interconnected nature of Smart Cities involves an intricate web of systems, including but not limited to smart grids, transportation networks, healthcare services, and public safety infrastructure. The seamless functioning of these systems relies heavily on the continuous exchange of data and real-time communication, making them susceptible to cyber threats and unauthorized access. Recent instances of cyberttacks on city infrastructure underscore the urgency of addressing security concerns to safeguard the foundation of Smart Cities. This chapter will explore the multifaceted dimensions of security in Smart Cities, emphasizing the necessity of implementing robust measures to protect critical infrastructures. Drawing insights from exemplary cases around the globe, we will analyze successful security strategies. These case studies will highlight innovative solutions such as blockchain technology, advanced encryption protocols, and AI driven threat detection systems that fortify the security fabric of Smart Cities. In conclusion, as Smart Cities advance towards a more interconnected and intelligent future, the imperative to fortify security across various systems cannot be overstated. By examining real-world examples and emerging technologies, this book chapter aims to contribute valuable insights to the discourse on securing the digital backbone of Smart Cities, ensuring that the promise of a smarter, safer urban future is realized.

2 What is a Smart City?

A smart city is an urban environment that leverages technology and data-driven solutions to enhance the efficiency, sustainability, and quality of life for its residents. These cities integrate advanced technologies such as the Internet of Things (IoT), artificial intelligence, and data analytics to optimize various aspects of urban living, including transportation, energy consumption, waste management, and public services. By utilizing interconnected systems and real-time data, smart cities aim to streamline operations, reduce environmental impact, and improve overall urban resilience. The goal is to create more intelligent, responsive, and user-friendly urban spaces that address the challenges of rapid urbanization and contribute to a more sustainable and connected future.

2.1 Smart City Systems

The idea behind a smart city is to use data and technology to improve the efficiency, sustainability, and quality of life in metropolitan areas [1]. By attending to the requirements of its citizens, the main goals are to maximize resource use, reduce waste and pollution, and improve the general livability of cities. Smart cities use technology to enhance services for residents, with particular applications designed to meet the unique needs and goals of each city:

- **Intelligent Transportation Systems**: These systems aim to optimize traffic flow, reduce congestion, and improve public transportation.
- **Smart Building Management Systems**: Focused on reducing energy consumption and enhancing citizen comfort within buildings.
- **Smart Lighting Systems**: These systems dynamically adjust lighting levels based on the presence of people and vehicles.
- **Smart Parking Systems**: Designed to assist drivers in efficiently locating available parking spaces.
- **Smart Waste Management Systems**: Geared towards optimizing waste collection and disposal.
- **Smart Water Management Systems**: These systems aim to enhance the quality and distribution of water.

A smart city is an urban environment that leverages technology and data-driven solutions to enhance the efficiency, sustainability, and quality of life for its residents. These cities integrate advanced technologies such as the Internet of Things (IoT), artificial intelligence, and data analytics to optimize various aspects of urban living, including transportation, energy consumption, waste management, and public services. By utilizing interconnected systems and real-time data, smart cities aim to streamline operations, reduce environmental impact, and improve overall urban resilience. The goal is to create more intelligent, responsive, and user-friendly urban spaces that address the challenges of rapid urbanization and contribute to a more sustainable and connected The integration of Internet of Things (IoT) technology is a crucial component of smart cities. The main goal is to use data and technology to improve urban areas sustainability, efficiency, and quality of life. Smart cities can create more livable, sustainable, and resilient communities by improving their ability to respond to the demands of their citizens and by leveraging modern communication and IoT technologies to optimize resource utilization. Figure 1 depicts several additional domains of smart cities.

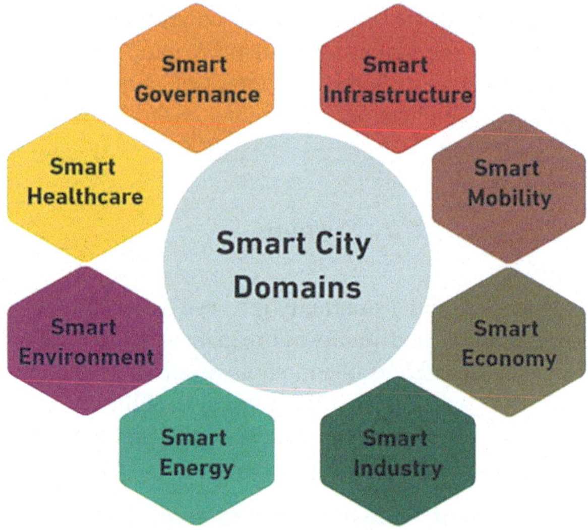

Fig. 1 Smart city domains [15]

2.2 *Internet of Things*

The term "Internet of Things" (IoT) describes how gadgets are connected to each other over the internet [1]. The Internet of Things (IoT) extends connectivity to real-world physical things, allowing them to exchange data, in contrast to the conventional usage of the internet to connect computers. It encompasses technologies like embedded systems, mobile computing, ubiquitous computing, wireless sensor networks, and machine-to-machine communication. Typically, the device, network, session, application, business, administration, and security layers make up the layers that make up an IoT architecture. Every layer contributes in a different way to the operation and administration of the Internet of Things. Sensors and other physical devices that gather data are part of the device layer. Data is then sent over the network layer to provide connection and transit functions. While the application layer is in charge of numerous IoT services like smart farming, smart cities, and smart homes, the session layer handles service connections. While the management layer facilitates device and network administration, the business layer specifies business logic and workflows. The security layer is essential for ensuring data confidentiality, integrity, and access control. It does this by putting security mechanisms in place at various levels. One of the most important aspects of the Internet of Things is interoperability, which is handled by middleware, layered system design, and standard protocols (such MQTT, CoAP, and HTTP). Regardless of device or technological variations, middleware ensures successful interoperability by facilitating data translation and communication between disparate systems. IoT system risks and vulnerabilities must be continuously identified and addressed through monitoring, assessment, and security procedures. Figure 2 summarizes the various processes involved in an IoT-based smart city.

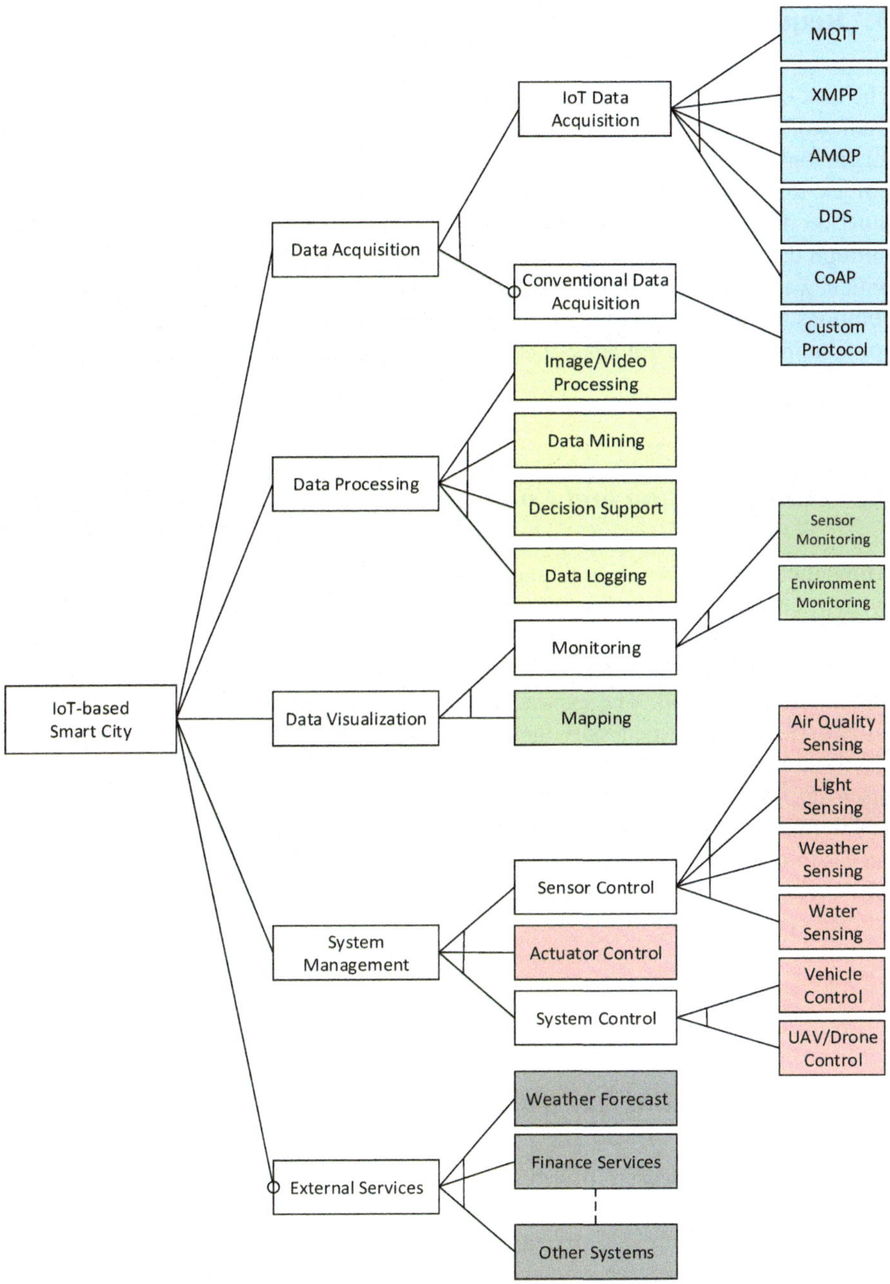

Fig. 2 Model for IoT based smart city [1]

3 Requirements and Challenges

The task of securing smart cities is complex and demands a complete cybersecurity framework in order to protect interconnected systems. Because of the vast array of technologies involved, it is imperative to proactively identify and mitigate vulnerabilities, underscoring the significance of standardized security protocols. A fine line must be drawn between privacy concerns and the optimization of civic functions through data analysis. Technology is always changing, making it harder to keep up with new threats. This calls for ongoing security monitoring and adaptable protection solutions. In order to fully realize the promise of Smart Cities and protect their digital ecosystems from cyberattacks, it is imperative that these requirements and obstacles are met.

3.1 Authentication and Confidentiality

Authentication stands as a fundamental requirement across various levels of a smart system, ensuring the verification of identities and restricting access to authorized customers within a heterogeneous environment [2]. In smart cities, IoT devices play a crucial role by authenticating messages originating from control stations, other nodes, and the network itself. The exponential growth in authentication data within smart cities underscores the necessity for cutting-edge technology to facilitate accurate and timely authentication processes. Confidentiality, aimed at preventing information disclosure to unauthorized sources or passive attacks, is paramount. In the realm of Internet of Things applications, there exists the concern that attackers may gain access to equipment or eavesdrop on conversations. To counteract this, encryption-based technologies are commonly employed to establish reliable communication and storage systems. These technologies play a pivotal role in securing the confidentiality of information transmission between nodes, ensuring the protection of sensitive data in smart systems.

3.2 Accessible and Integral

In a broad sense, accessibility refers to the capability of tools and services to be available as needed [2]. Within the context of our theme, intelligent programs or systems should be capable of maintaining normal operations even in the face of potential threats. Given the susceptibility of these devices to attacks, a smart system must be adept at recognizing unusual circumstances and taking proactive measures to prevent any potential system damage. Resilience, as defined by certain perspectives, represents a system's ability to endure attacks and withstand malfunctions and failures related to major disasters. Robust resilience and continuous adaptability are indispensable for protective mechanisms to effectively counter increasingly sophisticated threats. Equally critical is ensuring the integrity of IoT devices and the data shared between them and the cloud. In a comprehensive smart application, data is exchanged among numerous devices. If not adequately safeguarded during transmission, there is a risk of unauthorized alterations. Considering that most IoT devices possess limited processing power, specific techniques such as firewalls and protocols can regulate data traffic in IoT communications, although they may not guarantee integrity at endpoints. This underscores the importance of implementing comprehensive measures to safeguard the integrity of data exchanged between IoT devices and the cloud in smart systems.

3.3 Interoperability

Interoperability is a pivotal facet in securing Smart Cities, demanding a seamless integration of diverse technologies and systems. The requirements for interoperability in this context are multifaceted, necessitating standardized communication protocols, data formats, and cohesive frameworks. Achieving interoperability ensures that different components within a Smart City ecosystem can effectively communicate and share data, enabling real-time responses to security threats. However, the path to interoperability is fraught with challenges. The diverse array of devices, platforms, and technologies employed in Smart Cities often operate in silos, hindering effective communication and collaboration. Standardization becomes a formidable task due to the presence of proprietary systems and varying technological maturity levels across different sectors. Moreover, security protocols must be harmonized without compromising the functionalities of individual systems, posing a delicate balancing act. In the pursuit of interoperability, issues such as data privacy, scalability, and regulatory compliance loom large. Ensuring that sensitive information is securely exchanged while adhering to legal frameworks adds layers of complexity. Moreover, as Smart Cities expand and evolve, scalability becomes a pressing concern, demanding interoperable solutions that can adapt to the dynamic nature of urban environments.

3.4 Analysis of Security Systems

Analyzing security systems for Smart Cities poses a dual challenge, as outlined by Fan et al. [3]. Firstly, the complexity of Smart Cities necessitates security systems that are not only robust but also adaptable to urban dynamics. These systems must enable real-time threat detection, employ encryption protocols, and support resilient infrastructure to protect critical systems such as transportation, healthcare, and utilities. Moreover, ensuring interoperability among different security components is crucial for a comprehensive defense strategy. Secondly, the sheer scale and diversity of interconnected technologies present challenges in analyzing security systems for Smart Cities. Managing the vast data generated by IoT devices, integrating across domains, and safeguarding privacy and system integrity are key concerns. Additionally, the evolving nature of cyber threats requires security systems capable of swift adaptation and proactive risk management.

4 Smart City Security Issues

Smart cities, with their integration of advanced technologies and interconnected infrastructure, bring forth a host of security challenges that need careful consideration. One major concern is the vulnerability of the vast network of IoT devices and sensors that form the backbone of smart city systems. These devices, if not properly secured, can become entry points for cyberattacks, potentially compromising critical services such as transportation, energy, and healthcare. The reliance on data collection and analytics also raises privacy concerns, as citizens' personal information is constantly being gathered and analysed. Moreover, the interconnected nature of smart city components creates a domino effect; a breach in one area can potentially lead to the compromise of the entire system. Ensuring the security of smart cities requires robust cybersecurity measures, regular updates to software and firmware, and a proactive approach to identifying and addressing vulnerabilities. Balancing innovation and convenience with the need for stringent security protocols is a crucial aspect of developing resilient smart cities for the future.

4.1 IoT Infrastructure

By creating a network architecture that is in charge of collecting and processing data from dispersed sensors and smart devices, the Internet of Things (IoT) significantly influences the infrastructure of smart cities [4]. The network paradigm, in which tangible things such as sensor-based devices gather information on critical network interactions and exchange data via wired or wireless connections, is inextricably linked to the vulnerability of Internet of Things applications. After being uploaded, processed, and stored, the data could become vulnerable to denial-of-service and man-in-the-middle attacks. As a result, unless safety precautions are taken during the collection and transfer of data over IoT infrastructure, the security and privacy of smart cities face serious threats. The threat posed by IoT botnets to IoT systems is significant, as demonstrated by the Mirai botnet, which may infect a range of devices, such as routers, IP cameras, webcams, printers, DVRs, and cameras [2]. This botnet coordinates Distributed Denial of Service (DDoS) assaults against specific servers and disseminates infections among various IoT devices. People, devices, and sensors are becoming more and more interconnected, which raises privacy issues and highlights the need for strong data protection measures. Beyond privacy, information security protects against possible systemic breakdown of urban systems by embracing confidentiality, integrity, availability, and interoperable security. To evaluate and address risks in smart cities, effective risk management is necessary. This involves tackling issues like standards and gaps in technological competence. immaturity of data transfers and fluxes between IoT devices and network components The prevalence of online advertising, which has surpassed more conventional means such as broadcast advertising in radio, television, and newspapers, poses a number of privacy concerns for mobile advertising. Significant concerns are raised by instances of privacy violations, such as businesses selling consumers' personal information in spite of their pledges to keep it private [5]. In particular, if compromised or disclosed without consent, Personally Identifiable Information (PII), which is defined as "information that can be used to distinguish or trace an individual's identity," becomes susceptible to harm, embarrassment, inconvenience, or unfairness. In order to prevent such abuses, it is necessary to solve these difficulties and ensure responsible treatment of personal information in the context of mobile advertising.

4.2 Mobility Threats

In order to improve mobility, comfort, safety, and efficiency, intelligent transportation systems (ITS) combine technologies like sensing, control, analysis, and communication within the travel infrastructure and transportation [6]. The travel experience is being revolutionized by manufacturers' ongoing advances to vehicle intelligence in conjunction with better roads and infrastructure. Travel may be made more dependable and efficient by implementing new technology and continuing ITS research and development. The development of Autonomous Vehicles (AVs), involving substantial investments, seeks to mitigate traffic accidents and foster the evolution of a cleaner, smarter society [2]. However, this rapid advancement presents significant security risks, as hacking could compromise both safety and data privacy. Vulnerabilities in AV systems may enable hackers to execute remote attacks, manipulating steering, engine functions, and braking mechanisms. Additionally, concerns arise regarding the extensive collection of personal data by AV computer systems, posing potential privacy risks. AI-enabled autonomy is a key component of both present and future Intelligent Transportation Systems (ITS) [6]. AI and machine learning are key components of many ITS applications, ranging from driver assistance technology to analytics and planning at the municipal level. Recent studies, however, show how vulnerable different machine learning techniques—supervised, unsupervised, reinforcement learning, or hybrid approaches—are to vulnerabilities at every stage of the learning process, from testing and deployment to training. To ensure pedestrian and passenger safety, for example, control systems in autonomous vehicles rely on AI algorithms trained to detect things, such road signs. However, these algorithms are susceptible to adversarial instances, which are deliberately constructed inputs meant to trick the system into generating particular results. Adversarial examples are used in the field of classification to attempt to incorrectly classify inputs into groups that are distinct from their proper classes. In addition to AI-enabled autonomy, other attack vectors that affect ITS include vulnerabilities in tire-pressure monitoring systems, which expose in-car wireless sensor networks. The CAN bus is vulnerable to attacks because it lacks authentication, which allows hackers to access additional Electronic Control Units (ECUs), such as those connected to power steering, airbag control, engine control, and x-by-wire systems. Furthermore, complex systems like infotainment systems with GPS navigation could be the focus of an attack, which could result in the exfiltration of data from onboard sensors and systems and raise privacy issues.

4.3 Smart Power System

An essential component of a smart city's overall security and privacy architecture is its power infrastructure, which makes it possible for outside parties with access to the grid to track usage patterns and forecast consumer behavior as shown in Fig. 3. The different systems in smart cities that supply and control light and heat mainly depend on wireless network technology, which could lead to security flaws in the grid. By using bogus data injection techniques to cause breakdowns in the first transmission lines, attackers can take advantage of the power system [7]. Circuit breakers trip as a result of the operation center detecting an outage by injecting false data into the measurement system of the target lines. After a while, this first failure spreads to later links, and the redistribution process ends when there are no more failing nodes.

New technologies must be integrated at different power system levels in order to modernize the outdated electrical grid [8]. These initiatives include advanced metering infrastructure (AMI), local control and protection systems, distribution

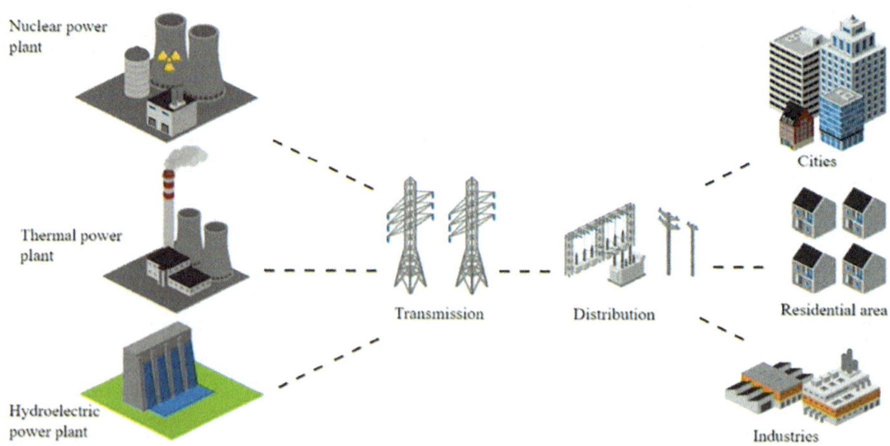

Fig. 3 Conventional power grid [16]

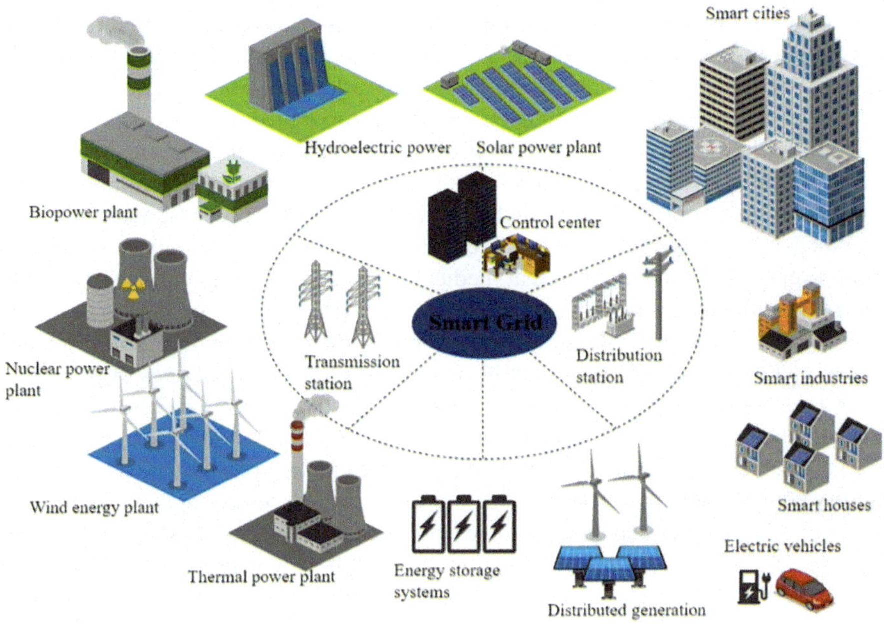

Fig. 4 Smart grid [16]

automation, and communication technologies at the distribution system level as visualized through Fig. 4. Although these improvements present chances for dispersed network protection and dynamic control, they also bring in fresh security holes that may be used to take down the electrical system.

In a similar study, we investigate a particular kind of attack on the power system called the load oscillation attack. The three stages of this attack are as follows: in the reconnaissance stage, estimate the load frequency swing factor (LFSF); in the design stage, optimize the load oscillation magnitudes and identify the target line; and in the execution stage, find the resonant frequency in real-time to maximize the attack's impact. This emphasizes how sophisticated attacks present changing risks and difficulties for smart city power systems.

4.4 *Smart Healthcare*

In order to minimize data disclosure and create a robust information security architecture, it is essential to secure the privacy and integrity of healthcare services and concepts within smart cities [4]. Over the past ten years, there has been an increasing integration of sensors in medicine and healthcare; recent study highlights the key security and privacy challenges inherent in developing Internet of Things (IoT) architecture for healthcare applications. The study highlights the serious risks associated with gathering personal health information, such as blood pressure and heart rate, and highlights the requirement for a system-wide integrated security solution. In the context of smart city healthcare services, this all-encompassing strategy seeks to safeguard private information and safeguard sensitive health information. Technological developments in artificial intelligence (AI) and the internet of things (IoT) have opened up previously unheard-of opportunities to improve community health, lower costs, and improve clinical and patient services [17]. However, modern healthcare management systems have many difficulties when it comes to the safe transfer and preservation of data. Given that medical information systems frequently handle enormous volumes of sensitive data and are connected to internal networks via the Internet, security is crucial. Due to their connectedness, hospitals are more vulnerable to malicious assaults and possible virus risks from outside devices. A prominent hazard is ransomware, a type of malicious software that obstructs access to networks, infrastructure, or device data and demands a fee to unlock. Ransomware attacks have been connected to medical process disruptions that impact patient care. Hackers can modify equipment to give false readings, give patients excessive dosages of medication, or perform other acts that endanger their health if they manage to take control. Because they are vital pieces of infrastructure, health- care institutions exchange and keep a lot of private data. They also rely a lot on IoT devices and electronic medical records to provide patient care. Because of this combination, healthcare facilities are more appealing targets for cybercriminals looking to demand large ransom payments. Identity theft is a serious issue since it can result in the theft of personal data such as bank and credit card information, insurance details, names, policy numbers, dates of birth, billing information, diagnosis codes, and birth dates. Cybercriminals might use this information to fabricate identification documents, market medical supplies, and make false insurance claims. The possible repercussions highlight how crucial it is for medical information systems to have strong security measures in place in order to protect patient data and guarantee the accuracy of healthcare services.

4.5 Social Networks

Information extracted from online social networks, including platforms such as Facebook, Instagram, and LinkedIn, provides valuable insights into social, economic, and cultural dynamics [4]. Government entities, policymakers, authorities, and commercial industries utilize this data to comprehend market trends and behavioral patterns that shape individual interactions through open data sources. While offering numerous benefits, the utilization of online social networks raises pertinent privacy concerns. The study delineates critical vulnerabilities linked to the use of online social networks, specifically outlining risks that jeopardize an individual's identity, anonymity, personal space, privacy, communication, and security. These vulnerabilities predominantly emanate from potential actions by third parties, underscoring the necessity for thoughtful consideration and protective measures to address privacy issues associated with online social networks.

5 Current Security Approaches

In the rapid evolution of Smart Cities, ensuring robust security approaches has become paramount to mitigate emerging threats and vulnerabilities. The intricate interconnectivity of diverse urban systems, from transportation and healthcare to energy grids, demands a comprehensive and adaptive security framework. Traditional security models are proving insufficient in the face of sophisticated cyber-attacks, prompting the exploration of innovative approaches. This introduction delves into the dynamic landscape of security in Smart Cities, exploring multifaceted strategies such as blockchain technology, artificial intelligence, and decentralized networks. The urgency to secure citizen data, critical infrastructure, and communication channels underscores the critical need for effective security paradigms. In the next subsections, we will discuss a few technologies that play an important role in the security of smart cities.

5.1 Cryptography

Cryptographic techniques provide the foundation for security and privacy protection in smart applications by thwarting unauthorized access to data while it is being stored, processed, and shared. In this section, we summarize some of the state-of-the-art technologies and offer an overview of the cryptographic tools used in smart systems today. For devices with limited resources, traditional algorithms and encryption standards are problematic because of their computational complexity and energy consumption [2]. As such, the usage of lightweight encryption has become essential for the practical use of cryptographic technology. User data has long been protected by cryptography during both transmission and storage (at-rest) [9]. Traditional methods like full disk encryption (FDE) and Transport Layer Security (TLS) form the foundation for protecting the majority of internet traffic and data stored on personal devices. Encrypting data from the point of creation to the point of destination is known as complete cryptography, and it guarantees protection against unauthorized access at every stage of the process. This method changes the way that data is secured by ensuring that, once encrypted, it is independent of any transport or storage system, protecting its validity, confidentiality, and integrity. A novel method based on the Blue Fish cipher in symmetric key cryptography is presented by the proposed Message Authentication technique for Sensor Nodes (MASN) [10]. In a Wireless Sensor Network (WSN), this solution handles scalable message authentication and counteracts multiple node threats. Intermediary nodes authenticate every communication from source to base station, a crucial element of the WSN network that includes the identification and rejection of faulty messages. Additionally, nodes guarantee message integrity when sending data to the base station. The results of the study show that total source anonymity and unforgeability against specific message attacks are achieved by the MASN system for sensor nodes. Concerns remain, though, about how the MASN technique affects sensor node computation overhead, energy usage, and other aspects. Researchers show that the protocol is safe, resilient against many types of attacks, such as replay, impersonation, and sensor node acquisition, through testing and simulation. This shows that, in the context of wireless sensor networks, the MASN strategy successfully strikes a balance between security, resilience, and efficiency.

5.2 Blockchain

Blockchain technology has emerged as a promising solution for enhancing privacy and security in Smart Cities, as evident in various research papers exploring its applications in this domain [11]. Researchers recognize the decentralized and tamper-resistant nature of blockchain as a means to address security concerns across different systems within Smart Cities. Blockchain utilizes cryptography to administer a write-only ledger, addressing issues of data integrity and authenticity associated with traditional techniques [9]. This ledger, shared by numerous users, consists of "blocks" added to the previous block, forming a "chain" of verified transactions. These transactions are validated by participants through proof-of-work calculations, with the decentralized system preventing forgery or attempts to alter existing data. While the most widely used blockchain implementation, Bitcoin, prioritizes public verifiability, it lacks confidentiality, and de-anonymization methods can identify users based on transactions. It's crucial to note that privacy and confidentiality are not the main objectives of most blockchain solutions, and assessing a technology's security qualities requires consideration of its real-world applications. Blockchain technology, with its distributed, unchangeable ledger, provides strong data integrity qualities. Blockchain is leveraged significantly in securing Internet of Things (IoT) devices within Smart Cities. Research papers emphasize how blockchain establishes a transparent and secure ledger for managing IoT device identities, transactions, and data exchanges. By decentralizing control over IoT networks, blockchain minimizes the risk of single points of failure and unauthorized access, thereby fortifying overall security. Smart Contracts, a feature of blockchain technology, are extensively studied for their potential in automating and securing various processes within Smart Cities [12]. Studies demonstrate that in Software Defined Networking (SDN) and smart contract-enabled municipal smart cities, innovative architectures based on authentication and authorization for limited contexts are essential for collaborative tasks.

5.3 Artificial Intelligence

Artificial Intelligence (AI) plays a pivotal role in securing Smart Cities, employing predictive analytics for threat detection, facial recognition in surveillance, and autonomous systems for monitoring and response. AI-driven cybersecurity measures identify and respond to potential cyber threats, while behavioral analysis tools enhance security in crowded areas. Smart traffic management, optimized by AI, contributes to accident prevention and efficient emergency response. Additionally, AI aids in the development of autonomous security devices such as drones and robots. Its role extends to optimizing emergency response, predicting incident spread, and coordinating response units during crises. The use of AI reflects a paradigm shift toward intelligent, proactive, and adaptive security strategies, ensuring the safety and resilience of Smart Cities in the face of evolving challenges. Using either anomaly-based or signature-based techniques, intrusion detection systems (IDS) are made to automatically recognize occurrences suggestive of hostile adversary attacks [13]. While anomaly-based techniques concentrate on spotting departures from the ordinary, signature-based techniques seek out and match attack signatures. Algorithms for machine learning, such as ensemble learning algorithms like random forest and gradient boosting, classical ML models like Na¨ıve Bayes, decision trees, and support vector machines, and deep learning models like MLP, CNN, and RNN, identify anomalies that indicate possible attacks or malfunctions. While there isn't a unanimous agreement on the optimal classifier, these methods allow for quick action and reduced risk. Adversaries possess the capability to target Machine Learning (ML)-based intrusion detection systems to mislead the classifier [14]. In the realm of cybersecurity, intrusion detection systems utilize machine learning algorithms to identify patterns and anomalies indicating a security threat. However, attackers can intentionally manipulate or provide misleading data to these systems, attempting to deceive the machine learning classifier. This type of attack, known as adversarial attacks on machine learning models, involves techniques such as injecting malicious input or modifying features to evade detection by the intrusion detection system.

6 Future Work

The future of security in smart cities is poised for significant advancements, driven by a nuanced understanding of the challenges and opportunities that arise in the complex urban landscape. One critical aspect that requires attention is the need for a holistic perspective on security. Beyond the traditional focus on technological solutions, future research is likely to delve deeper into the human-centered factors that influence the adoption and acceptance of smart city technologies. This includes considerations of citizen perspectives, the" lived in" experience of smart cities, and the factors that contribute to building trust among the population. As smart cities continue to evolve, the role of legal and institutional dimensions in ensuring security becomes increasingly pivotal. Research should explore how the legal system can effectively address trust challenges within smart cities, providing a robust framework for governance and accountability. This broader perspective encompasses not only the technological aspects but also the regulatory and legal structures that underpin smart city initiatives. Blockchain technology is expected to play a crucial role in enhancing security and privacy in smart cities. Ongoing and future research will likely focus on the development of new government and industry regulations to facilitate the integration of blockchain, ensuring secure and transparent transactions. Pilot projects will be essential to evaluate the cost implications of deploying blockchain-based systems, and comprehensive cost-benefit analyses will guide informed decision-making by governments. Interoperability and standardization are key challenges that the future of smart city security must address. Heterogeneity, communication protocol issues, and proprietary infrastructure currently hinder effective communication between smart cities. Future efforts will likely concentrate on establishing global standards that foster interoperability, enabling seamless collaboration and communication across diverse smart city ecosystems. In the realm of network security, the Internet of Things (IoT) will be a focal point of research. The complex network of interconnected devices in smart cities demands innovative technologies to address evolving security challenges. Understanding malware propagation characteristics, modeling information spread patterns in wireless sensor networks, and developing effective prevention strategies are areas ripe for exploration. Fog-based systems, emerging as a vital component of smart cities, introduce new security challenges due to their distributed nature. Future research will likely prioritize developing security measures tailored to the unique vulnerabilities of Fog systems. Protecting smart devices in Fog-based smart systems and preserving consumer privacy at the edge of the network will be critical considerations.

User-centric approaches are anticipated to gain prominence in the future of smart city security. Providing individuals with the ability to control their data, including the right to delete or move data between service providers, will be a key focus. User-friendly protection assistants that enhance both security and user comfort across various smart applications are expected to be developed, aligning with individual preferences and requirements. Data minimization strategies will become increasingly important in smart cities to address privacy concerns. Ensuring that only necessary data is collected, used, and stored will require technical guarantees and support from governance and political structures. Striking a balance between data utility and privacy preservation will be a central theme in the future of smart city security. The necessity for lightweight security solutions is underscored by the resource constraints of sensors and devices in smart cities. Future research will likely prioritize the development of efficient, low-overhead security measures that align with the dynamic and flexible nature of smart city environments.

Lastly, theoretical studies will form the foundation for comprehensive security frameworks in smart cities. As smart applications continue to proliferate globally, establishing a uniform concept of a smart city, including its definition and architecture, will be essential. Theoretical studies will bridge the gaps in understanding and facilitate the incorporation and sharing of security mechanisms and network protocols across the entire smart city landscape.

7 Summary

The book chapter comprehensively explores the challenges and requirements associated with securing smart cities. It emphasizes the need for a robust cybersecurity framework, highlighting the complexity of the task due to diverse technologies. The delicate balance between privacy concerns and optimizing civic functions through data analysis is underscored. The text delves into key areas such as authentication, confidentiality, accessibility, and integrity, elucidating their significance in smart city security. Interoperability emerges as a pivotal facet, demanding standardized protocols for effective communication. Security issues in specific domains like IoT infrastructure, mobility, power systems, healthcare, and social networks are thoroughly examined, emphasizing the potential vulnerabilities and threats. The challenges posed by IoT infrastructure, mobility threats, smart power systems, health care security, and social networks are discussed, offering insights into potential risks and mitigations.

The book chapter also examines contemporary security strategies, emphasizing the importance of blockchain, artificial intelligence, and cryptography in protecting smart cities. A thorough examination is conducted of these technologies' roles in guaranteeing data integrity, confidentiality, and proactive threat detection. The final section of the text anticipates future developments in the areas of user-centric strategies, data minimization, blockchain integration, interoperability standards, IoT network security, fog-based systems, user-centered security approaches, and lightweight security solutions. It also highlights how crucial theoretical research is to developing a unified understanding of smart cities and all-encompassing security frameworks.

References

1. Tekinerdogan, B., Köksal, Ö., Çelik, T.: System architecture design of IoT-based smart cities. Appl. Sci. **13**(7), 4173 (2023)
2. Cui, L., Xie, G., Qu, Y., Gao, L., Yang, Y.: Security and privacy in smart cities: challenges and opportunities. IEEE Access **6**, 46134–46145 (2018)
3. Fan, J., Yang, W., Liu, Z., Kang, J., Niyato, D., Lam, K.Y., Du, H.: Understanding security in smart city domains from the ANT-centric perspective. IEEE Internet of Things J. (2023)
4. Ismagilova, E., Hughes, L., Rana, N.P., Dwivedi, Y.K.: Security, privacy and risks within smart cities: literature review and development of a smart city interaction framework. Inform. Syst. Front. 1–22 (2020)
5. Ullah, I., Boreli, R., Kanhere, S.S.: Privacy in targeted advertising on mobile devices: a survey. Int. J. Inf. Secur. **22**, 647–678 (2023)
6. Hahn, D., Munir, A., Behzadan, V.: Security and privacy issues in intelligent transportation systems: classification and challenges. IEEE Intell. Transp. Syst. Mag. **13**(1), 181–196 (2019)
7. Nguyen, T.N., Liu, B.H., Nguyen, N.P., Chou, J.T.: Cyber security of smart grid: attacks and defenses. In: ICC 2020–2020 IEEE International Conference on Communications (ICC), pp. 1–6. IEEE (2020)
8. Alanazi, F., Kim, J., Cotilla-Sanchez, E.: Load oscillating attacks of smart grids: vulnerability analysis. IEEE Access (2023)

9. Stromire, G., Potoczny-Jones, I.: Empowering smart cities with strong cryptography for data privacy. In: Proceedings of the 1st ACM/EIGSCC Symposium on Smart Cities and Communities, pp. 1–7 (2018)
10. Sundarrajan, M.: Authentication Scheme Based on Blow Fish Cryptography in Categorized Sensor Networks (2020)
11. Mora, O.B., Rivera, R., Larios, V.M., Beltrán-Ramírez, J.R., Maciel, R., Ochoa, A.: A use case in cybersecurity based in Blockchain to deal with the security and privacy of citizens and smart cities cyberinfrastructures. In: 2018 IEEE International Smart Cities Conference (ISC2), pp. 1–4. IEEE (2018)
12. Siddiqui, S., Hameed, S., Shah, S.A., Khan, A.K., Aneiba, A.: Smart contract-based security architecture for collaborative services in municipal smart cities. J. Syst. Architect. **135**, 102802 (2023)
13. Schmitt, M.: Securing the digital world: Protecting smart infrastructures and digital industries with artificial intelligence (AI)-enabled malware and intrusion detection. J. Ind. Inf. Integr. **36**, 100520 (2023)
14. Jmila, H., Khedher, M.I.: Adversarial machine learning for network intrusion detection: a comparative study. Comput. Netw. **214**, 109073 (2022)
15. Bellini, P., Nesi, P., Pantaleo, G.: IoT-enabled smart cities: a review of concepts, frameworks and key technologies. Appl. Sci. **12**(3), 1607 (2022)
16. Ibrahim, M.S., Dong, W., Yang, Q.: Machine learning driven smart electric power systems: current trends and new perspectives. Appl. Energy **272**, 115237 (2020)
17. Almalawi, A., Khan, A.I., Alsolami, F., Abushark, Y.B., Alfakeeh, A.S.: Managing security of healthcare data for a modern healthcare system. Sensors **23**(7), 3612 (2023)

A Deep Learning Framework Based on Convolutional Neural Network for Automatic Detection of Cyberattacks in IoT Use Cases

Sivananda Hanumanthu, G. Anil Kumar, Amjan Shaik, K. Naga Jyothi, Christine Günther, Ingrid Haas, Frank Holzwarth, Anna Kramer, Leonie Kunz, Nicole Sator, Erika Siebert-Cole, and Peter Straßer

Abstract Internet of Things (IoT) technology has enabled unprecedented smart applications made possible. However, it also brought increased security risks as the technology is the amalgamation of heterogeneous artefacts and lack of global standards yet. In this context, it is indispensable to have mechanisms to safeguard IoT applications from ever increasing cyberattacks. The emergence of artificial intelligence (AI) paved way for solving many real world problems. In this paper, we proposed a deep learning based framework known as Learning based Cyberattack Detection Framework (LbCADF) which exploits an enhanced Convolutional Neural Network (CNN) for automatic detection of cyberattacks in IoT use cases. The framework is capable of detecting attack traffic flows from benign ones. We proposed an algorithm known as Enhanced CNN for Attack Detection and Classification (ECNN-ADC). Our algorithm exploits feature selectin and hyperparameter tuning for leveraging quality of training. We configured early stopping criterion to get rid of overfitting. The proposed framework is evaluated using a benchmark dataset known as CICIDS2017. Our empirical study has revealed that the ECNN-ADC outperforms many state of the art models such as MLP and baseline CNN with highest accuracy 95% in cyberattack detection. The abstract should summarize the contents of the paper and should contain at least 70 and at most 150 words. It should be set in

S. Hanumanthu (✉) · C. Günther · I. Haas · F. Holzwarth · A. Kramer · L. Kunz · N. Sator · E. Siebert-Cole · P. Straßer
Department of CSE, BESTIU, Gowniváripalli, Andhra Pradesh, India
e-mail: siva.phd1984@gmail.com

G. Anil Kumar
Scient Institute of Technology, Ibrahimpatnam, Telangana, India
e-mail: anil_deva@yahoo.com

A. Shaik
St. Peters Engineering College, Maisammaguda, Telangana, India

K. Naga Jyothi
BEST Innovation University, Gorantla, Andhra Pradesh, India
e-mail: research.director@bestiu.edu.in

S. N. Mohanty et al. (eds.), *Explainable IoT Applications: A Demystification*, Information Systems Engineering and Management 21,
https://doi.org/10.1007/978-3-031-74885-1_25

9-point font size and should be inset 1.0 cm from the right and left margins. There should be two blank (10-point) lines before and after the abstract. This document is in the required format.

Keywords Cybersecurity · Internet of Things · Deep learning · Convolutional neural network · Cyberattack detection

1 Introduction

With ever increase in the usage of computer and other networks in cyberspace and the emergence of Internet of Things (IoT) use cases, security is to be given paramount importance. With IoT networks integrated with other networks, global brain is being realized. However, there is ever increase in cyberattacks due to significant vulnerabilities in real life networks. The vulnerabilities are due to usage of resource constrained devices and lack of global standards for security for IoT integration. With the emergence of Artificial Intelligence (AI) and its related techniques along with big data, security in complex networks like IoT is made possible with learning based approaches [1]. As discussed in [2–4] cyber security threats can be detected using deep learning models that are designed based on neurons in human brain. Industrial cyber physical systems, as discussed in [5–10] could be protected from cyberattacks with deep learning based approaches. Various attacks including the ones that are distributed in nature could be detected using deep learning approaches. From the literature, it was observed that deep learning has potential to learn complex patterns and features so as to improve the attack detection process.

Gani et al. [1] observed that IoT's growth raises security concerns. Deep learning and big data show promise in addressing vulnerabilities, requiring novel integrated solutions. Al-Abassi et al. [6] proposed a deep learning-based ICS cyber-attack detection model, outperforming conventional classifiers in accuracy and F1-score. Rathore and Park [7] proposed DeepBlockIoTNet that integrates secure DL and blockchain for IoT, addressing centralized control, security, and privacy challenges, enhancing accuracy. The study done in [11–16] addressed escalating cybersecurity threats by proposing a distributed deep learning system for attack detection in IoT/Fog networks. The research study done in [17–20] found that the IoT's growing influence prompts the exploration of data extraction and knowledge utilization. Deep learning stands out for its transformative potential in managing massive IoT datasets. From the review of literature, it is found that there is need for improved CNN models for leveraging detection of cyberattacks.

2 Preliminaries

It focuses on the architecture of baseline CNN model. We extended this model and used in the proposed deep learning framework. CNN is one of the popular deep learning models being used in various application domains. Of late it is widely used for intrusion detection research as well. With its architecture consisting of convolution and pooling layers, it is possible to learn very complex patters from high-dimensional space. Therefore, CNN is found to be suitable for training with network traffic data. For instance, CICIDS2017 dataset used in this paper has 78 features and this kind of complexity can be handled by CNN model.

As presented in Fig. 1, architecture of a typical CNN model is provided with 8 × 8 matrix input size. The model is made up of three components such as convolutional layer, pooling layer and fully connected layer. The first layer extracts feature maps from input while the second one optimizes them. Final classification is carried out by the third layer. The feature map generation process in the convolutional layer is expressed as in Eq. 1.

$$feature\ map\{i\} = input\quad feature\ detector\{i\};\ i \in K \tag{1}$$

Once feature maps are generated, an activation function is used to break model linearity. In the pooing layer, optimization of feature maps is carried out resulting in reduction of dimensions. Once feature maps are pooled, there is flattening procedure that converts data into 1D which is suitable for third layer. The third layer has input neurons, hidden neurons (optional) and output neurons. Eventually output

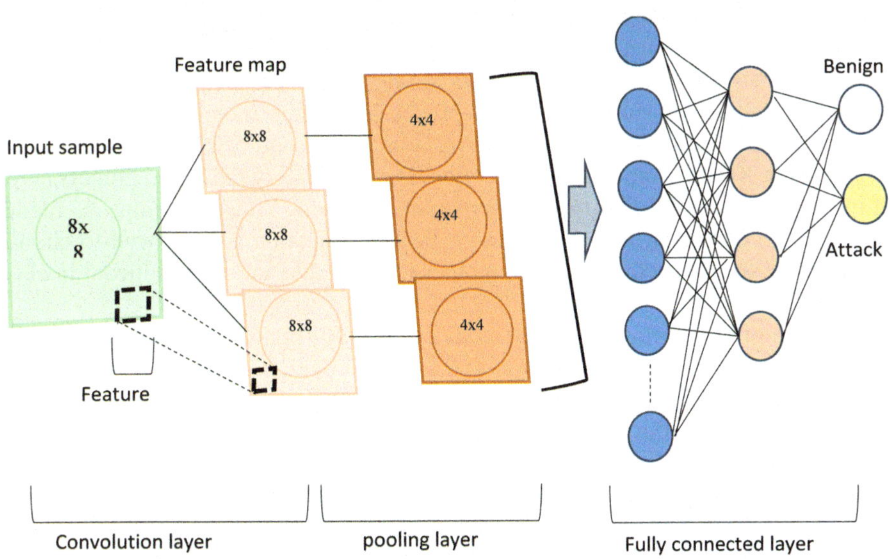

Fig. 1 Architecture of a typical CNN

neurons determine class labels. With the case study of this paper, output classes are nothing but different kinds of cyberattacks. CNN is characterized by its shared weights which makes the model efficient with fewer optimized parameters. It also makes an optimizer to converge faster. It also has potential to get rid of overfitting. The constraints associated with model weights can help in generalization of tasks. Thus CNN is capable of detecting cyberattacks if it is used as IDA. Moreover, CNN is efficient in learning complex patterns in the training data. In this paper, we introduced pre-processing in our deep learning framework to balance data for multi-class classification using CICIDS2017 dataset.

3 Materials and Methods

3.1 *Problem Definition*

Provided unlabelled network flows as input, proposing a deep learning framework which can automatically detect and classify cyberattacks is the problem considered. However, the proposed system is designed to be more efficient in attack detection and classification when compared with existing models.

3.2 *Proposed Framework*

A deep learning based framework is proposed as shown in Fig. 2. The framework has mechanisms to learn from training data and detect different kinds of cyberattacks automatically. The framework has strong pre-processing methodology. The given dataset [21] is subjected to pre-processing which includes dimensionality reduction using Principle Component Analysis (PCA) and t-SNE, hyperparameter tuning using XGBoost and feature selection. Dimensionality reduction process helps in reducing number of features. Hyperparameter tuning is used to improve model performance based on the given dataset. Feature selection is employed to choose contributing features while discarding other features that are not significant in class label prediction.

Once feature importance is estimated, a threshold is used for selecting features. In our framework the threshold is set to 0.1 and it could be adjusted with empirical study. We adjusted it to average feature importance *0.1. Then the dataset is split into two parts known as training set (T1) and test set (T2) with 80% and 20% instances respectively. An enhanced CNN model as shown in Fig. 3 (described in Sect. 3.3) is configured and compiled. The resultant model is trained with training data. The learned model is then persisted for future usage. Afterwards, the test data (unlabelled data) is given the learned model. The advanced CNN model is thus used to detect cyberattacks and classify them.

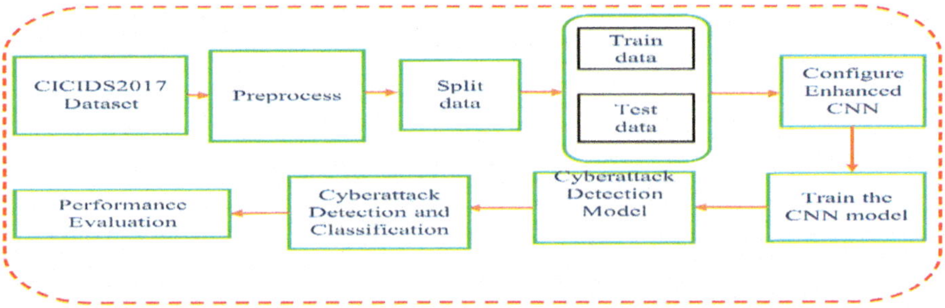

Fig. 2 Proposed deep learning based framework for cyberattack detection and classification

3.3 Enhanced CNN Model

We enhanced baseline CNN model presented in Sect. 3 by configuring layers with empirical study towards higher performance in cyberattack detection. The proposed model is shown in Fig. 3. Python 3 is used for its implementation and PyTorch is the framework used to configure layers in the enhanced CNN model. The proposed model has number of layers implemented appropriately. Empirical study is made to configure layers with the influence of the work in [21–23]. Based on the complexity of the task in hand, it is important to have more layers in CNN. This is evident in ResNet50 model. However, there is caveat that more layers do not guarantee better performance always as discussed in [23]. Therefore, in this paper, we fixed the layers with empirical study. The model takes batches of matrices as input where 9 × 9 is the size of each matrix comprising 77 features along with 4 zero pads. The enhanced model is made up of two convolutional layers, two max pooling layers and two fully connected layers. Exhaustive grid search is carried out to fix kernel size, number of layers, hyperparameters and number of neurons. Hyperparameters are chosen from a subset of values and model is iteratively evaluated to fix best hyperparameters. The first convolutional layer in the architecture (Fig. 3) kernel size is set to 3 × 3 while padding and stride are set to 1. The input feature maps set is 16 and ReLU is the activation function. Therefore, 16 feature maps of size 9 × 9 are generated by the model for each matrix linked to the training batch. Each feature map is subjected to ReLU. Afterwards, pooling of feature maps is done at the max pooling layer where 2 × 2 is the kernel size, padding is 1 and stride is 2. This will result in 5 × 5 sized 16 feature maps. Then second convolutional layer followed by second max pooling layer are executed that have similar padding, stride and kernel size as that of preceding layers.

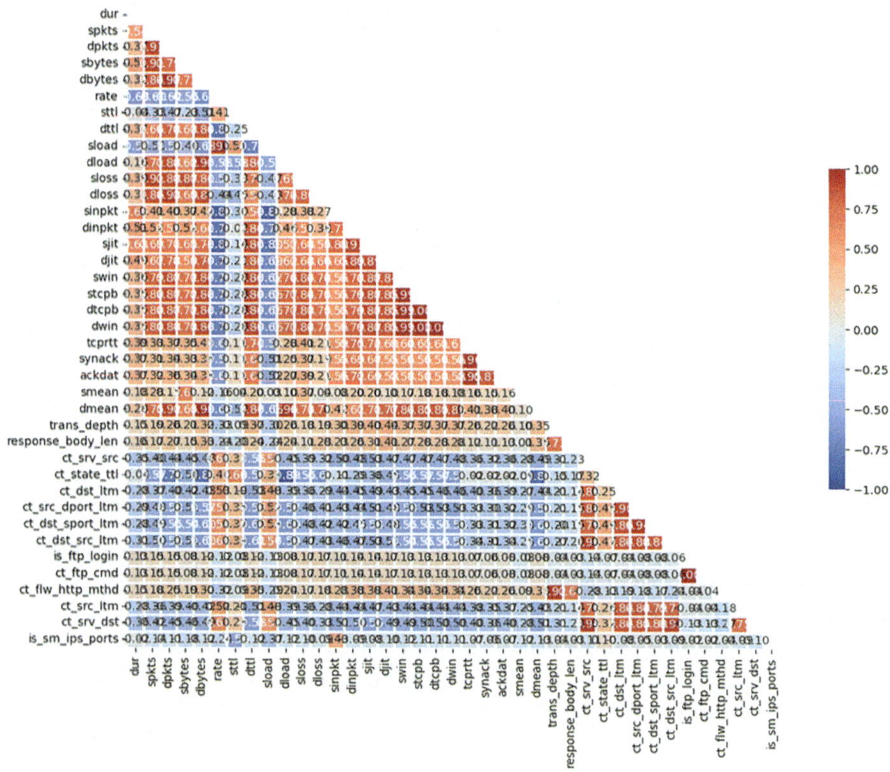

Fig. 3 Feature correlation analysis

4 Experimental Results

This section presents our experimental results using CICIDS2017 dataset is collected from [21]. The proposed enhanced CNN model used in our algorithm is compared against baseline CNN and Multilayer Perceptron (MLP) models. Enhanced CNN model is configured as in Fig. 3 with number of epochs 100, batch size 64 and early stop call back. Learning rate is set to 0.001 while loss function used is sparse categorical loss entropy. The model converged at epoch 50 as we used the early stopping criteria to eliminate overfitting possibilities.

As presented in Fig. 3, the features in the dataset are used to visualize a heatmap reflecting feature correlation.

As presented in Fig. 4, the result of dimensionality reduction using t-SNE is provided reflecting the fact that malicious and normal network flows are overlapped and not clearly grouped into two clusters.

As presented in Fig. 5, feature importance of each feature is computed using XGBoost model. Then the average feature importance is multiplied with 0.1 to arrive at a feature importance threshold dynamically.

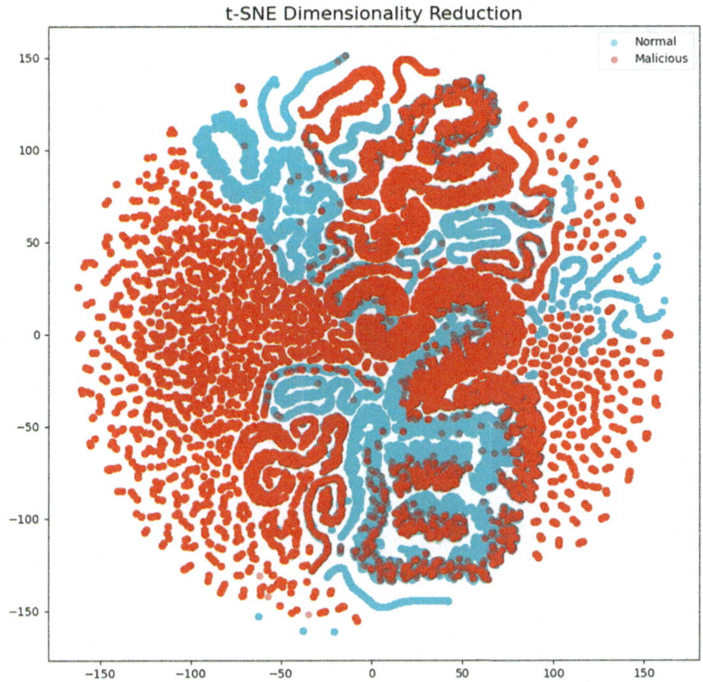

Fig. 4 Dimensionality reduction using t-SNE

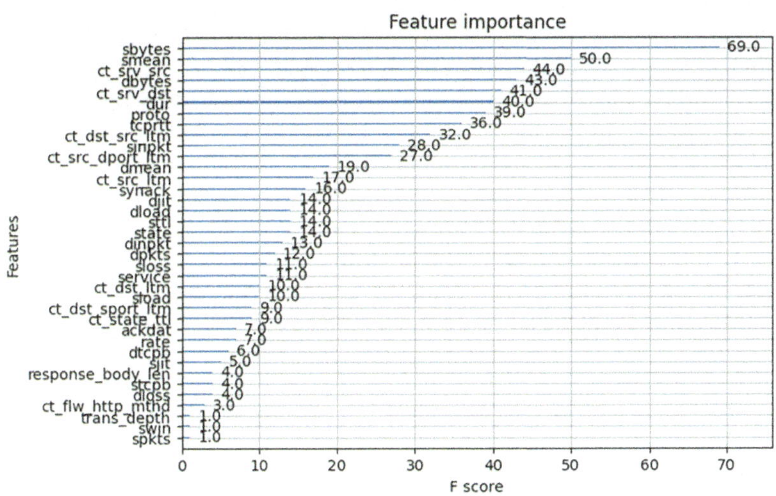

Fig. 5 Feature importance visualization

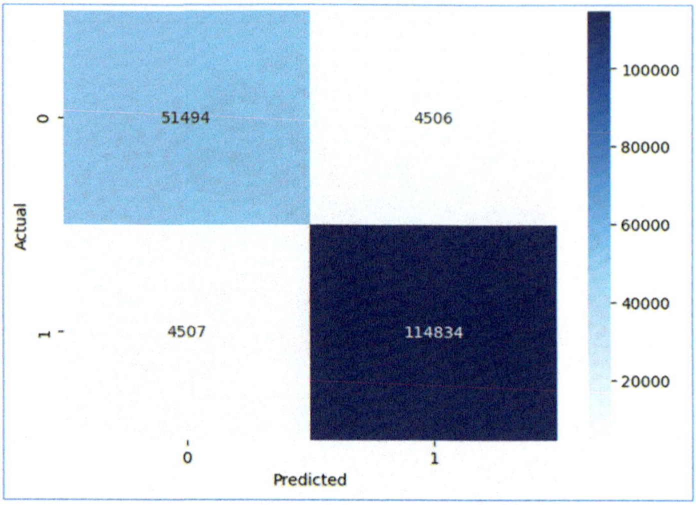

Fig. 6 Confusion matrix of the enhanced CNN model

As presented in Fig. 6 confusion matrix is visualized to compare algorithm predictions with ground truth towards arriving at the performance statistics.

As presented in Table 1, performance of the proposed algorithm ECNN-ADC which is based on enhanced CNN model is compared against existing models such as MLP and baseline CNN.

As presented in Fig. 7, performance of the proposed ECNN-ADC algorithm is compared against state of the art models. The precision of MLP is 89%, baseline CNN 89% and the proposed algorithm 94%. Highest precision is observed with the proposed algorithm. In the same fashion, ECNN-ADC could have achieved highest recall with 94% while MLP and baseline CNN could achieve 87% and 92% respectively. As the F1-Score is the harmonic mean of precision and recall, this measure showed similar trends as that of precision and recall. F1-Score of MLP is 88%, baseline CNN 90% and the proposed algorithm achieved 94%. When it comes to accuracy, the proposed algorithm could have achieved highest accuracy with 95% while MLP and baseline CNN could show 89% and 91% accuracy respectively.

Table 1 Performance comparison among cyberattack detection models

Cyberattack detection model	Performance (%)			
	Precision	Recall	F1-score	Accuracy
MLP	89	87	88	89
Baseline CNN	89	92	90	91
ECNN-ADC (proposed)	94	94	94	95

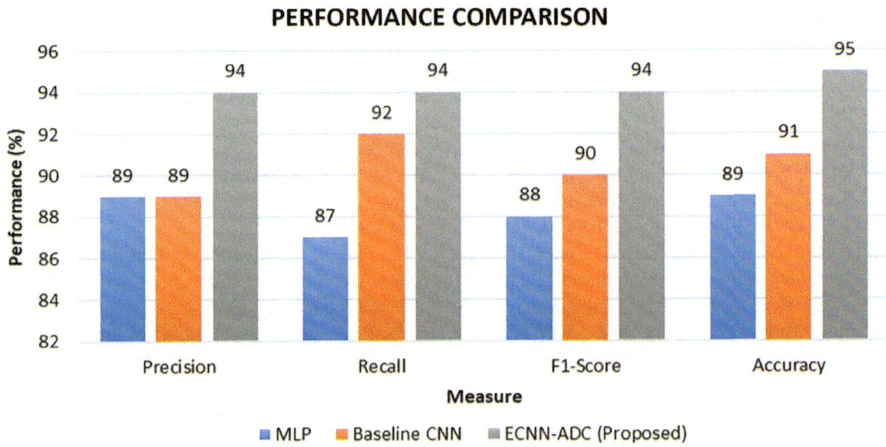

Fig. 7 Attack detection performance evaluation

5 Conclusion and Future Work

In this paper, we proposed a deep learning based framework known as Learning based Cyberattack Detection Framework (LbCADF) which exploits an enhanced Convolutional Neural Network (CNN) for automatic detection of cyberattacks in IoT use cases. The framework is not only capable of detecting attack traffic flows from benign ones but also classify them. The framework has strong pre-processing methodology. The given dataset is subjected to pre-processing which includes dimensionality reduction using Principle Component Analysis (PCA) and t-SNE, hyperparameter tuning using XGBoost and feature selection. Dimensionality reduction process helps in reducing number of features. Hyperparameter tuning is used to improve model performance based on the given dataset. Feature selection is employed to choose contributing features while discarding other features that are not significant in class label prediction. We proposed an algorithm known as Enhanced CNN for Attack Detection and Classification (ECNN-ADC). Our algorithm exploits feature selectin and hyperparameter tuning for leveraging quality of training. We configured early stopping criterion to get rid of overfitting. The proposed framework is evaluated using a benchmark dataset known as CICIDS2017. Our empirical study has revealed that the ECNN-ADC outperforms many state of the art models such as MLP and baseline CNN with highest accuracy 95% in cyberattack detection. In future, we intend to improve our framework with feature engineering and model scaling towards detection of unknown attacks as well.

References

1. Amanullah, M.A., Habeeb, R.A.A., Nasaruddin, F.H., Gani, A., Ahmed, E., Nainar, A.S.M., Akim, N.M., Imran, M.: Deep learning and big data technologies for IoT security. Comput. Commun. S0140366419315361 (2020). https://doi.org/10.1016/j.comcom.2020.01.016
2. Ullah, F., Naeem, H., Jabbar, S., Khalid, S., Latif, M.A., Al-Turjman, F., Mostarda, L.: Cyber Security Threats detection in Internet of Things using deep learning approach. IEEE Access, 1–1 (2019). https://doi.org/10.1109/ACCESS.2019.2937326
3. Ferrag, M.A., Friha, O., Maglaras, L., Janicke, H., Shu, L.: (2021). Federated deep learning for cyber security in the Internet of Things: concepts, applications, and experimental analysis. IEEE. **9**, 138509–138542. https://doi.org/10.1109/ACCESS.2021.3118642
4. Aversano, L., Bernardi, M.L., Cimitile, M., Pecori, P.: A systematic review on deep learning approaches for IoT security. Comput. Sci. Rev. (2021). https://doi.org/10.1016/j.cosrev.2021.100389
5. Gumusbas, D., Yldrm, T., Genovese, A., Scotti, F.: A Comprehensive survey of databases and deep learning methods for cybersecurity and intrusion detection systems. IEEE Syst. J. 1–15 (2020). https://doi.org/10.1109/JSYST.2020.2992966
6. Al-Abassi, A., Karimipour, H., Dehghantanha, A., Parizi, R.M.: An ensemble deep learning-based cyber-attack detection in industrial control system. IEEE Access 1–1 (2020). https://doi.org/10.1109/ACCESS.2020.2992249
7. Rathore, S., Park, J.H.: A blockchain-based deep learning approach for cyber security in next generation industrial cyber-physical systems. IEEE Trans. Ind. Inform. (2021). https://doi.org/10.1109/TII.2020.3040968
8. Atitallah, S.B., Driss, M., Boulila, W., Ghézala, H.B.: Leveraging deep learning and IoT big data analytics to support the smart cities development: review and future directions. Comput. Sci. Rev. **38**, 100303 (2020). https://doi.org/10.1016/j.cosrev.2020.100303
9. Khalil, R.A., Saeed, N., Masood, M., Fard, Y.M., Alouini, M.-S., Al-Naffouri, T.Y.: Deep learning in the industrial Internet of Things: potentials, challenges, and emerging applications. IEEE Internet of Things J. (2021). https://doi.org/10.1109/jiot.2021.3051414
10. Sarker, I.H.: Deep cybersecurity: a comprehensive overview from neural network and deep learning perspective. SN Comput. Sci. (2021). https://doi.org/10.1007/s42979-021-00535-6
11. Abeshu, A., Chilamkurti, N.: Deep learning: the frontier for distributed attack detection in fog-to-things computing. IEEE Commun. Mag. **56**(2), 169–175 (2018). https://doi.org/10.1109/MCOM.2018.1700332
12. Koroniotis, N., Moustafa, N., Sitnikova, E.: A new network forensic framework based on deep learning for Internet of Things networks: a particle deep framework. Future Gener. Comput. Syst. **110**, 91–106 (2020). https://doi.org/10.1016/j.future.2020.03.042
13. Diro, A.A., Chilamkurti, N.: Distributed attack detection scheme using deep learning approach for Internet of Things. Future Gener. Comput. Syst. S0167739X17308488 (2017). https://doi.org/10.1016/j.future.2017.08.043
14. Abbasi, M., Shahraki, A., Taherkordi, A.: Deep learning for network traffic monitoring and analysis (NTMA): a survey . Comput. Commun. (2021). https://doi.org/10.1016/j.comcom.2021.01.021
15. Rahman, A., Hossain, M.S., Alrajeh, N.A., Alsolami, F. (2020). Adversarial examples—security threats to COVID-19 deep learning systems in medical IoT devices. IEEE Internet of Things J. 1–1. https://doi.org/10.1109/JIOT.2020.3013710
16. Dixit, P., Silakari, S.: Deep learning algorithms for cybersecurity applications: a technological and status review. Comput. Sci. Rev. **39**, 100317 (2021). https://doi.org/10.1016/j.cosrev.2020.100317
17. Popoola, S.I., Adebisi, B., Hammoudeh, M., Gui, G., Gacanin, H.: Hybrid deep learning for botnet attack detection in the Internet of Things networks. IEEE Internet of Things J. 1–1 (2020). https://doi.org/10.1109/jiot.2020.3034156

18. Lakshmi S.K., Thenmozhi, M., Vijayakumar, K., Kohli, R.: An intensive healthcare monitoring paradigm by using IoT based machine learning strategies. Multimed. Tools Appl. (2021). https://doi.org/10.1007/s11042-021-11111-8
19. Ahmad, R., Alsmadi, I.: Machine learning approaches to IoT security: a systematic literature review. Internet of Things (2021). https://doi.org/10.1016/j.iot.2021.100365
20. Fernandez Molanes, R., Amarasinghe, K., Rodriguez-Andina, J., Manic, M.: Deep learning and reconfigurable platforms in the Internet of Things: challenges and opportunities in algorithms and hardware. IEEE Ind. Electron. Mag. **12**(2), 36–49 (2018). https://doi.org/10.1109/MIE.2018.2824843
21. Intrusion Detection Evaluation Dataset (CIC-IDS2017). Retrieved from https://www.unb.ca/cic/datasets/ids-2017.html
22. Kim, J., Shin, Y., Choi, E.: An intrusion detection model based on a convolutional neural network. J. Multim. Inf. Syst. **6**(4), 165–172 (2019)
23. Vinayakumar, R., Soman, K.P., Poornachandran, P.: Applying convolutional neural network for network intrusion detection. In: Proceedings of International Conference on Advances in Computing, Communications and Informatics, pp. 1222–1228 (2017)

Digital Attack Identification for the Internet of Things Using Machine Learning

Saswati Chatterjee and Suneeta Satpathy

Abstract Millions of devices are connected by the Internet of Things (IoT), a quickly expanding field that enables little user participation in device interaction. Nevertheless, because the internet is public, IoT is vulnerable to several kinds of cyberattacks. To get around this, practical security measures like network detection algorithms are essential. Traditional security solutions are insufficient given the storage and computing power limits of IoT devices. Therefore, machine learning (ML)-based intelligent network-based solutions are essential. Although ML techniques for attack detection have been the subject of numerous studies in this paper, we evaluate different machine learning (ML) techniques for IoT network threat detection to close this gap. The authors of this study evaluate different machine learning (ML) techniques for IoT network alerting to close this gap. They apply six alternative machine learning algorithms and achieve high accuracy by using a recently generated dataset named Bot-IoT for evaluation. They also extract additional characteristics from the collected data that perform better than the features from earlier research. The study adds to the body of knowledge by showcasing successful ML-based strategies for IoT network security.

Keywords Cyberattacks · Bot-IoT dataset · Machine learning · Internet of Things (IoT)

S. Chatterjee (✉)
Parul University Vadodara, Vadodara, Gujarat, India
e-mail: cshiv68@gmail.com

S. Satpathy
Centre for Cyber Security, Siksha 'O' Anusandhan (Deemed to Be) University, Bhubaneswar, Odisha, India

S. N. Mohanty et al. (eds.), *Explainable IoT Applications: A Demystification*, Information Systems Engineering and Management 21,
https://doi.org/10.1007/978-3-031-74885-1_26

1 Introduction

Globally, there is a growing concern about security and privacy when it comes to computer networks because information technology is being used so widely in daily life. There have been more attempts to compromise network and system integrity as a result of the growth of web-based technologies.

The advanced technologies cover interconnected devices capable of autonomous communication, eliminating the need for human intervention. This connectivity extends to various objects equipped with sensors, spanning sectors such as healthcare, agriculture, transportation, etc. IoT applications streamline tasks, conserve resources, and revolutionize both work and lifestyle practices. IoT nodes, unlike traditional networks, possess limited capacity and resources and lack manual controls. Better and improved assault screening methods are therefore essential. One method that appears to be promising for integrating intelligence into the Internet of Things is machine learning or ML. Network traffic analysis, intrusion detection, and botnet detection are just a few of the network security tasks for which machine-learning approaches have proven effective Despite its myriad benefits, IoT also poses security challenges, as it opens avenues for potential breaches [1]. Nonetheless, IoT offers vast opportunities for knowledge exchange, innovation, and economic growth. Machine Learning (ML) is an integral component of IoT solutions, enabling intelligent devices to adaptively modify or automate their behavior based on acquired knowledge. ML algorithms leverage data generated by devices or humans to infer valuable insights and are commonly employed in regression and classification tasks [2]. Similarly, ML can enhance security in IoT networks by providing advanced protection services.

Within cybersecurity, the use of machine learning for recognizing attacks is a quickly developing field. Though ML techniques for attack detection have been the subject of a lot of research in the literature, there hasn't been much study on effective detection strategies designed for IoT contexts. Two main approaches that machine learning may for identifying malware jobs are signature-driven [3–7].

This method's ability to detect unknown attacks is advantageous, but it may lead to high false alarm rates as legitimate but unusual behaviors can be flagged as anomalies.

Hybrid techniques combine both anomalous and identification of signatures methods to leverage their respective strengths. For example, a hybrid approach presented in [8] enhances detection rates for known attacks while reducing false positive rates for unknown attacks. Overall, while signature-based techniques excel in detecting known attacks with minimal false alarms, anomaly-based methods offer the advantage of detecting unknown attacks [9]. Hybrid techniques aim to leverage the strengths of both approaches to enhance overall detection effectiveness. In this work, we add to the body of knowledge by investigating how well machine learning techniques identify Internet of Things network intrusions. Next, we apply seven different machine learning techniques and obtain significant performance gains. Through this investigation, we provide valuable insights into enhancing defense mechanisms against IoT network attacks.

2 Relevant Work

Many scholarly articles and a great deal of research have been done in the field of algorithmic learning with an emphasis on machine intelligence and data mining approaches for detecting intrusions, especially in conventional networks. We reviewed several articles and gathered them into Table 1 to deliver a scenario of current trends in attack analysis. Every study outlines the datasets utilized, the detection strategies applied, and the machine learning techniques used. Diversity in the datasets and learning methods used was a key component of our selection criteria. The examined research shows that machine learning methods are capable of efficiently identifying intrusions in Internet of Things networks. Detection techniques can be divided into two categories based on the literature addressing the application of machine learning for Internet security: supervised and unsupervised techniques. Unsupervised methods include studies such as [9–14], while supervised methods encompass works like [9, 15–17]. In particular, unsupervised machine learning methods have been applied to identification issues; auto-encoders are a prominent example of an unsupervised technique that has been used in numerous publications. For example, Kitsune is an unsupervised network security system that uses autoencoders to efficiently identify network attacks. Mirsky et al. advocated using autoencoders to improve cyber threat detection. Similarly, one author suggested and assessed a unique recognition technique [18, 19]. However, unsupervised machine learning algorithms face challenges in network traffic analysis, where anomalies like attacks are rare compared to normal flows, impacting detection success rates negatively. On the contrary, supervised learning algorithms, trained on datasets labeled with attack instances, offer better results. In a recent study, Moustafa et al. assessed the Bot-IoT dataset using different models [20]. Our work, on the other hand, focuses on assessing various models on the Bot-IoT dataset and extracting new features from it. Furthermore, a further investigation used the Bot-IoT dataset [21]. All things considered; this research demonstrates how machine learning approaches may be used to improve threat detection in Internet of Things networks. Both supervised and unsupervised learning techniques provide insightful information about how to strengthen network security [22, 23].

Table 1 The list of features

Flow IAT Mean	Flow Duration	Flow Pkts/s
Flow IAT Max	Fwd Pkt_Len_Mean	TotLen Fwd Pkts
Fwd Pkt Len Mean	Tot Bwd Pkts	Fwd IAT Tot
Flow IAT Std	Flow Bytes	Tot Fwd Pkts
Flow IAT Min		

3 Proposed Approach

This section describes the dataset that was used and our recommended technique for spotting dangers in IoT networks. To effectively identify abnormalities, our approach applies machine learning algorithms after a series of initial processing stages. First, CICFlowMeter is used to take out flow-based attributes from the raw dataset [24]. Next, we preprocess the data to make sure machine learning algorithms work with it. The data must be transformed into a format that is appropriate for analysis at this stage. After pre-processing we choose the qualities that the machine learning algorithms will use throughout the feature selection phase. The application of machine learning methods for attack detection brings our strategy to a close.

Figure 1 gives a visual representation of the processes involved in the detection process and summarises our suggested approach.

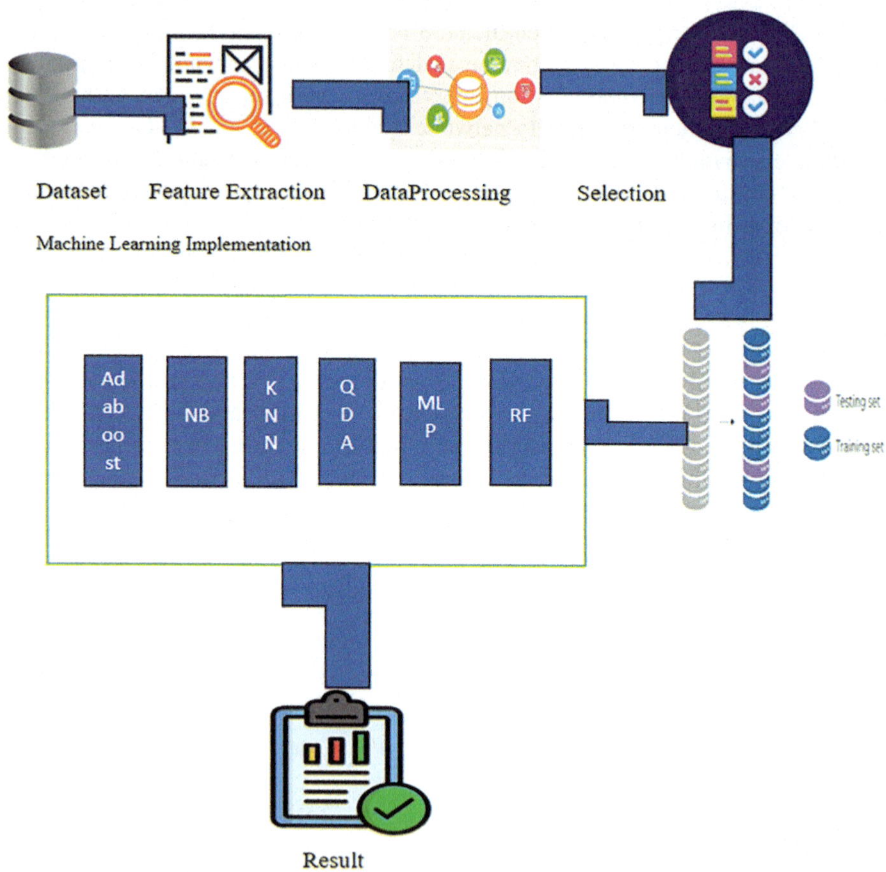

Fig. 1 Proposed architecture

For several important reasons, we chose to conduct our studies using the Bot-IoT dataset. To begin with, the dataset is updated frequently to ensure that it remains relevant to today's IoT security issues. Secondly, it provides an extensive range of attack types, including common botnet behaviors like information theft, denial of service (DoS), and snooping. Thirdly, the dataset is extremely pertinent to our investigation of IoT network security since it contains traffic produced by IoT devices. Finally, the Bot-IoT dataset broadens the scope of research by enabling the creation of additional characteristics from the raw data.

3.1 Implementation

This section focuses on our studies' main goal, which is to assess how well machine learning algorithms identify IoT network assaults. We start by outlining the dataset that we used in our research. We then go over the machine learning algorithms that we used in this research. Lastly, we outline the measures we took to put our experiments into practice. To prepare readers for the in-depth analysis and results presentation that follows, this part offers a thorough summary of the major elements of our evaluation approach.

4 Dataset

When it comes to computer security activities that use predictive techniques, big datasets are essential for properly analyzing and differentiating between regular and anomalous traffic. Many experiments have been carried out to produce network datasets over the years. Much research that uses algorithmic learning has evaluated their methods using real or simulated network data, as. Although there are many samples available, there hasn't been much work done in creating realistic datasets that include IoT-generated traffic and novel Botnet scenarios. Certain datasets don't include traffic created by IoT devices, or they don't produce new features. Figure 2 is the graphical picturization of the dataset.

Additionally, certain datasets might have imbalanced class distributions or a lack of variety in attack scenarios. For instance, a relatively big and clean N-BaIoT dataset has problems with class imbalance, as evidenced by the substantially lower ratio of typical information to targeted data.

We used the Bot-IoT dataset, which was created by Moustafa et al. to overcome these issues in our studies. The Bot-IoT sample comprises modeled and real-world IoT network traffic with different kinds of assaults categorized as data theft, denial of service (DoS), and sniffing attacks [25, 26]. This dataset provides a more thorough and realistic testbed for testing, making it an invaluable tool for assessing techniques that use machine learning to locate connected device assaults.

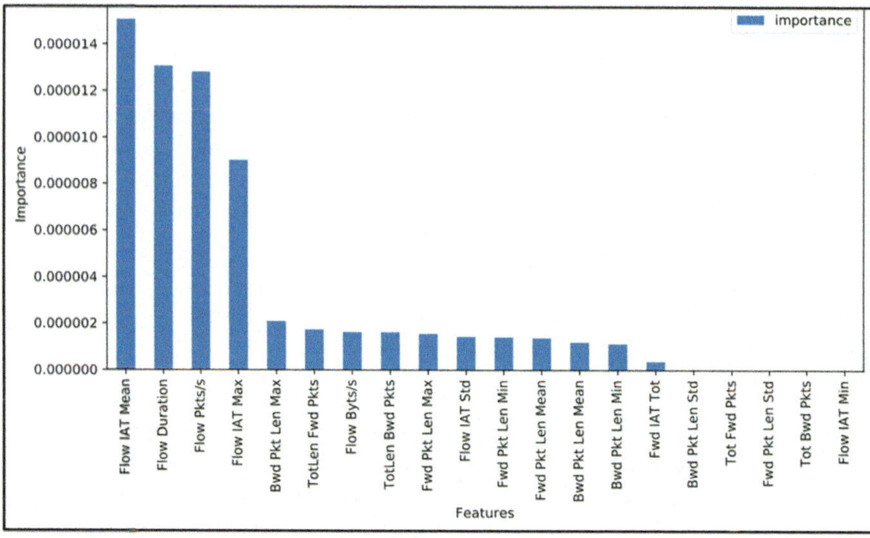

Fig. 2 Graphical representation of the entire dataset

5 Machine Learning Approach

The neural networks were chosen to cover a wide variety of techniques with unique features. In a nutshell, the selected classifiers are: A non-parametric technique called K-Nearest Neighbours (KNN) uses the largest class of each data point's k nearest neighbors to classify the data points. To maximize information acquisition, the ID3 (Iterative Dichotomiser 3) decision tree algorithm chooses the optimal characteristic at each stage to divide the data into groups.

Random Forest: An approach to ensemble learning that minimizes overfitting AdaBoost: An interactive learning method that builds a strong classifier by repeatedly training weak classifiers on the incorrectly categorized data and giving them greater weights. Quadratic Discriminant Analysis (QDA): A classification technique that uses a quadratic decision boundary to simulate each class's distribution. An artificial neural network with numerous layers of nodes that can recognize intricate patterns in data is called a multilayer perceptron (MLP). Naive Bayes (NB): A probabilistic classifier that relies on the separateness properties and is based on the Bayes theorem. These classifiers offer a thorough evaluation of machine learning methods for IoT network attack detection; they were selected based on their popularity and efficacy in a range of classification tasks [27–30].

K-Nearest Neighbors (KNN):

A straightforward reliable supervised learning algorithm is KNN. It searches through the available data and links incoming information with similar existing points in the dataset. KNN is quick while training but comparatively slow during estimating, and it performs well with multiple dimensions.

Quadratic Discriminant Analysis (QDA):

For supervised classification tasks, QDA is appropriate. It's a statistical method for allocating measurable data to a single group out of many groups. When there is little data available to characterize a category, QDA works best.

Random Forest (RF):

It puts together numerous distinct decision tree structures to create a "forest". When contrasted with alternative techniques, RF is lightweight, robust against noise and outliers, and can handle big datasets with efficiency.

Adaptive Boosting (AdaBoost):

AdaBoost concentrates on classification problems and strives to make inadequate classifiers stronger. When combined with different learning algorithms, it enhances performance. AdaBoost is capable of efficiently handling errors in datasets.

Multilayer Perceptron (MLP):

One type of feedforward artificial neural network (ANN) is the MLP class. It consists of invisible layers, the result, and the source.

Naive Bayes (NB):

NB is a popular supervised algorithm that is renowned for its ease of use. It is used for tasks like traffic categorization in intrusion detection Although NB treats features individually, it is more user-friendly but has less capacity to capture feature interactions.

These algorithms were chosen for their popularity and effectiveness in addressing various aspects of classification tasks, providing a diverse set of techniques for detecting IoT network attacks.

5.1 Implementation Steps

Feature Extraction:

To fetch the flow-based attributes from actual internet traffic data in pcap format, we used CICFlowMeter.Distributed by CIC, CICFlowMeter yield 84 network traffic features. The objective of this procedure was to extract novel features from the dataset to improve the classifiers' prediction power.

Data Pre-processing:

The process of preparing a dataset for machine learning is known as data pre-processing. To increase productivity, it contains processes to clean the dataset by eliminating erroneous or unnecessary data that could compromise its accuracy.

Splitting Data:

Data are needed for training and testing throughout the machine-learning process.

Eighty percent of the Bot-IoT dataset was used for training, and the balance of the data was used for validation. This enables us to analyse the algorithm's effectiveness and determine how effectively it operates with unknown data.

Feature Selection:
Limiting the number of attributes and using only those required for the development and evaluation of the algorithms is crucial to developing a lightweight security solution appropriate for IoT systems. Since the Random Forest Regressor algorithm has been demonstrated to be an efficient means of reducing dataset dimensionality, we used it for feature selection.

5.2 Implementation of Machine Learning Algorithms

Using well-known machine-learning packages all experiments were carried out in Python. Three stages comprised our organization of the dataset's machine learning algorithm assessment process:

1. Dealing with each assault in the information set independently using the suggested algorithms.
2. Using a set of features that combines the finest attributes for every assault.
3. Using the seven top features found during the feature selection phase, apply the algorithms to the complete dataset.

During these stages, we were able to thoroughly examine the machine learning algorithms' performance and determine how well they detected IoT network assaults. These steps collectively form the foundation of our approach, ensuring that the dataset is appropriately processed and partitioned for effective machine learning analysis of IoT network attacks.

6 Result Analysis

Choosing the right performance metrics for the job at hand is crucial when assessing how well predictive models perform. The following KPIs were employed by us for our assessment:

Accuracy: The percentage of occurrences in the dataset that are properly classified out of all instances.

Precision: The percentage of actual positive predictions among all the model's positive predictions. It assesses how well the model avoids producing false positive results.

Recall: The percentage of accurate positive forecasts among all real positive examples found in the collection. It assesses how well the model can account for every positive example.

F-measure: A balanced indicator of a classifier's efficiency, calculated as the harmonic mean of precision and recall.

These performance metrics enable us to evaluate how well our machine-learning algorithms identify and classify connected device threats. Table 2 describes the various attacks. We divided the process of evaluating machine learning techniques for the given data set into three stages:

Phase 1: Applied machine learning techniques to every assault in the information set independently. Table 3 provides a summary of the findings. Except for Naive Bayes (NB) and Quadratic Discriminant Analysis (QDA), all algorithms successfully detected the majority of kinds of attacks over ninety of the time. Because the ID3 algorithm has a short processing time, it prioritizes speed and demonstrated an outstanding success rate in six out of ten tests. Despite having the lowest overall F-measure, Naive Bayes showed quicker processing speeds.

Phase 2: Utilized machine learning techniques throughout the complete dataset, combining feature sets derived from the most effective characteristics for every attack.

Table 2 The type of attack

Attack Names
DDOS HTTP
DDOS UDP
DDOS TCP
DOS HTTP
DOS UDP
DOS TCP
Data exfiltration
Keylogging
Service Scan
OS Scan

Table 3 Performance analysis

ML algorithm	Accuracy	Precision	Recall	F-measure
NB	0.79	0.85	0.79	0.77
QDA	0.87	0.89	0.87	0.86
RF	0.97	0.97	0.97	0.97
Adaboost	0.97	0.97	0.97	0.97
MLP	0.84	0.87	0.84	0.83
KNN	0.99	0.99	0.99	0.99

The findings, which indicate that Adaboost is the method that performs the best, While Naive Bayes demonstrated the lowest score but the fastest execution, KNN demonstrated slower processing in comparison to other algorithms, despite having a higher performance score.

Phase 3: Utilizing the seven top features found from feature selection, machine learning methods were applied to the complete dataset. Although the algorithms' F-measure performance remained unchanged, the execution times significantly decreased as a result of the feature count reduction, from 13 attributes in Phase 2 to 7 attributes in Phase 3.

In the evaluation process, we structured our approach into three distinct phases:

Phase 1: We utilized distinct machine-learning algorithms for every assault present in the dataset. The findings, which are compiled in Table 3, showed that the majority of algorithms identified various kinds of assaults with over 90% accuracy rate. except for Naive Bayes (NB) and Quadratic Discriminant Analysis (QDA). The ID3 algorithm demonstrated the highest success rate across six out of ten tasks, emphasizing its efficiency due to low processing time. However, Naive Bayes exhibited faster processing times despite having the lowest overall F-measure.

Phase 2: Utilizing feature sets created from the most effective characteristics associated with each attack, machine learning methods were applied to the whole dataset. Notably, Naive Bayes demonstrated faster execution while having a lower score than KNN, which showed slower processing when compared to other algorithms.

Phase 3: We used the seven best features from feature selection to apply machine learning methods to the complete dataset. The execution times significantly lowered as a result of the feature count being reduced from 13 attributes in Phase 2 to 6 characteristics in Phase 3, even though there were no significant changes in performance as assessed by the F-measure. Together, these stages offered a thorough grasp of how predictive algorithms function in identifying Internet of Things network threats, emphasizing the trade-off between processing speed and performance.

7 Conclusion

This article used the Bot-IoT dataset, which is regularly updated, has a wide variety of attacks, and includes a range of network protocols, to detect IoT attacks on networks using machine learning techniques. Using CICFlowMeter, movement-based characteristics were obtained from raw traffic traces to create 84 network-related features that defined the network flow. The Random Forest Regressor algorithm was utilized to calculate importance values throughout implementation to identify the characteristics that were utilized in machine learning techniques. To discover common key qualities, two approaches were used: one included computing weights independently for each sort of attack, while the other involved calculating weights for a group that included all attacks. We employed seven popular machine learning algorithms, and their performance was measured using F-measure scores:

Subsequent research endeavors may aim to improve the accuracy of detection by assessing the efficacy of unsupervised algorithms and integrating diverse learning algorithms into a multi-layered model.

References

1. Deogirikar, J., Vidhate, A.: Security attacks in iot: a survey. International Conference on I-SMAC (I-SMAC), pp. 32–37 (2017)
2. Bodstrom, T., Hämäläinen, T.: State of the art literature review on network anomaly detection with deep learning. Internet of Things, Smart Spaces, and Next Generation Networks and Systems, pp. 64–76 (2018)
3. Pattnaik, L., Satpathy, S., Paikaray, B.K., Swain, P.K.: DDoS analysis using machine learning: survey, issues, and future directions. Int. J. Bus. Contin. Risk Manag. **14**(1), 57–76 (2024)
4. Du, M., Li, F., Zheng, G., Srikumar, V.: Deeplog: anomaly detection and diagnosis from system logs through deep learning. Proceedings of the 2017 ACM SIGSAC Conference on Computer and Communications Security, pp. 1285–1298 (2017)
5. Radford, B.J., Richardson, B.D., Davis, S.E.: Sequence aggregation rules for anomaly detection in computer network traffic (2018). arXiv preprint arXiv:1805.03735
6. Lambert, I., Glenn, M.: Security Analytics: Using Deep Learning to Detect Cyber Attacks (2017)
7. Stevanovic, M., Pedersen, J.M.: Detecting bots using multi-level traffic analysis. IJCSA **1**(1), 182–209 (2016)
8. Sedjelmaci, H., Senouci, S.M., Al-Bahri, M.: A lightweight anomaly detection technique for low-resource iot devices: a game-theoretic methodology. IEEE International Conference on Communications (ICC), pp. 1–6 (2016)
9. Koroniotis, N., Moustafa, N., Sitnikova, E., Turnbull, B.: Towards the development of realistic botnet dataset in the internet of things for network forensic analytics: Bot-iot dataset. Futur. Gener. Comput. Syst. **100**, 779–796 (2019)
10. Mirsky, Y., tshman, T., Elovici, Y., Shabtai, A.: Kitsune: An ensemble of autoencoders for online network intrusion detection. arXiv preprint arXiv:1802.09089 (2018)
11. Yuan, X., Li, C., Li, X.: Deepdefense: identifying ddos attack via deep learning. IEEE International Conference on Smart Computing (SMARTCOMP), pp. 1–8 (2017)
12. Meidan, Y., Bohadana, M., Mathov, Y., Mirsky, Y., Shabtai, A., Breitenbacher, D., Elovici, Y.: N-baiot—network-based detection of iot botnet attacks using deep autoencoders. IEEE Pervasive Comput. **17**(3), 12–22 (2018)
13. Putchala, M.K.: Deep Learning Approach for Intrusion Detection System (ids) in the Internet of Things (iot) Network Using Gated Recurrent Neural Networks (gru) (2017)
14. Ferrag, M.A., Maglaras, L.: Deepcoin: a novel deep learning and blockchain-based energy exchange framework for smart grids. IEEE Transactions on Engineering Management (2019)
15. Meidan, Y., Bohadana, M., Shabtai, A., Ochoa, M., Tippenhauer, N.O., Guarnizo, J.D., Elovici, Y.: Detection of Unauthorized iot Devices Using Machine Learning Techniques. arXiv preprint arXiv:1709.04647 (2017)
16. Soe, Y.N., Feng, Y., Santosa, P.I., Hartanto, R., Sakurai, K.: Rule generation for signature-based detection systems of cyber-attacks in iot environments. Bull. Netw. Comput. Syst. Softw. **8**(2), 93–97 (2019)
17. Bezerra, V.H., da Costa, V.G.T., Junior, S.B., Miani, R.S., Zarpelao, B.B.: One-class classification to detect botnets in iot devices. Anais do XVIII Simposio Brasileiro em Seguranc¸a da Informação de Sistemas Computacionais, pp. 43–56 (2018)
18. Hodo, E., Bellekens, X., Hamilton, A., Dubouilh, P.-L., Iorkyase, E., Tachtatzis, C., Atkinson, R.: Threat analysis of iot networks using artificial neural network intrusion detection system.

International Symposium on Networks, Computers and Communications (ISNCC), pp. 1–6 (2016)
19. Summerville, D.H., Zach, K.M., Chen, Y.: Ultra-lightweight deep packet anomaly detection for internet of things devices. In: IEEE 34th International Performance Computing and Communications Conference (IPCCC), pp. 1–8 (2015)
20. Yavuz, F.Y.: Deep learning in cyber security for internet of things. Ph.D. Dissertation (2018)
21. Ibitoye, O., Shafiq, O., Matrawy, A.: Analyzing adversarial attacks against deep learning for intrusion detection in iot networks. arXiv preprint arXiv:1905.05137 (2019)
22. Cvitic, I., Perakovíc, D., Perísa, M., Botica, M.: Novel approach for detection of iot generated ddos traffic. Wireless Networks, pp. 1–14 (2019)
23. Baig, Z.A., Sanguanpong, S., Firdous, S.N., Nguyen, T.G., So-In, C., et al.: Averaged dependence estimators for dos attack detection in iot networks. Futur. Gener. Comput. Syst. **102**, 198–209 (2019)
24. Satpathy, S., Swain, P.K., Mohanty, S.N., Basa, S.S.: Enhancing Security: Federated Learning against Man-In-The-Middle Threats with Gradient Boosting Machines and LSTM. In: 2024 IEEE International Conference on Advanced Video and Signal Based Surveillance (AVSS), pp. 1–8. IEEE (2024, July)
25. Yu, S.: Study on the internet of things from applications to security issues. International Conference on Cloud Computing and Security, pp. 80–89 (2018)
26. Kotsiantis, S.B., Zaharakis, I., Pintelas, P.: Supervised machine learning: a review of classification techniques. Emerging Artificial Intelligence Applications in Computer Engineering **160**, 3–24 (2007)
27. Satpathy, S., Pradhan, S.K., Ray, B.B.: A digital investigation tool based on data fusion in management of cyber security systems. Int. J. Inf. Technol. Knowl. Manag. **2**(2), 561–565 (2010)
28. Panda, M., Patra, M.R.: Network intrusion detection using Naive Bayes. Int. J. Comput. Sci. Netw. Secur. **7**(12), 258–263 (2007)
29. Basahel, S.B., Bajaba, S., Yamin, M., Mohanty, S.N., Lydia, E.L.: Teamwork optimization with deep learning based fall detection for IoT-enabled smart healthcare system. Comput. Mater. Continua. **75**(1), 1353–1369 (2023). ISSN: 1546-218. https://www.techscience.com/cmc/v75n1/51539
30. Chatterjee, S., Satpathy, S., & Nibedita, A. (2023). Digital investigation of network traffic using machine learning. *EAI Endorsed Transactions on Scalable Information Systems, 11*(1). https://doi.org/10.4108/eetsis.4055

IoT Applications and Cyber Threats: Mitigation Strategies for a Secure Future

Pratik Kumar Swain, Lal Mohan Pattnaik, and Suneeta Satpathy

Abstract The rapid evolution in development and implementation of IoT (Internet of Things) devices has transformed the way we interact with the digital world which offers unparalleled challenges. This research provides a comprehensive study of IoT technology, its various applications and exploring into the IoT cyber threat landscape, this paper serves as a roadmap for navigating the complex and dynamic IoT cyber threat landscape. The study begins by explaining the fundamental principles and architecture of IoT along with explanation of the complex web of interconnected devices in IoT ecosystems. It then proceeds to explore the multilayered applications of IoT across various domains like smart cities, agriculture, smart homes, industrial automation, etc. Along with expansion of IoT ecosystems, cyber threats and challenges are also increasing. An in-depth study of these potential cyber threats, vulnerabilities and risks related with IoT devices and IoT ecosystems has been done in the manuscript. Strategies for mitigating these cyber threats are discussed which encompasses both technical and policy-oriented approaches and also highlights the evolving regulatory landscape surrounding IoT security, emphasizing the need for standardized frameworks and best practices to ensure a resilient and secure IoT infrastructure. The study of emerging trends and technologies shaping the future of IoT such as edge computing, artificial intelligence, machine learning and blockchain along with their implementation for enhancing security and mitigation strategies. The interdisciplinary domain of IoT requires collaboration between researchers, manufacturers, stakeholders and policymakers.

Keywords IoT · Cybersecurity · Cyber threats · IoT architecture · Machine learning

P. K. Swain · L. M. Pattnaik
Faculty of Engineering and Technology, Sri Sri University, Cuttack, Odisha, India
e-mail: lalmohan.p@srisriuniversity.edu.in

S. Satpathy (✉)
Center For Cyber Security, SoA Deemed to be University, Bhubaneswar, Odisha, India
e-mail: suneeta1912@gmail.com

S. N. Mohanty et al. (eds.), *Explainable IoT Applications: A Demystification*,
Information Systems Engineering and Management 21,
https://doi.org/10.1007/978-3-031-74885-1_27

1 Introduction

Connectivity is the foundation of technological evolution in today's generation, the IoT (Internet of Things) has developed as a model-shifting force in which it helps in connecting the bridge between the physical and digital world. The beginning of the IoT ecosystem builds a monumental shift in our technological landscape, where it consists of a variety of smart devices and from everyday smart machines to industrial machinery are linked and generating vast streams of data. This interconnected IoT ecosystem provides an unparalleled advancement in efficiency, productivity, user experience, changing the way we work, interact and live with surroundings. From resource optimization and utilization in smart cities to healthcare devices revolutionizing patient's health care, the impact of IoT echoes across many industries, reshaping conventional methods and enhancing the decision-making capabilities [1]. But as we admire the promise of a connected future, the inherent challenges that are embedded in the IoT landscape cannot be overlooked or ignored, particularly those challenges that are related to security. The vulnerabilities or loopholes of the interconnected devices raise concerns about the confidentiality, integrity and availability (CIA triad) of the data, unauthorized access and the threat of cyber-attacks. Understanding the distinctions of IoT security is dominant in safeguarding the IoT ecosystems of this interconnected world.

1.1 Internet of Thing (IoT)

IoT refers to a network of interconnected smart devices, objects and systems that are embedded with sensors, hardware, software and other technologies that are embedded in them which it collects, processes and exchanges data over the internet. These IoT devices can range from everyday devices such as smartphones, smart wearables and home appliances to industrial machinery, automobiles and infrastructure components. The key components of IoT are:

Devices and Sensors. IoT devices are equipped with various sensors to collect data from the surrounding environment. These sensors can measure parameters like temperature, pressure, humidity, motion, etc.

Connectivity. IoT devices depend on connectivity technologies like Wi-Fi, Bluetooth, cellular networks, Zigbee and others to transmit data to the cloud or other devices within the network or over the network.

Data Processing and Analytics. The data collected by IoT devices is processed and analysed to extract meaningful information. This involves techniques like data filtering, machine learning and predictive analytics.

Cloud Computing. It plays a vital role in the IoT ecosystem by providing scalable storage, computing power and services for data processing, analysis and management. It also facilitates remote monitoring, control and management of IoT devices.

Edge Computing. Data processing and analysis are executed closer to the source of data (i.e., at the edge of the network) rather than in centralized cloud servers in edge computing. It reduces latency, conserves bandwidth and enhances the privacy and security in IoT ecosystems and deployments.

Security and Privacy. Its objective is to protect the IoT ecosystem against data breaches, unauthorized access and other cyber-attacks. Defensive measures include encryption, authentication, secure firmware updates, access control and intrusion detection systems.

1.2 IoT Architecture

The IoT architecture includes a network of interconnected devices, sensors, software applications and actuators that work simultaneously in collecting, transmitting and processing of the data. This architecture typically consists of several layers, each layer serves as a distinct function in the IoT ecosystem [2]. The following key layers of an IoT architecture are shown in Fig. 1.

Perception Layer. This layer comprises hardware embedded with sensors and actuators to observe the physical environment. Sensors collect data from the surrounding environment, while actuators enable devices to interact with the physical world.

Network Layer. This layer helps in creating and maintaining communication between IoT ecosystems by enabling data transmission. It makes use of many networking technologies like Bluetooth, Wi-Fi, etc.

Middleware Layer. This layer acts as a middle-man between perception layer and application layer. Its objective is to preprocess data, protocol translation, device management and ensure the continuous communication between diverse devices and applications.

Application Layer. This layer consists of software applications that utilizes data which are generated by the IoT ecosystem in order to provide value added services and can perform some specific tasks ranging from data visualization to complex analytics.

Edge Computing Layer. This layer assists IoT devices by giving computational resources which allows data processing and analysis, reducing latency, conserving bandwidth and enhancing real-time responsiveness.

Cloud Computing Layer. This layer provides scalability for processing, analysing and storing large volumes of IoT data through the help of services like data storage, analytics and application hosting.

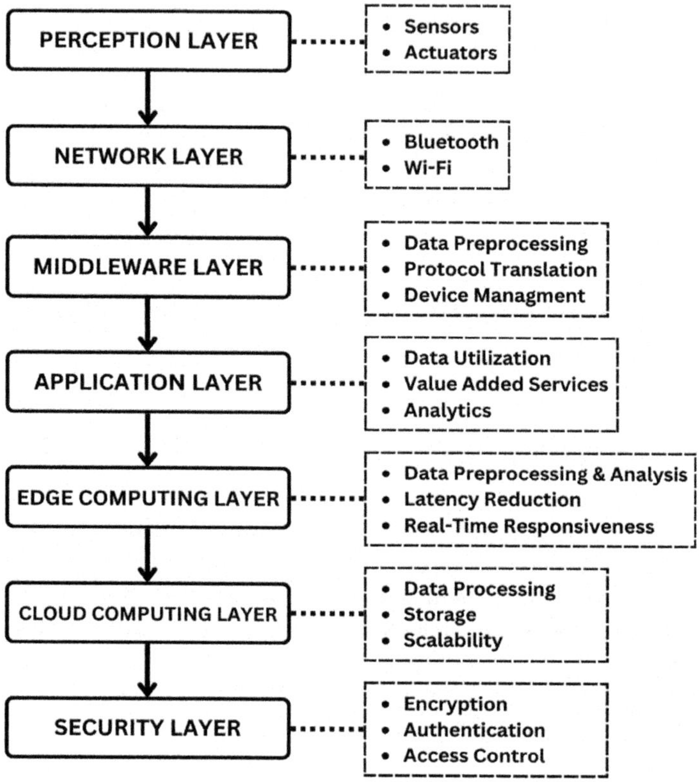

Fig. 1 IoT architecture

Security Layer. This layer implements security techniques which includes encryption, authentication, access control and intrusion detection to protect the CIA triad of IoT devices, data and communication from cyber threats.

These layers work together as one in which it enables seamless integration of physical devices with the digital ecosystem. Each layer plays a vital role in collecting, processing and analysing data to deliver valuable insights and services while addressing cybersecurity concerns through robust security measures and protocols.

2 Literature Survey

The literature encompassing the relation between the cybersecurity and IoT applications offers valuable insights into the multilayered challenges faced by the interconnected devices in the IoT ecosystem. Many research scholars and practitioners have extensively explored various parameters of IoT technology like its benefits in enhancing efficiency of the security mechanism and detection of security vulnerabilities in IoT devices.

Thakkar and Lohiya [3] studied many IDS (Intrusion Detection System) tools for IoT networks and their study highlighted the need for advanced defence mechanisms to mitigate the IoT threats. Their research was about the different strategies for deployment of IDS within the IoT ecosystem with the combination of Machine Learning and Deep Learning algorithm in detecting cyber-attacks within the network. They highlighted many security challenges in the IoT ecosystem and the scope of the IoT cyber threats.

Da Costa et al. [4] highlighted the challenges faced by intrusion detection in the IoT networks. They presented on recent developments in IDS and the integration of other defence techniques for securing the IoT networks. Their literature survey mainly discussed the development of security protocols by the community and industries that balances energy consumption and protection of the IoT devices. They studied the advanced defence techniques that are used in network security specifically in IDS and aimed for enhancing the positive mitigation rates while minimizing the false positive rates.

Ashraf et al. [5] worked on a comprehensive study of the application of Deep Learning and Machine Learning algorithms in IDS particularly developed for safekeeping of the IoT ecosystem. They presented the overview of the architecture of IoT, protocols, vulnerabilities and potential protocol based cyber threats. They also reviewed a variety of researches that are related to the IDS techniques and attack classification particularly for the IoT network. The authors discussed the emerging challenges and showed that current IDS tools used in IoT networks are not perfect and it requires further development.

Ahmad and Alsmadi [6] presented their in-depth study after analysing research papers about emerging IoT security trends from the year 2019–2020. They mainly focused on 3 domains: Information Security, Machine Learning (ML) and IoT in which it led to 6 research questions and the aim of their study was on finding out mitigation strategies that are implemented to classify cyber-attacks like DDoS attacks, unauthorized access, etc. on IoT ecosystems.

Hussain et al. [7] worked on the ML and DL application in the domain of IoT in terms of security purposes. They highlighted the security challenges, attack methodology and need of security within the IoT ecosystem. They discussed various ML and DL algorithms and their performance during securing IoT devices together with the limitation of traditional Machine Learning algorithms. They also researched about the existing security solution and pointed out research gaps. They suggested enhancing the main components of Deep Reinforcement Learning and DL and it must be examined and tested based on computational ability and effectiveness in learning the pattern of the attacks.

Liang et al. [8] showed the pros and cons during the integration of Machine Learning for IoT and CPS and also discussed the advantages of ML implementation to enhance the security in IDS and CPS while underscoring the challenges and security issues. They discussed the ML loopholes that existed across various phases like data collection, data pre-processing, validation, training, testing and implementation, and also alerts on misusing Machine Learning in deploying cyber-attacks and invented mitigation strategies to dissolve these issues.

Chaabouni et al. [9] reviewed various researches related to Network IDS for IoT implementing multiple learning techniques. Their review consists of existing datasets that are related to Network IDS and implementation of open-source tools which are strictly evaluated. The author discussed the architecture of NIDS and the methodology in the IoT ecosystem encompassing traditional Machine Learning based mitigation techniques. They proposed a state-of-the-art learning technology about the NIDS in the IoT ecosystem. This proposed model showed high detection accuracy and less false positive rates.

Tahsien et al. [10] proposed ML base security mechanism for the protection of IoT devices and introduced various IoT architecture layers and showed types of security challenges faced by each of the layers together with different types of cyber threats on the IoT threat landscape. Their review presented various ML approaches and how these approaches can be implemented to mitigate these cyber-attacks on IoT ecosystems. They showed an in-depth study of security related classification solutions for the IoT systems particularly focused on ML algorithm integration in the IoT system architecture layers and also highlighted the limitations and challenges of ML based approaches for the IoT system.

Wu et al. [11] had done a comprehensive evaluation of the unique features and polices of the IoT security protection and also researched about the AI methodologies including ML and DL which can be implemented in IoT security. Their study has presented a comprehensive overview of AI based solutions encompassing various algorithms and technologies to mitigate four emerging security threats: Unauthorized access, Intrusion Detection System, Malware and against DDoS attack. For the future purpose, the authors can explore the impact of the emerging challenges faced during implementation of AI in IoT security.

Bibi et al. [12] studied the major cyber-attacks that are targeting IoT devices at the network and physical layers. They have discussed the evolution of the botnet of IoT devices, architecture of an IoT botnet and compared it with traditional architecture of botnets. They also presented Mirai Botnet Attack case study and also discussed the tools and techniques that are used to detect botnet networks. Their work aimed to enhance the defence mechanism and the security configuration of these IoT devices by analysing the cyber-attacks that are targeting these devices.

Aamer [13] explained various types of security threats related to the IoT environment and its layers including network attacks, encryption attacks, firmware attacks and physical attacks. They also presented about the applications of IoT in various sectors like Smart Cities, Healthcare, Agriculture, Smart Homes, etc. The main contribution of the author was to examine various security challenges on the IoT architecture layers and design a model to gather and provide information to manufacturers and IoT researchers in which it helps in enhancing the security of the IoT devices.

Gerodimos et al. [14] showed an IoT model along with the security threats faced by other research scholars like privacy issues and data mining problems. They described the proposed IoT model and various types of security attacks that target each layer of the IoT model that includes application layers, transportation layers and perception layers. They have done comparisons of existing various models and the techniques implemented were also presented.

Tsiknas et al. [15] presented the security threats and risks related to the IoT ecosystem and introduced a ML based model to classify and predict IoT attacks. The aim of this study was to provide the methodology of IoT security threats so that the cyber security researchers can easily understand the relation between IoT network and general network security. They introduced an IoT system and examined the data collected by it over nine months, then evaluated the effectiveness of the classification models based on ML and proposed a framework for integration in the IoT ecosystem.

Ahmad et al. [16] proposed two IoT architectures: 3-layered and 5-layered architecture in which 3-layer architecture includes perception, network and application layer and the 5-layer architecture consists of perception, network, middleware, application and business layer. They highlighted some of the features like connectivity, intelligence, sensors, etc. and advantages of implementing the IoT ecosystem. They listed out the vulnerabilities of the IoT devices and various threats that can harm in every layer of the architectures.

Wheelus and Zhu [17] studied the IoT architecture, interpretations of its layers and applications of IoT in various domains. They did a literature review on IoT security and its challenges and also mentioned various mitigation methods to mitigate these challenges and many technological challenges faced due to limited resources given to the IoT system. They proposed a technological innovation that can address these challenges and it enhances the performance of the IoT devices and also enhances the security model of the IoT.

Borcherding et al. [18] discussed some security issues related to IoT like threat to integrity of the data, encryption and decryption, privacy, frameworks and many other challenges based on IoT. Their study highlighted the networks and inter connection of IoT that have already been researched by researchers. The author proposed a security model for the IoT network that has 6 main layers: coding layer, perception layer, network layer, middleware layer, application layer and business layer and the author also discussed various communication protocols, components and standards that builds a robust architecture for the IoT ecosystem.

Le et al. [19] examined the recent security state in the domain of IoT and discussed the security challenges related to the IoT ecosystem. They described the IoT applications in the domain of healthcare services and industries and also discussed the security threats faced in various layers in IoT healthcare architectures. Malware types are presented and its relation to IoT like viruses, spyware, keyloggers, worms, Trojan horses and some of the real-life case studies of malware impacts and its attack methodology like Mirai botnet attack, reaper and echo Bot.

The complex relation between IoT applications and cybersecurity highlights both the huge potential for innovation and the crucial need for the development of robust IoT security defensive measures through analysing the results and its insights from existing research.

3 IoT Application Landscape

The development of Internet of Things (IoT) technologies has resulted in evolving changes across various industries, revolutionizing business operation and interaction of the individuals with their surroundings [20, 21]. The multilayered applications of IoT across various domains that leveraging interconnected devices to drive innovation and efficiency are shown in Fig. 2.

3.1 Smart Cities

Implementation of IoT technologies enables the urban landscape in reshaping and in development which provides municipalities unparalleled abilities to enhance infrastructure resilience, optimize resource allocation and improve quality of life for residents [22, 23]. Smart city initiatives encompass a wide array of applications, including:

Traffic Management. IoT sensors implanted in roads, traffic lights, traffic cameras and vehicles enable real-time monitoring of traffic flow, allowing dynamic adjustments to signal timings and routing to lessen traffic congestion.

Public Safety and Security. Integrated surveillance systems united with IoT ecosystem and analytics platform which allows law enforcement agencies to proactively identify and respond to security threats, enhancing situational awareness and enhancing quick emergency response capabilities.

Energy Efficiency. IoT-enabled energy management system is implemented to monitor the pattern of the energy consumption, helps in optimizing energy distribution networks and helps in assisting in reducing waste and mitigate environmental impact.

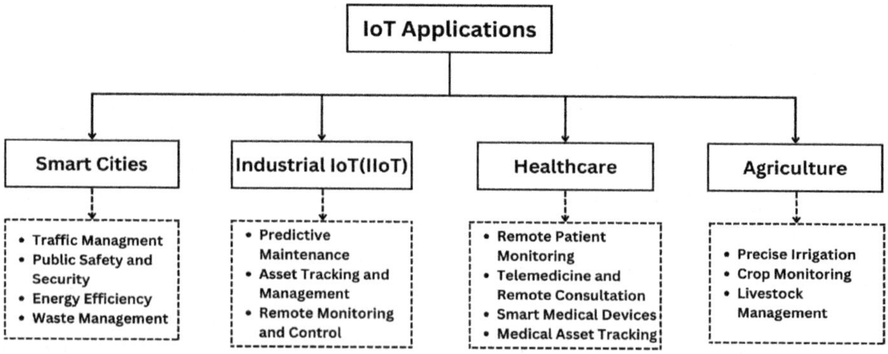

Fig. 2 Applications of IoT

Waste Management. IoT-enabled waste collection vehicles helps in streamline collection of wastes and also provides a real-time fill level monitoring, route planning optimization and minimizing operational costs while encouraging sustainability goals.

3.2 *Industrial IoT (IIoT)*

In industries, IoT technologies are changing traditional manufacturing processes into hybrid processes which are more productive and efficient. It enables industries in predictive maintenance of machines to minimize downtime and optimize asset utilization [24]. Key application of IIoT includes:

Predictive Maintenance. IoT sensors installed on industrial machines collect real-time performance data and it enables predictive analytics algorithms to predict equipment failures. It provides the schedule for the proactive maintenance of industrial equipment in which it helps in reducing unplanned downtime and maintenance costs optimization.

Asset Tracking and Management. IoT-enabled industrial asset tracking solutions leverage GPS, RFID tags and wireless connectivity to monitor the condition, location and status of these assets throughout the supply chain and streamlined logistics operations.

Remote IoT Monitoring. IoT ecosystem enables remote managing and monitoring of all industrial IoT equipment. It assists operators in optimizing production parameters, monitoring environmental conditions and responding swiftly to any anomalies.

3.3 *Healthcare*

In the healthcare sector, application of IoT technologies is revolutionizing by improving patient care, enhancing clinical workflows, enhancing operation efficiencies and enabling remote patient health monitoring [25]. Key applications include:

Remote Patient Monitoring: Smart wearable devices equipped with biosensors and wireless connectivity enable continuous monitoring of the health parameters and vital signs like heartbeat monitor, oxygen monitor, BP monitor, etc. in which it facilitates early detection of any health issues and enables timely treatment.

Telemedicine and Remote Consultation: IoT-enabled telemedicine platforms helps in connecting patients with healthcare providers remotely, enabling virtual consultations and treatment planning, thereby improving access to healthcare services and reducing the load on traditional healthcare infrastructure.

Smart Medical Devices: IoT-enabled medical equipment like smart insulin pumps, connected inhalers and digital pill dispensers in which it helps in improving medication adherence, dosage accuracy and treatment effectiveness. This equipment can also automatically transmit data to healthcare providers for analysis and decision-making.

Medical Asset Tracking: IoT integration for medical asset management helps in tracking location and utilization of critical medical equipment like infusion pumps and defibrillators, etc. In which it improves inventory management, enhancing equipment utilization and ensuring regulatory compliance.

3.4 Agriculture

In the agricultural sector, integration of IoT technologies is driving the emergence of precision agriculture practices in which it enables farmers to optimize resource allocation, monitor crop health, analyse soil quality, observe climate conditions and enhance productive yield [26]. Key applications include:

Precise Irrigation. IoT sensors that are deployed in fields in which it measures moisture levels of the soil, weather conditions, pest detections and water requirements for the crops. It enables precision irrigation systems to deliver the correct amount of water at the right time, minimizing water waste, maintaining sustainable development goals and maximizing crop yields.

Crop Monitoring. IoT integrated drones and ground-based sensors capture data on crop health, growth patterns and environmental conditions in which it provides farmers with insights to optimize planting strategies, detect pests and diseases early and apply quick response.

Livestock Management. IoT integration in livestock management helps in monitoring and tracking anime health, their behaviour and real-time locations. It assists farmers in optimizing feeding schedules, detecting any sign of illness or distress in livestock and improving overall livestock management practices.

The diverse applications of IoT across industries, smart cities, healthcare, agriculture and other domains provides opportunity in boosting innovation, efficiency and sustainability. However, the expansion of interconnected devices and the IoT ecosystem have introduced evolving cybersecurity risks, demanding robust mitigation strategies to safeguard sensitive data, protect critical infrastructure and ensure the protection of the CIA triad of the IoT ecosystems.

4 Cyber Threat Landscape in IoT

The growth of Internet of Things (IoT) devices has accompanied unparalleled connectivity and convenience across diverse domains, ranging from smart homes and healthcare to industrial automation and critical infrastructure like PowerStation, Hospitals, etc. However, this exponential growth in IoT integration has also given rise to a complex and evolving cyber threat landscape, categorized by diverse attack vectors, sophisticated malware and significant security vulnerabilities. Understanding this multilayered nature of cyber threats in the IoT ecosystem is crucial for inventing effective mitigation strategies to safeguard interconnected devices and IoT networks [27, 28].

4.1 Attack Vectors

IoT ecosystems consist of a wide range of interconnected endpoints which includes sensors, actuators, wearables and industrial machinery. Each presenting unique attack surfaces for malicious actors to exploit. Common attack vectors in the IoT landscape are shown in Fig. 3

Weak Authentication and Authorization. Many IoT devices are delivered with default credentials, making them vulnerable to credential stuffing attacks, guessing attacks and unauthorized access. Weak authentication mechanisms coupled with careless configuration of authorization controls enables rivals to compromise devices and gain unauthorized control over critical functions.

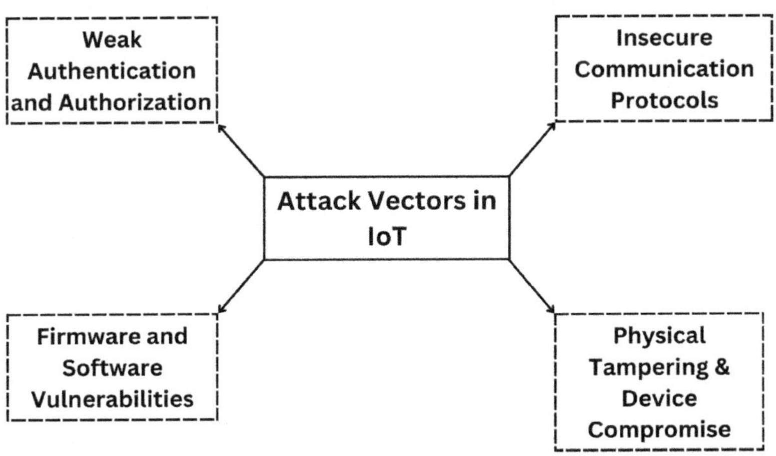

Fig. 3 IoT attack vectors

Insecure Communication Protocols. IoT devices often rely on insecure communication protocols like Bluetooth and Wi-Fi, which may lack encryption or authentication mechanisms, leaving data transmissions vulnerable to interception, modification, fabrication and tampering. Man-in-the-middle attacks and packet sniffing techniques can be leveraged to eavesdrop on sensitive information exchanged between IoT devices and backend systems.

Firmware and Software Vulnerabilities. Firmware and software updates to address security vulnerabilities and enhance device functionality are released by the manufacturers. However, IoT devices deployed in real-world environments may remain unpatched or outdated resulting in exposing them to exploitation by known vulnerabilities. Attackers can exploit these vulnerabilities to execute remote code execution (RCE) attacks, buffer overflows and other exploits targeting vulnerable firmware and software components which can disrupt the entire IoT ecosystem.

Physical Tampering and Device Compromise. Physical unauthorized access to IoT devices can enable tampering, reverse engineering or hardware-level attacks in which it aims in extracting sensitive data, modifying device configuration or injecting malicious code. Compromised devices may be reused as botnet networks which are used to launch DDoS attacks or to incorporate into larger-scale cyber-attacks targeting critical infrastructure or government networks.

4.2 Emerging IoT Threats

The evolving landscape of the IoT ecosystem and their integration into many diverse environments have given rise to evolving and emerging threats which exploit IoT vulnerabilities. Some emerging IoT threats are shown in Fig. 4

Botnets and DDoS Attacks. Botnets are the networks of compromised devices controlled by hackers in which it poses a threat on IoT networks. These botnets

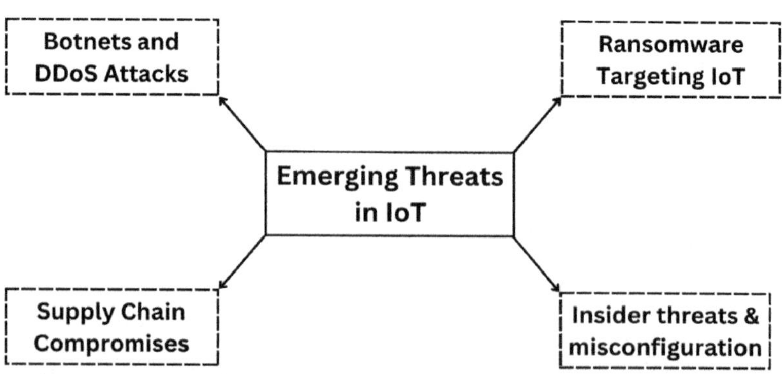

Fig. 4 Emerging IoT threats

leverage vulnerable IoT devices like routers, IP cameras and smart home appliances to conduct large-scale Distributed Denial of Service (DDoS) attacks. By harnessing the collective computing power of thousands of compromised devices and with the help of botnet operators can overwhelm targeted networks and services, disrupting operations and causing widespread outages [29].

Ransomware Targeting IoT. Ransomware is a malicious software that encrypts files or locks users out of their devices until a ransom is paid in the form of cryptocurrency which are untraceable. It has evolved and is targeting IoT devices with increasing frequency where the attackers exploit vulnerabilities in IoT firmware configurations to gain unauthorized access to devices, encrypting confidential data and disrupting device functionality. This poses significant risks, particularly in critical sectors such as healthcare and manufacturing where ransomware attacks on IoT devices can lead to operational disruptions, data loss, data leaks and financial extortion [30].

Supply Chain Compromises. It poses a significant threat and often overlooked in IoT security, it involves attackers infiltrating into manufacturing and distribution processes to implant malicious hardware or software within IoT devices. These compromised devices may contain firmware vulnerabilities, hidden backdoors and malicious codes that can be remotely exploited by malicious actors to gain unauthorized access, exfiltrate sensitive data and can have persistence connection later can launch further attacks within targeted networks.

Insider threats and misconfiguration. It represents a persistent threat in IoT security. Inadequate security practice such as failure to change default credentials, misconfiguring device settings or neglecting software updates can expose vulnerability of IoT devices to exploitation by malicious actors. Additionally, insider threats present a significant risk as disgruntled employees or insiders with privileged access may deliberately abuse their permissions to compromise IoT infrastructure or steal sensitive data.

4.3 Vulnerabilities in Critical Infrastructure

The integration of the IoT ecosystem into critical infrastructure of a nation which includes power stations, transportation, healthcare, etc. introduces unique security challenges due to the potential impact of cyber-threats on public safety, economic stability and national security. Vulnerabilities in Industrial Control Systems (ICS), Supervisory Control and Data Acquisition (SCADA) systems and Operational Technology (OT) networks can expose critical infrastructure assets to operational disruption, physical damage and compromise of safety–critical functions in which it can result in creating chaos.

SCADA and ICS Vulnerabilities. SCADA and ICS environments are increasingly interconnected with IoT devices and IT networks expanding the attack surface and enhancing the risk of cyber threats. Vulnerabilities in SCADA systems, insufficient access controls and lack of network segmentation create opportunities for rivals to disrupt industrial processes, manipulate control systems and can cause physical harm to infrastructure assets.

Healthcare IoT Risks. Due to the sensitive nature of medical data and healthcare IoT devices face challenges during safe keeping of these devices. Vulnerabilities or loopholes in networked medical devices like patient monitors, infusion pumps, etc. can result in exploitation, unauthorized access, data breaches and compromise of patient safety through tampering or manipulation of the IoT devices.

4.4 Privacy and Data Security Concerns

IoT devices collect vast quantities of personal and sensitive data that raises significant privacy and data security concerns particularly regarding transmission, data storage and access control. Weak security measures like unencrypted data storage, default credentials, etc. exposes IoT ecosystems to identity theft, data breaches and unauthorized access.

Data Breaches and Privacy Violations. Unauthorized access to IoT data archives or cloud storage services can result in exposing confidential information like personal identifiers, behavioural patterns and health records to unauthorized parties or get auctioned in dark webs. Privacy violations from IoT devices with built-in cameras, microphones, or location tracking capabilities raise ethical and legal concerns regarding user consent and transparency in data processing practices.

Regulatory Compliance and Legal Implications. Regulatory Compliance with privacy regulations such as the General Data Protection Regulation (GDPR), Health Insurance Portability and Accountability Act (HIPAA) and California Consumer Privacy Act (CCPA) are crucial for organizations deploying IoT devices particularly in regulated sectors like healthcare and finance. Non-compliance with data protection laws may result in heavy fines, reputational damage and legal sanctions for organizations found responsible for data breaches or privacy violations.

4.5 Future Trends and Challenges

IoT technologies continues to develop, evolve and expand across various industries, several key trends and challenges that are controlled in order to maintain the future of the IoT cyber threat landscape:

5G Connectivity and Edge Computing. The deployment of 5G networks and advancements in edge computing are likely to accelerate the integration of IoT devices to enable low-latency, high-bandwidth communication for real-time applications. However, the increased complexity and surface area introduced by 5G and edge computing integration may intensify security risks, necessitating enhanced security measures to protect against emerging threats.

Artificial Intelligence and Machine Learning. The implementation of AI and ML into IoT devices for enhancing automation, anomaly detection and predictive analytics. AI-powered IoT ecosystems also developed new attack vectors such as machine learning attacks, model inversion techniques and data poisoning in which malicious actors can exploit it to manipulate AI-enabled decision-making ability and evade detection [31, 32].

Blockchain and Distributed Ledger Technology (DLT). It offers potential solutions for enhancing IoT security by providing tamper-resistant data integrity, secure peer-to-peer communication and decentralized identity management. By leveraging blockchain-based mechanisms and smart contracts, IoT ecosystems are able to mitigate the risk of single points of failure, unauthorized access and data manipulation essential in centralized architectures [33].

Regulatory and Standards Frameworks. It provides frameworks like NIST (National Institute of Standards and Technology), IoT Cybersecurity Improvement Act European Union Agency for Cybersecurity (ENISA) guidelines and IoT cybersecurity framework provide valuable guidance and documentation for organizations seeking to enhance their regulatory landscape and implement best practices for securing IoT ecosystem.

The IoT cyber threat landscape consists of emerging cyber threats and attack vectors across critical infrastructure. Addressing these challenges requires a novel approach encompassing frameworks and industry collaboration to mitigate cyber threats and enhance the security of interconnected IoT ecosystems. By proactively mitigating these cybersecurity threats in the IoT ecosystem, organizations can harness the potential of IoT technologies while safeguarding against malicious exploitation and ensuring a secure future for connected devices and networks.

5 Mitigation Strategies

Safeguarding IoT devices against cyber threats requires a proactive mitigation techniques and defence strategies, following mitigation techniques highlights key measures that organizations can adopt to enhance the security posture of their IoT devices and safeguard against potential vulnerabilities.

5.1 Strong Authentication Mechanisms

Unauthorized access to IoT devices and its networks can be mitigated with the implementation of robust authentication mechanisms by utilizing strong and unique passwords in which it will be difficult for the hackers to crack the passwords by using guessing attacks or brute force attacks and minimizes the time taken to crack due to the default credentials. Organizations should implement advanced authentication methods like multi-factor authentication (MFA) which uses biometric authentication, voice authentication, one time password (OTP), etc. to add an extra layer of security before gaining access to IoT devices.

5.2 Encryption Protocols

To protect confidential information from interception, modification, fabrication and man-in-the-middle attacks that are transmitted between IoT devices and backend servers, various encryption protocols are employed. By utilizing advance cryptographic algorithm and protocols like Transport Layer Security (TLS) and Datagram Transport Layer Security (DTLS) ensures end-to-end encryption and safeguards the data confidentiality and integrity during the transmission process. Also, encryption techniques can be used in storage devices for IoT devices to prevent unauthorized access in the event of device theft or compromise [34].

5.3 Patch Management and Regular Software Updates

By regularly updating software upgrades and security patches can be critical in addressing known vulnerabilities and minimize or mitigates the exploitation of these vulnerabilities by the malicious actors. Organizations should integrate comprehensive patch management procedures to swiftly identify and install security patches across their IoT ecosystem and by automating the patch management system it can streamline the installing process in which it ensures timely upgrading without disrupting any operational workflows.

5.4 Network Segmentation and Access Controls

By implementing network segmentation, it can isolate IoT devices from critical systems and divide the network based on criticality of those IoT devices in which it helps in quarantining potential security breaches and limit the impact of compromised devices. Segmenting IoT networks into sub-networks combined with access controls can mitigate lateral movement by malicious actors. With the combination of firewalls, intrusion prevention system (IPS) and network access control (NOC) helps in regulating and monitoring the network traffic in which it enhances overall network security.

5.5 Intrusion Detection/Prevention System (IDS/IPS) and Threat Intelligence

Integrating IPS/IDS helps in detecting and mitigating malicious packets indicative of IoT related threats based on the rules configured. By leveraging threat intelligence gathered from SIEM (Security Information and Event Management), it enhances threat detection abilities and provides real-time monitoring of network traffic, system logs and quick response action against emerging threats within IoT ecosystems [35].

5.6 Employee Awareness Training

Providing proper cybersecurity training and awareness programs to the employees can help in minimizing the human errors done by mistake in which it ultimately assists in thwarting cyber threats effectively and also encourages employees to follow the established security practice and policies like maintaining good password hygiene and works with IoT devices with caution in which it enhances overall security infrastructure within the organization.

5.7 AI and ML-Driven Security Solutions

Integrating Artificial Intelligence (AI) and Machine Learning (ML) capabilities into IoT security frameworks can significantly enhance threat detection and incident response. By implementing AI and ML algorithms to analyse vast amounts of data generated by IoT devices can reveal anomalous patterns, identify emerging threats and adapt their defence mechanisms in real-time [36, 37]. The integration of AI and ML-driven security solutions into IoT environments offers several key benefits:

Anomaly Detection. AI and ML algorithms has ability to learn patterns and behaviours of IoT data and can detect potential security threats and any malicious behaviours like unusual network traffic patterns, unauthorize login attempts and triggered alerts by monitoring and analysing the huge number of data and logs generated by IoT devices.

Predictive Analytics. AI and ML models can predict emerging security threats and vulnerabilities based on historical data and background information by analysing past security events and logs. These models can predict threats and help in inventing proactive security measures to mitigate this threat before they start affecting organisations. It empowers organizations to implement a pre-emptive approach to IoT security enabling them to stay one step ahead of evolving threats and exploitation of vulnerabilities.

Behavioural Profiling. AI and ML helps in creating behavioural profiles for IoT devices in which it allows organizations to establish patterns of device activity and identify anomalies from normal behaviour thorough continuously monitoring device behaviour and comparing it against established profiles, model can detect unauthorized access, data exfiltration or malicious activity which are indicative of compromised IoT devices.

Adaptive Response. AI and ML integrated security solutions enable quick response mechanisms that dynamically adjust security controls based on evolving threat landscapes and changing environmental conditions through automation techniques in isolating compromised devices, blocking malicious traffic and updating access control policies in real-time. It can mitigate security incidents and minimize their impact on IoT ecosystems.

Threat Intelligence Integration. AI and ML algorithms integrate external threat intelligence and security information to enhance threat detection and quick response capabilities. It analyses a wide range of threat intelligence sources including known malware signatures, suspicious IP addresses and emerging attack vectors, these algorithms can compare security events and prioritize incident response actions based on the severity level of the threats.

The implementation of these mitigation strategies is essential for an organization in addressing the evolving emerging threats that can affect the whole IoT ecosystem. With the combination of access controls, security frameworks and threat management practices can enhance the fortification of the IoT ecosystem against cyber threats and also ensures a secure foundation for the future IoT innovations and its applications.

6 Case Studies

Case studies are vital for understanding the real-world implications and challenges in the domain of IoT security. Below, three distinct case studies based on true events that had happened in the IoT ecosystem. Each case study dives into a specific incident, analysing vulnerabilities exploitation, impact on affected systems and mitigation efforts taken in that case studies.

6.1 Mirai Botnet Attack

The Mirai botnet attack occurred in 2016, it is an important event in the domain of IoT security, highlighting the vulnerabilities essential in connected devices. In this attack default usernames and passwords are exploited to compromise a vast number of IoT devices and harness them into a massive botnet network capable of launching Distributed Denial of Service (DDoS) attacks. Mirai primarily targeted IP cameras, routers and CCTV in which their computing power was used to overwhelm targeted servers and networks with huge amounts of malicious traffic.

Mirai's modus operandi was relatively simple but it was highly effective. It used brute-force attacks to guess login credentials commonly used default usernames and passwords that are hardcoded into many IoT devices. Once a device was compromised, it would become part of the Mirai botnet and awaiting commands from its command and control (C&C) server.

Impact. It disrupted numerous critical services including important websites and online services by rendering them inaccessible to users. Notable targets included a major Domain Name System (DNS) provider in which it experienced severe downtime as a result of the DDoS attack. This incident highlighted the potential for IoT devices to be commanded in mass and used as a devastative weapon in large-scale cyber-attacks. The Mirai botnet attack had far-reaching consequences causing widespread disruption to internet services and IT infrastructure. Some of the major impacts include:

DDoS Attacks. Mirai Botnet was primarily used to launch Distributed Denial of Service (DDoS) attacks by flooding targeted websites and online services with overwhelming amounts of traffic. These attacks disrupt the targeted services and make them inaccessible to legitimate users causing financial losses.

Internet Outages. The strength of Mirai botnet powered DDoS attacks resulted in internet outages affecting millions of users worldwide. It affected high profiled websites including Netflix, Twitter, GitHub, etc. experienced extended periods of downtime as the result of the DDoS attacks done by the botnet.

Infrastructure Vulnerabilities. The ability of Mirai botnet is to compromise critical infrastructure devices such as routers, network switches and other networking devices in which it raised concerns about the security of internet infrastructure. The attack reveals the need for enhanced defence measures to safeguard against similar threats in the future.

Lessons Learned. The Mirai botnet attack highlighted the importance of securing the IoT devices and its network against exploitation. It revealed the threats and risks that exist due to weak credentials and focused on the need for advanced authentication mechanisms in IoT ecosystems. The incident encouraged awareness regarding the security consequences of IoT expansion and it encouraged efforts to enhance IoT device security standards.

Mitigation Strategies. To mitigate the similar types of cyber-attacks in the future, organizations must prioritize security in IoT devices through implementing defensive measures like using strong login credentials, access controls and regularly updating device firmware to address known vulnerabilities and doing network segmentation to isolate IoT devices from critical systems. The development of intrusion detection and prevention systems helps to detect malicious behaviour that indicates botnet activity can aid in pre-emptively identifying and mitigating these potential attacks. The Mirai botnet attack highlighted the importance of proactive cybersecurity measures to mitigate the threats of similar incidents. Some key defence strategies for mitigating the impact of Mirai botnet like attacks include:

Credential Management. Employees should adopt strong security policies and avoid using easily guessable credentials. By regularly changing passwords on a time-to-time basis and integrating MFA (Multi-Factor Authentication) can minimize the risk of unauthorized access.

Device Hardening. IoT device manufacturers should prioritize security in the product design and development by implementing robust security features like secure boot, remote update and firmware validation abilities to minimize vulnerabilities and mitigate cyber threats.

Network Segmentation. Dividing IoT devices networks into sub networks from main network infrastructure through network segmentation can contain the spread of malware and limit the impact of compromised devices of critical services.

Anomaly Detection. Implementing network monitoring tools and anomaly detection models can help identify malicious patterns of traffic that indicates botnet activity and it allows organizations to respond swiftly to potential threats.

This case study serves as a reminder for the risks, threats and vulnerabilities comes with IoT devices and understanding the attack methodology and developing effective mitigation strategies can protect the IoT ecosystem and ensures integrity of the inter connected devices against similar cyber threats in future.

6.2 *Stuxnet Worm Attack*

The Stuxnet worm discovered in 2010, was targeting industrial infrastructure and its IoT devices particularly the SCADA (Supervisory Control and Data Acquisition) system used in critical infrastructure like nuclear facilities, etc. Stuxnet particularly used to sabotage Iran's nuclear enhancement program by compromising PLCs (Programmable Logic Controllers) used in centrifuge operations in the nuclear facility.

Impact. Stuxnet worm attack showed the potential of the sophisticated cyber weapons that can inflict physical damage by exploiting the vulnerabilities in ICS (Industrial Control Systems). It first compromised the PLCs and then it caused centrifuges to malfunction resulting in significant operational disruption of Iran's nuclear facility. The incident marked as critical cyber threats that closes the limits between traditional cyber espionage and kinetic warfare. Some major impacts include:

Physical Damage to Critical Infrastructure. Stuxnet particularly targeted ICS components including PLCs and other IoT components that are used in centrifuges for uranium enhancement. Stuxnet infiltrated this system and then manipulated their operation causing disruption and caused physical damage. This demonstrated the potential for cyber-attacks that can disrupt critical infrastructure and industrial processes leading to significant operational and financial losses.

Disruption of Nuclear Program. Stuxnet's primary target was Iran's nuclear facilities where it destroyed thousands of centrifuges. The attack significantly disrupted Iran's nuclear program, delaying its progress and causing extensive setbacks. This highlighted the use of cyber-attacks as a tool for espionage and geopolitical agendas with the implications for international relations and its security.

Awareness of ICS Vulnerabilities. After Stuxnet worm attack, it raises the awareness of the vulnerabilities that exist in ICS components. It demonstrated the potential for malicious actors to exploit security weaknesses in industrial control systems emphasizing the need for enhanced cybersecurity defensive measures in critical infrastructures.

Lessons Learned. The Stuxnet attack revealed the lack of isolated systems that can safeguard the critical infrastructure from these types of cyber threats. It emphasized the need for robust and advanced defence mechanisms that are capable of detecting and mitigating these advanced malwares targeting ICS. Stuxnet served as a wake-up call regarding the potential consequences of state-sponsored cyber operations prompting increased caution among governments and organizations worldwide.

Mitigation Strategies. In order to enhance the defences against Stuxnet-like threats in future, organizations should integrate a multi-layered defence approach encompassing network segmentation, intrusion detection and prevention systems and role-based access controls. Regularly doing security assessments and vulnerability scans on a weekly basis can help identify and fix potential vulnerabilities within industrial control systems. Identifying and mitigating the threats and risks of Stuxnet-like attacks requires a comprehensive and proactive approach to cybersecurity in industrial environments. Key mitigation strategies include:

Patch Management. Timely patching of software vulnerabilities and loopholes are essential for preventing malware exploitation like Stuxnet. Organizations must prioritize patch distribution for ICS components and regularly update firmware and software to address known vulnerabilities.

Network Segmentation. Dividing ICS networks from external networks into isolated sub networks and implementing strict access controls can limit the spread of malware within industrial IoT components. This reduces the attack surface and minimizes the impact of cyber-attacks on critical infrastructure.

Intrusion Prevention Systems (IPS). Implementing IPS mechanism allows an organization to detect and block suspicious activity targeting ICS devices and its networks. Most organisations prefer IPS over IDS because Intrusion detection can only detect suspicious behaviour that are indicators of cyber-attacks while intrusion prevention mechanisms can automatically respond to and mitigate threats in real-time.

Security Awareness Training. Training and educating the employees about the risks of social engineering attacks, phishing attacks and malware infections are crucial for enhancing cybersecurity infrastructure within an organization. Employees should be trained to recognize and report suspicious activity while following best security practices and obey established protocols for incident response.

Cybersecurity Standards and Best Practices. Following the cybersecurity standards like the ISA/IEC 62443 series provide guidelines for implementing robust cybersecurity controls in industrial automation and control systems and best practices based on industries can help organizations mitigate the risk of Stuxnet-like attacks.

Incident Response Planning. Developing and testing incident response plans allows organizations to effectively respond to cyber-attacks targeting ICS components. A well-defined incident response plan must include procedures for identifying, containing, eradicating and recovering from security events or incidents and minimizing disruption to critical operations.

6.3 Healthcare IoT Breaches

The use of IoT devices in healthcare sectors has increased the efficiency of healthcare workflows but it has also introduced unparalleled security challenges and threats. Healthcare IoT breaches ranging from unauthorized access into medical IoT devices to data exfiltration from networks poses a significant risk to patient safety and privacy of their healthcare data.

Impact. Healthcare IoT breaches can have comprehensive consequences, threatening confidentiality of patient data, disrupting medical operations and compromising the integrity of healthcare data. Cases of unauthorized access to medical devices, like infusion pumps, etc. raises concerns about the manipulation of treatment parameters and potentially endangering patient lives. The theft and leaks of sensitive healthcare information can lead to identity theft, insurance fraud, trust and reputational damage for healthcare providers.

Lessons Learned. The interconnected medical devices require robust defence mechanisms to protect against unauthorized access and data breaches. The merging of healthcare and IT (Information Technology) dictates interdisciplinary collaboration between medical professionals, cybersecurity experts and device manufacturers in order to develop and implement comprehensive security protocols.

Mitigation Strategies. Mitigating the risks and threats associated with healthcare IoT devices requires a novel approach encompassing technical controls, security policy frameworks and ongoing risk assessment. Implementing encryption protocols to secure data transmission between IoT devices and backend servers can help safeguard the confidential information from man-in-the-middle attacks. Healthcare organizations must establish clear guidelines for the secure deployment and maintenance of IoT devices including regular security updates and vulnerability management. Training medical staff about cybersecurity best practices and incident response procedures can enhance preparedness and flexibility during any events of emerging threats.

These case studies of the Mirai botnet attack, Stuxnet worm targeting ICS and healthcare IoT breaches highlights the diverse range of cyber threats confronting IoT ecosystems across various domains. After analysing and studying these case studies an organization can gather valuable insights about the evolving threat landscape and implement effective mitigation strategies to strengthen their defences and secure the security and integrity of connected systems.

7 Conclusion and Future Scope

The integration of Internet of Things (IoT) devices has increased the connectivity and innovation across diverse domains and revolutionizing the industries ranging from healthcare and manufacturing to transportation and smart cities. However, alongside the numerous benefits offered by IoT technology, there exists a parallel landscape of cybersecurity threats, risks and challenges. Through the analysis of case studies of the Mirai botnet attack, Stuxnet worm targeting ICS and healthcare IoT breaches, the vulnerabilities that were present in interconnected IoT ecosystems poses a significant threat to CIA triad of the data. The intelligence gathered from these case studies highlights the critical vulnerabilities of the organizations and implement various advance mitigation strategies to prioritize IoT security and defence in order to mitigate the evolving cyber threats. Integrating proactive and multi-layered cybersecurity approaches encompassing access controls, security policy frameworks and collaborative partnerships with stakeholders can enhance the defences of IoT ecosystems and protect against vulnerability exploitation and breaches. Nurturing cybersecurity awareness and education in employees is crucial in allowing individuals and organizations to identify and respond effectively to emerging threats in the IoT landscape.

Looking ahead into the future of IoT security presents both challenges and opportunities for innovation and advancement as IoT ecosystems continue to expand and evolve fuelled by advancements in edge computing, AI and ML and 5G connectivity. Future research in the domain of IoT security may explore novel approaches like the integration of blockchain technology to enhance data integrity and decentralized authentication mechanisms. Additionally, the development of standardized security protocols and certification frameworks personalized specifically for IoT devices can enable trusts across heterogeneous IoT ecosystems. Furthermore, interdisciplinary collaboration between cybersecurity experts, IoT manufacturers and end-users will be the driving force to address the multilayered challenges in the IoT cyber threat landscape. With the development of an ecosystem of innovation and best security practices helps in securing the future for IoT ecosystems ensuring the continued transformative potential of connected technologies while mitigating the IoT cyber threats.

References

1. Whig, P., Velu, A., Nadikattu, R.R., Alkali, Y.J.: Role of AI and IoT in intelligent transportation. In: Artificial Intelligence for Future Intelligent Transportation, pp. 199–220. Apple Academic Press (2024)
2. Huangpeng, Q., Yahya, R.O.: Distributed IoT services placement in fog environment using optimization-based evolutionary approaches. Expert Syst. Appl. **237**, 121501 (2024)
3. Thakkar, A., Lohiya, R.: A review on machine learning and deep learning perspectives of IDS for IoT: recent updates, security issues, and challenges. Archiv. Computat. Methods Eng. **28**, 3211–3243 (2020)
4. Costa, K., Papa, J., Lisboa, C., Munoz, R., Albuquerque, V.: Internet of Things: a survey on machine learning-based intrusion detection approaches. Comput. Netw. **151** (2019). https://doi.org/10.1016/j.comnet.2019.01.023
5. Ashraf, J., Moustafa, N., Khurshid, H., Debie, E., Haider, W., Wahab, A.: A review of intrusion detection systems using machine and deep learning in Internet of Things: challenges, solutions and future directions. Electronics **9** (2020). https://doi.org/10.3390/electronics9071177
6. Ahmad, R., Alsmadi, I.: Machine learning approaches to IoT security: a systematic literature review. Internet of Things. **14**, 100365 (2021). https://doi.org/10.1016/j.iot.2021.100365
7. Hussain, F., Hussain, R., Hassan, S.A., Hossain, E.: Machine learning in IoT security: current solutions and future challenges. IEEE Commun. Surv. Tutorials **22**(3), 1686–1721 (2020)
8. Liang, F., Hatcher, W.G., Liao, W., Gao, W., Yu, W.: Machine learning for security and the internet of things: the good, the bad, and the ugly. IEEE Access **7**, 158 126–158 147 (2019)
9. Chaabouni, N., Mosbah, M., Zemmari, A., Sauvignac, C., Faruki, P.: Network intrusion detection for IoT security based on learning techniques. IEEE Commun. Surv. Tutorials **21**(3), 2671–2701 (2019)
10. Tahsien, S., Karimipour, H., Spachos, P.: Machine learning based solutions for security of Internet of Things (IoT): a survey. J. Netw. Comput. Appl. **161**, 102630 (2020). https://doi.org/10.1016/j.jnca.2020.102630
11. Hui, W., Han, H., Wang, X., Sun, S.: Research on artificial intelligence enhancing Internet of Things security: a survey. IEEE Access, pp. 1–1 (2020). https://doi.org/10.1109/ACCESS.2020.3018170
12. Bibi, N., Iqbal, F., Akhtar, S., Anwar, R., Bibi, S.: A survey of application layer protocols of Internet of Things, **21**, 301–311 (2021). https://doi.org/10.22937/IJCSNS.2021.21.11.41
13. Aamer, S.: Internet of Things security threats and key technologies. J. Discrete Math. Sci. Crypt. **24**, 1–7 (2021). https://doi.org/10.1080/09720529.2021.1957189
14. Gerodimos, A., Maglaras, L., Ferrag, M.A., Ayres, N., Kantzavelou, I.: IoT: communication protocols and security threats. Internet of Things Cyber-Phys. Syst. **3**, 1–13 (2023)

15. Tsiknas, K., Taketzis, D., Demertzis, K., Skianis, C.: Cyber threats to industrial IoT: A survey on attacks and countermeasures. Informatics **2**, 163–188 (2021). https://doi.org/10.3390/iot2010009
16. Ahmad, I., Niazy, M.S., Ziar, R.A., Khan, S.: Survey on IOT: Security threats and applications. J. Robot. Control. (JRC) **2**, 38–49 (2021)
17. Wheelus, C., Zhu, X.: IoT network security: Threats, risks, and a data-driven defense framework. IoT **1**(2), 259–285 (2020)
18. Borcherding, A., Feldmann, L., Karch, M., Meshram, A., Beyerer, J.: Towards a better understanding of machine learning based network intrusion detection systems in Industrial Networks. In: Proceedings of the 8th International Conference on Information Systems Security and Privacy, Online, 9–11 February (2022)
19. Le, K.-H., Nguyen, M.-H., Tran, T.-D., Tran, N.-D.: IMIDS: an intelligent intrusion detection system against cyber threats in IoT. Electronics **11**, 524 (2022). https://doi.org/10.3390/electronics11040524
20. Mu, X., Antwi-Afari, M.F.: The applications of Internet of Things (IoT) in industrial management: a science mapping review. Int. J. Prod. Res. **62**(5), 1928–1952 (2024)
21. Sakthivel, C.R., Sundararajan, S., Periyasamy, P., Shathik, A., Fenitha, J.R.: A novel performance enhancement of real-time iot applications using big data analytics. In: AIP Conference Proceedings, vol. 2742, No. 1. AIP Publishing (2024)
22. Ahmad, A.Y.B., William, P., Uike, D., Murgai, A., Bajaj, K.K., Deepak, A., Shrivastava, A.: Framework for sustainable energy management using smart grid panels integrated with machine learning and IOT based approach. Int. J. Intell. Syst. Appl. Eng. **12**(2s), 581–590 (2024)
23. Priyadarshini, I.: Anomaly detection of IoT cyberattacks in smart cities using federated learning and split learning. Big Data Cogn. Comput. **8**(3), 21 (2024)
24. Quy, V.K., Nguyen, D.C., Van Anh, D., Quy, N.M.: Federated learning for green and sustainable 6G IIoT applications. Internet of Things **25**, 101061 (2024)
25. Nadhan, A.S., Jacob, I.J.: Enhancing healthcare security in the digital era: safeguarding medical images with lightweight cryptographic techniques in IoT healthcare applications. Biomed. Signal Process. Control **88**, 105511 (2024)
26. Kumbhare, S., Ubale, S.A., Dharmale, G., Mhala, N., Gandhewar, N.: IoT-enabled agricultural waste management for sustainable energy generation. Int. J. Intell. Syst. Appl. Eng. **12**(13s), 477–482 (2024)
27. Yazdinejad, A., Kazemi, M., Parizi, R.M., Dehghantanha, A., Karimipour, H.: An ensemble deep learning model for cyber threat hunting in industrial internet of things. Digit. Commun. Netw. **9**(1), 101–110 (2023)
28. Satpathy, S., Pradhan, S.K., Ray, B.B.: A digital investigation tool based on data fusion in management of cyber security systems. Int. J. Inf. Technol. Knowl. Manag. **2**(2), 561–565 (2010)
29. Pattnaik, L.M., Swain, P.K., Satpathy, S., Panda, A.N.: Cloud DDoS attack detection model with data fusion & machine learning classifiers. EAI Endorsed Trans. Scalable Inform. Syst. **10**(6) (2023)
30. Satpathy, S., Swain, P.K., Mohanty, S.N., Basa, S.S.: Enhancing Security: Federated Learning against Man-In-The-Middle Threats with Gradient Boosting Machines and LSTM. In: 2024 IEEE International Conference on Advanced Video and Signal Based Surveillance (AVSS), pp. 1–8. IEEE (2024, July)
31. Walia, G., Kumar, M., Gill, S.S.: AI-empowered fog/edge resource management for IoT applications: a comprehensive review, research challenges and future perspectives. IEEE Communications Surveys & Tutorials, pp. 1–1 (2023). https://doi.org/10.1109/COMST.2023.3338015
32. Usoh, M., Asuquo, P., Ozuomba, S., Stephen, B., Inyang, U.: A hybrid machine learning model for detecting cybersecurity threats in IoT applications. Int. J. Inf. Technol. **15**(6), 3359–3370 (2023)
33. Lepping, A., Pham, H.M., Mons, L., Rueb, B., Grulich, P.M., Chaudhary, A., Markl, V.: Showcasing data management challenges for future IoT applications with NebulaStream. Proceed. VLDB Endowment **16**(12), 3930–3933. (2023)

34. Cherbal, S., Zier, A., Hebal, S., Louail, L., Annane, B.: Security in internet of things: a review on approaches based on blockchain, machine learning, cryptography, and quantum computing. J. Supercomput. **80**(3), 3738–3816 (2024)
35. Sharma, B., Sharma, L., Lal, C., Roy, S.: Explainable artificial intelligence for intrusion detection in IoT networks: a deep learning-based approach. Expert Syst. Appl. **238**, 121751 (2024)
36. Dhinakaran, D., Sankar, S.M., Selvaraj, D., Raja, S.E.: Privacy-preserving data in IoT-based cloud systems: a comprehensive survey with AI integration (2024). arXiv preprint arXiv:2401.00794
37. Ayasrah, F.T.M., Abu-Alnadi, H.J., Al-Said, K., Shrivastava, G., Mohan, G.K., Muniyandy, E., Chandra, U.: IoT integration for machine learning system using big data processing. Int. J. Intell. Syst. Appl. Eng. **12**(14s), 591–599 (2024)

Internet of Things and OpenCV-Based Smart Posture Recognition Chair

Swarna Prabha Jena, Mangaldeep Chakraborty, Jayanta Mondal, Shaikh Habibur Rehaman, Pratik Ranjan Dash, Bijay Kumar Paikaray, and Sujata Chakravarty

Abstract The Internet of Things (IoT) technology is expanding quickly and makes it possible to connect actual devices to the Internet. In our daily lives, we spend a lot of time sitting in incorrect postures, which over time leads to a host of health issues such back discomfort, stiffness, cervical pain, etc. Here's a creative way to use IoT and OpenCV to create a smart posture detection chair that will solve the posture problem. In order to address these health concerns and aid us with our posture, this approach seeks to design a smart chair. We used various sensor interfaces to gather and process real-time data in order to identify posture. We were able to get a satisfactory result that indicates if the position is good or terrible in the Boolean value. Here, 86% accuracy was achieved in classifying the good and bad posture using the machine learning model YOLOV5.

Keywords Smart chair · Smart posture · Real-time data · Sensors · Deep learning · Object detection · Video processing · YOLOV5

S. P. Jena (✉)
Department of ECE, Centurion University of Technology and Management, Bhubaneswar, Odisha, India
e-mail: prabha.jena@gmail.com

M. Chakraborty
Department of Electronics and Telecommunications Engineering, Bishnupur Public Institute of Engineering (Polytechnic College), Bishnupur, West Bengal, India
e-mail: mangaldeepchakraborty73914@gmail.com

J. Mondal
School of Computer Engineering, KIIT Deemed to Be University, Bhubaneswar, India
e-mail: jayanta.mondalfcs@kiit.ac.in

B. K. Paikaray
Centre for Data Science, Department of Computer Science and Engineering, Siksha 'O' Anusandhan (Deemed to be University), Bhubaneswar, Odisha, India
e-mail: bijaypaikaray87@gmail.com

S. H. Rehaman · P. R. Dash · S. Chakravarty
Department of CSE, Centurion University of Technology and Management, Bhubaneswar, Odisha, India

S. N. Mohanty et al. (eds.), *Explainable IoT Applications: A Demystification*, Information Systems Engineering and Management 21,
https://doi.org/10.1007/978-3-031-74885-1_28

1 Introduction

IoT and ML have recently been widely used in human poster recognition systems. This system demonstrates effective use in computer–human interaction (HCI). Generally speaking, a human posture is crucial to recognize. The way we hold our bodies when we sit, stand, or lie down is known as our posture. It's essential to have proper posture for general wellness. It offers numerous advantages, including lowered back discomfort and boosted confidence and energy.

It has been used in a variety of industries, including senior health care, HCI, environmental awareness, and physical training. To recognize posture, one might use a variety of techniques. The conventional approaches are sensor-based and computer vision-based. Both strategies offer benefits and drawbacks based on the quantity, size, and expense. But both have issues with robustness in challenging environments, illumination and backdrop, and identification accuracy. Nonetheless, most current research concentrates on action identification instead of posture recognition, with a growing emphasis on routine actions. Furthermore, there are presently few datasets and posture recognition-based algorithms available. To achieve more accurate posture recognition, this research suggests a human posture recognition algorithm using real-time datasets encompassing a variety of postures.

After reviewing roughly 20 articles, we abstracted the information as follows: [1–3]. Here, the authors began utilizing pressor sensors to measure pressure to recognize posture. Even though the number of pressure sensors used varied [4–6]. Pressure sensors are used for posture recognition, although a distinct textile pressure sensor is employed [2, 4, 7–10]. They employed machine learning methods to determine if the posture was good or poor in each case we looked at. For classification, they employed various machine learning techniques such as ANN, CNN, Logistic Regression, Naïve Bayes, etc. Additionally, the outcome is more accurate [7]. We investigated a novel type of sensor called a flex sensor, which is composed of plastic flake and conductive particles in a polymer ink. Depending on the individual's bending technique [11]. A combined sensor system was employed [12, 13]. We investigate piezoelectric and piezoresistive sensors' applications, the two piezo sensors [14]. There is a wearable gadget with a notch sensor. Overall, these experiments provided us with information on the cost and accuracy of each model. Thus, we attempted to enhance the posture identification technique by employing a piezoelectric sensor to lower our costs and boost model accuracy. We tested our model using real-world cases and achieved a high accuracy of 98%.

1.1 Comparison Sheet

This table explains the different papers we have read. It includes the paper title, the published year, the sensors used, and their quantity; the microcontroller used if the connection is wired or wireless; the ML model used; their result, including accuracy percentage; and their estimated costs.

1.2 Research Gap

We have researched through many papers and acknowledged that most of the documents came to have some or other drawbacks, which first of all include the overall cost of the device (chair), which was high for maximum cases, also not implementing an ML Model [20, 21] was another drawback in many instances, many models were using a heavily wired system, and some models did not use any microcontroller [22, 23]. Also, some models lag because they do not have a notifying application or system (Table 1).

1.3 Contribution

Therefore, we are attempting to overcome these shortcomings in this study. To lower the cost of the real-world model, we looked over every component previously utilized for the model, compared their costs and performance, and ultimately selected the most affordable and effective product. We experimented with a variety of platforms for data storage, including AWS, Influx DB, and Google Sheets, using app scripts. Ultimately, we settled on MySQL and deployed a deep machine-learning model, YOLOv5, for real-time object detection. We also offer a simply accessible, satisfactorily maintained notification webpage for chair users. The overall arrangement of the work is as follows: The technological features of the smart chair and a comparison of earlier studies are highlighted in Section I. Section II discusses the suggested system, Section III presents the findings and conclusions, and section.

2 Proposed Methodology

The flow chart for this study is divided into three sections shown in Fig. 1: Sensing, Recognition, and Output. The flow chart for this study is shown in Fig. 1 and is divided into three sections: Sensing, Recognition, and Output. First, In the Sensing section, the user sits on a chair where Piezoelectric and ultrasonic Sensors are implemented. We obtain all the user's live data from these sensors, which goes to the microcontroller. Then, in the Recognition section, our code analyses the posture from a combination of both sensors. It matches the posture level, orientation and, finally, posture composition. The object detection and classification model is implemented in the recognition section to detect good and bad posture through the camera. Then, all the sensors' data and machine learning model output are combined, giving

Table 1 Comparison sheet of the previous works

Ref	Sensor	Quantity	Micro-controller	Connection	Model	Results	Cost
[1]	Pressure sensor sheet	64 Pressure sensor in a sheet	No description	Wired	No description	93%	2700*
[2]	Pressure sensor	9	No description	Bluetooth	Random forest classifier, decision tree classifier	98.93%	3500*
[4]	Textile sensor	1	Arduino	Bluetooth	No description	93.30%	–
[14]	Load cell	4	Arduino	Wired	SVM	97.20%	1700*
[15]	FSR	6	STM32F411RARM cortex-M4 core	Bluetooth low energy (BLE)	ANN	96.40%	7000*
[16]	A 52 by 44 piezoresistive sensor array	1	No description	Wired	ANN	0.0098 MSE	
[7]	Flex sensors	6	FPGA	No description	ANN	97.43%	3100*
[3]	8 by 8 pressure sensor array	1	Arduino Mega 2560	Wired	ANN	99%	2200*
[11]	Force sensing resistors, infrared reflective sensors	6 each	No description	Wired	KNN classifier	0.92	3500*
[8]	Pressure sensor	4	No description	Wired	LDA classifier	0.783	1000*
[17]	Force sensor	6	Arduino	Wired	Not defined	0.99	1000*
[12]	Piezoresistive pressure sensor	9	ATmega32u4	Wired	Not defined	96%	1000*
[9]	Pressure sensor, ultrasonic sensor	2	Arduino	Wired	CNN	96%	1000*

(continued)

Table 1 (continued)

Ref	Sensor	Quantity	Micro-controller	Connection	Model	Results	Cost
[18]	Flexible sensor	4	Arduino	Bluetooth	KNN	93.30%	1600*
[5]	Pressure sensor	8	Arduino	Bluetooth	Not defined	96.20%	–
[19]	Kinect (IR depth camera)	1	No description	No connection	PHP and MySQL used not ML		20,000*
[10]	FSR sensor	64	No description	No connection	CNN	97.5%	1500*
[14]	Notch sensor	1	No description	Wired	KNN	98%	
[13]	Piezoresistive sensors	13	No description	Wired	DT, SVMs, KNN classifiers and feed-forward neural networks (NN)	98%	1000*
[6]	Textile sensor	1	Arduino dev	Wired	Used calibration curve, not ML	85.9%	2300*

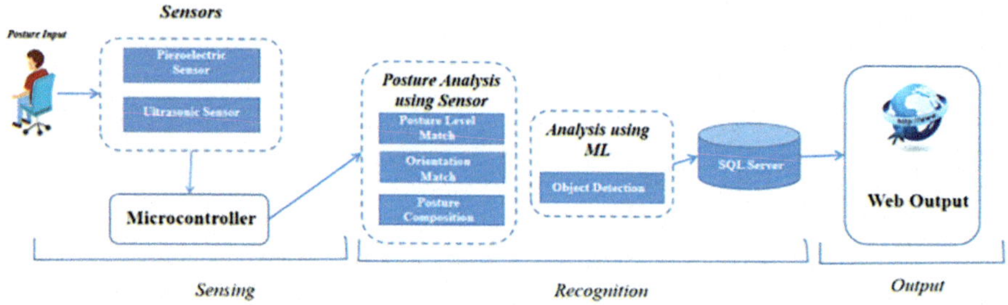

Fig. 1 Shows the block diagram of how the work will process

binary results, such as good or bad posture. Then, all the information is uploaded to the MySQL database using a PHP API. And lastly, in the Output section, all the outputs are shown on the website, including the individual data of sensor in tabular form can also be observed which was previously stored in MySQL database.

2.1 *Hardware Required*

Table 2 explains the specification, quantity and price of the components we used in our model.

Table 2 Cost sheet of the hardware used in the proposed system

S.No.	Components	Specifications	Qty	Price
1	ESP32 Dev Kit	Built-in RF, power amplifier, voltage 3V3, and antenna switches Built-in Bluetooth and Wi-Fi		
2	Piezoelectric sensor	Impedance: $\leq 500\ \Omega$	8	120/—
		Voltage: ≤ 30 Vp-p		
		Operating temperature: $-$ 20 °C ~+ 60 °C		
		Material: Quartz (mostly)		
3	Ultrasonic sensor	Power supply: 3.3 V – 5 V. Operating current: 8 mA. Working frequency: 40 Hz. Ranging distance: 3 cm – 350 cm/3.5 m. Resolution:1 cm\sTrigger Input Pulse width: 10uS TTL. Dimension: 50 mm × 25 mm × 16 mm	1	80/–

2.2 Software Required

Several software packages have been integrated to carry out the task. MySQL is an Oracle relational database management system (RDBMS) built on structured query language (SQL). An organized collection of data is called a database. It could be anything, such as an intuitive shopping list, an image gallery, or a location to store a large amount of data on a business network.

A local host or server verifying clients or sites precede file transfers to a remote web server. On a local PC with the XAMPP server software installed, MYSQL, PHP, Apache, and Perl projects can all be tested in a suitable environment.

The Arduino Software (IDE), also known as the Arduino Integrated Development Environment, has various menus, a message box, a text console, a toolbar with buttons for commonly performed tasks, and a text editor for writing code. It establishes a connection with the Arduino hardware to upload and communicate with software. It is an open-source platform that works with open-source hardware.

TensorFlow is an open-source software package that may be used for various dataflow programming applications. It is a symbolic math library in neural networks and other machine-learning applications. At Google, it's utilized for both production and research.

Anaconda is an open-source software collection that is used in many applications for dataflow programming. Neural networks and other machine learning applications use this symbolic math library. Google makes use of it for both research and production.

Brackets is a web development-focused source code editor. Adobe Inc. made it and is currently being maintained by open-source developers under the MIT License on GitHub. The software is open-source and free. PHP, HTML, CSS, and JavaScript are used in its writing.

2.3 Block Diagram

2.4 Interfacing with Sensors

Interfacing with Piezoelectric Sensor. Piezo sensors are used in many devices and sensors. They are used to convert a physical property, such as pressure or acceleration, into an electrical signal. Piezo sensors are used to sense pressure, acceleration, or strain changes by converting them into electrical charges.

In Fig. 2 the interfacing of piezoelectric sensor is shown. A esp32 devkit microcontroller is used along with a breadboard, a led and jumper wires for connection purpose. A required code is uploaded to microcontroller using Arduino IDE and executed. The glowing LED indicates that the interfacing is successful.

Fig. 2 Piezoelectric sensor

Figure 3 shows the output of the piezoelectric sensor. Each line shows the result in a charge proportional to pressure.

Interfacing with Ultrasonic Sensor. Ultrasonic sensors emit sound waves at a frequency that is too high for human hearing. They then wait for the sound to be reflected before timing it to determine the distance. This is comparable to how radar measures the time elapsed between a radio wave striking an item and returning.

There are four pins in all. (1) VCC: The VCC pin provides power to the sensor. (2) Trig: The trigger pin is an input pin. (3) Echo: The echo pin is an output pin. (4) GND: The system's ground is connected to this pin.

In Fig. 4 the interfacing of ultrasonic sensor is shown. An esp32 devkit microcontroller is used along with a breadboard and jumper wires for connection purpose. A required code is uploaded to microcontroller using Arduino IDE and executed. The glowing LED of the Microcontroller indicates that the interfacing is successful (Fig. 5).

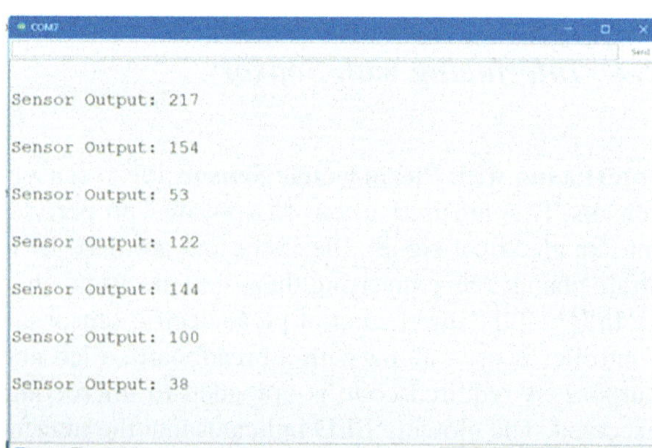

Fig. 3 Serial monitor of piezoelectric sensor

Fig. 4 Interfacing with ultrasonic sensor

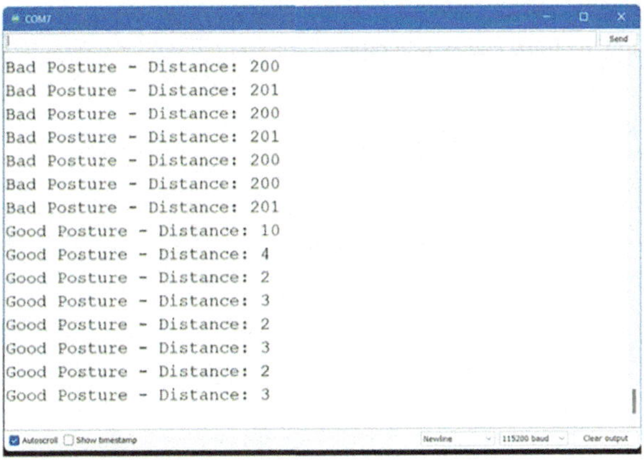

Fig. 5 Serial monitor of ultrasonic

Figure 9 shows the output of the ultrasonic sensor. In each line the output is shown as distance in meters.

2.5 Final Circuit Diagram

Figure 6 shows the connection diagram of our model. In this model six piezoelectric Sensor and one ultrasonic Sensor is used and to control these sensors we have used a micro controller ESP32 (Fig. 7).

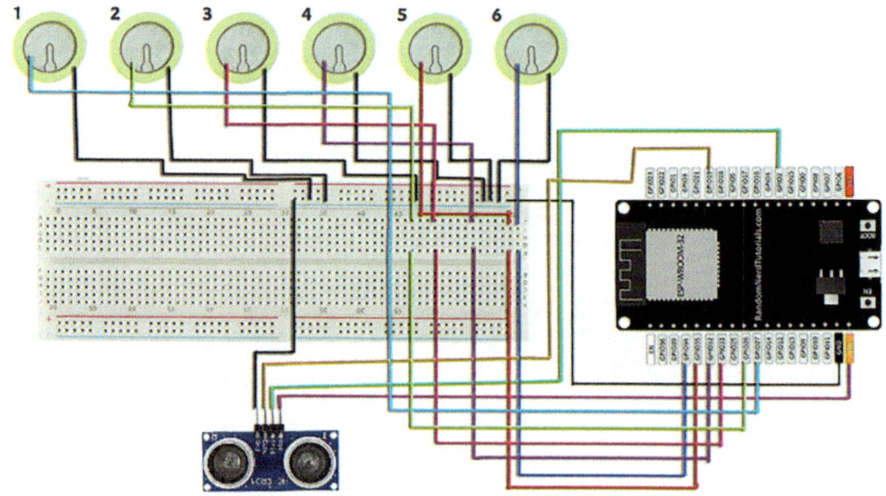

Fig. 6 Final circuit diagram

Fig. 7 Designed model

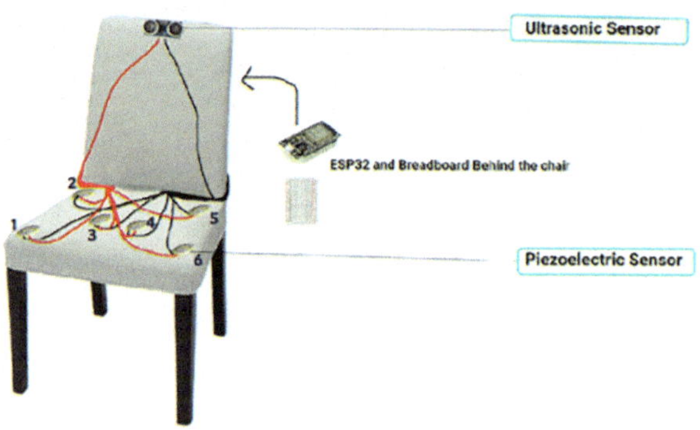

3 Result and Discussion

3.1 Real Model

See Fig. 8.

Figure 12 shows the accurate model developed with the sensors equipped with this chair.

Fig. 8 Front and backside of the model

3.2 *Output*

Case-1. No person is sitting on the chair (Figs. 9 and 10).

When no person is sitting on a chair, the piezoelectric sensor gives zero or garbage value. Here the posture is showing as "Bad Posture" in Serial monitor and 0 in website.

Case-2. Person is sitting with a correct posture (Figs. 11 and 12).

Fig. 9 Output of the sensors at the initial state

```
13:28:39.544 -> Distance: 97
13:28:39.544 -> Sensor 1:0
13:28:39.544 -> Sensor 2:0
13:28:39.544 -> Sensor 3:0
13:28:39.544 -> Sensor 4:144
13:28:39.544 -> Sensor 5:24
13:28:39.544 -> Sensor 6:0
13:28:39.544 -> value_Avg_L:0
13:28:39.544 -> value_Avg_R :56
13:28:39.544 -> Bad Posture -
```

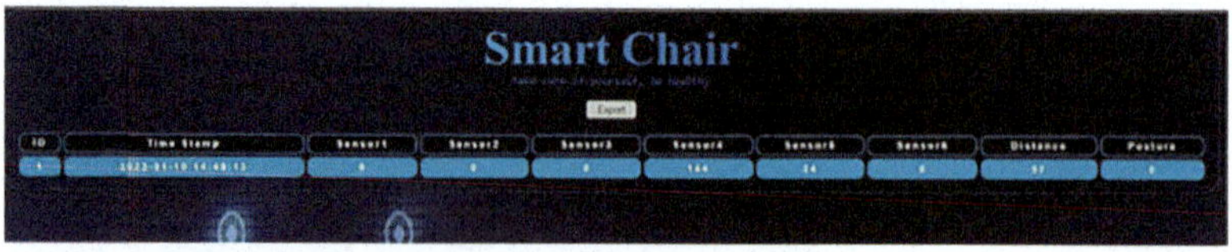

Fig. 10 Output of visualize in website at the initial state

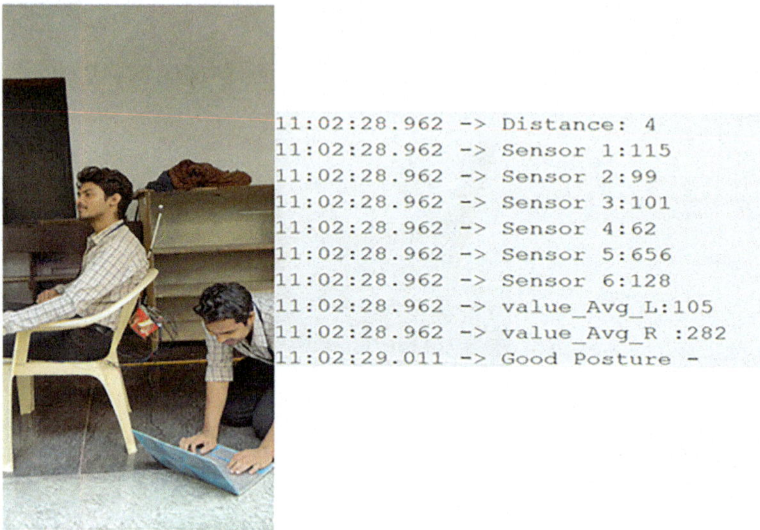

Fig. 11 Output showing the correct posture of a person

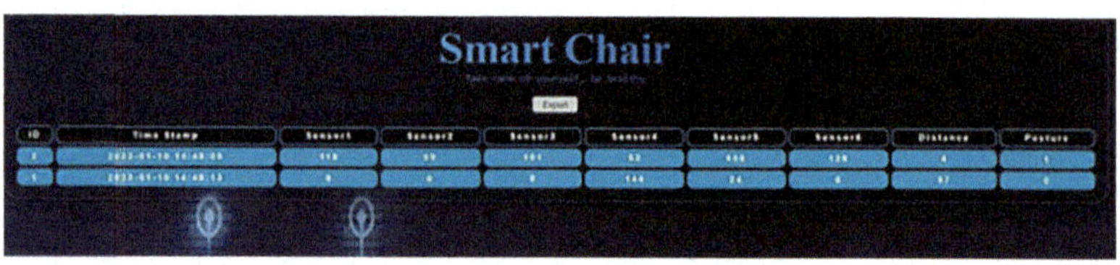

Fig. 12 Output of visualization of correct posture of a person in the web output

When a person is sitting on the chair with a correct posture, the piezoelectric sensor gives value in between 0 to 500. Here the posture is showing as "Good Posture" in Serial monitor and 1 in website.

Case-3. Person is sitting with a incorrect posture (Figs. 13 and 14).

When a person is sitting on the chair with an incorrect posture, the piezoelectric sensor gives value in between 1000 to 4000. Here the posture is showing as "Bad Posture" in Serial monitor and 0 in website.

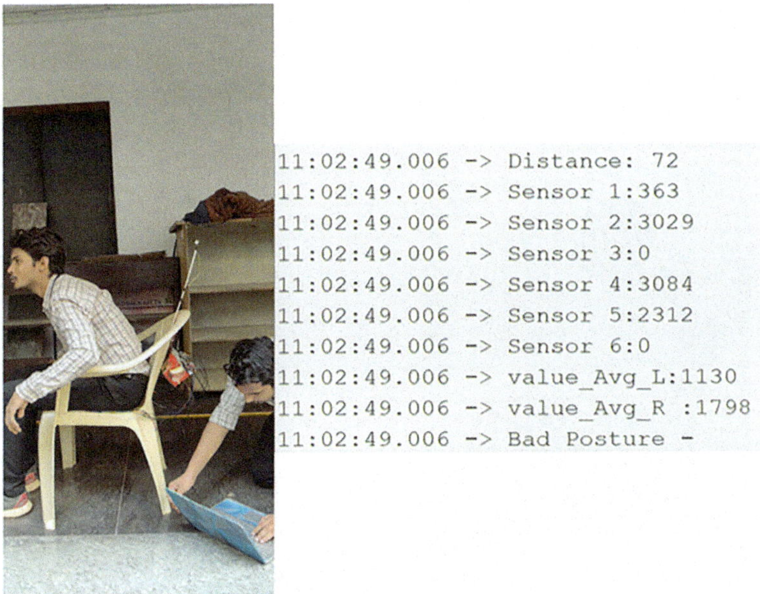

Fig. 13 Output showing the bad posture of a person

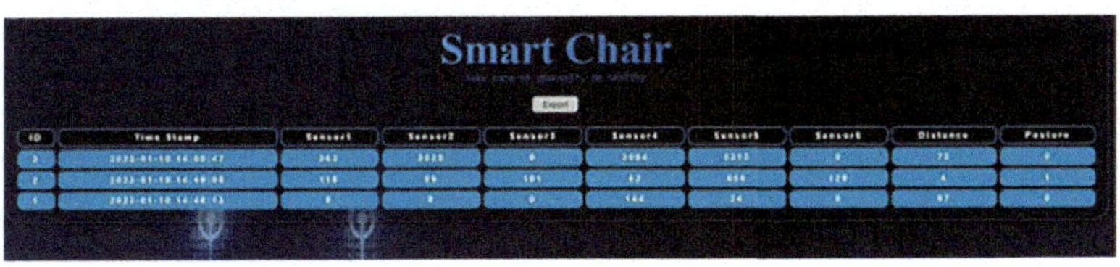

Fig. 14 Output of visualization of bad posture of a person in the web output

Here, the YOLOv5 model recognizes the posture, and the machine learning model uses computer vision to identify good and bad posture. An object recognition technique that effectively finds objects in pictures and videos is the YOLOv5 model. It can handle a range of sizes and aspect ratios and splits the image into grids, predicting bounding boxes and class probabilities for each grid. YOLOv5 is built on a backbone network and has a simplified architecture. It supports a wide range of applications and is simple to use. We record live video, extract individual frames, assign coordinates to each frame, and produce an excel label dataset. Next, in order to achieve better results, we processed various data cleaning procedures, including light correction and noise reduction. Following the entire procedure, we obtained the annotated data, which we then used to train the posture recognition model.

Figure 15 [13] represents the good position of a person, through the camera model classified as good posture, same as in Fig. 16 [6] represents the bad posture

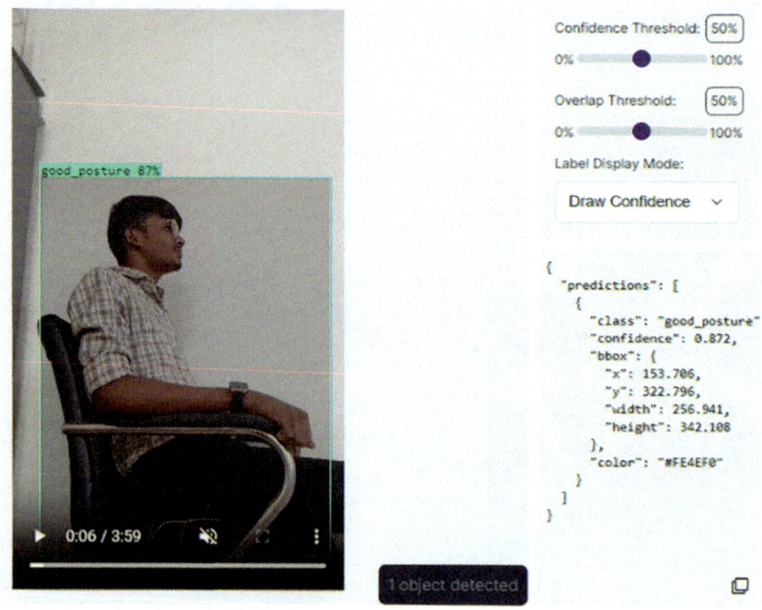

Fig. 15 Good posture classified by machine learning model

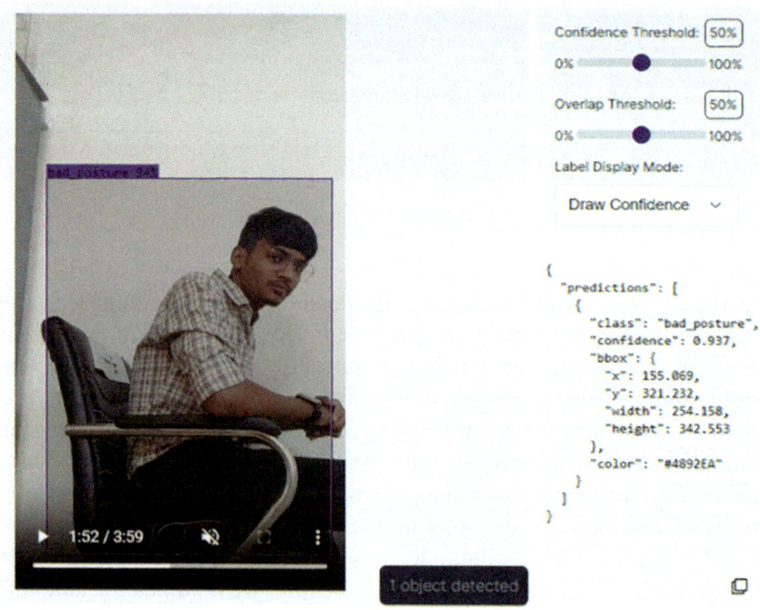

Fig. 16 Bad posture classified by machine learning model

4 Conclusion

With the help of IoT and OpenCV, a model has been developed to classify an individual's posture. This work provides a learning platform where IoT hardware devices, their features, efficiency, capacity etc. are covered. We surfed the Internet, read research papers, and compared them to choose the best components for our real-world model. Overall, the machine learning model was built successfully with an accuracy of 86%, which showed us if our posture was good or bad in our website, which we had already created. Furthermore, in the future, we are trying to implement this model for commercializing purposes in IT sectors, where it will benefit office workers who constantly sit in front of computers.

References

1. Roh, J., Park, H.J., Lee, K.J., Hyeong, J., Kim, S., Lee, B.: Sitting posture monitoring system based on a low-cost load cell using machine learning. Sensors **18**(1), 208 (2018)
2. Anwary, A.R., Cetinkaya, D., Vassallo, M., Bouchachia, H.: Smart-cover: a real time sitting posture monitoring system. Sens. Actuators, A **317**, 112451 (2021)
3. Tlili, F., Haddad, R., Bouallegue, R., Mezghani, N.: A real-time posture monitoring system towards bad posture detection. Wirel. Pers. Commun. **120**(2), 1207–1227 (2021)
4. Ran, X., Wang, C., Xiao, Y., Gao, X., Zhu, Z., Chen, B.: A portable sitting posture monitoring system based on a pressure sensor array and machine learning. Sens. Actuators, A **331**, 112900 (2021)
5. Martinaitis, A., Daunoraviciene, K.: Low cost self-made pressure distribution sensors for ergonomic chair: are they suitable for posture monitoring? Technol. Health Care **26**(S2), 655–663 (2018)
6. Kulikajevas, A., Maskeliunas, R., Damaševičius, R.: Detection of sitting posture using hierarchical image composition and deep learning. Peer J. Comput. Sci. **7**, e442 (2021)
7. Cha, Y., Nam, K., Kim, D.: Patient posture monitoring system based on flexible sensors. Sensors **17**(3), 584 (2017)
8. Tlili, F., Haddad, R., Ouakrim, Y., Bouallegue, R., Mezghani, N.: A survey on sitting posture monitoring systems. In: 2018 9th International Symposium on Signal, Image, Video and Communications (ISIVC), pp. 185–190. IEEE (2018)
9. Anwary, A.R., Bouchachia, H., Vassallo, M.: Real time visualization of asymmetrical sitting posture. Procedia Comput. Sci. **155**, 153–160 (2019)
10. Luna-Perejón, F., Montes-Sánchez, J.M., Durán-López, L., Vazquez-Baeza, A., Beasley-Bohórquez, I., Sevillano-Ramos, J.L.: IoT device for sitting posture classification using artificial neural networks. Electronics **10**(15), 1825 (2021)
11. Huang, M., Gibson, I., Yang, R.: Smart chair for monitoring of sitting behavior. In: DesTech 2016: Proceedings of the International Conference on Design and Technology, pp. 274–280. Knowledge E (2017)
12. Kim, W., Jin, B., Choo, S., Nam, C.S., Yun, M.H.: Designing of smart chair for monitoring of sitting posture using convolutional neural networks. Data Technol. Appl. **53**(2), 142–155 (2019)
13. Hu, Q., Tang, X., Tang, W.: A smart chair sitting posture recognition system using flex sensors and FPGA implemented artificial neural network. IEEE Sens. J. **20**(14), 8007–8016 (2020)
14. Nayak, S.K., Swain, S.K., Mohanta, B.K., Paikaray, B.K.: Secure framework for data leakage detection and prevention in IoT application. In: 2022 IEEE 2nd International Symposium on Sustainable Energy, Signal Processing and Cyber Security (iSSSC), pp. 1–6. IEEE (2022)

15. Jung, H.Y., Ji, J.K., Min, S.D.: Real-time sitting posture monitoring system using pressure sensor. Trans. Korean Inst. Electr. Eng. **64**(6), 940–947 (2015)
16. Birsan, J., Stavarache, D., Dascalu, M.I., Moldoveanu, A.: SpiMO-sitting posture monitoring system. In: RoCHI, pp. 143–146 (2017)
17. Shin, D.J., Kim, M.S., Song, W., Min, S.D., Hong, M.: Implementation of sitting posture monitoring system with kinect. In: Advanced Multimedia and Ubiquitous Engineering, pp. 144–150. Springer, Singapore (2017)
18. Kim, M., Kim, H., Park, J., Jee, K.K., Lim, J.A., Park, M.C.: Real-time sitting posture correction system based on highly durable and washable electronic textile pressure sensors. Sens. Actuators, A **269**, 394–400 (2018)
19. Xu, W., Huang, M.C., Amini, N., He, L., Sarrafzadeh, M.: Ecushion: a textile pressure sensor array design and calibration for sitting posture analysis. IEEE Sens. J. **13**(10), 3926–3934 (2013)
20. Jena, S.P., Chakravarty, S., Sahoo, S.P., Nayak, S., Pradhan, S.K., Paikaray, B.K.: Automatic leaf diseases detection and classification of cucumber leaves using internet of things and machine learning models. Int. J. Web Grid Serv. **19**(3), 350–388 (2023)
21. Basahel, S.B., Bajaba, S., Yamin, M., Mohanty, S.N., Lydia, E.L.: Teamwork optimization with deep learning based fall detection for IoT-enabled smart healthcare system. Comput. Mater. Continua. **75**(1), 1353–1369 (2023). ISSN: 1546-218. https://www.techscience.com/cmc/v75n1/51539
22. Jena, S.P., Chakravarty, S., Paikaray, B.K.: Internet of things-based remote monitoring and classification of Spinacia oleracea leaf disease using deep learning approach. Int. J. Web Grid Serv. 20(2), 159–187 (2024)
23. Shaik, R., Jena, S.P., Pramanik, J., Paikaray, B.K., Samal, A.K.: Implementing swarm intelligence for image enhancement: a comparative study. Int. J. Reason. Based Intell. Syst. 16(2), 160–169 (2024)

Security Concerns in Low Power Networks for Internet of Things (IoT)

Muhammad Zunnurain Hussain, Muhammad Zulkifl Hasan, and Zurina Mohd. Hanapi

Abstract With the continuous growth of the Internet of Things (IoT), LPWAN becomes more meaningful in order to have sustainable and extremely distributed connectivity while being energy efficient. In this review article, the security challenges that create serious risks in LPWAN deployments are examined, which are important in different IoT applications like urban infrastructure and industrial systems. By means of a critical summary of the recent research we disclose the multifaced security difficulties of such networks, including data integrity breaches, encryption weaknesses and unauthorized access. Review covers a lot, from the current security solutions and protocols to the weaknesses and hence the need for the innovative security frameworks that take into consideration the unique features of the LPWANs. We propose a novel security framework which inherently enhances the fault tolerance and accommodates the operational constraints that characterize the low-power networks. The paper outlines the current security issues while projecting possible future threats and provides a roadmap for the improvement of LPWAN protocols that will ensure the strength of the next generation IoT networks.

Keywords Internet of everything (IoE) · Low power and lossy networks (LPLN) · IoT · Security · LPWAN · LoRaWAN · NB-IoT · DBPSJ · GFSK · BPSK

M. Z. Hussain (✉) · M. Z. Hasan · Z. Mohd. Hanapi
Department of Communication Technology and Network, Faculty of Computer Science and Information Technology, Universiti Putra Malaysia, 43400, Serdang, Selangor, Malaysia
e-mail: gs58270@student.upm.edu.my

M. Z. Hasan
e-mail: gs58279@student.upm.edu.my

Z. Mohd. Hanapi
e-mail: zurinamh@upm.edu.my

S. N. Mohanty et al. (eds.), *Explainable IoT Applications: A Demystification*,
Information Systems Engineering and Management 21,
https://doi.org/10.1007/978-3-031-74885-1_29

1 Introduction

In recent years, communication technology has undergone massive changes centered on the internet of things (IoT) leading to numerous disruptions across several domains. Along with that, the Internet of Things (IoT) plays a crucial role, as it helps to build up an enhanced connectivity and automation across sectors such as intelligent cities, healthcare, industrial systems, and more. The IoT statistics, which showed the presence of more than 35 billion active devices before the end of 2021, and expected more than 50 billion devices in 2025, demonstrate the tremendous role of IoT and its fast growth [4, 6].

The IoT expansion leads to many security challenges, especially within the low-power wide-area networks (LPWANs) that make up the foundation of most IoT infrastructures. LPWANs are prone to these threats due to the limitation of power, very low processing capabilities, and lack of security mechanisms. These networks are challenged by complex network security issues because of the number of communication channels needed, which are leading to serious power consumption, data throughput, and coverage problems [7–9]. Data breach, ransomware, phishing attacks are some of the security threats that are common in these environments, and they lead to the compromise of integrity and functionality of IoT implementations [13, 14].

The security of IoT devices and networks is becoming a crucial issue because they are integrated into essential infrastructure sectors. The weaknesses of the low-power networks made them a key target of cyberattacks, putting both data privacy and operational safety and effectiveness of the connected systems under the risk. Such networks are plagued with several specific security challenges:

- **Configuration Issues**: Wireless devices often have configuration errors due to improper setup and can be the source of unauthorized access and network compromise [15].
- **Denial of Service (DoS)**: LPWANs are vulnerable to DoS attacks in which attackers inundate their network with excessive traffic and in doing so they prevent legitimate users from accessing the network [15].
- **Passive Capturing**: The likelihood of passive data interception is high, as attackers have very easy chances to obtain data transmitted through these networks, thus exposing sensitive information. An important feature of wireless IoT protocols that amplifies this risk is their inherent openness [15].
- **Rogue Access Points**: Devices may be spoofed to join unauthorized access points and traffic can be eavesdropped or altered, which can result in very serious security vulnerabilities [15].
- **Stolen Wireless Devices**: The theft or unauthorized use of these devices may result in the loss of critical data and cybercriminals may use these as a gateway into the network in case they are not properly secured [15].

This paper examines the existing theories, empirical findings, and cryptographic techniques for network security of low-power IoT networks in a complete literature review. By the use of a systematic analysis of all the existing research this review highlights the most critical gaps and proposes a consolidated view a security strategy in LPWANs and as a result, contributes to the development of more secure, efficient, and reliable IoT systems.

2 Literature Review

Speaking of which, the network design which has a lightweight solution to topological vulnerabilities in a low power and lossy environment is defined as LPLN. Exploiting the flaws in the ranking algorithm, etc. is studied. According to the author, there is a set of model oracle release dates that are in agreement researcher research. This system which is proposed can tackle the risks efficiently to ensure that it is updated timeously and that spoofing is not a threat to the network.

AES has transformed. Thus, this engenders an increase in its energy efficiency and a boost in dynamic capabilities. Continually changing the AES S-Box and keys for all devices interoperable with LoRaWAN was what they incorporated The research discovered, that the LoRaWAN devices, that use modified AES operations, expend only about 26.2% less energy compared to those with the jobs of LoRaWAN doing the traditional way [21].

The nested nature of IoT devices in their connected environment makes the single authentication task much more complex regardless of the number of nodes [22]. The researchers put on themselves new techniques of multi-key to the elliptic curve encryption. The sessions' key of nodes for encryption can be changed at the same time. Only the destination node will be able to decrypt its message with the key that was previously shared. The authors accordingly recommend this method where the handling of up to 40 sessions is involved. There was the best performance strategy for IoT applications to work well. Moreover, this study's recommended methods worked well in the study. Despite being deployed, it still did not succeed in achieving the targets set as some time, the data still suffered from high processing activities.

Since low-power wireless area networks are in the energy-efficient zone, they can be easily subjected to identity theft node simulation. These kinds of assaults might be as hazardous as disturbing the network, compromising the most profound nodes, which in the end, bring about the end-user's data conversion impossibility as well. Hence, a decent way on how to inspect and act in case of such kind of attacks are required therefore. One of the most important tasks faced to damage the network was to identify and perceive infected devices the attacker had used. Furthermore, billing system engaged these devices to authenticate their identity through biometrics, such as fingerprints and passwords.

In order to develop a corresponding fingerprint, the authors embellished the fact of packet and of the nodes timing difference ability. For this purpose, the fingerprint node was suggested in order to accomplish node authentication as well as security in networks. Also the option that is available is the creation the fingerprints that can be used for the authentication or identification of networks that belongs to our neighbours. It was announced to attribute nodes based on the physical uncolorability feature in order to enhance the security (PUF). On behalf of the intrinsic lack of reliability in the course of electrical chip manufacturing, PUFs are being allocated some value. The 96 percent to be right on it off the bat has been an expected result of the proposed model. A parallel level of quality shown in this performance which did not display any kind of shortcomings [23–25]. Use our AI to write for you about any topic! You can use Artificial Intelligence to write content on any given topic easily with our software. Our writing tool is capable of creating high-quality, unique content for you that is optimized for SEO. Whether you need blog posts, product descriptions.

The authors suggest the possibility of low-power file synchronization over the networks of a dedicated IoT-type. The research objective aims at achieving energy efficiency through decreasing data transfers, increasing the life span of IoT devices and showcasing the device synchronization potential for the most effective efficiency. The increasing battery life is likely to be achieved only when the data transfer rates of IoT devices are lower.

It is GTIDs, that the authors introduce' for devices authentication. They do a good job there. The authenticating device paradigm is examined by four major components. A feature, print, divide by a similarity index, and finally, key in the are phases of a process. This traffic-rate fingerprint (DR) was generated as the distributor of the nodes distribution during an inter-arrival of packets. ANN, a neural network which works on the principle of an Artificial Neural Network, was used for ANN to store the patterns for the network device once fingerprints were solved. Node fingerprints were elaborated around the way to spot imposter nodes and became [28].

They employed the advantages of direct sequence code division multiple access (CDMA), together with low power consumption and wide coverage of thin and light devices, and changed the medium access control (MAC) architecture. RA-CDMA is thought to be a supersonic related protocol since its computations is done at the physical layer with the same approach. With a network topology stripped to the simplest star-like shape, the star structure. This scheme not only provides the advantage of low power and miniature complexity compared with the previous MAC protocols but also high data rates. Thus, the presented one is aimed at the use of IoT sensors in a compact multilayer network structure and improvement of the minimal power network design [29].

LoRa nodes were encouraged to join gatherings that were already in place thanks to the approach. As the authors tried to overcome the problem of node joining by the mono-key joining one, they created a new method of node joining at a dual-key node formation procedure due to the problematic side-effects related to node joining. Due to the fact that the very same key used in the first time of shared building, was updated and designed to enhance security. The method suggested uses more energy as each of the two keys in the lock have to be unlocked before the mechanism can row. We obtain our security from the nature of this key: its generation is done for every sessions individually. Nevertheless, the design will involve only LoRa-based network of nodes which, supposedly, would be sufficient for the proper operation.

Peer-to-peer communication between LoRaWAN devices that will be based on the key generation concept. The AES128 keys for LoRaWAN devices is generated using seven stages of processes separately including. And the key, again the study says, was changed every three hours to ensure security.

Since it made full-blown verification, Messages Authentication Codes (MAC) command was added. Instead of pre-share key, the key for the asymmetric key was employed in that way. The session key and user authentication are set in four phases of their algorithm (Step-by-step approach). These two are considered each time an ID is generated for the new key. In regard to the research, long as it took to register a node in eight weeks, the authentication method deployed was streamlined and could be carried out via a simple and affordable mechanism [33].

Through a scheme, they were able to illustrate a key pairs generation mechanism by HD wallet and server using BIP32. Having access to the given data, the server performs the operation of creating and saving a couple of root keys for communication. Using this approach has increased the laundry device's integrity because [33].

For these reasons, the roots of LoRaWAN network keys will convert to the plan of using alternative pupation form. Spot Rabbit stream cipher was used as the TRNG and as a two-reeds key method in order to create and produce the last two-needed key steps. Following this, root keys are necessary for data transmission and session keys generation, respectively. The analysis shows that the described solution is much more economical of resources than HBK (hash-based key) mechanism used in LoRaWAN now (paragraph 3, 43).

Topologies of LoRa devices can be enhanced by introducing a new security key management approach that involves both the device and the network exchange on a regular basis. The contract relied ECC-DH agreement that is a cryptographic tool used in the strengthening of a wave diffraction system for the tunnelling implications. Here, hierarchical deterministic (HD) wallets were a method of choice.

Low power (IoT) sensor technology for waste management in Smart Cities. Thus, in their opinion, the work brought into question an sustainable energy way to deal with waste in smart cities. Furthermore, architectural design using low-power, low-cost sensor elements was developed [37].

The LPWAN networks have been explored by the network security mechanisms used in the Low Power Warning Area Network devices. To begin with, we switched to an unknown key to transmit the power strength codes. Besides, because it is a one-way communication that transmits data to the end-user, the node must be connected [29]. The research, however, did not provide any data or explaining analysis of energy consumption more thoroughly.

To pursue the exploration of a unique method, proxy-based encryption was involved and the creation of a system relies on proxy nodes to encrypt and deliver encrypted data was completed such a setting, the end node and the proxy node have secure node communication and hence data privacy and safety because reliable and authentic encryption technique was needed for the data's integrity and trustworthiness, getting along with a reliable node was compulsory for this [38].

Pre-computational method was applied, where CBE or Cypher specification was defined so as to communicate with the public. The data collection procedures leveraged pre-computations to save up the acquired information. The model was run by using the data from operations and speeding up EEL for calculating. This low cost strategy although it came in handy saving on hardware costs, it needed a large memory store [39].

It was proposed to monitor the remote network of appliances and many IoT devices who use little electricity as the IBSMC system. Intelligence monitor behaviors constituted the safeguards of IBSMS system. Coding scheme Swapped Huffman was suggested known as an encoder employed an algorithm of encryption based on a secret key [40] for the protection of the data. Ciphers can be simple and shown as an elegant solution to data encryption. Still, there is an opportunity for an attacker to crack the system after he/she has enough data to decode the data pattern and as a result decrypt the scheme and retrieve the plain text.

Researchers, as described in their method, have found that side-channel attacks are employed. South Korea was the first country in the world to consider approving a Lightweight Encryption Algorithm (LEA) for Internet of Things in 2013 (IoT). Encryption reinforced the trust and awe in LEA as it took them ten hours to escape. XOR and addition rotation are typically used in AES instead of S-box lookups for the same reason, to avoid the ALE side-channel leakage. If the attack was carried out by an unauthorized code, the original pattern would change to a bit pattern to prevent a side-channel attack and save the updated data. The side-channel attack was defeated and which seems to be a secure way of encrypting the Wi-Fi networks became soon after. S-box must be removed if the algorithm has problems in the encryption part [42].

It differs through the use of a suite of subscriber appliance to build a crypto net device. The core energy-adoptive system applies a collection of public key and symmetric cryptography modes depending on situations such as node energy consumption. Implementing the node having power from the sun for instance, the solar-powered node, for cryptography purpose like public-key encryption, is an example of the consideration of power consumption which is less relevant issue in this case.

2.1 Low Power Devices

The diodes in this setup consists of personal communication network (PAN) and local communication network (LAN) devices. ZigBee and Bluetooth are referred to direct short range technologies which connect personal equipment. Local area network administrators can create a wireless communication system using Wi-Fi for sharing. The low-power wireless area will be seen in more details in the following sections section especially with Bluetooth and Zigbee wireless technologies and Wi-Fi discussed.

2.2 ZigBee

ZigBee joins neighboring devices to form a network. This is mostly used in low-power applications. It has the added advantage to be used on network of minimal size but use as little power as possible. Working within the area network that has adopted the use of open standard, including the low power function and minimal environmental impact, makes the IEEE 802.15.4 to follow the rules and protocols set by IEEE 802.15.4. Such technology gives the possibility to work the frequency range of 2.4 GHz all over the world. ZigBee is very different from Bluetooth, which only covers an approximated distance of 10 to 100 m, and there is networks with a range of up to 100 m. The channel access method "Career-Sense for Multiple Access with Collision Avoidance" is employed in CSMA/CA protocol. In the same way QPSK and BPSK modulation is used during modulation process. Since the data rate supported by ZigBee is at a maximum 250 Kbps (Kilobyte per second), the data transfer speed is quite slow and becomes a bottleneck. It uses those sixteen frequencies that work through RF. There are 255 different items that may be linked to the ZigBee device, a thing that brings its responsibility. Despite the fact that it has only small bandwidth requirements, it usually overcome Bluetooth that is low bandwidth. Similarly it is not unusual to buy ZigBee devices with a higher price. As a result, implementing ZigBee technologies in the mass production is more likely to be neglected at the early stage. There is a major area which is connecting ZigBee networks to satisfy the requirement of monitoring and controlling to implement a wireless area network. It is done in the areas of residential, commercial and agricultural worlds with a very spin of the time. Hence, also, for the commercial and industrial land it may hold out to the best choice [44–50].

2.3 Bluetooth

Short distances wireless technology standards like Bluetooth become common and well-established in our life. The end-user targeted area networks with power consumption constraints is what it is for. The IEEE 802.15.4 standard for access security is used in it and it operates in the 802.15.4 PAN through the 802.15.4 MAC protocol. The Bluetooth standard operates with a frequency of 2.4 GHz and maximum data rate of 1 MHz, which is sufficient for small data packets exchange across a range of a few meters. The maximum range of Bluetooth and ZigBee is a few feet. When these tow devices are compared, they will be compared to a wide blast range. Universally less common topologies of the type that was mainly used in Bluetooth devices is Star architecture. We are able to achieve a distance of 30 m with maximum speed of 3 Mbps (MegaBytes per second). The number of the devices allowed in one network is 1000. FDMA/TDMA is a popular media access scheme available in the Bluetooth systems. The modulation phase web utilizes 8DPSK (8 DPSK) and GFSK (GFSK).

Most of the gadgets, like smartphones and other individual units, that are used for communication must be rented nowadays [51–59].

2.4 Wi-Fi

The term Wi-Fi is a short form for wireless fidelity. 802.11ac uses IEEE 802.11b and 802.11g standards for a better management, increased speeds, and better performance. Wireless communications can provide connection of mobile devices and laptops within the small world for instance corporate areas, school classrooms and houses by using wireless communication technology. It works at 2.4 and 5 GHz and transferring the data between the two is one of its key strengths. whereas the ZigBee and Bluetooth networks length is limited to 10 m, the HomePAN technology takes over in this field with respect to the range it reaches. Moreover, the rate of ZigBee devices suffers on network performance in the same way it does when it comes to reliability. For sure, despite the fact that wired devices are more energy-efficient than those connected via PAN. There is a collision in network segments which Wi-Fi detects using the Carrier Sense Multiple Access and Collision Detection methods. It could be required to have much more bandwidth when setting the frequencies across a wide range. In contrast to its rivals like PAN, it achieves a data rate of 7 Gbps which is the highest. 23. Just in check the range and data rates of WAN devices compared to Bluetooth and ZigBee, you find the differences that give WAN advantage over PAN. An analyzes of three wireless communication protocols Bluetooth, ZigBee, and Wi-Fi is carried on Table 1 through their technical features [60–64].

Table 1 Technical features & security issues of PAN and LAN

Technical features of ZigBee, Bluetooth, and Wi-Fi							
	Frequency	Channel access	Modulation	Maximum data rate	Maximum range	Maximum devices support	Security issues
Bluetooth	2.4 GHz	FDMA/ TDMA	GFSK, 8DPSK	3 Mbps	30 m	1000	MAC spoofing, man in the middle attack
ZigBee	2.4 GHz	CSMA/ CA	BPSK/ QPSK	250 Kbps	10–100 m	255	Eavesdropping, DOS, Node compromise, sink hole, warm hole, physical attacks
Wi-Fi	2.4 GHz, 5 GHz	CSMA/ CA	Various	7 Gbps	100 m	255	Limited range, data protection, connectivity issues

As far as data speeds, power consumption and the area range are considered, PAN and LAN wired technology also come into play to connect the various devices. Nevertheless, these devices generally focus on some variables over others, what we call constraints. Due to this bandwidth limitation, they still don't manage to meet the efficiency, capacity, and network range one can see in the existing networks. On the other hand, this deficiency dictates you create these things with the goal of solving this specific issue. Following low data rate, long-range supporting and low power consumption, an important factor during the production is to fabricate devices with low power commands [65–67].

With the examples of NB-IoT, LoRaWAN, DASH7, Weightless and Sigfox, low-power wireless area networks (LPWANs) available are just some of the ones today.

2.5 NB-IoT

3GPP (Third Generation Partnership Project) is responsible for creating and standardizing LPWA (Low Power Wide Area) wire type of communication known as NB-IoT. Discussion on the LTE (Long Term Evaluation) spectrum as used for data transmission in the NB-IoT. In the end, our list of release standards for version 13 includes all the key areas that need to be addressed. The Next Generation the Mobile was also labeled as the 5G technology the next year. LPWANs (Low Power Wide Area Networks) are its recent and most prevailing superlative. Due to the fact that among its features it can be regarded as very versatile, it may be used for a number of new applications in the industrial aspects and smart parking. Polar communications capabilities for cheaper batteries and equipment features would be our technology priority. In addition, the NB-IoT has used the LTE protocols of authentication and authorizing for its devices' security. The perception layer performs as a shield to block any destruction of data using unauthorized methods. This technology utilizes the frequencies of G-Smart wave and low energy transmission (LET). If you are required to send or receive messages over air EM air up to a distance of 5 km, this communication technique ensures a great range in terms of transmission speed. QPSK modulation of NB-IoT signals is utilized in the delivery of these signals. The NET B platform maintains NB-IoT specs with 128-bit of AES encryption and a star topology as its scaffolding. The size of a huYonaw datagram is 2048 bytes. In addition to the other compliance clauses, IoT's NB-security is still the only approach that does not fit any standard on the planet [68–76].

2.6 LoRa

One of the latest proven wireless technologies is LoRaWAN, or Long-Range Wireless Area Network, which is an output of the company Semtech Corporation and calls for commercialization. More so, just like the hiring of engineers at Lora corporation, the organization has got a macro-layer protocol for wireless devices that use batteries in operation. The LoRa comes in 1GHZ unlicensed frequency band and possesses a mobile wide area network characteristic that make it the most intriguing. Also, the modem supports all these frequencies (868 MHz) no matter where the border frequency (433 MHz) is. Uplink and downlink are hence the method of the communications utilizing frequency shift keying (FSK) along with Chirp Spread Spectrum (CSS) spread spectrum modulation. The highest data rate is 50 kbps per second and a packet size of 2047 kB, which allow radio can span maximum 20 km in distance. In each LoRaWAN correct node we have the AES 128-bit encryption. Internet protocols are the backbone of a conventional backbone, gateways, and star and mesh topologies being in development. At the same year, the LoRaWAN network was pushed out. Nodes in the encrypted network communicate to the hub first. Afterwards, servers get the information that nodes are allowing or refusing the request. Without encryption, the emergency messages were both sent and received in a timely manner in order to comply with the said network rules. If there is a message broadcast through the server, there should be no worries since the applaud will make sure the message does not get corrupt. LoRaWAN is the long distance communication alternative we have been looking for. Write 3 paragraphs answering the following question: Describe the significance of the arts and culture in shaping a nation's identity [30, 77–80].

2.7 Sigfox

Most often the Sigfox is used as the broadcast mode (LPWAN) for the low-power network (LPWAN). Its enables the LPWAN type connect which very important to solve the problem of different network connection. In 2009, the French mobile network company named Sigfox developed a communications system, called Sigfox, especially for those devices that have few or no power. The frequencies provided are UNB and the broadcast can be streamed in 200 kHz. The BTU RTK PPK in DGPS-compatible form provides data transition from UL to DL speeds, up to 100 kbps, and is compatible with the USB-OTG port as a simple application interface, and it has a range of up to 50 km. The delivery time would become 24 bytes for a 12 byte message up-linking and 8 for down linking. Currently, a communication average of (2 s). On 868/902 MHz, modulation is performed in DBPSKJ and GFSK methods. With the use of ultra-narrowband spectrum, the limited noise level existence and the easement of such a signal demodulation are ensured. An important feature is that in Sigfox, message signing as a default is not activated. As set forth by their needs customers can choose an end-to-end encryption or a solution from Sigfox [81–84].

2.8 Weightless

Bs. low power networks Weightless-P, Weightless-W, and Weightless-N make it possible to deploy flangeless, which is a network. A non-profit group introduced this method, which was already around since 2008. Neul, who initiated this technology in the UK, has also signed agreements with Huawei to facilitate its rollout. Home wrecking indeed is one of the newest which belongs to 802.11n family and involves USNBband, one of the most cutting gambling technologies (1 GHz spectrum, 24 uplink channel access). With the doubling of rate and the increase of range from 5 to 9 km, the current combo of BPSK, QPSK and DBPSK can support a transmission rate of 10 Mbps. Keeping the nodes encrypted and authenticated by AES 128-bit method, we strive to offer an extra-security and protection status for participants among other choices. employing a star topology, an avenue of safe data circulation is accessible. The effect achieved by the use of AES 128 by the nodes is that the nodes. If a physical attack results in the hijack of session key, the private key of the node can also be exposed [85, 86].

2.9 Dash7

Likewise, open-source LPWAN radio protocol DASH7 is forcefully meant for sensor work and this protocol operates in the 433/868/915MHz frequency range. The original coupon was released back in 2003 (ISM). Authentication for the node is determined by 128-bit AES symmetric key encryption. It just assumes the network is safe and offensives its confidence upon that. On last, key is preserved on the gate versus independent node to save power. Last but not the least, defense vulnerabilities for the low power wide Area Networks (LPWANs) supplying end devices with tiny power, compute, and memory (for example, wearables) escalated significantly from future tech advancement and the growth of IoT devices. LPWAN network DASH7 has a tremendous power saving potential by providing the base stations nearly endless battery life for the devices for many years. 128-bit one time, higher end to end data encryption is applied throughout the data transaction. Transmitting 167 kbps, DASH7 will be taking huge share of bits and bytes by areas much larger than that. Forwarding of the 3 succinct GFSK modulation channels is also be allowed. By Ethernet cable or wireless network, a tree or star are the network's configuration. DASH7 contains sensor portals, hubs, and nodes. Two-way flow direction is considered storing it in the server if end-to-end data is actively collected by the gateway. Just like a gateway it has got its own sleep cycle to control power usage. DASH7 is recognized as energy-efficient quality [87]. Table 2 shows the technical specifications for the Low Power (LoF) Wireless Domestic Network (DoN).

Table 2 Technical features of LPWAN

Technical features and security issues of low power wide area network					
	NB-IoT	LoRa	Sigfox	Weightless	DASH7
Frequency	LET and GSM, USA	433/868/915 MHz	868/902 MHz	1 GHz	433/868/915 MHz
Channel access	Multiple	10 (EUR), 8 (DL)	360	24 (UL)	3
Modulation	QPSK	CSS/FSK	DBPSK and GFSK	BPSK, QPSK, DBPSK	GFSK
Maximum data rate	UL (158.5 kbps), DL (106 kbps)	50 Kbps	UL (100 kbps), DL (600 kbps)	10 Mbps	167 kbps
Maximum range	0–5 km	5–20 km	10–50 km	5 km	0–5 km
Encryption	AES 128 bit	AES 128 bit	AES	AES 128 bit	AES 128 bit
Topology	Star	Star/Mesh	Star	Star	Star/Tree
Packet size	2047 B	2047 B	UL (12 B), DL (8 B)	> 10 Kb	256 B
Security	In development	Developed	Partially Afforested	Developed	N/A
Founded	2016	2015	2009	2012	2013
Security issues	Insufficient authorization/ authentication Poor application and end-point security Lack of physical security	MTM Attacks, Frame payload attacks,Network Flooding attacks, Physical attacks. RF Jamming Attack	Replay attack, DOS attack, POC Attack	Key attack	Authentication

2.10 *Confidentiality*

Data exchange and reception area should be restricted and encrypted in order to prevent communication by unnecessary parties only the sender may use the network node to access the data for this purpose. With so many compromising facts, data security becomes crucial. No party should feel exposed or violated. Since data is more than a crucial resource, it cannot be treated as an expendable asset and must not be in jeopardy under any circumstances. Man-in-the-middle attacks commonly use two common assaults: the make-or-break situation of the SCT, and worst of all is the SDGT. In the first instance, appending scenery from different countries rise to attacker's theft of a vital key before they start an attack. Let's imagine someone stealing the user's private key could not get the data once entered which they could change or postpone and send it to the receiver. When two nodes both consider access to their data or password or code as secured connections or they are simply making assumptions, but a third party accesses the data and the password or the code. This is called a "man in the middle" attack. Many literary works dealing with them during an efficiency design on the networks with a low power level being added to the list.

2.11 Integrity

Virtual private networks (VPN) share a common goal of keeping the network out of the hands of diverse attackers. Integrity of the network is the greatest lost. Generating precise information is indeed one of the key elements of data integrity. Means of data manipulation may range from accidental to purposeful by exploiting this loophole. For the elimination of deletions, versions of the code, and changes, the organization of the network's integrity is important. Since the jamming is the vector given to wormhole or replay attacks in case of low-power networks, integrity is an important thing to consider. Bearing in mind that the strike can be mounted without Geneva jammers and announce sniffer should be ensured. The sniffer ascends to the data packet then the deciding system operator decides whether or not to jam the packet. Problems of low-power wide-area networks are demonstrated in Table 3.

Table 3 Attacks on LPWAN

Attacks on low power wide area network					
	NB-IoT	LoRa	Sigfox	Weightless	DASH7
Replay attack	-	Frame counter	Sequence number	Data frame counter	N/A
Possible attacks	Jamming	Jamming	Replay attack	Jamming	Jamming
Possible attacks	Physical attacks Port scanning ARP spoofing DNS spoofing Malicious UE attack	Network flooding attacks Packet forging	POC attack	Key attack	–

3 Methodology

3.1 Defining the Research Question and Objectives

The methodology begins with a precise definition of the research question, which is central to guiding the entire review process. For this survey, the primary question is: "What are the prevailing security challenges and proposed solutions within Low Power Wide Area Networks (LPWAN) for Internet of Things (IoT) applications?" The objectives include identifying existing security threats, reviewing proposed security measures, and assessing the efficacy of these solutions within the constraints of LPWAN technologies.

3.2 Literature Search Strategy

A structured literature search strategy is employed to ensure comprehensive coverage of the topic. This includes:

Databases and Sources: Selection of databases such as IEEE Xplore, ScienceDirect, and Google Scholar for their strong repository of technology and security-related academic articles.

Search Terms: Development of a list of keywords and phrases like "LPWAN security", "IoT network vulnerabilities", "cryptographic methods in IoT", and "security protocols in LPWAN". These keywords are combined using Boolean operators to refine the search results.

Time Frame: The search is limited to documents published in the last ten years to focus on the most current research and developments.

Inclusion and Exclusion Criteria: Inclusion criteria are set to select studies that specifically address security aspects of LPWANs within IoT contexts, peer-reviewed articles, and papers in English. Exclusion criteria exclude non-peer-reviewed articles, articles not directly related to LPWAN technologies, and redundant publications.

3.3 Screening of Literature

The identified literature is first subjected to a title and abstract screening, assessing relevance based on the predefined inclusion and exclusion criteria. This initial screening is followed by a full-text review of selected articles to ensure they provide substantial information relevant to the research questions.

3.4 Data Extraction and Categorization

Relevant data from each article are extracted, including author details, publication year, study focus, methodologies employed, main findings, and security solutions discussed. This data is organized in a matrix format to facilitate easy comparison and synthesis. Categories are formed based on common themes such as types of security threats (e.g., physical, network, encryption), types of LPWAN technologies discussed (e.g., LoRaWAN, Sigfox), and the nature of proposed solutions (e.g., cryptographic, protocol-based).

3.5 Quality Assessment

To ensure the credibility of the literature review, a quality assessment of the selected papers is conducted. This assessment focuses on methodological rigor, relevance to the research question, and the impact factor of the publication sources. The assessment helps in weighting the conclusions drawn from the various studies in the final synthesis.

3.6 Thematic Synthesis

This step involves synthesizing the information extracted from the literature to identify common findings, contradictions, and gaps in existing research. The summary is arranged according to main theses, which are only recognized as the process of classification meets. This topic analysis aids in understanding the broader context of security of LPWAN and its solutions within the domain of IoT.

3.7 Development of the Security Framework

The security framework for LPWANs in IoT is designed based on the analyzing synthesized data. This framework integrates those strategies that had been effective in their implementation and fills in the gaps of the existing research. It puts forward integrated solutions customized to the particular requirements and limitations of the LPWAN networks.

3.8 Drafting the Review

The outcomes and analyses are composed into a consistent review paper. The paper is designed to start with the problem area, followed by the methodology, the highlights from the literature review, the security framework, and finally the implications for future research and practice.

3.9 Conclusions and Future Work

The final part of the paper is the summary of the most important findings, the opportunities for practice of the proposed framework and the suggestions for the development of the research in the area of LPWAN security within IoT environments.

Figure 1 diagram represents the procedure of research methodology for the research on security issues belonging to Low Power Wide Area Networks (LPWAN) for Internet of Things (IoT) applications. It details the interactions between the researcher and various components such as databases, review processes, and analysis, leading to the development of a security framework. The iterative refinement process between the researcher and the review is also highlighted, emphasizing the dynamic nature of conducting a comprehensive literature review.

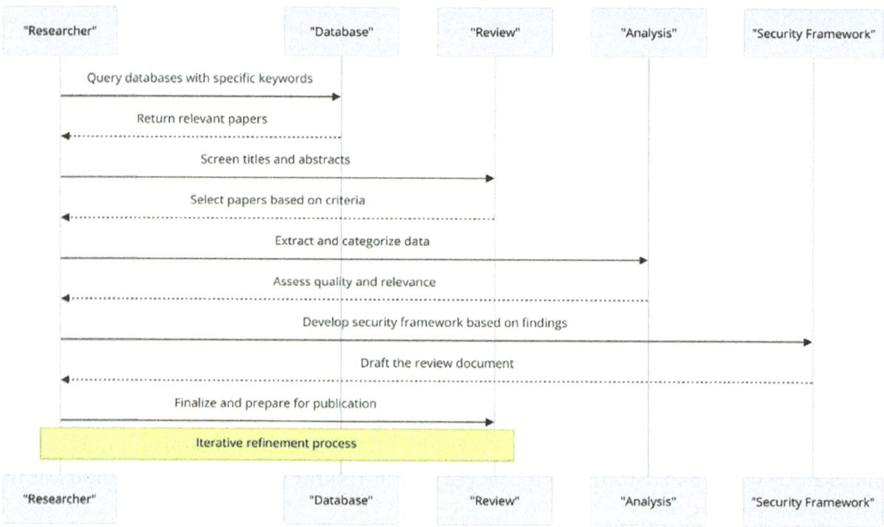

Fig. 1 Research methodology diagram for security framework

4 Discussion

The challenges of security in the low power networks are specific because they have a different architecture. While scientists changed the world in diverse ways, their common goal always remains the same—to move humanity forward. Mainly Personal area networks (PANs), very close area networks (VLAN) are key subfields of low-power networks designed for short-range communications (LAN). Bluetooth & ZigBee are PAN (Personal Area Network) devices that communicate each other via radio frequency. Next to this, Wi-Fi is an example of LAN (Local Area Network) devices that require setting up in order to communicate with each other. Low-power wireless radio frequency devices manifests a host of constraints. This is not just of joining the social interaction. It entails as well the focus on protecting ones' data. Low Power Wireless Aera Networks must also be considered to achieve better communication and therefore scientists constantly struggled to do this. The combination of various technologies ensures that services can be carried out across a high proportion of their vocal spectra and usage. In matters of the IoT devices, it is scarcely necessary to experience the problems of the battery. The issue of energy solution has been proposed to tackle IoT devices' consumption problems in terms of use of solar power. The actors are explained along as well. Security on the Internet of Things (IoT) is on the list of top problems. Security analysis has been addresses in many studies and gives solution for it. The upkeep of high standard of confidentiality and trustworthiness is undoubtedly the crucial element in building relationships with financial institutions. References [22, 32, 39, 41, 90] raises the problem of cleanness of IoT devices, however, other authors are also covering a similar topic. In [22, 25], we bring up the problem of integrity; in [30, 32], we highlight the topic of integrity. Every aspect of this trial to be analyzed in the bibliography study, and a precision for fixation is also given.

5 Conclusion

The low-power wireless, area networks technology offers inexpensive possibilities in areas where the need is wireless. The rapid pace of technology is accompanied by an ever-increasing amount of research into the technology known as the Internet of Things. Along with this trend, sometimes WLANs further trouble these IoT devices. This research targets the collation of information for both the known and it will also identify the potential security challenges that they will arise while using low-wireless LANs as the local networking. Studies on various IoT technologies specifically put emphasis on the real parts ensuring that wireless technology functions as it should. Academicians are considering bettering standards for IoT devices and security to prevent possible vulnerabilities and cybercrimes. Researchers that come into women studies to take a look at things that are broken with women are pulled to the windows like moths to the flame. Well key distribution and authentication mechanism, encryption is already getting solved, but more research is needed to address those for low-power networks.

Acknowledgements This work was supported by Geran Putra Berimpak Universiti Putra Malaysia, Vote Number 9659400. Our sincere thanks to Geran Putra Berimpak Universiti Putra Malaysia for their support.

References

1. Cho, J.: Roles of smartphone app use in improving social capital and reducing social isolation. Cyberpsychol. Behav. Soc. Netw. **18**(6), 350–355 (2015)
2. Tyagi, A.K., Nair, M.M.: Internet of Everything (IoE) and Internet of Things (IoT): threat analyses, possible opportunities for future. J. Inf. Assur. Secure. **15**(5) (2020)
3. Bogatinoska, D.C., Malekian, R., Trengoska, J., Nyako, W.A.: Advanced sensing and internet of things in smart cities. In: 2016 39th International Convention on Information and Communication Technology, Electronics, and Microelectronics (MIPRO), pp. 632–637 (2016)
4. Mills, M.P.: Mines, Minerals, And 'Green' Energy: A Reality Check (2020)
5. Namasudra, S., Deka, G.C., Johri, P., Hosseinpour, M., Gandomi, A.H.: The revolution of blockchain: state-of-the-art and research challenges. Arch. Comput. Methods Eng. **28**(3), 1497–1515 (2021)
6. Ejaz, W., Anpalagan, A.: Dimension reduction for big data analytics in Internet of Things. In: The Internet of Things for Smart Cities, pp. 31–37. Springer (2019)
7. Yaacoub, E., Alouini, M.-S.: A key 6G challenge and opportunity—connecting the base of the pyramid: A survey on rural connectivity. Proc. IEEE **108**(4), 533–582 (2020)
8. Konte, K.: Mobile Ad Hoc Networks in Transportation Data Collection and Dissemination. Rowan University (2019)
9. Shen, H., Bai, G., Hu, Y., Wang, T.: P2TA: privacy-preserving task allocation for edge computing enhanced mobile crowdsensing. J. Syst. Archit. **97**, 130–141 (2019)
10. Himma, K.E.: Internet Security: Hacking, Counter Hacking, and Society. Jones & Bartlett Learning (2007)
11. Pant, V.K., Prakash, J., Asthana, A.: Three-step data security model for cloud computing based on RSA and steganography. International Conference on Green Computing and Internet of Things (ICGCIoT) **2015**, 490–494 (2015)
12. Choubey, S.D., Namdeo, M.K.: Study of data security and privacy-preserving solutions in cloud computing. International Conference on Green Computing and Internet of Things (ICGCIoT) **2015**, 1101–1106 (2015)
13. Mohan, N.R., Kumar, N.P.: Predicting and Analysis of Phishing Attacks and Breaches in E-Commerce Websites (2020)
14. Cheng, L., Liu, F., Yao, D.: Enterprise data breach: causes, challenges, prevention, and future directions. Wiley Interdiscip. Rev. Data Min. Knowl. Discov. **7**(5), e1211 (2017)
15. Khan, S., Loo, K.-K., Naeem, T., Khan, M.A.: Denial of service attacks and challenges in broadband wireless networks **8**, 7 (2008)
16. Akyildiz, I.F., Wang, X.: A survey on wireless mesh networks. IEEE Commun. Mag. **43**(9), S23–S30 (2005)
17. Li, X., Li, D., Wan, J., Vasilakos, A.V., Lai, C.-F., Wang, S.: A review of industrial wireless networks in the context of industry 4.0. Wirel. Netw. **23**(1), 23–41 (2017)
18. Yaqoob, I., et al.: Internet of things architecture: recent advances, taxonomy, requirements, and open challenges. IEEE Wirel. Commun. **24**(3), 10–16 (2017)

19. Akinyele, D.O., Rayudu, R.K.: Review of energy storage technologies for sustainable power networks. Sustain. Energy Technol. Assess. **8**, 74–91 (2014)
20. Nikravan, M., Movaghar, A., Hosseinzadeh, M.: A lightweight defense approach to mitigate version number and rank attacks in low-power and lossy networks. Wirel. Pers. Commun. **99**(2), 1035–1059 (2018). https://doi.org/10.1007/s11277-017-5165-4
21. Tsai, K.-L., Huang, Y.-L., Leu, F.-Y., You, I., Huang, Y.-L., Tsai, C.-H.: AES-128 based secure low power communication for LoRaWAN IoT environments. Ieee Access **6**, 45325–45334 (2018)
22. Tsai, K.-L., Huang, Y.-L., Leu, F.-Y., You, I.: TTP based high-efficient multi-key exchange protocol. IEEE Access **4**, 6261–6271 (2016)
23. Sieka, B.: Active fingerprinting of 802.11 devices by timing analysis. In: CCNC 2006. 2006 3rd IEEE Consumer Communications and Networking Conference, vol. 1, pp. 15–19 (2006)
24. Xu, Q., Zheng, R., Saad, W., Han, Z.: Device fingerprinting in wireless networks: challenges and opportunities. IEEE Commun. Surv. & Tutorials **18**(1), 94–104 (2015)
25. Xing, K., Liu, F., Cheng, X., Du, D.H.C.: Real-time detection of clone attacks in wireless sensor networks. In: 2008 The 28th International Conference on Distributed Computing Systems, pp. 3–10 (2008)
26. Petroni, A., Cuomo, F., Schepis, L., Biagi, M., Listanti, M., Scarano,G.: Adaptive data synchronization algorithm for IoT-oriented low-power wide-area networks. Sensors (Switzerland) **18**(11) (2018). https://doi.org/10.3390/s18114053
27. Radhakrishnan, S.V., Uluagac, A.S., Beyah, R.: Gtid: a technique for physical device and device type fingerprinting. IEEE Trans. Dependable Secure Comput. **12**(5), 519–532 (2014)
28. Rehman, S.U., Sowerby, K.W., Chong, P.H.J., Alam, S.: Robustness of radiometric fingerprinting in the presence of an impersonator. In: 2017 IEEE 28th Annual International Symposium on Personal, Indoor, and Mobile Radio Communications (PIMRC), pp. 1–5 (2017)
29. Petrosky, E.E., Michaels, A.J., Ernst, J.M.: A low power IoT medium access control for receiver-assigned CDMA. Int. J. Interdiscip. Telecommun. Netw. **11**(2), 24–41 (2019). https://doi.org/10.4018/ijitn.2019040103
30. Kim, J., Song, J.: A dual key-based activation scheme for secure LoRaWAN. Wirel. Commun. Mob. Comput. (2017)
31. Ruotsalainen, H., Zhang, J., Grebeniuk, S.: Experimental investigation on wireless key generation for low-power wide-area networks. IEEE Internet Things J. **7**(3), 1745–1755 (2019)
32. Roselin, A.G., Nanda, P., Nepal, S.: Lightweight authentication protocol (LAUP) for 6LoWPAN wireless sensor networks. IEEE Trustcom/BigDataSE/ICESS **2017**, 371–378 (2017)
33. Wuille, P.: Bip32: Hierarchical Deterministic Wallets. https://github.com/bitcoin/bips/blob/master/bip-0032. mediawiki (2012)
34. Parhami, B.: "Data Longevity and Compatibility", in Encyclopedia of Big Data Technologies, pp. 559–563. Springer International Publishing, Cham (2019)
35. Xing, J., Hou, L., Zhang, K., Zheng, K.: An improved secure key management scheme for LoRa system. In: 2019 IEEE 19th International Conference on Communication Technology (ICCT), pp. 296–301 (2019)
36. Cerchecci, M., Luti, F., Mecocci, A., Parrino, S., Peruzzi, G., Pozzebon, A.: A low power IoT sensor node architecture for waste management within smart cities context. Sensors (Switzerland) **18**(4) (2018). https://doi.org/10.3390/s18041282
37. Han, B., Peng, S., Wang, X., Wang, B.: Distributed physical layer key generation for secure LPWAN communication. In: 2019 IEEE 25th International Conference on Parallel and Distributed Systems (ICPADS), pp. 225–232 (2019)
38. Naoui, S., Elhdhili, M.E., Saidane, L.A.: "Enhancing the security of the IoT LoraWAN architecture. International Conference on Performance Evaluation and Modeling in Wired and Wireless Networks (PEMWN) **2016**, 1–7 (2016)
39. Oualha, N., Nguyen, K.T.: Lightweight attribute-based encryption for the internet of things. In: 2016 25th International Conference on Computer Communication and Networks (ICCCN), pp. 1–6 (2016)

40. Kontogiannis, S.: An internet of things-based low-power integrated beekeeping safety and conditions monitoring system. Inventions **4**(3) (2019). https://doi.org/10.3390/inventions4030052
41. Jang, Y.S., Usman, M.R., Usman, M.A., Shin, S.Y.: Swapped Huffman tree coding application for low-power wide-area network (LPWAN). In: 2016 International Conference on Smart Green Technology in Electrical and Information Systems (ICSGTEIS), pp. 53–58 (2016)
42. Choi, J., Kim, Y.: An improved LEA block encryption algorithm to prevent side-channel attack in the IoT system. Asia-Pacific Signal and Information Processing Association Annual Summit and Conference (APSIPA) **2016**, 1–4 (2016)
43. Kim, J.M., Lee, H.S., Yi, J., Park, M.: Power adaptive data encryption for energy-efficient and secure communication in solar-powered wireless sensor networks. J. Sensors (2016)
44. Gill, K., Yang, S.-H., Yao, F., Lu, X.: A ZigBee-based home automation system. IEEE Trans. Consum. Electron. **55**(2), 422–430 (2009)
45. Di Francesco, M., Anastasi, G., Conti, M., Das, S.K., Neri, V.: Reliability and energy-efficiency in IEEE 802.15. 4/ZigBee sensor networks: an adaptive and cross-layer approach. IEEE J. Sel. areas Commun. **29**(8), 1508–1524 (2011)
46. Vishwakarma, D.: IEEE 802.15. 4 and ZigBee: a conceptual study. Channels **868**, 866–868 (2012)
47. Safaric, S., Malaric, K.: ZigBee wireless standard. Proceed. ELMAR **2006**, 259–262 (2006)
48. Georgakakis, E., Nikolidakis, S.A., Vergados, D.D., Douligeris, C.: An analysis of BlueTooth, ZigBee and BlueTooth low energy and their use in wbans. In: International Conference on Wireless Mobile Communication and Healthcare, pp. 168–175 (2010)
49. Hwang, S., Yu, D.: Remote monitoring and controlling system based on ZigBee networks. Int. J. Softw. Eng. Its Appl. **6**(3), 35–42 (2012)
50. Al-Adwan, I., Al-D, M.S.N.: The use of ZigBee wireless network for monitoring and controlling greenhouse climate. Int. J. Eng. Adv. Technol. **2**(1), 35–39 (2012)
51. Zhang, T., Lu, J., Hu, F., Hao, Q.: Bluetooth low energy for wearable sensor-based healthcare systems. IEEE Healthc. Innov. Conf. (HIC) **2014**, 251–254 (2014)
52. Shah, R.C., Nachman, L., Wan, C.: On the performance of Bluetooth and IEEE 802.15. 4 radios in a body area network. In: Proceedings of the ICST 3rd International Conference on Body area Networks, pp. 1–9 (2008)
53. Pengg, F., Barras, D., Kucera, M., Scolari, N., Vouilloz, A.: A low power miniaturized 1.95 mm 2 fully integrated transceiver with fast PLL mode for IEEE 802.15. 4/bluetooth smart and proprietary 2.4 GHz applications. In: 2013 IEEE Radio Frequency Integrated Circuits Symposium (RFIC), pp. 71–74 (2013)
54. Bhagwat, P.: Bluetooth: technology for short-range wireless apps. IEEE Internet Comput. **5**(3), 96–103 (2001)
55. Haartsen, J.: Bluetooth-The universal radio interface for ad hoc, wireless connectivity. Ericsson Rev. **3**(1), 110–117 (1998)
56. Gomez, C., Oller, J., Paradells, J.: Overview and evaluation of BlueTooth low energy: an emerging low-power wireless technology. Sensors **12**(9), 11734–11753 (2012)
57. Nieminen, J., et al.: Networking solutions for connecting Bluetooth low energy enabled machines to the internet of things. IEEE Netw. **28**(6), 83–90 (2014)
58. Esmailzadeh, R., Nakagawa, M.: TDD-CDMA for wireless communications. Artech House (2003)
59. Choi, Y., Lee, H.B., Park, S.-B., Hong, B.-H., Lee, S.-Y., Tchah, K.H.: A unified GFSK, $π$/4-shifted DQPSK, and 8-DPSK baseband controller for enhanced data rate Bluetooth SoC. Curr. Appl. Phys. **6**(5), 862–872 (2006)
60. Kaushik, S., Kaushik, M.: An overview of technical aspect for wireless fidelity for Wi-Fi and wireless networks. Int. J. Adv. Electr. Electron. Eng. **1**(2), 173–178 (2012)
61. Khan, J., Khwaja, A.: Building secure wireless networks with 802.11. Wiley (2003)
62. Dolińska, I., Jakubowski, M., Masiukiewicz, A.: Interference comparison in wi-fi 2.4 ghz and 5 ghz bands. In: 2017 International Conference on Information and Digital Technologies (IDT), pp. 106–112 (2017)

63. Lee, J.-S., Su, Y.-W., Shen, C.-C.: A comparative study of wireless protocols: Bluetooth, UWB, ZigBee, and Wi-Fi. In: IECON 2007–33rd Annual Conference of the IEEE Industrial Electronics Society, pp. 46–51 (2007)
64. Fairhurst, G.: Carrier sense multiple access with collision detection (CSMA/CD). http://www.erg.abdn.ac.uk/users/gorry/course/lan-pages/csmacd.html (2004)
65. Kharel, J., Reda, H.T., Shin, S.Y.: Fog computing-based smart health monitoring system deploying Lora wireless communication. IETE Tech. Rev. **36**(1), 69–82 (2019)
66. Chaudhari, B., Borkar, S.: Design considerations and network architectures for low-power wide-area networks. In: LPWAN Technologies for IoT and M2M Applications. Elsevier, pp. 15–35 (2020)
67. Chaudhari, B.S., Zennaro, M., Borkar, S.: LPWAN technologies: emerging application characteristics, requirements, and design considerations. Futur. Internet **12**(3), 46 (2020)
68. Anand, S., Routray, S.K.: Issues and challenges in healthcare narrowband IoT. International Conference on Inventive Communication and Computational Technologies (ICICCT) **2017**, 486–489 (2017)
69. Alagarsamy, G., Shanthini, J., Balaji, G.N.: A survey on technologies and challenges of LPWA for narrowband IoT. Trends Cloud-Based IoT, pp. 73–84 (2020)
70. Lauridsen, M., Kovács, I.Z., Mogensen, P., Sorensen, M., Holst, S.: Coverage and capacity analysis of LTE-M and NB-IoT in a rural area. In: 2016 IEEE 84th Vehicular Technology Conference (VTC-Fall), pp. 1–5 (2016)
71. Qadir, Q.M., Rashid, T.A., Al-Salihi, N.K., Ismael, B., Kist, A.A., Zhang, Z.: Low power wide area networks: a survey of enabling technologies, applications and interoperability needs. IEEE Access **6**(C), 77454–77473 (2018). https://doi.org/10.1109/ACCESS.2018.2883151
72. Nshimba, K.T.: Identification of security threats to data privacy posed by smart appliances in home area networks. North-West University (South Africa) (2020)
73. Sanchez-Gomez, J., Garcia-Carrillo, D., Marin-Perez, R., Skarmeta, A.F.: Secure authentication and credential establishment in narrowband IoT and 5G. Sensors **20**(3), 882 (2020)
74. del Peral-Rosado, J.A., López-Salcedo, J.A., Seco-Granados, G.: Impact of frequency-hopping NB-IoT positioning in 4G and future 5G networks. IEEE International Conference on Communications Workshops (ICC Workshops) **2017**, 815–820 (2017)
75. Ribeiro, L.E., Tokikawa, D.W., Rebelatto, J.L., Brante, G.: Comparison between LoRa and NB-IoT coverage in urban and rural Southern Brazil regions. Ann. Telecommun. **75**(11), 755–766 (2020)
76. Khan, I.: Suitability of LoRa, Sigfox, and NB-IoT for Different Internet-of-Things Applications (2019)
77. Mekki, K., Bajic, E., Chaxel, F., Meyer, F.: Overview of cellular LPWAN technologies for IoT deployment: Sigfox, LoRaWAN, and NB-IoT. IEEE International Conference on Pervasive Computing and Communications Workshops (Percom Workshops) **2018**, 197–202 (2018)
78. Sinha, R.S., Wei, Y., Hwang, S.-H.: A survey on LPWA technology: LoRa and NB-IoT. ICT Express **3**(1), 14–21 (2017)
79. Dasiga, S., Bhatia, A.A.R., Bhirangi, A., Siddiqua, A.: LoRa for the last mile connectivity in IoT. In: 2020 9th International Conference System Modeling and Advancement in Research Trends (SMART), pp. 195–200 (2020)
80. Drüagulinescu, A.M.C., Manea, A.F., Fratu, O., Drüagulinescu, A.: LoRa-based medical IoT system architecture and testbed. Wirel. Pers. Commun. 1–23 (2020)
81. Ikpehai, A., et al.: Low-power wide-area network technologies for internet-of-things: a comparative review. IEEE Internet Things J. **6**(2), 2225–2240 (2018)
82. Buurman, B., Kamruzzaman, J., Karmakar, G., Islam, S.: Low-power wide-area networks: design goals, architecture, suitability to use cases and research challenges. IEEE Access **8**, 17179–17220 (2020). https://doi.org/10.1109/ACCESS.2020.2968057
83. Aernouts, M., Berkvens, R., Van Vlaanderen, K., Weyn, M.: Sigfox and LoRaWAN datasets for fingerprint localization in large urban and rural areas. Data **3**(2), 13 (2018)
84. Poursafar, N., Alahi, M.E.E., Mukhopadhyay, S.: Long-range wireless technologies for IoT applications: a review. Eleventh International Conference on Sensing Technology (ICST) **2017**, 1–6 (2017)

85. Stoyanov, V., Poulkov, V., Valkova-Jarvis, Z.: Low power wide area networks operating in the ism band-overview and unresolved challenges. In: International Conference on Future Access Enablers of Ubiquitous and Intelligent Infrastructures, pp. 96–109 (2019)
86. Ismail, D., Rahman, M., Saifullah, A.: Low-power wide-area networks: opportunities, challenges, and directions. In: Proceedings of the Workshop Program of the 19th International Conference on Distributed Computing and Networking, pp. 1–6 (2018)
87. Weyn, M., Ergeerts, G., Wante, L., Vercauteren, C., Hellinckx, P.: Survey of the DASH7 alliance protocol for 433 MHz wireless sensor communication. Int. J. Distrib. Sens. Netw. **9**(12), 870430 (2013)
88. Ergeerts, G., et al.: DASH7 alliance protocol in monitoring applications. In: 2015 10th International Conference on P2P, Parallel, Grid, Cloud and Internet Computing (3PGCIC), pp. 623–628 (2015)
89. Mainsah, V.L., Hasnat, M.R., Manoharan, A.: Survey on DASH7: A New Open Source Standard for Wireless Sensor Networks
90. Pathak, G., Gutierrez, J., Rehman, S.U.: Security in low powered wide area networks: opportunities for software-defined network-supported solutions. Electron. **9**(8), 1–24 (2020). https://doi.org/10.3390/electronics9081195
91. Al-Kashoash, H.A.A., Kemp, A.H.: Comparison of 6LoWPAN and LPWAN for the Internet of Things. Aust. J. Electr. Electron. Eng. **13**(4), 268–274 (2016). https://doi.org/10.1080/1448837X.2017.1409920
92. Basahel, S.B., Bajaba, S., Yamin, M., Mohanty, S.N., Lydia, E.L.: Teamwork optimization with deep learning based fall detection for IoT-enabled smart healthcare system. Comput. Mater. Continua. **75**(1), 1353–1369 (2023). ISSN: 1546-218. https://www.techscience.com/cmc/v75n1/51539
93. Potluri, S., Mohanty, S.N.: An efficient scheduling mechanism for IoT based home automation system. Int. J. Electron. Bus. **16**(2), 147–156 (2021). https://doi.org/10.1504/IJEB.2021.115719. ISSN: 1470-6067

Comprehensive Review of Security Challenges and Issues in Wireless Sensor Networks Integrated with IoT

Lopamudra Prusty, Pratik Kumar Swain, Suneeta Satpathy, and Satyasundar Mahapatra

Abstract A wireless sensor network (WSN) is a distributed sensor network of wireless devices that can gather and communicate information through wireless links. The gathered information will be sent to the base station or sink for further processing. The Sensor in WSN communicates wirelessly, so device location can be changed at any time and network settings should be flexible. WSN are spatially distributed autonomous sensors to monitor physical environmental conditions such as temperature, sound, pressure etc. It started to connect with internet of things (IoT) through internet which has also an ability to connect the sensor nodes. Now, huge amounts of data which are collected by WSNs are being accessed by IoT. The objective of this paper is to identify security challenges and various issues in wireless sensor networks which are integrated with IoT. The wireless sensor network currently must deal with the main issues that are security. For certain type of attacks like eavesdropping, jamming, and spoofing. Finally, to address security and privacy issues in IoT domain of WSN a light has been shed on a few obstacles and potential future research avenues by using Machine Learning algorithms.

Keywords Wireless sensor network · Security · Internet of Things · Machine learning

---

L. Prusty (✉)
Computer Science Department, ABIT Engineering College, Cuttack, Odisha 753014, India
e-mail: lopa.lipina@gmail.com

P. K. Swain
Faculty of Engineering and Technology, Sri Sri University, Cuttack, India

S. Satpathy
Center For Cyber Security, Siksha 'O' Anusandhan Deemed to Be University, Bhubaneswar, Odisha 751030, India

S. Mahapatra
Pranveer Singh Institute of Technology, Kanpur, Uttar Pradesh 209305, India
e-mail: satyasundara.mahapatra@psit.ac.in

S. N. Mohanty et al. (eds.), *Explainable IoT Applications: A Demystification*, Information Systems Engineering and Management 21,
https://doi.org/10.1007/978-3-031-74885-1_30

1 Introduction

In recent years wireless communications and micro electro Mechanical systems (MEMS) advancements have made easier creation of wireless sensor networks (WSNs), in which sensor nodes gather meaningful data from the surroundings. Wireless sensor network is a device network for information communication from a monitored field through wireless links [1]. Connecting the devices in a network with an approach of using wireless link for transmitting data from sensors with proper observation. It can also be said collection of sensing devices that are communicated wirelessly. For example if in some condition people are not able to visit hospital because of some issues. So here the concept of WSN comes. Using this technology patient can interact with doctor and check him. In case of emergency, he/she may receive the message from doctor [2].

Information can be transferred without requiring a physical link between two or more sites wirelessly. Due to the lack of any physical infrastructure wireless communication has various benefits which would frequently involve distance or space compressing. Wireless communication is important for data transmission due to flexibility and cost effectiveness. A wireless network can be configured more flexible and adaptably than a conventional network [3]. As the wireless network is simple to setup and does not require cables so it is comparatively less expensive. The main purpose of WSNs is to gather sensor data and send it to a central processing unit for analysis and decision making. However a wireless network is a more general word which includes WSNs to cover a variety of wireless communication technologies. Wireless network can be used for data transport, internet access and mobile communication. Wireless sensor network are designed for collecting and monitoring sensor data which can support a variety of communication protocols and technologies.

Wireless sensor network in internet of things (IoT) interconnect various devices to exchange data and communicate with one another, improving automation and efficiency. For example in smart homes, sensors control lighting, heating and security systems. IoT application areas as shown in Fig. 1 will therefore grow steadily and dramatically for all facets of life if this goal is realized. Security is a major issue for WSNs and IoT, especially if they are being used for mission critical functions. For example, in tactical military applications where friendly forces could be casualties due to a security breach in the network a combat zone. Different types of benefits of IoT are discussed below:

- Efficient resource utilization—Utilization of our resources must be in a proper way or it should not be wasted which is helped by IoT.
- Save time—saving the time is equal to saving the life
- Human efforts and errors—It is near about same as resource utilization.
- Security—It is the most important factor in our surrounding.
- User friendly/Easy to use.

In this unit the article presents an overview of wireless sensor network integrated with IoT. Security in WSN and IoT are a major issue especially if they are collected

Fig. 1 Various applications of WSN [3]

for mission critical task [4]. For example health cares section in case of patient privacy if their confidential health data were exposed through the leakages from misbehaving nodes or due to system failure. So in computer science solution to defend against security attacks towards network comprises of three essential components that is prevention, Detection and mitigation.

This article narrates a brief explanation of the literature survey in Sect. 2. Subsequently, an overview of the challenges encountered and attacks by WSN for network security in the form of problem statement in sub point of Sect. 3 and explained some research questions of all articles. Further Sect. 4 details the discussion of types of solutions to defend attacks towards WSN followed by the concluding remarks in Sect. 5.

2 Literature Survey

For conducting a very thorough, complete and methodical review of relevant and on-point papers on a particular research issue, a systematic literature review methodology is a detailed procedure. The protocol of a systematic literature review plays a crucial role in ensuring that the review process is open and duplicable. For doing systematic literature review the major objective is to identify, evaluate and understand various information's which will guide the review process. From the final selection of applicable research papers the examination of the information acquired will next be used to research process.

In the first reference paper [5] of I. Butun there are two major steps to ensure security i.e. intrusion prevention and intrusion detection. Network is secured as a first line of defense by using intrusion prevention method such as authorization, authentication and access control. Butun et al. [6] have made a survey of the state-of-the-art in intrusion detection system that is purposed for WSN are presented. Detailed information about IDS and a brief survey of IDSs proposed for mobile Ad-Hoc networks are presented in this paper and applicability of those systems to WSN are discussed. V. Friedman has started to merge WSN with IoT [7] through the introduction of internet access capability in sensor nodes. Thereby IoT over the internet provide large amount of data collected by WSN. Then Lin et al. [8] particularly shows the relationship between cyber physical systems and IoT which play important roles in realizing an intelligent cyber physical world. This also investigate the relationship between IoT and fog computing and discuss issues in fog computing based IoT. M. Kocakulak and I. Butun has discusses detailed overview of WSNs [9]. It also accesses the technology as well as characteristics of WSNs and provides a review of WSN application integrated with IoT.

In the paper of Fang et al. [10] shows a novel IIS that combines IoT, cloud computing, Geo-informatics, geographical information system and global positioning system and e-science for environmental monitoring and management with a case study on regional climate change and its ecological effects. Abomhara and Køien [11] have discussed the IoT vision, existing security threats and open challenges in the domain of IoT. The current state of research on IoT security requirements is discussed and future research directions with respect to IoT security and privacy are presented. Then Gope and Hwang [12] show the major security requirements in BSN-based modern health care system. In addition to that a secure IoT based healthcare system using BSN called BSN-care is proposed which can efficiently accomplish those requirements. Li et al. [13], introduced various definitions of IoT and emerging techniques for the implementation of IoT. Here open issues are discussed related to the IoT applications are explored followed by some major challenges which need addressing by the research community and corresponding potential solutions are investigated. Zhu et al. [14] consider a typical threat in the latter category is known as the node replication attack. It also reproduces the node in large quantity. A.-S. K. Pathan, H.-W. Lee and C. S. Hong has investigated the security related issues and challenges in wireless sensor networks. Security threats are identified [15]. The holistic view of security is discussed for ensuring layered and robust security in wireless sensor networks [16]. Sharma and Ghose [17] focus on physical assault and issues in remote sensor systems. The methodology of security discovery against physical assaults are talked [4]. Kumar and Sanyal [18] has made a survey on the paper which has given importance on the security arising out of the information exchange technologies that is used in internet of Things. I. Butun, N. Pereira, and M. Gidlund provide the first importance into the security of LoRaWAN. Overview of the protocol and several threats to the new version of protocol are presented [19]. Then I. Butun, N. Pereira, and M. Gidlund clarify and review the security aspects of LoRaWAN for Comprehensive security risk analysis is provided by ETSI guidelines and discusses several points to the security risk [20]. The research paper of M. Eldefrawy, I. Butun, N. Pereira, and M. Gidlund concludes a formal study of LoRaWAN security is presented that is an increasingly popular technology that defines the structure and operation of LPWAN networks [21]. Then J. Haxhibeqiri, E. De Poorter, I. Moerman, and J. Hoebeke provides an overview of the research work on LoRa and LoRaWAN that has been published in between 2015 to 2018 [22]. R. S. Sinha, Y. Wei, and S.-H. Hwang provides a comprehensive survey on NB-IoT and LoRa which is an efficient solution to connecting devices [23].

This paper of I. Butun provides privacy and trust for IoT that is presented as a multidimensional relation which efforts in doing a new path for security in IoT, especially for the FIDO applications [24]. I.-R. Chen, J. Guo, and F. Bao developed a technique based on distributed collaborative filtering to select feedback using similarity rating friendship, community of interest relationships and social contact as the filter. The technique also supports service composition applications in SOA-based IoT systems [25]. Relevant results are presented in the paper written by D. Ott, C. Vishik, D. Grawrock, and A. Rajan from Intel's university research program on trust evidence, that include participants from the US and Europe [26]. A formal study of LoRaWAN security is presented by S. Sicari, A. Rizzardi, L. A. Grieco, and A. Coen-Porisini which is an increasingly popular technology that defines the structure and operation of LPWAN networks based on the LoRa physical layer [27]. The properties of IoT trust management's objective and a survey are provided by Z. Yan, P. Zhang, and A. V. Vasilakos on the current literature advances towards trustworthy IoT [28]. Now Shelby et al. [29] specifies simple protocol extension for CoAP that are specified enabling CoAP clients to observe resources [30]. Banks and Gupta as well as Yokotani and Sasaki purposes enhancements for better performance of MQTT. Here the performance of HTTP is compared with that of MQTT which is a type of name based protocol [31, 32]. Brand et al. as well as Zhang and Li has made an in-depth study on a popular implementation of the routing protocol for low power and loss network that provides insights and guidelines for the adoption of these standards [33, 34]. The behavior of the low power protocol is analyzed by V. P. Nikshepa and U. K. K. Shenoy that is called 6LowPan abbreviated as IPv6 over low power wireless personal area network under many types of topological scenarios [35]. A. Fabre et al. has done implementation and deployment of a system is focused in this research paper using contiki, 6LoWPAN over an 868 MHz radio network, together with a CoAP as a standard application layer protocol [36]. To combine the standardization effort for allowing low power wireless devices to communicate D. Dujovne, T. Watteyne, X. Vilajosana, and P. Thubert have used the internet protocol that become true fingers of internet and greatly simplifying their integration into existing networks [37]. T. Watteyne, P. Tuset-Peiro, X. Vilajosana, S. Pollin, and B. Krishnamachari discussed the way the 6TiSCH stack can be incorporated into existing and new way to teach the next generation of electrical engineering and computer science professionals for designing and deploying the networks [38]. T. Chang, T. Watteyne, K. Pister, and Q. Wang presented an adaptive synchronization technique in this article that allows a node to learn and predict the way its clock is drifting relative to its neighbors and the instants is coordinated at which the nodes re-synchronize [39].

The environment, problem statement is described by T. Watteyne, M. Palattella, and L. Grieco in their paper and goals for using the time-slotted channel hopping Medium Access [40]. The research focus on cyber manufacturing systems that evolve as a significant terms and also introduces the development of IoT along with digital factory and cyber-physical systems [41]. S. Firestorm, I. Butun, M. Eldefrawy, U. Jennehag, and M. Gidlund presented the challenges, needs and requirements of industrial applications in this research paper when it comes to securing IoT systems [42]. H. Song, D. B. Rawat, S. Jeschke, and C. Brecher, in their paper [43] fully describes on foundation, principle and applications of cyber physical systems. K. Xu, X. Wang, W. Wei, H. Song, and B. Mao purposed the core idea of SDN and the software defined smart home platform are referred in this article along with the core technology of SDSH and the application value of fore core technology are discussed [44]. A cross layer protocol for internet of medical things (IoMT) is presented in this paper [45]. The cross layer communication of physical, data link, and routing layers have considered in proposed methodology for multimedia applications. Y. Sun, H. Song, A. J. Jara, and R. Bie describes the concept of smart and connected communities that evolved from the idea of smart cities [46]. A security and privacy preservation scheme is purposed [47] to solve the issues related to face identification and resolution technology. An outline based on the fog computing face identification and resolution framework is given and summarizes the security and privacy issue. The research paper is focused on enhancing the secrecy of wireless communications in CPS using physical layer security techniques [48]. A amplify and forward relay network are studied where all devices are equipped with a single antenna. J. Zhu, Y. Song, D. Jiang, and H. Song purposed through multiple channels the problem of how to achieve the proper strategy to transmit packets of different buffers to maximize the system throughputs [49]. The integration of fog computing with IoT and its implications are surveyed in this paper [50]. This paper aims to find and emphasize problems; especially security related which arise with the employment of fog computing by IoT [51]. The basics of smart cities are provided in this paper [52] and examine the possible future trends of this technology. Cyber security and privacy challenges in TCPS, novel, transformative, multidisciplinary approaches are addressed in this paper at the confluence of cyber security, privacy and TCPS are needed [53]. KhaledTelli, OkbaKraa, Yassine Himeur, Mohamed Boumehraz, Shadi Atalla, Wathiq Mansoor, Abdelmalik Ouamane encompasses an experimental setup that includes the use of Pseudo-random binary sequence signals for excitations to show the efficacy of NARX-NN in predicting and controlling quad rotor dynamics [54]. A mobile application like BAdDrolds leveraging machine learning are introduced for detecting malware on resource constrained devices [55]. The economic affects the university research which is compared in the United States and japan-countries. In this paper that is similar in economic and technological capabilities but different in culture, tradition and institutional structure [56].

An introduction to the intersection of both the fields is provided in this paper with special emphasis on the techniques used to protect the data [57]. Mohammad Al-Rubaie and J Morris Chang gives an introduction of a wearable fingerprinting technique focused on Bluetooth classic protocol are done in this paper that is a common protocol used by the wearable's and other IoT devices [58]. A mutual information based algorithm is purposed in this paper [59] which analytically selects the optimal feature for classification. This mutual information can handle linearly and nonlinearly dependent features that are based on feature selection algorithm.

3 Problem Statements

For conducting a very thorough, complete and methodical review of relevant and on-point articles on a particular research issue or topic, a problem statement methodology is a detailed procedure. The protocol of a problem statement plays a crucial role in ensuring that the review process is open and duplicable. For doing problem statement the major objective is to identify, evaluate and understand various strategies which will guide the review process. From the final selection of applicable research papers the examination of the information acquired will next be used to answer the following objectives as a complete problem statement.

O1. Discover the challenges that are encountered in WSN for network security.
O2. Discover the solutions that defend against various attacks towards the WSNs on all the layers of OSI model.
O3. Attacks and security measures to be surveyed in the WSNs of industrial supervisory control and data acquisition (SCADA) systems.
O4. Challenges of the IoT to be discussed for security purpose.

For finding the solutions for the above objectives the following strategies are executed:

3.1 Search Strategy

A thorough search procedure is started that allow for the completion of systematic survey. A good search strategy must be developed for a systematic survey to be successful, which requires selection of number of relevant databases from which to draw relevant material. Between 2018 and 2023, for this investigation a two-step search process was used, with step 1 employing the digital libraries of the ACM, IEEE, springer, and science direct. The academic search engine Google scholar is used to make sure no relevant literature was overlooked in step two of the search. There are various classifications for attacks towards the WSNs in the literature. Among these the attacker's activity will consider as main categorization and open systems interconnection (OSI) model is targeted as sub classifications. Figure 2 depicts how several steps were carried out during the survey.

3.2 Study Selection Process

In this research the primary objective of study selection was to filter out irrelevant material that did not align with the specified research objectives. Sensor network are highly distributed, lightweight nodes, deployed in large number to monitor the environment or system sensor node consists of three things that are sensor subsystem,

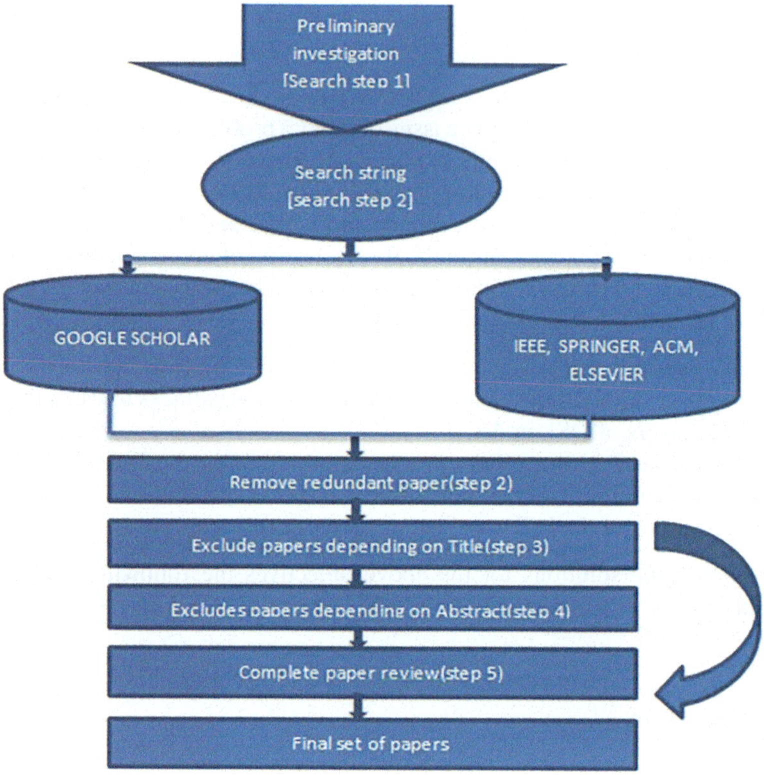

Fig. 2 SLR of attacks towards WSNs

processing system and communication system. In search step 1, a total of 4806 entries were formed by combining the first 1504 entries from search step 2 with the 3302 entries collected in the initial search phase. Following the removal of 302 duplicate items from the previous stage, articles were further eliminated on their title (3351), abstracts (868), and full texts (393). Ultimately, after the fifth round of screening, a total of 69 research publications were chosen for inclusion in the study.

4 Reference Checking

The references were evaluated after reviewing the entire text of the 34 papers to make sure no relevant materials had been missed that were selected for the study. Throughout the evaluation, 75 more publications were discovered and they were evaluated for conformity with the inclusion and exclusion criteria using their abstracts, titles and full texts. As a result of this approach 50 articles were eliminated based on abstracts, 12 articles were excluded based on full texts, and 11 articles were discarded based on titles. For reference checking and final selection leaving only two articles, 71 articles were removed after these filtering steps. Finally 54 papers were selected after elimination of 71 articles from the filtering steps.

4.1 Data Extraction

Each article was meticulously read together the necessary data in order to address the study's objective. In a pre-designed form relevant information was then extracted and recorded that includes various fields, methodology, such as article title, dataset utilized, number of features, identification of attack and legitimate classes, preprocessing strategy, experiment model performance optimization, strengths and weakness, performance metrics, along with a summary of the article. The research question and critical evaluation of the compiled article are investigated by utilization of these fields.

4.2 Inclusion Criteria

- Papers that were relevant were included in the selection process.
- A machine learning technique for recognizing active and passive attacks towards WSNs were outlined.
- The study's objectives were addressed in research findings.
- On earlier research in the field, and investigations were build that closely related but differed in some important ways are considered independent primary investigation.
- Our primary goal of this SLR is to study briefly of attacks towards WSNs and IoT that was released between 2018 and 2023.

4.3 Exclusion Criteria

- Insufficient information was gained from redundant research studies and articles.
- For writing papers other languages except the English language are used.
- Subjects, editorial pieces, critiques, discussions, brief communications, posters, encyclopedia entries, ongoing research, keynote presentations, and invited speeches unrelated to the study topic were studied.

Finally all the sub points of this section give a brief idea of threats and challenges of WSN integrated with IoT. The filtering steps are done to eliminate the duplicate entries.

5 Discussions

In this sub-section various types of solutions to defend attacks towards WSN are discussed. Based on the systematic literature review this section looks at work that looks into the kind of attacks and security measures that are studied in wireless sensor networks of industrial supervisory control and data acquisition systems are Malware attacks, DoS attacks and IoT vulnerabilities [60, 61]. A search approach strategy is used that entails eliminating plenty of unused papers in order to reduce the number of publications that exactly meet our study objective. A quality evaluation criterion is also employed to ensure that the selected papers provide results that have been synthesized. Due to different variant and attack patterns detecting the kind of attacks can be difficult, makes it as a challenge to differentiate them from normal traffic. Various attacks and security measures towards WSNs are discussed and there are some issues in applying stochastic process.

Security is needed in sensor networks because sensors deploy in critical networks [62]. The sensor networks are needed to secure if not the enemy will attack the sensor network and manipulate. Denial of service attacks are the major threat to the systems connected to the internet, especially for e-commerce, financial services and government services. Computer network majorly works on client server architecture [63]. There are number of clients sending request to the same server. Server offers various services like HTTP or FTP (file transfer protocol) with which clients can interact to or share information [64, 65]. When server cannot respond to client request DOS condition arises. Dos occur when there is a description in sending information or communication.

A. Defense against passive attacks

Defend against passive attacks towards WSN is presented in this subsection.

- Defense against passive information gathering
- Defense against code attack
- Defense against active attacks

This section summarizes the solutions to defend against active attacks.

- Defense on physical layer
- Defense on data link layer
- Defense against network layer
- Defense on transport layer
- Defense against Application layer.

Figure 3a presents list of attacks for security major in detection method and Fig. 3b presents majority percentage of the security in prevention and mitigation method according to OSI representation. Accordingly; in the detection part as shown in Fig. 3a, 87% of the solutions are provided for the network layer, and 7%, 4%, 2% are for the physical, MAC and application layers, respectively. In this figure 87% is considered as DDos attacks, 7% for the MITM (Man-in-the-middle) attack, 4%

for malware attack and 2% for computer worm attack. It is apparent that majority of the solutions are designed to work on network layer. In the prevention and mitigation partas shown in Fig. 3b, 47% of the solutions are provided for the network layer, and 17%, 12%, 10%, are for the all, physical, transport and MAC application layers, respectively. The methods that are used in detecting various attacks are stochastic process, machine learning, game theory and optimization theory respectively as shown in pie chart. Here, distribution of the solutions among OSI layers is compared to detection methods, yet still there is an accumulation on network layer.

There are also some open issues and solutions on cyber security issues of WSN presented in this work are solved by the vast amount of efforts in the research community [66]. Normally people use internet to get connected with each other or to share data with each other but there is a change in technology, through which all can connect with each other physically which is known as internet of things. For example in case of ordering food items one can track the delivery person. Some important points of IoT are described below [65, 67]:

- Describes the network of physical objects- "things".

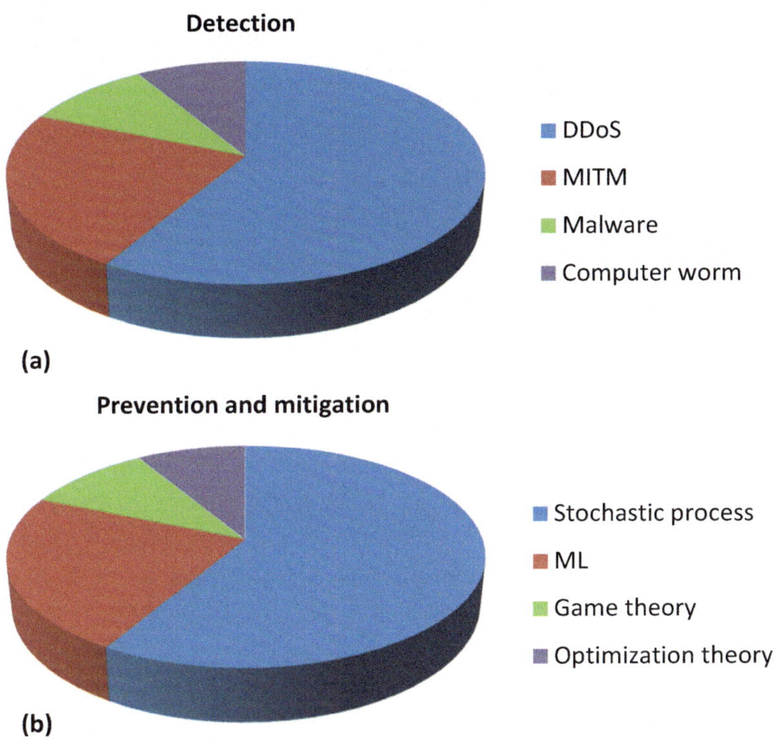

Fig. 3 **a** List of attacks for security in detection method. **b** Majority percentage of the security in prevention and mitigation method

- Internet of things [68, 69] are the devices embedded with sensors, software and other technologies for the purpose of connecting and exchanging data with other devices and systems over the internet.
- IoT is a sensor of network of billions of smart devices that connect people, systems and other applications to collect and share data.
- It may be ordinary household objects to sophisticated industrial tools.

The working of IoT is discussed below:

There are many examples of IoT, Smart watches are among the most popular example of the Internet of things [70, 71]. Fit bits and Apple Watches connect to other devices to share data which are the example of wearable IoT technology.

- Sensor in every device
- Every device emits data regarding working states of things
- IoT provides platform to all devices to dump their data
- IoT platform collect and analyses the data
- Valuable information is extracted
- And information is shared to other devices for decision making or information.

Benefits of IoT [72, 73]

Our lives have changed for IoT by making repetitious tasks more convenient, enabling wearable technology for greater health and wellness, enhancing home security systems and boosting connectivity.

- Real time information
- Help in monitoring the overall business process
- Help in improving the experience of the customer
- Save time and money
- Productivity of the employee will increase
- Business decisions can be made better by IoT
- Revenue generation can also be increased
- IoT encourages companies [74, 75] to rethink the ways they approach their business, industries, markets and can also improve their strategies.

5.1 Challenges Encountered by WSN for Network Security

As discussed in spite of highly practical usefulness of wireless sensor network there are some challenges. Different researches are categorized according to the methods and challenges are discussed below:

Scalability—It means how a wireless sensor network will be scaled over a large period of network. As the network is increased area the throughput decreases.

Quality of Service—If network of three nodes are there then the first node moves to such distance so that the distance should be checked whether the first distance between n1 and n3 (n1-first node, n2-second node, n3-third node) is less than second distance.

Energy Efficiency—It is constrained and don't have energy backup. Regular source of backup is not there.

Security—This is limited because if an attacker node comes in between three nodes then it will be connected with these three nodes and starts to send malicious packets.

5.2 Solutions to Defend Against Various Attacks Towards the WSNs on All of the Layers of OSI Model

The different types of solutions against active and passive security attack towards WSNs networks on all of the layers of OSI model are discussed below:

Defense Against Passive Attacks—In this subsection, defense against passive attacks towards IoT and WSN are discussed. Since the communication of WSN is achieved through the air, not the wires. There is no warranty according to the nature of wireless communications that the future planned packets arrive at the intended parties only. Routine physical checking of sensor nodes has been required by detection of the tampering attempts by eye or by special equipment like magnifier.

Defense Against Active Attacks—As discussed this section the solutions to defend against active attacks are discussed below. Time division technique that would provide dedicated time slots to each node for transmission of the packets In order to detect black hole attacks in WSN in active trust the author have created multi-detection routs in such areas which have remaining energy.

5.3 Attacks and Security Measures that Are Surveyed in the WSNs of Industrial Supervisory Control and Data Acquisition (SCADA) Systems

This segment describes various methods that are used in detecting various attacks. Various types of malicious attacks against trust model are presented in this paper. In constructing the network trust in Wireless sensor network play a key role. Based on this study the methods are divided in to four groups:

- Stochastic process
- Machine learning
- Game theory
- Optimization theory.

The attacks and security measures that are surveyed in WSNs of industrial supervisory control and data acquisition systems are Malware attacks, DoS attacks and IoT vulnerabilities. Tampering type of replication attack is detected. The replication attack of nodes in an environment is not secured for the replication of malicious nodes in to several clones.

5.4 Challenges of the IoT is Discussed for Security Purpose

This segment describes various challenges of the IoT that are used for security purpose.

Scalability: Addressing, naming, managing and servicing millions of devices are a unique challenge.

Communications: Various technologies are used by IoT devices, such as wired or wireless communications, e.g., Bluetooth, ZigBee and LPWAN.

Energy Consumption: This is one of the main challenging constraints of the IoT. Any kind of algorithm running on IoT devices needs to be designed with light-weight processing requirement.

Data Privacy: Privacy of the user data in the IoT can be an issue for some specific use cases.

Self-awareness: Smart objects of the IoT should self-organize themselves autonomously in order to fulfill some pre-determined specific tasks in responding real word environmental situations without too much human intervention.

Interoperability: In order for heterogeneous IoT devices to communicate, collaborate and share data with each other, there should be a pre-determined and standardized data exchange format.

From the discussions made throughout this paper along with its scalability the heterogeneity of the devices in IoT ecosystem causes several implications. While designing protocols for WSNs and IoT security is a key component. This paper shows all known types of security attacks towards WSN in IoT.

6 Conclusions and Future Scope

In IoT landscape research and development continues to grow in a fast pace as the application areas: cloud/fog based system, cyber physical systems, internet of drones, smart home appliance, industrial IoT and smart factories, are just namely a few prominent fields to mention [76]. Attacks by DDOs can hurt internet users in a number of ways which continues to be a severe danger to many big and small enterprises. To recognize DDOs attack various several machine learning methods are utilized [77–79]. The IoT is a large scale complex architectural design consist a variety of heterogeneous devices; therefore scalability, transparency, and reliability are most prominent issues to be solved. In the first place these issues need to consider security-related initiative. By designing lightweight security protocols and cryptography algorithm this can be achieved that are tailored according to specific needs of the resource-constrained devices of the IoT. From the discussions made throughout this paper, heterogeneity of the devices in IoT ecosystem can be deduced along with its scalability causes several implications, in terms of security. Although some of the new vulnerabilities can be discovered on time, related security patches cannot be installed to the end devices in a timely manner due to the mentioned IoT network implications above [80, 81]. Therefore IDS techniques become more important for IoT systems, as some of them are even efficient against zero-day-attacks. If the necessary IDS techniques require high processing power, gateway devices can be employed for this purpose. Going throughout the paper it is concluded that security is a key component when designing protocols for WSNs and IoT.

This paper has been studied with a hope that this will provide information about the field of WSNs and IoT, by leading them to produce more robust and secure network solutions along with the intrinsic relation between block chain and communication, network and computing are reveled. In future it's needed to think more about how to find solutions to defend the attacks towards WSNs and in case of open issues and solutions of cyber security, it requires further investigation, attention and deep research from the researchers.

References

1. Chen, C.-Y., Chao, H.-C.: A survey of key distribution in wireless sensor networks. Security Commun. Newt. **7**(12), 2495–2508 (2014)
2. Han, G., Jiang, J., Shu, L., Niu, J., Chao, H.-C.: Management and applications of trust in wireless sensor networks: a survey. J. Comput. Syst. Sci.Comput. Syst. Sci. **80**(3), 602–617 (2014)
3. Pathan, A.-S. K., Lee, H.-W., Hong, C.S.: Security in wireless sensor networks: Issues and challenges. In: Proceedings of 8th International Conference on Advanced Communication Technology (ICACT), vol. 2, p. 6 (2006)
4. Shabana, K., Fida, N., Khan, F., Jan, S.R., Rehman, M.U.: Security issues and attacks in wireless sensor networks. Int. J. Adv. Res. Comput. Sci. Electron. Eng. **5**(7), 81 (2016)
5. Butun, I.: Prevention and detection of intrusions in wireless sensor networks. Ph.D. Dissertation, Department of Electrical Engineering, University of South Florida, Tampa, FL, USA (2013)
6. Butun, I., Morgera, S.D., Sankar, R.: A survey of intrusion detection systems in wireless sensor networks. IEEE Communications Surveys & Tutorials, **16**(1), 266–282, 1st Quart (2014)
7. Friedman, V.: On the Edge: Solving the Challenges of Edge Computing in the Era of IoT (2018). Available: https://data-economy.com/on-the-edge-solving-the-challenges-ofedge-computing-in-the-era-of-iot/
8. Lin, J., Yu, W., Zhang, N., Yang, X., Zhang, H., Zhao, W.: A survey on Internet of Things: Architecture, enabling technologies, security and privacy, and applications. IEEE Internet Things J. **4**(5), 1125–1142 (2017)

9. Kocakulak, M., Butun, I.: An overview of wireless sensor networks towards Internet of Things. In: Proceedings of IEEE 7th Annual Computing and Communication Workshop and Conference (CCWC), pp. 1–6 (2017)
10. Fang, S., et al.: An integrated system for regional environmental monitoring and management based on Internet of Things. IEEE Trans. Ind. Informant. **10**(2), 1596–1605 (2014)
11. Abomhara, M., Køien, G.M.: Security and privacy in the Internet of Things: Current status and open issues. In: Proceedings of IEEE International Conference on Privacy and Security in Mobile System (PRISMS), pp. 1–8 (2014)
12. Gope, P., Hwang, T.: BSN-care: a secure IoT-based modern healthcare system using body sensor network. IEEE Sensors J. **16**(5), 1368–1376 (2016)
13. Li, S., Da Xu, L., Zhao, S.: The Internet of Things: a survey. Inf. Syst. Front. **17**(2), 243–259 (2015)
14. Zhu, W.T., Zhou, J., Deng, R.H., Bao, F.: Detecting node replication attacks in wireless sensor networks: a survey. J. Newt. Comput. Appl. **35**(3), 1022–1034 (2012)
15. Pathan, A.-S. K., Lee, H.-W., Hong, C.S.: Security in wireless sensor networks: Issues and challenges. In: Proceedings of 8th International Conference on Advanced Communication and Technology (ICACT), vol. 2, p. 6 (2006)
16. Bartariya, S., Rastogi, A.: Security in wireless sensor networks: attacks and solutions. Int. J. Adv. Res. Comput. Commun. Eng. **5**(3), 214–220 (2016)
17. Sharma, K., Ghose, M.: Wireless sensor networks: an overview on its security threats. Int. J. Comput. Appl.Comput. Appl. **1**, 42–45 (2010)
18. Kumar, U., Sanyal, S.: Survey of security and privacy issues of Internet of Things. Int. J. Adv. Netw. Appl. **6**(4), 2372–2378 (2015)
19. Butun, I., Pereira, N., Gidlund, M.: Analysis of Lora WAN v1.1 security. In: Proceedings of 4th ACM MobiHoc Workshop on Experiences with the Design and Implementation of Smart Objects, p. 5 (2018)
20. Butun, I., Pereira, N., Gidlund, M.: Security risk analysis of Lora WAN and future directions. Future Internet **11**(1), 3 (2019)
21. Eldefrawy, M., Butun, I., Pereira, N., Gidlund, M.: Formal security analysis of Lora WAN. Comput. Netw.. Netw. **148**, 328–339 (2019)
22. Haxhibeqiri, J., De Poorter, E., Moerman, I., Hoebeke, J.: A survey of Lora WAN for IoT: from technology to application. Sensors **18**(11), 3995 (2018)
23. Sinha, R.S., Wei, Y., Hwang, S.-H.: A survey on LPWA technology: Lora and NB-IoT. ICT Exp. **3**(1), 14–21 (2017)
24. Butun, I.: Privacy and trust relations in Internet of Things from the user point of view. In: Proceedings of IEEE 7th Annual Computing and Communication Workshop and Conference (CCWC), pp. 1–52017
25. Chen, I.-R., Guo, J., Bao, F.: Trust management for SOA-based IoT and its application to service composition. IEEE Trans. Services Comput. **9**(3), 482–495 (2016)
26. Ott, D., Vishik, C., Grawrock, D., Rajan, A.: Trust evidence for IoT: Trust establishment from servers to sensors. In: Proceedings of ISSE, pp. 121–131 (2015)
27. Sicari, S., Rizzardi, A., Grieco, L.A., Coen-Porisini, A.: Security, privacy and trust in Internet of Things: the road ahead. Comput. Netw.. Netw. **76**, 146–164 (2015)
28. Yan, Z., Zhang, P., Vasilakos, A.V.: A survey on trust management for Internet of Things. J. Netw. Comput. Appl.Netw. Comput. Appl. **42**, 120–134 (2014)
29. Shelby, Z., Hartke, K., Bormann, C.: The constrained application protocol (COAP). IETF, Fremont, CA, USA, Rep. RFC 7252 (2014)
30. Hartke, K.: Observing resources in the constrained application protocol (COAP). IETF, Fremont, CA, USA, Rep. RFC 7641 (2015)
31. Banks, A., Gupta, R.: MQTT Version 3.1.1, OASIS Standard 29 (2014)
32. Yokotani, T., Sasaki, Y.: Comparison with HTTP and MQTT on required network resources for IoT. In: Proceedings of International Conference on Control, Electronics, Renewable Energy and Communications (ICCEREC), pp. 1–6 (2016)

33. Brandt, A., Baccelli, E., Cragie, R., van der Stok, P.: Applicability statement: the use of the routing protocol for low-power and lossy networks (RPL) protocol suite in home automation and building control. IETF, Fremont, CA, USA, Rep. RFC 7733 (2016)
34. Zhang, T., Li, X.: Evaluating and analyzing the performance of RPL in CONTIKI. In: Proceedings of ACM 1st International Workshop on Mobile Sensing Computer and Communication, pp. 19–24 (2014)
35. Nikshepa, V.P., Shenoy, U.K.K.: 6LoWPAN—performance analysis on low power networks. In: Proceedings of International Conference on Computer Networks and Communication Technologies (ICCNCT), vol. 15, p. 145 (2018)
36. Fabre, A., et al.: Deploying a 6LoWPAN, COAP, low power, wireless sensor network. In: Proceedings of 14th ACM Conference on Embedded Network Sensor System (CD-ROM), pp. 362–363 (2016)
37. Dujovne, D., Watteyne, T., Vilajosana, X., Thubert, P.: 6TiSCH: deterministic IP-enabled industrial Internet (of Things). IEEE Commun. Mag.Commun. Mag. **52**(12), 36–41 (2014)
38. Watteyne, T., Tuset-Peiro, P., Vilajosana, X., Pollin, S., Krishnamachari, B.: Teaching communication technologies and standards for the industrial IoT? Use 6TiSCH! IEEE Commun. Mag.Commun. Mag. **55**(5), 132–137 (2017)
39. Chang, T., Watteyne, T., Pister, K., Wang, Q.: Adaptive synchronization in multi-hop TSCH networks. Comput. Netw.. Netw. **76**, 165–176 (2015)
40. Watteyne, T., Palattella, M., Grieco, L.: Using IEEE 802.15.4e time-slotted channel hopping (TSCH) in the Internet of Things (IoT): Problem statement. IETF, Fremont, CA, USA, Rep. RFC 7554 (2015)
41. Rawat, D.B., Brecher, C., Song, H., Jeschke, S.: Industrial Internet of Things: Cyber Manufacturing Systems. Springer, Cham, Switzerland (2017)
42. Firestorm, S., Butun, I., Eldefrawy, M., Jennehag, U., Gidlund, M.: Challenges of securing the industrial Internet of Things value chain. In: Proceeding of IEEE Workshop Metrology Industry IoT, pp. 218–223 (2018)
43. Song, H., Rawat, D.B., Jeschke, S., Brecher, C.: Cyber-Physical Systems: Foundations, Principles and Applications. Amsterdam, The Netherlands: Morgan Kaufmann (2016)
44. Xu, K., Wang, X., Wei, W., Song, H., Mao, B.: Toward software defined smart home. IEEE Commun. Mag.Commun. Mag. **54**(5), 116–122 (2016)
45. Rani, S., Ahmed, S.H., Talwar, R., Malhotra, J., Song, H.: IOMT: a reliable cross layer protocol for Internet of multimedia things. IEEE Internet Things J. **4**(3), 832–839 (2017)
46. Sun, Y., Song, H., Jara, A.J., Bie, R.: Internet of Things and big data analytics for smart and connected communities. IEEE Access **4**, 766–773 (2016)
47. Hu, P., Ning, H., Qiu, T., Song, H., Wang, Y., Yao, X.: Security and privacy preservation scheme of face identification and resolution framework using fog computing in Internet of Things. IEEE Internet Things J. **4**(5), 1143–1155 (2017)
48. Xu, Q., Ren, P., Song, H., Du, Q.: Security-aware waveforms for enhancing wireless communications privacy in cyber-physical systems via multipath receptions. IEEE Internet Things J. **4**(6), 1924–1933 (2017)
49. Zhu, J., Song, Y., Jiang, D., Song, H.: A new deep-Q-learning based transmission scheduling mechanism for the cognitive Internet of Things. IEEE Internet Things J. **5**(4), 2375–2385 (2018)
50. Butun, I., Sari, A., Österberg, P.: Security implications of fog computing on the Internet of Things. In: Proceedings of IEEE International Conference on Consumer Electronics (ICCE), pp. 1–6 (2019)
51. Song, H., Fink, G.A., Jeschke, S.: Security and Privacy in Cyber-Physical Systems: Foundations, Principles, and Applications. Wiley, Hoboken, NJ, USA (2017)
52. Song, H., Srinivasan, R., Sookoor, T., Jeschke, S.: Smart Cities: Foundations, Principles, and Applications. Wiley, Hoboken, NJ, USA (2017)
53. Sun, Y., Song, H.: Secure and Trustworthy Transportation Cyber-Physical Systems. Springer, Singapore (2017)

54. Telli, K., Kraa, O., Himeur, Y., Boumehraz, M., Atalla, S., Mansoor, W., Ouamane, A.: Quadrotor experimental dynamic identification with comprehensive NARX neural networks. In: 2023 6th International Conference on Signal Processing and Information Security (ICSPIS), pp. 167–172 (2023)
55. Aonzo, S., Merlo, A., Migliardi, M., Oneto, L., Palmieri, F.: Low-Resource Footprint, Data-Driven Malware Detection on Android. IEEE (2017)
56. Transactions on Sustainable Computing 3782, c (2017), 1–1. http://ieeexplore.ieee.org/document/8113505/
57. Brans comb, L.M., Kodama, F., Florida, R.L. (eds.). Industrializing Knowledge: University-Industry Linkages in Japan and the United States. MIT Press (1999)
58. Al-Rubaie, M., Morris Chang, J.: Privacy Preserving Machine Learning: Threats and Solutions. IEEE Security and Privacy Magazine (2018)
59. Hidayet Aksu, A., Uluagac, S., Bentley, E.: Identification of wearable devices with Bluetooth. IEEE Transactions on Sustainable Computing, 1–1 (2018)
60. Ambusaidi, M.A., He, X., Nanda, P., Tan, Z.: Building an intrusion detection system using a filter-based feature selection algorithm. IEEE Trans. Computer. **65**, 10 2986–2998 (2016)
61. Tournier, J., Lesueur, F., Le Mouël, F., Guyon, L., Ben-Hassine, H.: A survey of IoT protocols and their security issues through the lens of a generic IoT stack. Internet of Things **16**, 100264 (2021)
62. Douceur, J.R.: The Sybil attack. In: 1st International Workshop on Peer-to-Peer Systems (IPTPS "02) (2022)
63. Tun, Z., Maw, A.H.: Worm hole attack detection in wireless sensor networks. In: Proceedings of World Academy of Science, Engineering and Technology, vol. 36. ISSN 2070-3740 (2008)
64. El Kaissi, R., Kayssi, A., Chehab, A., Dawy, Z.: DAWWSEN: a defense mechanism against wormhole attack in wireless sensor network. In: Proceedings of the Second International Conference on Innovations in Information Technology (IIT"05) (2005)
65. Wood, A.D., Stankovic, J.A.: Denial of service in sensor networks. Computer **35**(10), 54–62 (2002)
66. Zorzi, M., Rao, R.R.: Geographic random forwarding (GeRaF) for Ad Hoc and sensor networks: Multihop performance. IEEE Trans. Mob. Comput.Comput. **2**(4), 337–348 (2003)
67. Ganesan, D., Govindan, R., Shenker, S., Estrin, D.: Highly-resilient, energy-efficient multipath routing in wireless sensor networks. Mob. Comput. Commun. Rev. **4**(5) (2001)
68. Raymond, D.R., Midriff, S.F.: Denial-of service in wireless sensor networks: attacks and defenses. IEEE Pervasive Comput.Comput. **7**(1), 74–81 (2008)
69. Karloff, C., Sastry, N., Wagner, D.: Tiny Sec: a link layer security architecture for wireless sensor networks. In: Proceedings of the 2nd International Conference on Embedded Networked Sensor Systems, Nov 03–05, Baltimore, MD, USA (2004)
70. I-SCOOP: Blockchain and the Internet of Things: The IoT Blockchain Opportunity and Challenge (2018) [Online]. Available: https://www.iscoop.eu/blockchain-distributed-ledger-technology/blockchain-iot/
71. Bit Fury, G.: Proof of stake versus proof of work. White Paper (2015)
72. Lamport, L., Shostak, R., Pease, M.: The byzantine generals problem. ACM Trans. Program. Lang. Sys. (TOPLAS), **4**(3), 382–401 (1982)
73. Christidis, K., Devetsikiotis, M.: Blockchains and smart contracts for the Internet-of-Things. IEEE Access **4**, 2292–2303 (2016)
74. Aminanto, M.E., Choi, R., Tanuwidjaja, H.C., Yoo, P.D., Kim, K.: Deep abstraction and weighted feature selection for Wi-Fi impersonation detection. IEEE Trans. Inf. Forensics Security **13**, 3 621–636 (2017)
75. Kang, J., Yu, R., et al.: Enabling localized peer-to-peer electricity trading among plug-in hybrid electric vehicles using consortium blockchains. IEEE Trans. Ind. Inf. **13**(6), 3154–3164 (2017)
76. Li, Z., Kang, J., et al.: Consortium blockchain for secure energy trading in industrial internet of things. IEEE Trans. Ind. Inf. **14**(8), 3690–3700 (2018)
77. Kotobi, K., Bilen, S.G.: Secure blockchains for dynamic spectrum access: a decentralized database in moving cognitive radio networks enhances security and user access. IEEE Veh. Technol. Mag.Veh. Technol. Mag. **13**(1), 32–39 (2018)

78. Pattnaik, L.M., Swain, P.K., Satpathy, S., Panda, A.N.: Cloud DDoS attack detection model with data fusion & machine learning classifiers. EAI Endorsed Trans. Scalable Inform. Syst. **10**(6) (2023)
79. Satpathy, S., Swain, P.K., Mohanty, S.N., Basa, S.S.: Enhancing Security: Federated Learning against Man-In-The-Middle Threats with Gradient Boosting Machines and LSTM. In: 2024 IEEE International Conference on Advanced Video and Signal Based Surveillance (AVSS), pp. 1–8. IEEE (2024, July)
80. Satpathy, S., Pradhan, S.K., Ray, B.B.: A digital investigation tool based on data fusion in management of cyber security systems. Int. J. Inf. Technol. Knowl. Manag. **2**(2), 561–565 (2010)
81. Zhumabekuly Aitzhan, N., Svetinovic, D.: Security and privacy in decentralized energy trading through multi-signatures, blockchain and anonymous messaging streams. IEEE Trans. Depend. Secure Comput. **15**, 5 840–852 (2018)

The manufacturer's authorised representative in the EU is Springer Nature Customer Service Centre GmbH, Europaplatz 3, 69115 Heidelberg, Germany. If you have any concerns regarding our products, please contact ProductSafety@springernature.com

Printed and bound by CPI Group (UK) Ltd, Croydon, CR0 4YY
30/09/2026
02225586-0017